Deforestation and Forest Degradation: Impacts, Mapping and Preventions

Deforestation and Forest Degradation: Impacts, Mapping and Preventions

Edited by **Aduardo Hapke**

R Callisto
Reference

New York

Published by Callisto Reference,
106 Park Avenue, Suite 200,
New York, NY 10016, USA
www.callistoreference.com

Deforestation and Forest Degradation: Impacts, Mapping and Preventions
Edited by Aduardo Hapke

© 2015 Callisto Reference

International Standard Book Number: 978-1-63239-146-9 (Hardback)

Printed in the United States of America.

Contents

Preface   VII

Part 1   Deforestation Impacts   1

Chapter 1   **The Climatic Effects of Deforestation
in South and Southeast Asia**   3
Rachindra Mawalagedara and Robert J. Oglesby

Chapter 2   **Impact of Deforestation on the Sustainability
of Biodiversity in the Mesoamerican Biological Corridor**   21
Vani Starry Manoharan, John Mecikalski,
Ronald Welch and Aaron Song

Chapter 3   **Impacts of Deforestation on Climate
and Water Resources in Western Amazon**   45
Ranyére Silva Nóbrega

Chapter 4   **Deforestation Dynamics: A Review and Evaluation
of Theoretical Approaches and Evidence from Greece**   59
Serafeim Polyzos and Dionysios Minetos

Chapter 5   **Deforestation and Water Borne Parasitic Zoonoses**   79
Maria Anete Lallo

Chapter 6   **Landslides Caused Deforestation**   93
Diandong Ren, Lance M. Leslie and Qingyun Duan

Chapter 7   **Dinaric Karst – An Example of Deforestation
and Desertification of Limestone Terrain**   121
Andrej Kranjc

Part 2   Mapping Deforestation   143

Chapter 8   **Sustainable Forest Management Techniques**   145
K.P. Chethan, Jayaraman Srinivasan, Kumar Kriti and Kaki Sivaji

Chapter 9   **Remnant Vegetation Analysis
of Guanabara Bay Basin, Rio de Janeiro, Brazil,
Using Geographical Information System**   171
Luzia Alice Ferreira de Moraes

Chapter 10   **Geospatial Analysis of Deforestation and Land Use
Dynamics in a Region of Southwestern Nigeria**   199
Nathaniel O. Adeoye, Albert A. Abegunde and Samson Adeyinka

Chapter 11   **Unsupervised Classification
of Aerial Images Based on the Otsu's Method**   225
Antonia Macedo-Cruz, I. Villegas-Romero,
M. Santos-Peñas and G. Pajares-Martinsanz

Chapter 12   **Bunjil Forest Watch
a Community-Based Forest Monitoring Service**   241
Chris Goodman

Chapter 13   **Deforestation and Waodani Lands in Ecuador:
Mapping and Demarcation Amidst Shaky Politics**   265
Anthony Stocks, Andrew Noss,
Malgorzata Bryja and Santiago Arce

Part 3   **Preventing Deforestation**   281

Chapter 14   **Preserving Biodiversity and Ecosystems:
Catalyzing Conservation Contagion**   283
Robert H. Horwich, Jonathan Lyon, Arnab Bose and Clara B. Jones

Chapter 15   **Agroforestry Systems
and Local Institutional Development
for Preventing Deforestation in Chiapas, Mexico**   319
Lorena Soto-Pinto, Miguel A. Castillo-Santiago
and Guillermo Jiménez-Ferrer

Chapter 16   **Efficiency of the Strategies to Prevent
and Mitigate the Deforestation in Costa Rica**   337
Óscar M. Chaves

Chapter 17   **Economic Models of Shifting Cultivation: A Review**   351
Yoshito Takasaki

Permissions

List of Contributors

Preface

The impacts, mapping and preventions of deforestation and forest degradation have been elucidated in this profound book. Forest degradation and deforestation constitute a considerable fraction of human-induced emission of greenhouse gases to the atmosphere across the globe every year, which is a primary cause of the loss of biodiversity and devastation of homes of millions of people. Regardless of local/regional causes, its impact is global. The book also includes analysis of the effects and causes of deforestation and the actions required for avoiding it along with a fundamental description about deforestation dynamics across the planet. Emphasis has been laid on distinct remote-sensing and mapping methodologies that could be employed as a tool for promoting forest conservation and preventing deforestation.

This book is the end result of constructive efforts and intensive research done by experts in this field. The aim of this book is to enlighten the readers with recent information in this area of research. The information provided in this profound book would serve as a valuable reference to students and researchers in this field.

At the end, I would like to thank all the authors for devoting their precious time and providing their valuable contribution to this book. I would also like to express my gratitude to my fellow colleagues who encouraged me throughout the process.

Editor

Part 1

Deforestation Impacts

The Climatic Effects of Deforestation in South and Southeast Asia

Rachindra Mawalagedara and Robert J. Oglesby
University of Nebraska, Lincoln
USA

1. Introduction

Deforestation is the removal of the existing natural vegetation cover, especially where the native cover is largely forest. The growth in the world population has increased the clearing of forests to obtain fuel and building material, to grow crops and to raise livestock. Over the past 300 years, 7-11 million km² of forest has been cleared (Foley et al. 2005). Deforestation can have a devastating impact on biodiversity as about 70% of land dwelling animals and plants are found in forests. Impacts such as land degradation in the absence of forest regrowth, soil erosion and sedimentation in rivers can have a negative impact on the environment. These impacts are discussed in greater detail in the other chapters of the book. Importantly, deforestation can also have strong effects on climate.

In the past it was assumed that the local climate determined the vegetation type in a region (Nobre et al. 1991) with the amount of incoming solar radiation, precipitation and soil type determining the vegetation cover of the region. But studies have shown that the atmosphere and the vegetation interact with each other, exchanging energy, moisture and momentum (Zeng et al. 1999) and are in a dynamic equilibrium (Nobre et al., 1991). Therefore any change in vegetation cover can potentially lead to a change in the climate. As deforestation is a pressing problem in most parts of the world it is important to understand the possible consequences of deforestation and the mechanisms by which the change in land cover can alter the climate.

The impacts of land cover changes on the atmosphere have been studied extensively using both observations and computer models (Suh and Lee, 2004; Lean and Warrilow, 1989; Kanae et al., 2001; Clark et al., 2001). Previous studies have shown that deforestation can change the surface albedo, surface roughness and the amount of evapotranspiration (evapotranspiration is the combined effect of evaporation from the surface and the transpiration from vegetation) (Gibbard et al., 2005; Oglesby et al., 2010; Hasler et al., 2007) thus, leading to a modification of the surface energy and moisture budgets.

In order to determine the full climatic impact of deforestation, it is necessary to understand the behavior of the surface energy and the moisture budgets, as deforestation interacts directly or indirectly with all the components of these budgets. The surface energy budget looks at all the possible sources and sinks of energy at the surface as well as any possible horizontal transport (fluxes) and storage of energy within the seasonally active layer just below the surface. Over land, incoming and reflected solar radiation (shortwave radiation), incoming and outgoing longwave radiation, sensible heat flux, latent heat flux and ground storage are the important terms that need to be considered for the surface energy budget. Unlike in the oceans where transport of energy (by the oceans) is significant to the climate system, horizontal transport of energy in the ground is negligible. The moisture budget is related to the hydrologic cycle and takes into account precipitation, evaporation, surface runoff (horizontal transport of water) and storage. If the period considered is a year or longer, the storage of energy and moisture can be considered negligible. The changes in the terms of the two budgets can be used to determine the changes in the climate.

To understand how deforestation impacts the climate it is important to understand the behavior of the terms that make up the surface energy and moisture budgets. The primary source of energy that contributes to the surface energy budget is the sun. The energy that sustains the Earth and drives the global circulation is acquired from incoming solar radiation but is not absorbed directly by the atmosphere. Instead, most of the incoming solar radiation is first absorbed by the surface. The amount of solar radiation absorbed by the surface depends on the surface albedo, which determines the fraction of solar radiation reflected from the surface. The albedo can be expressed either as a percentage or a fraction, ranging from 100% (1.0) to 0% (0.0) with the former value indicating that all the incoming radiation is reflected (no absorption) and the latter indicating that no reflection of radiation takes place (all incoming radiation is absorbed). The surface albedo is related to the texture and the colour of the surface, with dark rough surfaces (low albedo) absorbing more energy.

Once the energy is absorbed by the surface, radiative and non-radiative processes transfer it from the surface to the atmosphere. Part of the absorbed energy warms the surface and is then emitted as longwave radiation from the surface. The magnitude of the emission is determined by the temperature and the emissivity of the surface (emissivity depends on the type of the surface: vegetated, bare soil etc.) and can be determined by the Stephen-Boltzmann Law. Part of the longwave radiation emitted by the surface is absorbed and reemitted by the atmosphere (can be calculated using the emissivity and the temperature of the atmosphere). A portion of the reemitted longwave radiation then acts as a source of energy for the surface. The remaining energy is partitioned between sensible and latent heat fluxes which are the non-radiative terms in the surface energy budget. The sensible heat flux heats the atmosphere in contact with the surface and is a less efficient method of heat transfer compared to the latent heat flux. It is difficult to measure the sensible heat flux directly. But it can be calculated easily if the latent heat flux and the Bowen ratio are known. The Bowen ratio is a measure of the water availability in a region and is the ratio between sensible heat flux and the latent heat flux. The latent heat flux can be calculated using the rate of evaporation.

The latent heat exchange, a significant process in the surface energy budget, is proportional to the amount of evaporation, and thereby provides a link to the surface moisture budget. The magnitude of the latent energy flux depends both on the amount of moisture available at the surface and the energy available for evaporation. In the tropics energy is not usually a limiting factor and hence the dependency is on the water availability. Therefore the energy that is not emitted as longwave radiation or stored in the ground is partitioned between sensible and latent heat and this partitioning is determined by the amount of water available at the surface. The latent heat flux also provides a measure of cooling at the surface due to evapotranspiration (i.e. water absorbs energy from the surface and evaporates thus cooling the surface). Once the evaporated moisture condenses in the atmosphere the latent energy is released. The energy released contributes to convection and helps to drive the local circulation. Therefore the latent heat exchange is not only an important cooling mechanism but is an important measure of the energy available for regional circulation.

The surface moisture budget accounts for precipitation, evapotranspiration and surface runoff. The evapotranspiration term links this to the surface energy budget. Changes in availability and the partitioning of energy can have an impact on evapotranspiration and hence other terms in the moisture budget. The information from the surface moisture budget namely, precipitation and evaporation can be used to compute the atmospheric moisture convergence/divergence which gives the net amount of water vapor transported into or out of a region by the regional circulation. If the amount of precipitation is larger than the local evapotranspiration this indicates that the moisture transport into the region makes a significant contribution to precipitation. Thus the atmospheric moisture convergence can be used to determine the relative importance of an external moisture source to that of local evaporation.

As discussed above both surface albedo and emissivity are sensitive to the nature of the land surface and hence depend on the type of land-use (forest, water, urban etc). The magnitude of the latent heat flux also depends on the land-use category as the water available for evapotranspiration changes with land-use. For example a vegetated surface would have more moisture available for evapotranspiration than barren land, and, as described more fully later in the chapter, forested land will have higher values than grassland or shrubland. Therefore it is evident that any change in the land-use in a region will modify both the surface energy and the moisture budgets.

Deforestation alters the land surface properties and the interactions between the surface and the atmosphere. Two of the most important changes due to deforestation are the increase in surface albedo and the decrease in evapotranspiration. The significance of these changes is discussed in more detail under the methods section.

Deforestation results in two competing effects, warming due to the reduction in evapotranspiration and a cooling due to the increase surface albedo. Previous studies have shown that in most regions the magnitude of warming is much greater than that of cooling, resulting in warmer and drier conditions (Zhang et. al., 1996; Oglesby et al., 2010). But these impacts are further modulated or enhanced by the dominant circulation patterns and moisture sources of the considered region. For example the change in precipitation due to the decrease in evapotranspiration would be more dramatic in a region such as the Amazon basin where 50% of the moisture available for precipitation comes local evapotranspiration (Lean and Warrilow, 1989). But if the region is close to a water body such as an ocean or a lake and the local and/or regional circulations are favorable for moisture transport, the contribution from local evapotranspiration would not be as significant. Most coastal regions receive abundant moisture from the ocean, carried inland by onshore flow (i.e. sea breeze – local circulation) contributing to precipitation. This mechanism alone is not strong enough for moisture to be transported to inland regions far from the oceans. But a large scale circulation pattern such as the Asian monsoon can penetrate further inland supplying moisture to continental regions thus making the impact of reduced evapotranspiration on precipitation much smaller.

This study focused on South Asia, Southeast Asia and Sri Lanka, all regions where the Asian monsoon plays an important role, and all regions where deforestation is currently or potentially a major issue. In these regions, the monsoon flow brings moisture from the ocean over land and this together with moisture provided by evapotranspiration produce abundant rainfall. The wet season is mostly during the summer monsoon period, but some areas experience rainfall during the winter monsoon as well. Any changes in this established pattern of rainfall and associated climate could have devastating consequences as this is a heavily populated area where agriculture is of great importance.

The Hadley cell is one of the prominent features that make up the global circulation. The Hadley cell consists of two asymmetric cells extending between approximately 15o in the summer hemisphere and 30o in the winter hemisphere (Lu et al. 2007). This circulation is defined by a branch of rising air over the surface low pressure belt (Inter-Tropical Convergence Zone- ITCZ) in the tropics, a poleward flow in both hemispheres in the upper troposphere, a branch of subsiding air in the subtropics and equator-ward flow (in both hemispheres) at the surface that converges at the ITCZ (Mitas et al., 2005). This circulation transports both energy and angular momentum and gives rise to many of the climatic regions that are present today. For example regions that fall under the ITCZ receive abundant rainfall due to the strong convection associated with the rising branch of the Hadley cell and either receive rainfall throughout the year or have well defined wet/dry seasons. The subtropical deserts are in the regions where the subsiding branches of the Hadley cell are found as the sinking of air prevents the formation of clouds and hence precipitation. Any changes in the strength or the extension of the Hadley cell will therefore impact the climate within the region and have the ability to change the geographical boundaries of these climate zones. As the location of Southeast Asia is closely associated with the rising branch of the Hadley cell and the poleward transport of energy, deforestation can have consequences on both regional and global scales.

It is evident that tropical deforestation can have an impact on the climate by modifying the magnitudes and the spatial and the temporal patterns of temperature and precipitation, creating warmer and drier climatic conditions in most regions. These changes might then be modulated or enhanced by the dominant circulation patterns such as the Asian monsoon leading to additional changes in the climate. The local moisture sources of the considered region may also play a significant role in modifying the climatic effects of deforestation Therefore, it is important to understand the possible consequences of deforestation and the mechanisms by which the change in land cover can alter the monsoonal climate.

2. Methods

In order to understand the impacts of tropical deforestation on the climate in South and Southeast Asia, a widely-used regional climate model (WRF – Weather Research and Forecasting Model) was employed. The Weather Research and Forecasting (WRF) Model is a next generation mesoscale numerical weather prediction system that has been developed as a collaborative effort by the National Center for Atmospheric Research (NCAR), the National Oceanic and Atmospheric Administration (the National Centers for Environmental Prediction (NCEP) and the Forecast Systems Laboratory (FSL), the Air Force Weather Agency (AFWA), the Naval Research Laboratory, the University of Oklahoma, and the

Federal Aviation Administration (FAA). WRF can be used in a wide scope of spatial scales ranging from a few meters to thousands of kilometers and is suitable for both operational forecasting and atmospheric research (Skamarock et al., 2008; http://www.wrf-model.org/index.php).

The main focus of this study was to identify the impacts of deforestation on the monsoonal climate in South and Southeast Asia. Therefore a Regional Climate Model (RCM) was used for this study instead of a Global Circulation Model (GCM), so that it was possible to run the model at a high resolution. This allowed the model to include the regional features and predict the regional climate with more accuracy. The GCM which usually has a horizontal resolution of 100 – 250 km (McGuffie and Henderson-Sellers, 2005). can capture the features of large and synoptic scale atmospheric circulation, but is too coarse to include small scale features such as the effects of topography or land surface effects to simulate the climate on a regional scale accurately (Denis et al., 2002). Since GCM simulations are done for the entire globe it is not feasible to run it at very high resolution due to limitations in computing power. Therefore regional climate models (RCM) are used to study regional climate changes. The main difference is a RCM is focused on the region of interest and has a much higher resolution. For example WRF simulations can be done at a resolution of 4 km which allows many small scale features such as mountains, coastlines and, importantly for our purposes, land-use categories to be represented more accurately.

The model was forced at the lateral boundaries by NCEP/NCAR Reanalysis data (NNRP)[1] . Since the model is a regional model the simulations are done for a restricted region. Therefore, conditions at the boundaries of the specified domain need to be provided so that the model can properly simulate the climate within the domain. Reanalysis data is used to provide these boundary conditions as observations alone are not sufficient to describe the full state of the atmosphere due to missing or spatially non-uniform data. Reanalysis data solves this problem by combining actual observations with a global model of the atmosphere to produce a comprehensive data set that serves as a proxy for real observations, thus providing better boundary conditions for the regional model.

The model simulations focused on three specific domains: 1. South Asia (Southern India and Sri Lanka with a resolution of 12 km), 2. Southeast Asia (with a resolution of 12 km), 3. High resolution focus on Sri Lanka (with a resolution of 4 km). The WRF simulations were done for the years 1988, 1991 and 1993. These years represent a strong, weak and normal monsoon year with respect to South Asia. For each of these three years, a control run as well as two idealized runs (completely deforested and forested situations) were carried out and analyzed. The control runs are also compared to actual observations in order to identify model strengths and weaknesses, as well as any biases. In the deforested run all the land use categories (other than inland water) were replaced with grassland, which has a higher albedo than the tropical forests, but much less capability at extracting water from the soil. In the forested run evergreen broadleaf forest was used. These land-use changes, while extreme, provided the maximum possible range of impacts due to deforestation.

---

[1]The data for this study are from the Research Data Archive (RDA) which is maintained by the Computational and Information Systems Laboratory (CISL) at the National Center for Atmospheric Research (NCAR). NCAR is sponsored by the National Science Foundation (NSF). The original data are available from the RDA (http://dss.ucar.edu) in dataset number ds090.2.

In a WRF simulation each grid point has a land-use category (grassland, cropland, evergreen broadleaf forest, water etc.) assigned to it based on the land-use data set being used for the model run. The properties (surface albedo, surface emissivity, moisture availability, surface roughness length) of each land-use category depend on the land surface model used in the WRF run. The land surface model is the component that takes care of the processes involving land-surface interactions. For the WRF runs, the 5-layer thermal diffusion scheme was selected as the land surface model. USGS (winter) data set was used to specify land-use categories and their properties. To simulate deforested conditions, all the land-use categories other than water were replaced with grassland. Grassland has a higher albedo (23%) than most other land-use types and therefore absorbs less energy. The specified moisture capacity (0.30) is, however, also low. Water bodies have the lowest surface albedo (8%) and a specified moisture availability of 1.0 (saturated surface). Forested conditions were simulated by replacing all land-use categories other than water by evergreen broadleaf forest which has a very low surface albedo (12%), but much higher moisture availability (0.5) compared to grassland, with the former allowing the surface to absorb more incoming energy and the latter supporting larger evaporation amounts.

Before analyzing the results of the WRF simulations it is important to understand the possible impacts of the land-use changes made to the deforested and deforested runs (as mentioned above), on the surface energy and the moisture budgets and how the changes will ultimately affect the regional climate.

Tropical forests have a low surface albedo throughout the year. This allows the forests to absorb a large part of the incoming radiation. Most land-use categories have surface albedos that are higher than that of a tropical rainforest. Therefore tropical deforestation leads to an increase in the surface albedo, allowing the surface to reflect more radiation. As a result the surface absorbs less radiation creating a cooling effect. This also reduces the amount of energy available for evapotranspiration.

On the other hand, the trees found in tropical forests have the capacity to draw water from the soil and thereby add a large amount of moisture to the atmosphere via transpiration. The large leaf and stem area allows the trees to intercept a significant amount of the rainfall. The intercepted water is then evaporated into the atmosphere. Evapotranspiration can be an important moisture source for local precipitation, especially in regions that are not in the proximity of a water body. Also evapotranspiration helps to lower the surface temperature. Deforestation leads to a decrease in evapotranspiration. This removes or reduces the capacity of the local moisture source and the cooling effect of evapotranspiration. This also alters the energy partitioning between latent and sensible heat fluxes at the surface. Due to the reduction in evapotranspiration and hence the latent energy flux, the energy transfer between the surface and the atmosphere would be achieved mostly through the exchange of sensible heat. As this is a less efficient method of heat transfer (compared to the cooling by latent heat transfer), this would lead to an increase in the surface temperature. The reduction in the latent energy flux means the energy available for convection is reduced. This can potentially lead to a weakening of the local circulation that in turn can have a negative impact on the moisture convergence in the area. The reduction of moisture convergence and evapotranspiration result in a reduction in precipitation. Therefore the reduction in evapotranspiration can result in a warming of the surface and a decrease in precipitation. The decrease in precipitation then acts as a positive feedback to further reduce the evapotranspiration and enhance the warming effect. Thus, deforestation will result in a warmer and drier climate.

The surface energy and the moisture budgets were analyzed to see if the possible changes that are discussed above were present in the WRF output. Temperature, precipitation and evaporation anomalies (deforested – forested) were also calculated to identify the climatic impact of deforestation.

3. Results

In response to deforestation, for all three years, there is an increase in temperature over majority of the land areas whereas the changes over the oceans are more variable. Figure I show the changes in temperature (at 2 m) between the deforested and the forested runs for the year 1988. (For brevity, we show changes for one representative year.) Changes in the magnitudes in the spatial patterns of both the annual and the seasonal (JJA) temperatures are shown in the figure. The period of June, July and August (JJA) was selected to focus on the height of the summer monsoon season. Most regions show a clear warming, evident in both the annual and the seasonal values. The most prominent warming is seen along the west coast of India and Sri Lanka. There is a small region in the central highlands of Sri Lanka where the response is a cooling of temperature. Figure II shows the time series of temperature over land areas for 1988. The warming is evident in the monthly temperatures in all of the domains. The temperature at 2 m (for 1988) for the deforested and the forested situations and the corresponding changes are shown in table II.

The spatial pattern of precipitation over each of the three years shows a decrease over land whereas some regions over the oceans experience enhanced rainfall. The annual precipitation values show a larger reduction than just the values in the monsoon season. This indicates that while the amount of precipitation received during the summer monsoon is affected by deforestation so are the other mechanisms such as the winter monsoon and convection that provide rainfall during the rest of the year. Figure III shows the precipitation anomalies between the deforested and the forested simulations for the representative year 1988. Also the increase in Bowen ratio (over land) indicates that conditions become drier. The annual precipitation (for 1988) over South Asia decreases by 19% while the reduction over Sri Lanka is only 10% relative to the forested run. Southeast Asia experiences a 53% decrease in precipitation as a result of deforestation (All the percent changes provided in this chapter are calculated as the difference between the deforested and forested runs with respect to the forested run. i.e. [(D-F)/F]*100%).

Evaporation over land is much smaller after deforestation but the changes over the ocean in the domains over South Asia and Sri Lanka show an increase. (The increase over the ocean is likely due to the warmer temperatures due to deforestation. The amount of water vapor the atmosphere can hold strongly depends on the temperature with warm air being able to hold more moisture than cold air. As deforestation warms the atmosphere, the air over the adjacent ocean also warms, gaining the ability to hold more moisture. This implies that more evaporation can take place before the air is saturated thus enhancing evaporation over the ocean.) Overall, the annual evapotranspiration over all three domains is reduced. This can be seen in figure IV which shows the anomalies in evapotranspiration for 1988. The largest decrease is seen over Sri Lanka where the (area averaged) annual evaporation decreases by

465 mm (table I), which is 29% reduction compared to the forested run, over the year. The annual evapotranspiration over South Asia decreases by 420 mm (27%) whereas the decrease over Southeast Asia is 415 mm (25%). The reduction in evapotranspiration is a result of the lower transpiration, reduced interception of precipitation (due to reduced leaf and stem area) and smaller roughness length. The roughness length provides a measure of the surface friction and the exchange of moisture between the atmosphere and the surface, with smaller roughness lengths modulating the exchange.

The latent heat flux shows a decrease over all domains, consistent with the reduction in evaporation. As moisture becomes limited with the change in land-use, less energy is used for evapotranspiration. This means the evaporative cooling at the surface is reduced and the remaining energy goes into warming the surface. The sensible heat flux, which heats the air in contact with the surface, shows an increase due to the change in energy partitioning at the surface (i.e. in response to the decrease in latent energy flux). The changes in evaporation, latent heat flux and sensible heat flux support the warming observed in response to deforestation.

Precipitation is greater than the evapotranspiration over South Asia and Sri Lanka. This indicates that the moisture provided from a moisture source (i.e. moisture transported from the Indian Ocean into the area by the regional circulation) other than local evapotranspiration makes a significant contribution to the precipitation in the region, suggesting that the Indian monsoon plays a dominant role where precipitation is considered. Out of the two regions the moisture transport is more significant for Sri Lanka as precipitation over the region can be more than twice the amount of the local evapotranspiration.

The decrease in precipitation is much larger than that of evapotranspiration over Sri Lanka. The decrease in the local moisture recycling capacity alone cannot explain this strong reduction in precipitation. This indicates the moisture transported into the region has decreased, signaling a weakening or a change in the regional circulation leading to a reduced atmospheric moisture convergence into the area. This shows that while the moisture brought in with the monsoon flow may be a prominent factor in determining the amount of precipitation, deforestation can apparently weaken the moisture flux so that less water is available for precipitation in the region.

On the other hand, the reduction in precipitation over South Asia is less than that of evaporation. This indicates a stronger moisture convergence over the Indian subcontinent. As a result deforestation causes two competing effects on precipitation. The decrease in evapotranspiration has a negative impact on precipitation whereas the increased moisture convergence has a positive effect on it. But the magnitude of the latter is smaller than that of the evapotranspiration, therefore ultimately resulting in a decrease in precipitation.

The difference between precipitation and evapotranspiration is also a measure of surface runoff and storage. The changes in surface runoff can be very important for streamflow in the region. But further analysis of these terms are not possible because the land surface model used in the WRF runs does not have the capability to compute these terms separately. Clouds reflect part of the incoming solar radiation back out to space, hence the amount of incoming shortwave radiation at the surface can be used as a proxy for cloud cover. The incoming shortwave radiation shows a reduction for all three years over all domains. This is a indication of reduced cloud cover. The decrease in evaporation (and hence latent heat flux) means that there is less moisture.

Term	South Asia			Sri Lanka			Southeast Asia		
	Deforest D	Forest F	Change D-F	Deforest D	Forest F	Change D-F	Deforest D	Forest F	Change D-F
SW_{down}	249.26	242.46	6.80	225.36	219.44	5.92	277.88	270.59	7.29
SW_{up}	57.33	29.10	28.23	51.83	26.33	25.50	63.91	32.47	31.44
SW_{net}	191.93	213.36	-21.43	173.53	193.11	-19.58	213.97	238.12	-24.15
LW_{down}	387.23	390.36	-3.13	414.85	415.26	-0.41	366.21	371.19	-4.98
LW_{up}	405.92	416.00	-10.08	418.60	428.21	-9.61	402.41	411.88	-9.47
LW_{net}	-18.69	-25.64	6.95	-3.75	-12.95	9.20	-36.20	-40.69	4.49
R_{net}	173.24	187.72	-14.48	169.78	180.16	-10.38	177.77	197.43	-19.66
LE	89.72	122.96	-33.24	88.77	125.58	-36.81	96.02	133.84	-37.82
SH	35.14	27.11	8.03	39.41	27.63	11.78	44.30	38.53	5.77
G_E	48.38	37.65	10.73	41.60	26.95	14.65	37.45	25.06	12.39
P (mm/yr)	1444.39	1775.57	-331.18	4700.86	5206.67	-505.81	172.93	370.34	-197.41
E (mm/yr)	1134.89	1555.33	-420.44	1122.83	1588.47	-465.64	1214.58	1629.90	-415.32
P-E (mm/yr)	309.50	220.24	89.26	3578.03	3618.20	-40.17	-1041.65	-1259.56	217.91
Bowen Ratio	0.39	0.22	0.17	0.44	0.22	0.22	0.46	0.29	0.17

Table 1. The terms of the surface energy and the moisture budgets for deforested (D) and forested (F) WRF simulations as well as the difference between the two simulations (D-F) over South Asia, Sri Lanka, and Southeast Asia . These are the areas averaged values for 1988 and are computed only over land areas. Precipitation and evaporation are yearly totals. The units are Wm^{-2} unless noted otherwise in the table. The Bowen ratio is unitless. Here SW – shortwave radiation, LW – longwave radiation, R_{net} – net radiation, LE – latent energy flux, SH – sensible heat flux, G_E – storage, P – precipitation, E – evaporation. The subscripts (up/down) give the direction of the radiation. SW_{net} and LW_{net} are calculated as the difference between downward and upward radiation. A positive net radiation value indicates that the radiation is aimed towards the surface (downwards).

Term	South Asia			Sri Lanka			Southeast Asia		
	Deforest D	Forest F	Change D-F	Deforest D	Forest F	Change D-F	Deforest D	Forest F	Change D-F
Temperature (at 2 m) (K)	296.56	296.31	0.25	298.82	298.47	0.35	295.72	295.47	0.25
Surface Skin Temperature (K)	296.88	296.35	0.53	299.22	298.54	0.68	296.13	295.53	0.6
Surface Emissivity	0.92	0.95	- 0.03	0.92	0.95	- 0.03	0.92	0.95	- 0.03
Surface Albedo	0.23	0.12	0.11	0.23	0.12	0.11	0.23	0.12	0.11

Table 2. The area averaged annual temperature at 2 m and surface skin temperature (surface temperature) are shown here. Temperature values are in Kelvins (K). Surface emissivity and surface albedo for both deforested and forested simulation are included here. All the values in the table are for 1988

Annual Temperature Anomalies

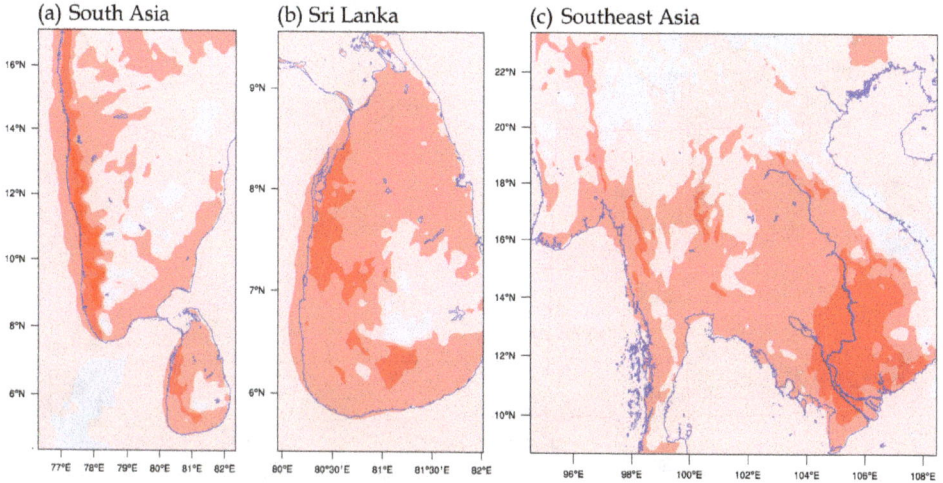

(a) South Asia   (b) Sri Lanka   (c) Southeast Asia

Seasonal Temperature Anomalies

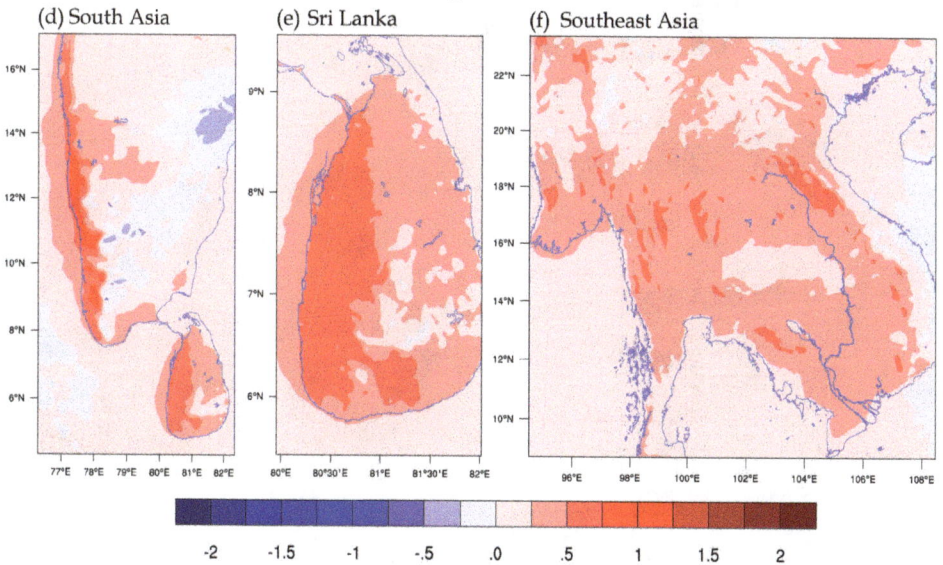

(d) South Asia   (e) Sri Lanka   (f) Southeast Asia

Fig. 1. The annual and the seasonal (June, July and August) temperature anomalies (i.e. deforest –forest) for South Asia (a,d), Sri Lanka (b,e) and Southeast Asia (c,f) for 1988. Temperature is the surface air temperature 2m above the surface and is expressed in Kelvins (K) with a contour interval of 0.25 K.

Monthly Temperatures

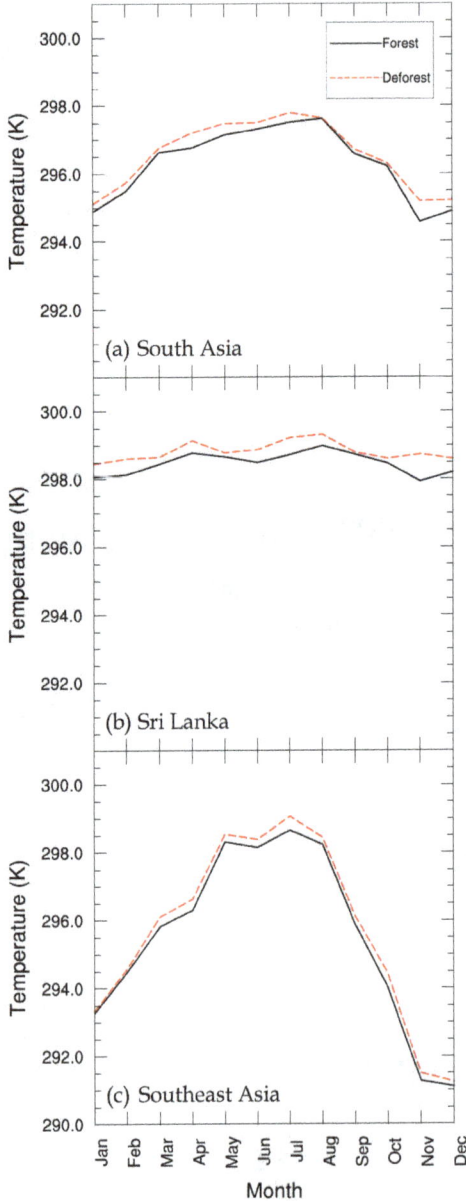

Fig. 2. The time series of temperature for 1988. The monthly temperature values over land regions (excluding oceans) for South Asia (a), Sri Lanka (b) and Southeast Asia (c) are shown here. Temperature is the air temperature 2m above the surface and is measured in Kelvin (K)

Annual Precipitation Anomalies

(a) South Asia   (b) Sri Lanka   (c) Southeast Asia

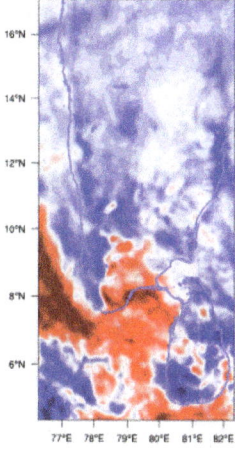
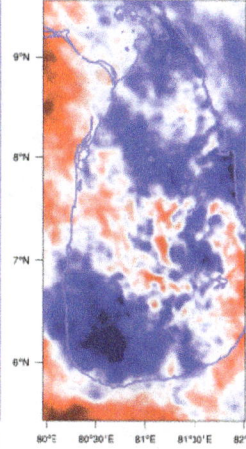
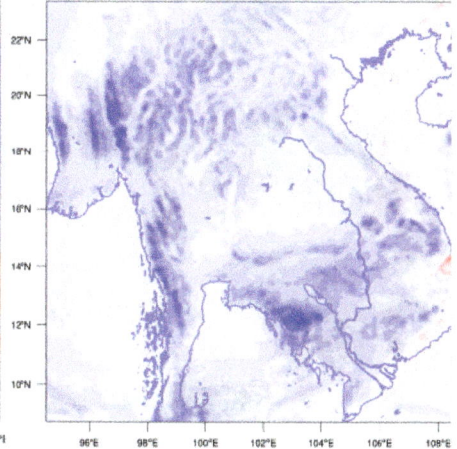

Seasonal Precipitation Anomalies

(d) South Asia   (e) Sri Lanka   (f) Southeast Asia

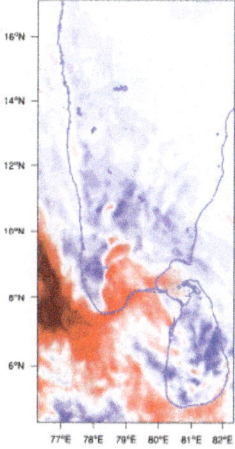
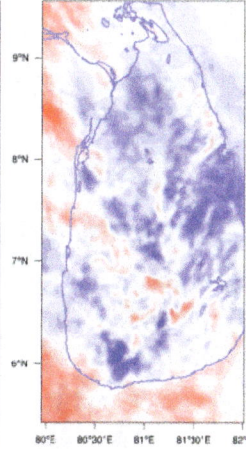
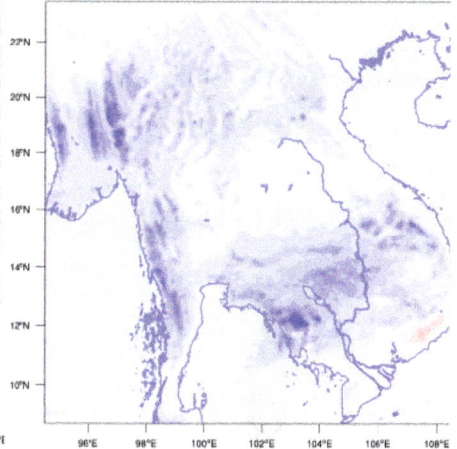

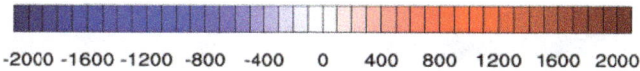

-2000 -1600 -1200 -800 -400   0   400 800 1200 1600 2000

Fig. 3. The total annual and the seasonal (June, July and August) precipitation anomalies (i.e deforest –forest) for South Asia (a,d), Sri Lanka (b,e) and Southeast Asia (c,f) for 1988. Precipitation is in millimeters (mm) with a contour interval of 100 mm.

Annual Evaporation Anomalies

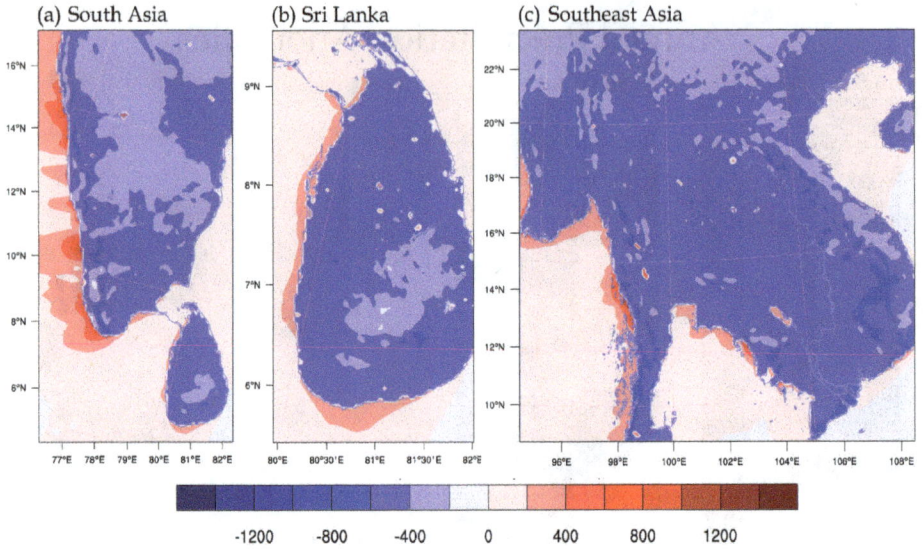

Seasonal Evaporation Anomalies

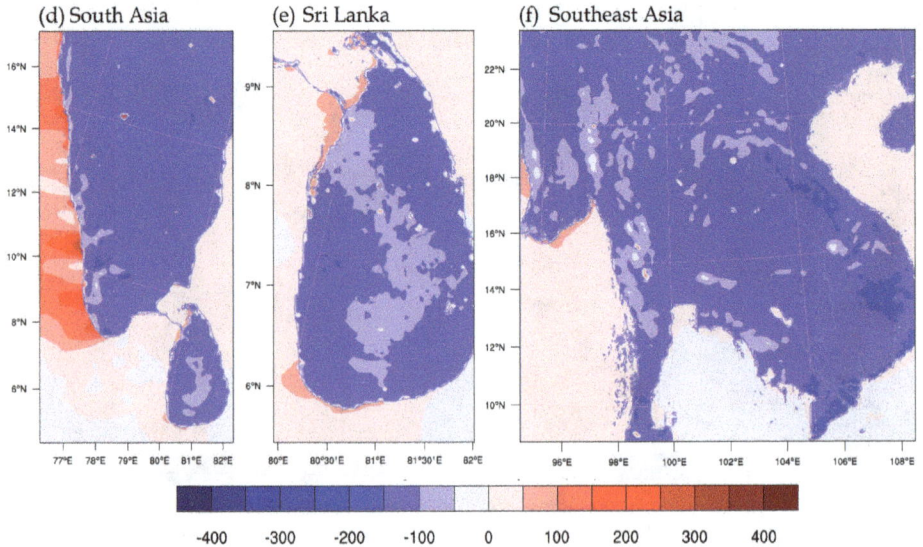

Fig. 4. The annual and the seasonal (June, July and August) evaporation anomalies (i.e deforest –forest) for South Asia (a,d), Sri Lanka (b,e) and Southeast Asia (c,f) for 1988. Evaporation is in millimeters (mm) with a contour interval of 200 mm for annual plot and 50 mm for the seasonal plot.

4. Discussion

Deforestation leads to warmer and drier climatic conditions. As a result of deforestation less moisture is available at the land surface, leading to a reduction in evapotranspiration. This in turn leads to an increase in temperature and a decrease in precipitation over land. These changes then work together to alter the moisture convergence in the region. These changes are seen not only during the monsoon season but through the entire year.

The monsoon is primarily driven by the pressure gradient created by the differential heating of the ocean and land. Land regions especially the Tibetan plateau warms up more than the Indian Ocean during the summer due to the small heat capacity of land. This creates a thermal low pressure over land and the resulting pressure gradient between ocean and land initiates the monsoon flow. Subsequently, the monsoon is sustained by the release of latent heat into the atmosphere. Therefore warmer temperatures should intensify the monsoon, increasing precipitation over land. But this is not seen in the results. The reasons for this are regionally specific. Over Sri Lanka it is likely that the warming of the land itself is reducing the amount of precipitation. The initial step in the formation of precipitation is the rising of parcels of warm moist air. The air being warm and moist by itself is not sufficient to cause rising motion. For air to rise the density of a parcel of air has to be lower than that of the surrounding air, that is the parcel temperature must be warmer than that of the surrounding (cooler) air. When deforestation warms the air over land, it reduces this difference, creating a more stable atmosphere which inhibits the rising of air and hence also weakens the atmospheric moisture convergence. Therefore even if the atmosphere otherwise holds plenty of moisture this process reduces the amount of precipitation that can form. Therefore the weakened moisture convergence together with the reduced evaporation causes the reduction in precipitation. The conditions are different in the Indian subcontinent where the summer monsoon is at its' strongest. The warming of the land region now results in a stronger moisture convergence indicating the possibility that the monsoon might become stronger over the region. It is possible that the warming over the Indian subcontinent (which has a smaller magnitude than the warming in Sri Lanka) is insufficient to stabilize the atmosphere. But as discussed in the results section the reduced evapotranspiration at any rate plays the dominant role in reducing the annual and the seasonal precipitation.

The changes in the climatic conditions due to deforestation can have a strong impact on society. Warmer and drier conditions can have far reaching consequences in many different ways. The elevated temperatures alone can cause life loss especially if the number of heat waves increases. India would be more susceptible to this as parts of the country experience high summer temperatures prior to the onset of the monsoon, a time during which people are vulnerable to the excess heat. The warm temperatures can also kill livestock and destroy stored goods.

The reduction in precipitation in combination with the warmer temperature can have a negative impact on agriculture in the region. These climatic changes can result in reduced crop yields, a shift in the life cycle of the crop and change the length of the growing season and the timing of the harvest. Also the regions that are the best suited for growing crops may migrate. As the warm and dry conditions act as positive feedbacks, deforestation can lead to persistent droughts. This may then result in desertification making some regions no longer viable as agricultural land and marginally habitable regions unsuitable for living. In addition to this the changes in temperature and precipitation may change the habitats of animals and expand or alter the spread of diseases such as malaria.

Some regions depend largely on rainfall and/or stream flow for drinking water. If the decrease in precipitation is large enough, the lack of access to clean drinking water can become a major problem. In addition to this the water level in the rivers are important for transport along the rivers, generation of hydropower and irrigation of crops. If the regions that feed the river flow experience a decrease in precipitation it would lower the water levels regardless of what happens downstream causing many economic problems.

All these changes would lead to a shortage in food and a disruption in the well established livelihoods in the region leading to poverty and famine in extreme cases. They would also cause many health and safety issues to which the very young, the old and the poor would be the most sensitive.

Deforestation in South and Southeast Asia has an impact on the monsoonal climate. These climatic impacts then in turn will cause social, economical, environmental and health problems in the region. It is important to notice that Southeast Asia still has a significant amount of forest whereas the situation in south Asia is closer to the conditions of the deforested simulations. Therefore it is imperative to set in place, policies that would prevent or slow down deforestation in the region as well as to implement steps to deal with issues that have already arisen due to deforestation.

This study illustrates the maximum possible range of changes possible due to deforestation and shows that the signal due to deforestation can override even a strong monsoon. Therefore it is worthwhile to continue this study, focused on more realistic land-use changes and to examine the magnitude of the changes. Also future studies would include the changes to the Hadley cell and the occurrence of extreme events, especially floods, under both realistic and idealized situations.

5. Conclusion

Deforestation impacts the climate by modifying surface energy and moisture budgets. These modifications are mostly due to the decrease in evapotranspiration and increase in the surface albedo. The changes in surface albedo and evapotranspiration have competing effects on the temperature, but as seen from the results of the WRF runs the warming effect due to the reduced evapotranspiration dominates over the cooling effects of the reduced albedo.

Results show that majority of the land areas would become warmer and drier in response to deforestation, with precipitation, evapotranspiration and cloud cover all showing a decrease. Atmospheric moisture convergence shows regionally specific changes. These changes are seen in both the annual and monsoon seasonal values, suggesting that the changes that take place due to deforestation have the ability to override even a strong monsoon signal. The changes over the oceans are more variable, with an increase in evaporation seen in both seasonal and annual values in the domains over South Asia and Sri Lanka.

These changes due to deforestation can have far reaching social, economic and environmental impacts as well as cause serious health issues. Warm temperatures can cause illness and even heat related deaths. The decrease in precipitation can lead to the drying of natural springs and reduced stream flow, cutting of access to clean drinking water in some regions. The habitats of plants and animals can change resulting in spread of diseases. Further the warm dry conditions can reduce or destroy crops, kill livestock and decrease the output from hydro powered electricity plants.

The climatic changes due to deforestation have the very real potential to impact the economic and the social structures of a country. As deforestation has been an ongoing problem, a climatic signal may already be present and the consequences of the changes already affecting us to some extent. If such a signal is present further deforestation may amplify these changes to a level that they would be clearly evident. Therefore it is important to understand the impacts of human activity on the climate, and set in place policies not only regarding deforestation but also implement steps to understand and deal with the consequences of a climatic signal that can result from current land-use practices.

6. References

Clark D. B., Xue Y., Harding R. J. & Valdes P.J. (2001). "Modeling the impact of land surface degradation on the climate of tropical north Africa". Journal of Climate, vol 14 1809-1822

Denis B., Laprise R., Caya D. & Côté J. (2002). "Downscaling ability of one-way nested regional climate models: the Big-Brother Experiment". Climate Dynamics 18:627-646

Foley J. A., DeFries R., Asner G. P., et al., (2005). "Global consequences of land use". Science, 309, 570-574.

Gibbard S.G., Caldeira K., Bala G., Phillips J.J. & Wickett M., (2005). "Climate effects of globa land cover change". Geophysical research letter.

Hasler, N. & Avissar R. (2007). "What controls evapotranspiration in the amazon basin?" Journal of Hydrometeorology 8(3): 380-395.

Kanae S., Oki T. & Musiake K. (2001). "Impacts of deforestation on regional precipitation over the Indochina peninsula". Journal of Hydroclimatology, vol 2 51-70

Lean J. & Warrilow D.A. (1989). "Simulation of the regional climatic impact of Amazon deforestation". Nature 342: 411-413

Lu J., Vecchi G. A. & Reichler T., (2007). "Expansion of the Hadley cell under global warming". Geophysical Research Letters, Vol 34

McGuffie K. & Henderson-Sellers A. (2005). A climate modeling primer (third edition). John Wiley & Sons, ISBN: 9780470857502

Mitas C. M. & Clement A., (2005). "Has the Hadley cell been strengthening in recent decades?". Geophysical Research Letters, Vol 32

Nobre C.A., Sellers P.J. & Shukla J., (1991). "Amazonian deforestation and regional climate Change". Journal of Climate vol 4: 957-988

Oglesby R. J., Sever T. L., Saturno W., D. J. Erickson, III & Srikishen J. (2010), "Collapse of the Maya: Could deforestation have contributed?", J. Geophys. Res., 115, D12106, doi:10.1029/2009JD011942.

Skamarock W.C., Klemp J.B., Dudhia J., Gill D.O., Barker D.M., Duda M.G., Huang Xiang-Yu, Wang W. & Powers J.G., (2008) "A Description of the Advanced Research WRF Version 3"

Suh M. S., & Lee D. K. (2004). " Impacts of land use/cover changes on Surface climate over east Asia for extreme climate cases using RegCM2". J. Geophy. Res.,109

Zeng N. & Neelin J. D., (1999). "A land-atmosphere interaction theory for the tropical deforestation problem". Journal of Climate Vol12: 857-872

Zhang H. & Henderson-Sellers A. (1996). "Impacts of tropical deforestation. Part 1: Process analysis of local climatic change". Journal of Climate vol19: 1497-1517

Impact of Deforestation on the Sustainability of Biodiversity in the Mesoamerican Biological Corridor

Vani Starry Manoharan[1], John Mecikalski[2],
Ronald Welch[2] and Aaron Song[2]
[1]Environmental Sciences Division, Argonne National Laboratory, Argonne
[2]Department of Atmospheric Sciences, University of Alabama in Huntsville,
USA

1. Introduction

Tropical rain forests play an essential role in housing global biodiversity, and they accommodate more than 50% of all species in the world while occupying only ~10% of the surface land of the Earth [*Myers*, 1992; *Pimm*, 2001]. However, during the last 10,000 years humans have significantly influenced land surface characteristics by altering the vegetation to include plant species more suitable for their consumption, a process that includes converting forests to agricultural lands, livestock grazing, and building settlements [*DeFries et al.*, 2004]. Current rates of deforestation are extremely high and are known to have significant impact on regional and global atmospheric and climate changes [*Laurence et al.*, 2004] in addition to the direct local effects of deforestation. The scale and speed of global habitat loss and fragmentation is alarming, with only about half of the pre-industrial forest areas remaining as forests. These forest fragments are becoming the only refuge for most of the global tropical wildlife, but as these habitats fragment into smaller and smaller pieces and become more isolated, local extinction rates accelerate [*Bennett*, 1999].

Central America or Mesoamerica, which occupies only 0.5% of the global land area, provides habitat for more than 7% of the world's species [*Mittermeier et al.*, 2000]. This region has a population growth rate of > 2% per year with high levels of poverty, unsustainable exploitation of natural resources, soil erosion and one of the world's highest rates of deforestation, losing 2.1% of forests/year [*FAO*, 1999]. This scale and speed of habitat loss and forest fragmentation in one of the earth's biologically richest regions has led conservationists to propose the Mesoamerican Biological Corridor (MBC) project, an integrated regional initiative intended to conserve biological and ecosystem diversity in a manner that also provides sustainable economic development [*Carr et al.*, 1994; *Miller et al.*, 2001].

The MBC is an ambitious effort intended to connect large existing isolated parks, forest fragments and reserves with new protected areas through an extensive network of biological corridors within the five southern states of Mexico and the Central American countries (Guatemala, Belize, El Salvador, Honduras, Nicaragua, Costa Rica and Panama). The intent is to establish an environment that provides better prospects for the long-term survival of the native species, provides migratory pathways for the others, and addresses the region's socioeconomic needs. This would stem and reverse the erosion of biodiversity in the existing forest fragments in Mesoamerica. Ideally these proposed connecting corridors would contain the biological communities that were originally present. However, most of the current and proposed connecting corridors do not contain their original forest, but are instead occupied by agricultural landscapes containing croplands, grassland and various forms of degraded woodlands. Additionally, these regions have a varied topography of altitudes ranging between sea level and 3000 m. The topography and land characteristics of the corridors vary widely, from lowlands to high mountain peaks and dense-forest regions to croplands and woodlands respectively. The establishment of fully functional corridors will depend upon the regrowth of forests in many areas. However, the extent of deforestation within Central America may already have had climatic consequences that affect the stability and sustainability of currently protected areas and the proposed corridor regions [*Laurence et al.*, 2004].

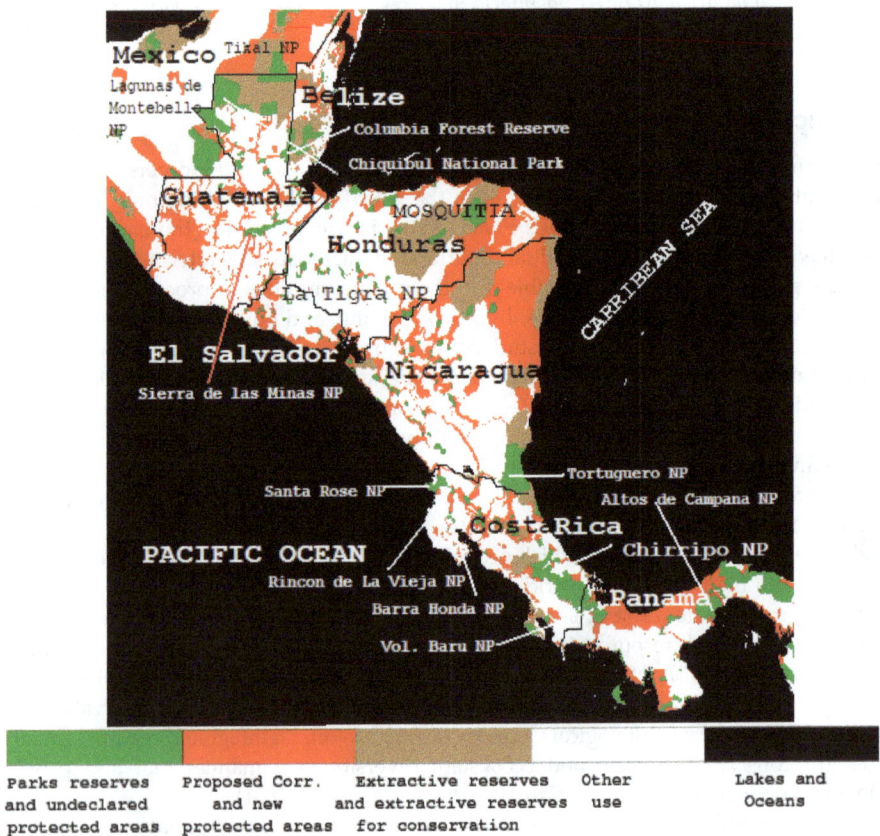

Fig. 1. The map of the Mesoamerican Biological Corridor.

According to the Intergovernmental Panel for Climate Change (IPCC) 3rd Assessment report [*IPCC*, 2001], there are high chances of adverse impacts on the existing natural global ecosystems, biodiversity and food supply due to the projected climate changes in the future.

Observations show that there have been decreases in frog population and other small mammals that may be related to climate change [*IPCC*, 2001].

Previous studies show that changes in land use impact regional climate which in turn may enhance and sustain these changes [*Hansen et al.*, 2001; *Hansen and Rotella*, 2002; *DeFries et al.*, 2002]. Deforestation changes the surface energy budgets and generally decreases latent heat (LH) fluxes from surface to the atmospheric boundary layer and increases the sensible heat (SH) fluxes [*Lawton et al.*, 2001; *Nair et al.*, 2003; *Ray et al.*, 2006]. The net result is hotter and drier air over the deforested and forested regions leading to increased dryness. However due to the continuous deforestation throughout Central America, there is concern about the long-term stability and even the sustainability of the MBC. This chapter is built upon studies to understand the environmental stability of forests in the corridors and their surroundings. The map of the proposed MBC is shown in Figure 1. The regions in red color in the map are the proposed corridor regions. National parks and protected areas are shown in green color. The extractive reserves shown in brown color are proposed for conservation. Some hunting and logging are permitted in these reserves. However, loggings in the extractive reserves are followed by reforestation. The guiding rules for the extractive reserves vary from country to country, but the basic requirement of reforestation is the same.

There are 600 protected areas as a part of the MBC network [*Herrera*, 2003]. The number and percent of territory covered by these protected areas are given in Table 1.

Country	Number of protected areas	% of territory covered by the protected areas
Southern Mexico	33	18.8%
Belize	59	44.8%
Guatemala	104	26.3%
El Salvador	3	1.6%
Honduras	106	19.0%
Nicaragua	76	21.7%
Costa Rica	151	24.6%
Panama	69	29.5%

Table 1. The number of protected areas in each territory and the percentage of area covered by them.

According to Central American Protected Areas System (SICAP), 29 percent of the protected areas which had been legally designated by 1998 cover less than 1,000 hectares and 67 percent of protected areas cover less than 10,000 hectares. Only 22 are larger than 100,000 hectares and only four of these cover more than 500,000 hectares.

2. Impact of deforestation on regional hydrometeorology

The clearing of forest results in an immediate reduction of intercepted water storage capacity followed by a decrease in interception and transpirational loss of water to the atmosphere [*Swift et al.*, 1975; *Eshleman*, 2004]. This is in turn followed by increase in runoff and/or overall decrease in water yield capacity [*Whitehead and Robinson*, 1993]. Further, deforestation leads to decrease in the water storage capacity of the soil [*Shukla et al.*, 1990] as well as the fraction of available soil moisture [*Ray et al.*, 2003; *Manoharan et al.*, 2009]. Additionally, the deforestation reduces the overall roughness of the surface [*Gash and Nobre*,

1997] and increases the albedo above the region [*Costa et al.*, 2007]. These factors have high impact on the aridity of these places. Moreover, deforestation leads to warming in the Tropics and cooling in the temperate regions [*Bonan*, 2004]. The changes in temperature can range from a couple of degrees to tens of degrees Celsius [*Gash and Nobre*, 1997; *Ray et al.*, 2003; *Manoharan et al.*, 2009]. Thus, deforestation changes land cover. Thus influencing the albedo, surface temperature, soil fertility, surface roughness, soil moisture and changes in the magnitude of thermal energy (LH and SH) fluxes. The above changes, in turn, influence the regional boundary layer depth, cloudiness, cloud optical properties and local rainfall [*Nair et al.*, 2003; *Ray et al.*, 2003; *Pielke*, 2001]. In a stable atmosphere the clouds tend to form earlier over the moist surfaces, whereas in a less stable environment, the clouds tend to form earlier over the drier surfaces [*Wetzel et al.*, 1996]. In a forested region the boundary layer air tends to be moister than in a deforested region. An average difference of 1g kg^{-1} difference in specific humidity has been observed between the forested and deforested forests in the Amazon [*Bastable et al.*, 1993]. *Pielke* [2001] provides a detailed review of the influence of vegetation and soil characteristics on cumulus cloud formation and precipitation, and notes in particular that the alteration in heat fluxes as a result of deforestation modifies the convective activity by modifying the environment for thunderstorms, which are an effective conduit of heat, moisture, and momentum to higher latitudes and exerts major impacts on global weather and climate.

MBCs are regions of contrasting vegetation types such as forests adjacent to deforested regions. There are several observational and modeling studies [*Segal et al.*, 1988; *Avissar and Liu*, 1996; *Weaver and Avissar*, 2001] that have shown that such contrasting vegetation types lead to differential heating which in turn results in sea-breeze-like mesoscale circulations that increase cloudiness. *Avissar and Liu* [1996] noted that the updrafts created by surface heterogeneity are much stronger than those created as a result of mechanical turbulence. Climatic feedbacks from deforestation can also alter rainfall patterns, initially increasing [*Avissar et al.*, 2002] then followed by drastic decreases in precipitation [*Avissar et al.*, 2004]. For example, *Lyons et al.* [1993] showed that in southwest Australia a substantial clearing of native vegetation led to a 20% decrease in local winter rainfall. Also *Ray et al.* [2003] in their study over western Australia along the bunny fence (a rabbit proof fence running 750 km separating the native vegetation from the farmland) area observed a higher frequency of low level cumulus cloud cover over the native vegetation side than over the farmland. The land use changes led to differences in soil moisture availability and hence the surface energy fluxes which in turn enhanced the cumulus cloudiness over the native vegetation than the agricultural land. However other studies by *Otterman* [1990], *Sud et al.* [1993], *Pielke et al.* [1998] show that deforestation leads to decreases in rainfall. Since the mid 1970s, tropical forest regions have experienced declines in precipitation at a rate of 1.0±0.8 % per decade with sharp declines in northern Africa (3 % to 4 %/decade), marginal declines over Asia, and no significant trend in Amazonia.

Forest clearing tends to increase SH fluxes and reduce LH fluxes. This, in turn, causes the development of deeper turbulent convective boundary layers, with widespread cumulus clouds more likely to occur over the deforested regions than over the pristine forests (e.g., *Avissar et al.*, [2002]). Tendencies of increased convective available potential energy (CAPE), increased updraft widths and increased cloud base heights also have been reported (e.g., *Negri et al.*, [2004]). CAPE is a measure of the energy available for convection. It is directly related to the vertical speed of the updrafts and thus higher values of CAPE indicate greater potential for severe weather. In tropical soundings, parcel temperatures of 1 to 2 K excesses may occur at a depth of 10 - 12 km. A typical value of CAPE is then 500 Jkg^{-1}. However, for a mid-latitude thunderstorm environment, the value of CAPE often may exceed 1000 Jkg^{-1}, and in severe weather cases it may exceed 5000 Jkg^{-1}. This small value of CAPE in the tropical environment is the major reason that the updraft velocities in tropical cumulonimbus are observed to be much smaller than those in mid-latitude thunderstorms [*Holton*, 2004].

Convective activity is strongly influenced by surface characteristics, with changes in land cover producing changes in local surface temperatures and precipitation rates. From satellite observations *Rabin et al.* [1990] and *Cutrim et al.* [1995] reported increased cloudiness over deforested areas in Amazonia and attributed this to land surface heterogeneity. Using GOES (Geostationary Operational Environmental Satellite), TRMM (Tropical Rainfall Measuring Mission) and Special Sensor Microwave Imager (SSM/I) satellite data, *Negri et al.* [2004] found that enhanced dry season surface heating created a thermal circulation which increased shallow cumulus clouds, and the precipitation resulting from deep convection over deforested relative to those over forested regions in Amazonia. However, observations are not entirely consistent. Based upon ten years of 3-hourly infrared window channel observations taken during the International Satellite Cloud Climatology Project (ISCCP) over a 2.5°x2.5° grid, *Durieux et al.* [2003] reported no significant differences in dry season cloud cover between forested and deforested regions of Amazonia.

Modeling studies likewise are inconsistent. *Eltahir* [1996] reported that large-scale deforestation could weaken large-scale circulation patterns which would lead to reduced rainfall. At the mesoscale, *Eltahir and Bras* [1994] reported that deforestation will lead to a reduction in precipitation, but would have no effect on large-scale circulations. However, at small scales on the order of 10-100 km, *Wang et al.* [2000] reported that the organization of rainfall reflects land cover patterns and that there is enhanced cumulus cloud cover and enhanced deep convection over deforested patches. As expected they found no relationship between shallow clouds and land cover patterns in the wet season. During the period between the dry and wet seasons, they suggested that deforestation may enhance afternoon cloudiness while contributing nothing to precipitation. However, during the dry season they found that the organization of rainfall does reflect land cover patterns, with enhanced shallow cumulus over deforested patches leading to enhanced deep convection. In contrast recent modeling studies by *Costa et al.* [2007] and *Sampaio et al.* [2007] came to divergent results. In these studies Amazonian forest was replaced by pasture and soybeans. *Costa et al.* [2007] used the Community Climate Model coupled to the Integrated Biosphere Simulator (CCM3-IBIS) climate model on a 2.81° grid and *Sampaio et al.* [2007] used the Centro de Previsao do Tempoe Estudos Climaticos do Instituto Nacional de Pesquisas Espaciais (CPTEC-INPE) global model at 2° spatial resolution. Both results showed that replacement of forest by pasture leads to higher values of albedo, SH flux, and surface temperature with decreased values of roughness, turbulence, Leaf Area Index (LAI), root depth, LH flux, evapotranspiration, cloud cover and precipitation. Intriguingly, precipitation was decreased even further as pasture was replaced by soybeans. This was attributed to the albedo of soybean plantation being higher than that of the pasture. *Sampaio et al.* [2007] found that cloud cover was significantly decreased by about 12% over deforested areas converted to pastures and by about 16% for soybean croplands, while precipitation decreased by 18% over pastures and 25% over soybeans. However, these studies utilized synoptic-scale grids that may not be representative of smaller-scale processes. Indeed, *Sampaio et al.* [2007] suggested that fragmented forest patches may create local circulations which in turn may enhance precipitation over the deforested regions. Table 2 summarizes the above discussed literature survey.

S.N.	Reference	Data & Study Area	Key conclusions/Remarks
1	*Otterman, 1990*	Landsat, Israel	Afforestation leads to reduced surface albedo and reduced soil heat flux and hence increases in precipitation.
2	*Rabin et al., 1990*	AVHRR from NOAA-7, Oklahoma	Clouds form earliest over regions of high SH and high albedo and are suppressed over regions of high LH fluxes. Convection enhancement at mesoscale results due to land surface heterogeneity.
3	*Lyons et al., 1993*	Field experiments, AVHRR, Southwestern Australia	Substantial clearing of native vegetation lead to a 20% decrease in local winter rainfall. Cumulus cloud frequency is thus high over areas with high LH fluxes and high CAPE.
4	*Eltahir and Bras, 1994*	Amazonia	Small scale deforestation (~250 km) may lead to reduction in precipitation, but would have no effect on large-scale circulations.
5	*Cutrim et al., 1995*	GOES, Amazonia	Enhanced dry season cumulus clouds frequency over forest cleared regions.
6	*Avissar and Liu, 1996*	RAMS	Contrasting vegetation types lead to differential heating which in turn results in sea-breeze-like mesoscale circulations that increase cloudiness. Updrafts created by surface heterogeneity are much stronger than those created as a result of turbulence.
7	*Eltahir, 1996*	Amazonia	Large-scale deforestation (~2500 km) could weaken large-scale circulation patterns which would lead to reduced rainfall.
8	*Rabin and Martin, 1996*	GOES, Central United States	Slightest change in elevation can modulate the cumulus cloud frequency. And cumulus frequency is inversely associated with plant cover and available soil moisture.
9	*Pielke et al., 1998*	RAMS, South Florida	During the past 100 years 11% decrease in summer deep cumulus rainfall due to land cover change and this climate change is irreversible due to permanent land cover change.
10	*Wang et al., 2000*	MM5V2, Rondônia, Amazonia	At scales of 10 km observed enhanced cloud cover and enhanced and enhanced deep convection over deforested patches during dry season.
11	*Weaver and Avissar 2001*	RAMS, ARM-CART, US Central Plains	Enhancement in mesoscale convection arising due to land cover heterogeneity.

S.N.	Reference	Data & Study Area	Key conclusions/Remarks
12	*Avissar et al., 2002*	Satellite observations and Model simulations, Amazonia	Climatic feedbacks from deforestation can also alter the rainfall patterns, initially increasing followed by drastic decrease in precipitation.
13	*Ray et al., 2003*	ASTER, MODIS, GMS5, Southwestern Australia	High frequency of cumulus cloud cover over regions of high LH heat flux and high available energy.
14	*Durieux et al., 2003*	ISCCP, GPCP, TRFIC, Amazonia	More wet season rainfall in deforested regions and less dry season rainfall than the forested regions.
15	*Nair et al., 2003*	GOES 8, Landsat MSS, RAMS, Costa Rica	Deforestation leads to warmer, drier air upwind of the mountain and increasing the cloud base heights.
16	*Malhi and Phillips, 2004*	Tropics	Since mid 1970s, significant decline in precipitation at a rate of 1.0 ± 0.8 % per decade with sharp decline in northern Africa (3 to 4 %/decade), marginal declines over Asia, and no significant trend in Amazonia.
17	*Negri et al., 2004*	TRMM & SSMI, Southwest Brazil	During dry season there is enhanced shallow cumulus cloudiness and deep convection over deforested areas than dense forests.
18	*Ray et al., 2006*	MODIS, GOES, RAMS, Guatemala	High frequency of cumulus cloud cover, high rainfall rate over drier deforested areas than pristine forests.
19	*Costa et al., 2007*	CCM3-IBIS climate model (2.8°), Amazonia	Forests to pasture lead to reduced cloud cover and rainfall; forest to soybean lead to increase in albedo and further decrease in cloud cover and precipitation.
20	*Sampaio et al., 2007*	CPTEC-INPE (2°), Amazonia	Observed similar to *Costa et al.* [2007].
21	*Nepstad et al., 2002*	Brazil's Tapajo´s National Forest, in east-central Amazonia	The forest leaves were quite tolerant to the soil moisture reduction provoked by throughfall exclusion. Instead of a pulse of leaf shedding, the exclusion treatment have inhibited the formation of new leaves, leading to a decline in fine litter production and, eventually, a thinning of the leaf canopy.

Table 2. Literature survey on satellite remote sensing and modeling studies of impact on land cover change on regional cumulus cloud cover and precipitation.

Other significant and indirect impacts of land use changes include loss of spiritual and cultural benefits from ecosystems, both for the indigenous people and others enjoying recreational opportunities [*Ramakrishnan*, 2001]. Deforestation could also result in disease emergence [*Patz et al.*, 2000; *Patz et al.*, 2004] such as dengue and malaria [*Tauil*, 2001], diarrhea [*De Souza et al.*, 2001] and other respiratory diseases [*D'Amato et al.*, 2001].

3. Impact of deforestation on flora and fauna due to hydrometeorological disturbances

Temperature, precipitation, wind/storms, solar radiation, long-wave radiation, atmospheric concentration of carbon dioxide and ozone influence biogeochemical cycles, greenhouse gas fluxes and surface energy balance, which in turn impact the biodiversity of ecosystems. Species tend to be attracted to their optimum climate. Therefore, if temperature and precipitation changes, species would be expected to either expand or contract their range depending on favored conditions [*Peters and Darling*, 1985; *Ford*, 1982].

Forests that are undisturbed tend to be dark under the canopy, humid, with stable temperature, and light wind [*Laurence et al.*, 2002]. Deforestation creates forest edges with increased temperatures, reduced humidity and increased sunlight. This "edge effect" can penetrate 40 to 60 m deep in the forest [*Kapos*, 1989; *Didham and Lawton*, 1999]. These changes in the hydrometeorology parameters impact the flora and fauna found in the forest fragments. Many tropical animals require large areas of native vegetation for their survival, and isolation of forest fragments impacts their survival due to lack of water and food. Studies by *Dale et al.* [1994] over the tropical forests show that some species in small and isolated patches of forests do not cross even relatively small deforested areas. The survival of species in isolated forest patches strongly depends on suitable habitats of sufficient spatial extent to support their population. Decreases in the movement of animals across an ecosystem can limit/reduce the nutrient exchange between forest patches [*Saunders*, 1991].

For plants, hydrometeorological changes are impacted by decreased evapotranspiration, and soil moisture depletion. Fragmented forests are more vulnerable to lateral shear force exerted by increased wind speed, turbulence and vorticity [*Bergen*, 1985; *Miller et al.*, 2001]. This increases the mortality rate of trees and damage within 100 to 300 m of the edges of forest fragments [*Ferreira and Laurence*, 1997; *Laurence et al.*, 1998]. *Laurence et al.* [2000] found that in Amazonia trees die three times faster near the edges than those at the interior. Some trees simply drop leaves and die at the forest edges due to sudden changes in temperature, moisture and sunlight [*Lovejoy et al.*, 1986; *Sizer and Tanner*, 1999]. Tree mortality impacts canopy dynamics [*Ferreira and Laurance*, 1997], which in turn alters forest structure, composition and diversity. New trees growing near the edges adapt to the new environment, but these are dissimilar to the forest-interior trees [*Viana et al.*, 1997]. Thus, rare and localized species whose range becomes unsuitable tend to be threatened to extinction unless dispersion and colonization is possible.

4. Outline of this study

Although there are many studies as discussed above [Table 2], from both observations and models [*Negri et al.*, 2004; *Wang et al.*, 2000; *Costa et al.*, 2007; *Sampaio et al.*, 2007] to quantify the impact of deforestation on regional weather and climate, many of these studies exclusively focus on large homogeneous, pristine forests of Amazonia that were replaced by farmlands and pastures. The results presented in the literature are not only inconsistent, but in many cases are opposing in their conclusions. Furthermore, it is unknown to what degree results reached for large, relatively homogeneous regions such as Amazonia are applicable to small regions with strong oceanic contributions to weather and climate.

Ray et al. [2006] utilized satellite observations and regional atmospheric model simulations in Guatemala and adjacent areas and reported that deforestation locally intensifies the dry season, increasing the risk of fire along the long corridor regions connecting the protected areas. They showed that the deforested habitats in dry season have higher day-time temperatures, less cloud cover, less soil moisture and low values of Normalized Difference Vegetation Index (NDVI) than the forests in the same life zone.

A detailed analysis of the climate parameters, such as change in land cover, surface temperature, NDVI, soil moisture, albedo, cloud formation, and precipitation over the protected reserves and the corridors connecting these reserves for different seasons, is required to determine the regions in the MBC that are potentially "*stable*" and "*unstable.*" This research work examines these issues in detail over samples of forested and deforested regions in Guatemala and the stability and sustainability of the MBCs in the region. analyzes whether relatively small-scale forested regions adjacent to deforested regions in Guatemala create meaningful differences in convective initiation and precipitation, which would be of importance to the sustenance of the proposed MBC.

5. Study area

Figure 1 shows the parks, reserves and protected regions within the study area along with the narrow corridors proposed to connect patches of similar habitat. The largest protected region is the Maya Biosphere Reserve in the northern Petén region which contains more than a million hectares of tropical forest. *Cochrane* [2003] showed that long corridors with long perimeters relative to their area may be particularly threatened by fire because forest edges usually are drier and more vulnerable. *Ray et al.* [2006] suggested that deforestation may be lengthening and locally intensifying the dry season which would adversely impact second-growth forest regeneration.

Guatemala is composed of the mountainous highlands, the Pacific coast region south of the highlands and the hot, humid tropical and relatively flat Petén lowlands to the north. Approximately one-third of Guatemala is forested with about half of that total is composed of primary forests. The choice of this site is dictated in part because *Ray et al.* [2006] showed that at the height of the dry season (March), many of the forests in the Petén region have estimated rainfall deficits of up to 25mm which would have potentially serious consequences for the corridor regions. The *Ray et al.* [2006] results were generated from cloud cover rainfall regression statistics, and a more extensive analysis of the region is necessary to estimate the potential vulnerability of the proposed MBC corridors.

6. Model

The GEMRAMS atmosphere-vegetation coupled model [*Eastman et al.*, 2001; *Beltrán* 2005; *Beltrán et al.*, 2008] was used for the modeling experiments. It is comprised of the Colorado State University Regional Atmospheric Modeling System (RAMS) 4.3 [*Pielke et al.*, 1992; *Cotton et al.*, 2003] and the General Energy and Mass Transport Model [GEMTM, *Chen and Coughenour*, 1994]. RAMS is a general-purpose, atmospheric simulation model that includes the equations of motion, heat, moisture and continuity in a terrain-following coordinate system. It is a fully three-dimensional and non-hydrostatic model. RAMS also includes a soil-vegetation-atmosphere transfer scheme, the Land Ecosystem-Atmosphere Feedback model, version 2 (LEAF-2) [*Walko et al.*, 2000] that represents the storage and exchange of heat and moisture associated with the vegetation and canopy air and soil.

GEMTM is an ecophysiological process-based model that can be used to simulate the dynamic interactions between the atmosphere and the growing canopy [*Chen and Coughenour*, 1994]. Several of the GEMTM components were coupled to RAMS: canopy radiation transfer, plant and root growth, soil water dynamics, biomass production and soil respiration submodels. These components require an additional set of parameters, mostly vegetation dependent, to characterize these biological processes.

In GEMRAMS the near-surface atmosphere and biosphere are allowed to dynamically interact through the surface and canopy energy balance. Precipitation, canopy air and soil temperature, humidity, winds and surface LH and SH fluxes are computed by RAMS. Photosynthesis at the leaf-level is calculated for sunlit and shaded leaves as a function of photosynthetic active radiation (PAR) and temperature. Water stress effect on the assimilation rate also is considered. At the canopy level, photosynthesis and conductance are calculated by scaling-up from the corresponding sunlit and shaded leaves using sunlit and shaded LAI using light extinction coefficients from a multi-level canopy radiation model [*Goudriaan*, 1977]. The available photosynthate is allocated to leaves, stems, roots and reproductive organs with variable partition coefficients which are functions of soil water conditions. As water stress increases, the fraction allocated to root growth increases. The root profile is updated daily through the processes of branching, extension and death [*Chen and Lieth*, 1993].

Cumulus parameterization in the model is accomplished using the Kain-Fritsch scheme [*Kain and Fritsch*, 1993]. GEMRAMS is a state-of-the-art soil-vegetation-atmospheric simulation approach which offers much increased realism in modeling complex forest-deforestation circulation patterns. The original GEMRAMS model was built based on RAMS v4.3. The RAMS model has evolved to the latest version v6.0, and one of the major improvements includes the use of independent satellite observations of vegetation greenness represented by the Normalized Difference Vegetation Index (NDVI).

6.1 Model configuration

Two nested domains are used. The outer domain, with a horizontal grid of 40 km x 40 km and covering an area about 2000 km on a side, is centered at the point of 16.0N and -90.0W in Guatemala. This domain is used mainly to incorporate as much real-time weather as possible within the chosen regional scale domain. Real-time weather data are obtained from National Centers for Environmental Prediction Final gridded analysis datasets (NCEP FNL), which is global at 1 x 1 degree, 6-hourly. The inner domain in this study has a horizontal grid of 10 km x 10 km and covers an area of 620 km x 620 km, centered at the same point. The inner and outer domains are shown overlaid onto the map of vegetation types compiled from MODIS data by the University of Maryland in Figure 2a. Figure 2b shows the topography of the study region. The Petén study area is characterized by relatively low relief (< 300 m) and is bounded roughly from 16N to 18N and 89W to 91.5W.

Thirty-five vertical levels are used in the model, with the lowest level 20 m from ground, increasing with a ratio 1.15 up to a maximum of 1200 m near the model top at 18.7 km. The fine vertical grid limited the model time step to 30 seconds for the outer and 10 seconds for the inner domain. It is noted that the model was run with the "full" dynamic features (topography, various landscape features, etc.) over the two nested domains.

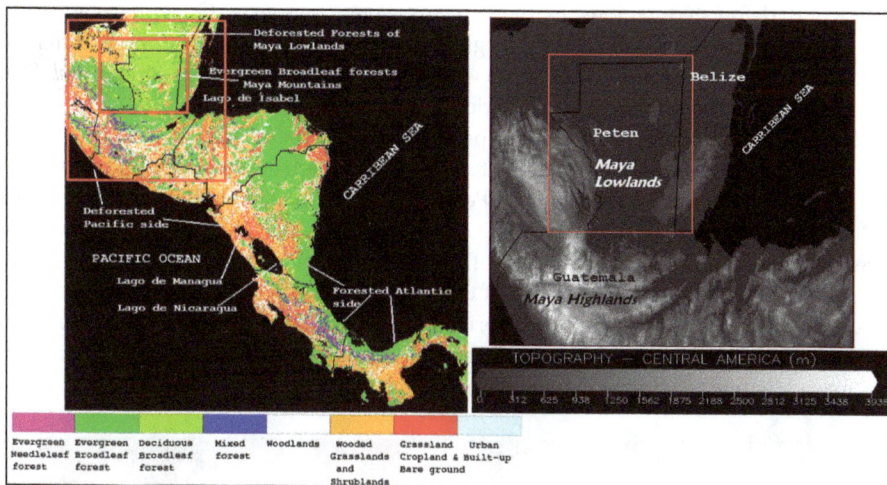

Fig. 2. (a) The University of Maryland ecosystem map of Central America with outer and inner domains overlaid (red boxes), (b) The two nested domains used for model simulation shown over the topography map of Guatemala. The outer domain has a horizontal grid of 40 km x 40 km and covering an area about 2000 km on a side and is centered at the point of 16.0N and -90.0W in Guatemala. The inner domain has a horizontal grid of 10 km x 10 km and covers an area of 620 km x 620 km, centered at the same point. The major vegetation is the Evergreen Broadleaf Trees, which shows indeed a quite "smoothed" distribution. However, there are small-scale patchy, heterogeneous land covers which represent partial- or non-forested land covers.

The northern half of the Petén region tends to be characterized by protected regions and tropical forests, while the southern and western regions are heavily deforested. LAI in these dense tropical rainforests are ~ 6 while values in the range of ~2 are associated with the small-scale, patchy, heterogeneous land surface conditions in the non-forested areas. Roughness heights are approximately 2 m in the forests and a few centimeters in the pastures. Albedo values of about 14% are taken for the forests and approximately 16.5% for the deforested regions.

6.2 Soil initialization

The model soil parameters are more difficult to initialize, due to the lack of satisfactory data in most regions. *Reichle et al.* [2004] discussed a three-way comparison regarding soil modeling (in-situ measurement, remote satellite observation, and numerical modeling) and pointed out that there is as yet no "consensus" among the three. In the current study, soil moisture is initialized using the NCEP FNL 2-layer soil data, in which vertical variation is incorporated into the model. Horizontally, the data over Guatemala is averaged. Soil temperature is initialized with "zero-offset" from the NCEP FNL lowest air temperature. Also, soil (3D) texture is initialized following the same procedure of LAI and vegetation fraction as provided by RAMS v6.0.

7. Model results

The objective of this study is to determine environmental and climatic differences in forested and deforested regions in the lowland Petén region of Guatemala as well as circulation patterns both within these regions and along the borders. Large differences between the forested and deforested areas would support the hypothesis of *Ray et al.* [2006] who proposed that the proposed MBC corridors are potentially unstable.

A 30-day simulation of the dry season is performed in the Petén region with the northern half forested and the southern half deforested. This is in rough agreement with the primarily forested conditions found in the northern Petén today and the largely deforested conditions found in the southern region.

Figure 3 shows the model configuration of vegetation assumed in this study. The vertical north-south line (AB) through the center of the region at 90.25W shows the location at which most of the results are described. The horizontal east-west line through the center of the region at 17N divides the primarily evergreen broadleaf forested region to the north from the primarily deforested pasture region to the south.

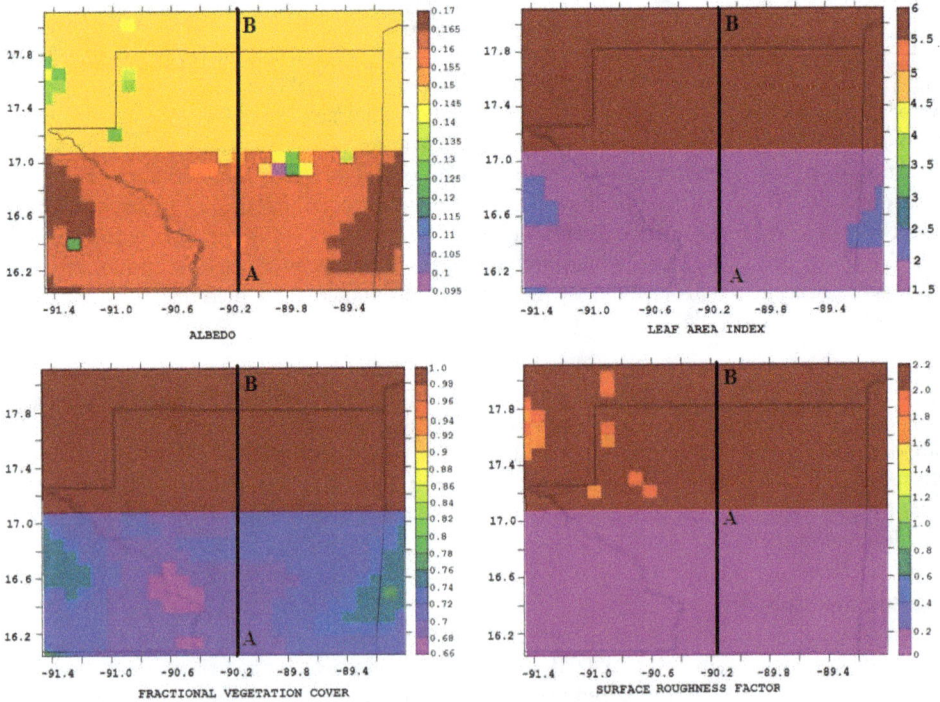

Fig. 3. Model Configuration: (a) Albedo of the Petén. (b) LAI field over the Petén on March 8th, 2003 (dry day) at 12 noon local time (1800 UTC). (c) and (d) Similar to (b) but for fractional vegetation cover and surface roughness factor respectively.

Based upon satellite images of the Petén region, it is assumed that fractional vegetation cover is about 98% in the forested regions and about 70-75% in the deforested regions. As seen in Figures 3 a, b, c and d the LAI is ~6 for the Evergreen Broadleaf Forests, surface roughness length is about 2m and surface albedo is about 14%. Note that once the vegetation type is chosen, the GEMRAMS model initializes the surface parameters. The value of 14% for the forests is consistent with satellite remote sensing estimates and the literature (e.g., *Pielke*, 1984, Table 11-8). Note that coniferous forests have much lower albedo on the order of ~10% and LAI values of 2-4, but these are not prevalent in the study region. In general, forests may have albedo values ranging from about 10% to 20%, depending upon forest type and season. Whereas, the LAI values for the deforested regions range from 1.5 - 2.5 with very small roughness lengths (nearly zero) appropriate for pastures and with albedo ~ 16.5%.It is important to note that the GEMRAMS model is fully dynamic. Therefore, the surface parameters of albedo, LAI, fractional vegetation cover and roughness factor change over time in response to environmental conditions such as surface temperature, humidity and precipitation. However, for the short term study conducted here over a one month period, the surface parameters varied little.

Within a typical dry season, there may be some days which are hot, dry and cloudless, others that have fair-weather cumulus and others that experience unstable atmospheric conditions, deep convection and heavy precipitation. Figure 4 shows daily precipitation averaged over the forested and deforested regions for March 2003. Note that there was no precipitation in first half of the month over both the regions. About a third of the days experienced small precipitation events such as showers, and about six of the days experienced unstable conditions with deep convection and heavy precipitation. The environmental conditions necessary for the formation of deep convection are destabilization of the air parcels and lifting the destabilized air to the level of free convection [*Wallace and Hobbs*, 2006]. The destabilization is associated with the lifting of the air parcel and the low level convergence influenced from large-scale forcings such as an extratropical cyclone, which are generally anticipated a day or more before. In this decade, 2003 was year of many extratropical cyclones. Central America was highly influenced by the troughs associated with the cold fronts coming from north [*IPCC*, 2007].And the lifting of air parcels that initiates the deep convection is associated with localized, short-lived and less predictable forcings such as the sea-breeze or outflow of any pre-existing convective storms [*Wallace and Hobbs*, 2006].

Figure 5a shows SH fluxes peak values tend to occur at 1400LT, which is the time of maximum convection on days in which it occurs. SH fluxes range in value from about 400-500 Wm-2 for about the first two weeks of March and then again on the dry days of 21-22 March. SH flux decreased to about 300 Wm-2 on 16 March the day convective rains. Then slowly increasedas the ground dried out. The day of 23 March had strongly unstable conditions and heavy precipitation and SH fluxes < 200 Wm-2 with similar low values at the end of the month. Note that SH fluxes are within 10% of each other in the forested and deforested regions throughout this dry season on dry and convective days.

Figure 5b shows the corresponding LH fluxes during this month at the same time of the day. LH fluxes range from < 200 Wm-2 on the very dry days to > 600 Wm-2 on the very convective days during the later half of March. High LH release associated with warm surface temperatures acts to maintain the horizontal temperature gradient thereby increasing the supply of potential energy to build up the deep convection [*Wallace and Hobbs*, 2006]. Note in particular that the LH fluxes are up to twice as large in the forested regions during the "dry" days.Differences in LH fluxes between the forested and deforested regions are much smaller during the convective days.

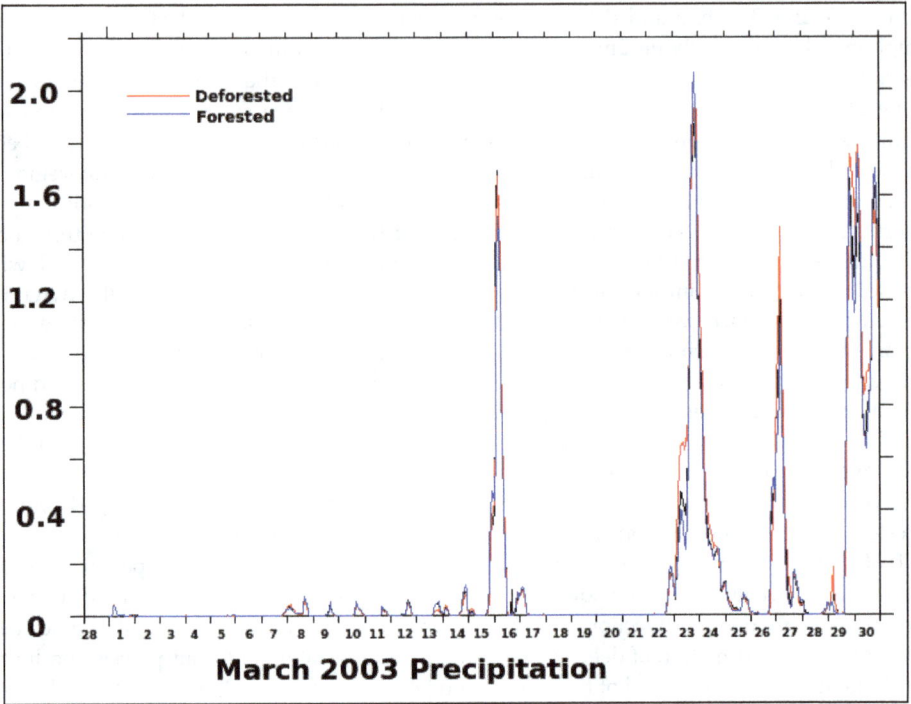

Fig. 4. Diurnal (hourly) precipitations for March 2003 averaged over forested and deforested regions.

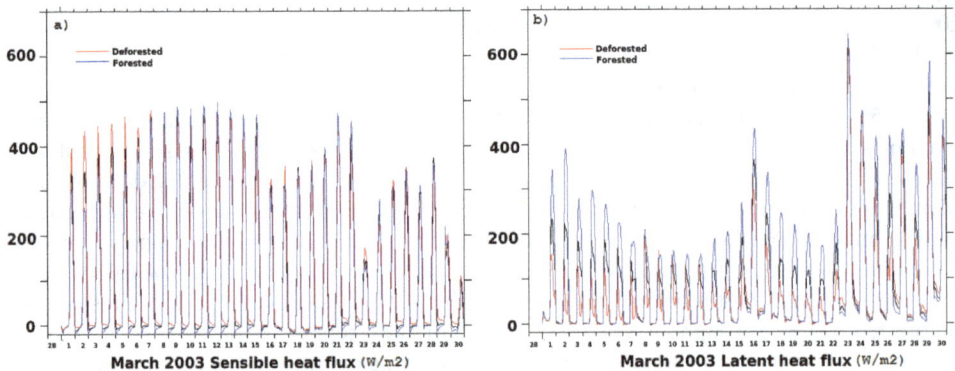

Fig. 5. (a) Diurnal (hourly) SH flux for March 2003 averaged over forested and deforested regions, (b) same as (a) but for LH flux.

It might be expected that the forests with their larger LH fluxes would produce greater cloud cover on the drier days, but this is not found to be the case during this study period. The following case studies examine representative dry, showery and convective days, comparing conditions in the forested and deforested regions in each case.

7.1 Dry day

Table 3a,b shows temperature, LH, SH, PBL height, precipitation and cloud cover for the south (pasture) and north (forest) sides for both a representative dry day (March 8) and a representative convective day (March 23) as a function of time (0600, 1000, 1200, 1400 and 1800 LT). These are values averaged over each of the two sides, forest to the north and pasture to the south.

a) Dry day						Convective day					
Time	Temperature (K)		LH (Wm-2)		SH (Wm-2)		Temperature (K)		LH (Wm-2)		SH (Wm-2)	
	Pasture	Forest	Pasture	Forest	Pasture	Forest	Pasture	Forest	Pasture	Forest	Pasture	Forest
6am	296.18	293.69	12	8.23	12.47	15.31	299.11	297.57	100.95	31.87	14.44	49.56
10am	305.92	303.56	163.96	54.02	309.58	439.61	302.12	299.42	618.114	719.05	87.52	47.08
noon	310.49	306.44	37.61	149.31	471.17	485.66	302.73	300.89	670.557	476.62	122.63	82.4
2pm	310.79	306.66	28.03	138.75	373.34	333.41	302.83	299.08	574.53	482.88	117.29	32.73
6pm	303.73	300.45	0.5044	0.7795	7.12	-26.14	298.42	295.94	128.29	60.37	-5.57	-10.197

b) Dry day						Convective day					
Time	Planetary Boundary layer (m)		Precipitation (mm)		Cloud Cover (%)		Planetary Boundary layer (m)		Precipitation (mm)		Cloud Cover (%)	
	Pasture	Forest	Pasture	Forest	Pasture	Forest	Pasture	Forest	Pasture	Forest	Pasture	Forest
6am	164.38	491.38	0.32	0.77	13.95	13.76	465.27	293.49	37.62	11.91	21.24	16.46
10am	1104.23	1267.63	0.59	0.87	16.72	13.8	927.65	1199.39	44.97	15.58	34.17	22.68
noon	1658.03	1811.46	0.59	0.87	13.85	13.76	1193.71	1456.6	50.73	18.12	33.08	22.49
2pm	2017.77	2065.17	0.6	0.88	12.62	12.62	1272.75	1727.46	55.29	22.82	37.57	21.72
6pm	1442.22	1088.76	0.6	0.88	10.88	12.14	834.66	994.18	61.39	25.52	20.67	18.85

Table 3. Temperature (K), LH flux (Wm-2), SH flux (Wm-2), PBL height (m), precipitation (mm) and cloud cover (%) for the pasture (south) and forest (north) sides for dry day (March 8th, 2003) and convective day (March 23rd, 2003) at 6 am, 10 am, 12 noon, 2 pm and 6 pm.

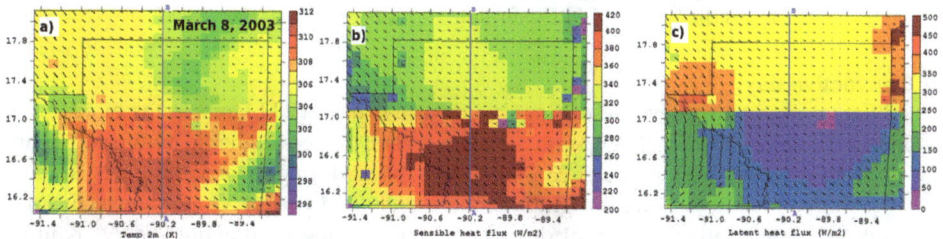

Fig. 6. Spatial distribution of temperature at 2 m (K), SH flux and LH flux on the dry day (March 8th, 2003) at 1400 LT

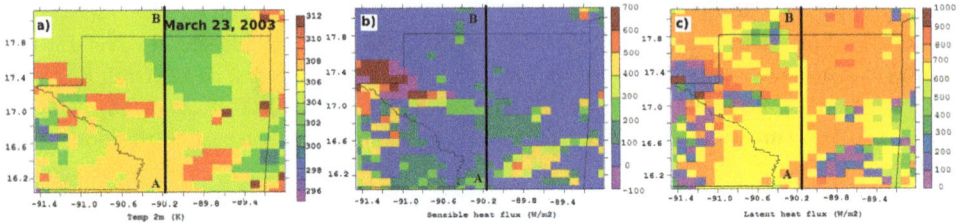

Fig. 7. Spatial distribution of temperature at 2 m (K), SH flux and LH flux at 1400 LT on the convective day (March 23rd, 2003)

For the dry day, Table 3a shows that temperature in the pasture is about 2K warmer than the forest in the early morning, increasing to about 4K warmer during mid-day. Indeed, the literature and the satellite observation studies discussed in Chapter 3 show that during dry season deforested regions tend to be hotter than forested ones in virtually all regions worldwide. Also we see that the values of LH and SH fluxes were very small in the early morning, with SH values approaching 500 Wm^{-2} in both pastures and forests by mid-day and then decreasing again by evening. Sensible heat values were relatively similar in both pastures and forests.

Under these very dry conditions, values of LH from the pastures spiked to values of about 160 Wm^{-2} in mid-morning (10am LT) due to "burnoff of early morning dew while forests had values only a third as large. By mid-day (noon LT) LH fluxes decreased to about 30 Wm^{-2} in the deforested regions while values in forests increased to about 140 – 150 Wm^{-2}. Note that during mid-day SH fluxes were about three times larger than the LH fluxes in both pastures and forests.

Cloud cover ranged from about 12% to 14% over both regions during this dry day, and precipitation was minimal, averaging less than 1 mm in both regions. The PBL height was <200 m over the pastures in the early morning and about three times that height (about 500 m) over the forests.

The boundary layer height grew rapidly during the morning to about 1800 m over the forests and about 1650 m over the pastures by noon and then to about 2000 m by 1400 LT. The PBL remained slightly higher over the forests for most of the day, except during the evening hours.

Figures 7a, b, c show the spatial distribution of surface temperature, SH and LH fluxes, respectively, for the study area at 1400 LT for this dry day. On the dry day there is clear distinction between the forests and deforested regions. 2m surface temperatures over the southern deforested regions are 4 to 8 K warmer than forests in the north. Similar differences are seen in the SH and LH fluxes spatial plots.

7.2 Convective day

Very different conditions prevailed during the convective day of 23 March. In the more humid conditions, Table 3a shows that temperatures were several degrees higher in both pastures and forests in the early morning hours of 23 March than found during the dry day. This is due to higher radiative cooling during the nighttime hours over the dry areas. On the convective day, temperatures were relatively constant during the day, ranging from about 298-299 K in the morning to only about 302-303 K during the early afternoon hours, a variation of about 4 K. Note that on the dry day temperatures increased by 12-14 K from morning to early afternoon.

Under the relatively wet conditions of the convective day, a reversal in the behavior of the LH and SH fluxes is found, as compared to the dry day. On the convective day the SH fluxes are relatively small, in the range from 50 – 125 Wm^{-2} in both pastures and forests, while LH fluxes reached 600-700 Wm^{-2} in the late morning hours before decreasing somewhat during mid-day. Note that LH fluxes were significantly larger over pastures during mid-day than over forests and that SH fluxes were somewhat higher over pastures than forests.

Cloud cover is significantly higher over pastures than over forests on the convective day, with values of about 23% over forests and up to 38% over pastures during the early afternoon. These differences in cloud cover also are reflected in differences in precipitation. The precipitation increased from about 12 mm in the early morning hours to about 25 mm by evening over the forests, an increase of about 13 mm during the day. However, precipitation increased from about 38 mm in the early morning over the pastures to about 61 mm by late afternoon, an increase of 23 mm during the day. Overall, precipitation was more than double over the pastures. On an average the rain rate of the forested region was 0.114 mm/hour and deforested region was 0.151 mm/hour. The results suggest that under sufficiently convective conditions, pastures generate significantly higher precipitation rates than do forests

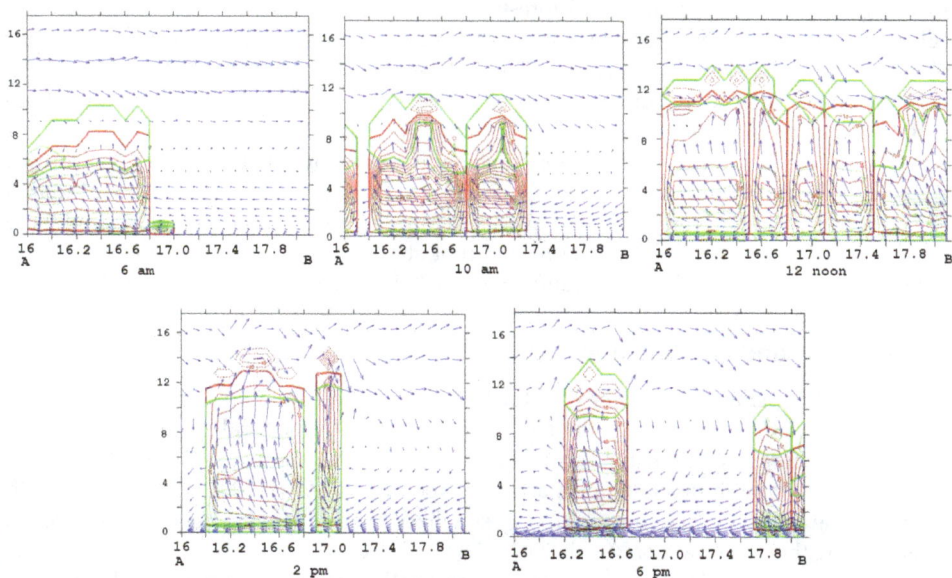

Fig. 8. Development of circulation patterns along the AB line shown in Figure 7 from early morning (6 am) to late afternoon (6 pm) on March 23rd, 2003 (convective day). The horizontal axis shows the latitude (in degrees) and the vertical axis is the height from the ground (in km). The red lines represent convective heating (^{o}C day^{-1}) (solid line – heating and dashed line – cooling) and the green lines represent convective moistening (g kg^{-1}day^{-1}) (solid line – moistening and dashed line – drying).

The height of the PBL was higher over the pastures (~450 m) than over the forests (~300 m) in the early morning hours. By 1400 LT the PBL grew to about 1300 m over the pastures and to about 1700 m over the forests, a value only slightly smaller than during the dry conditions. The height of the PBL over forested regions is not highly sensitive to wet and dry conditions. However, note that the height of the PBL over pastures is very sensitive to wet and dry conditions, reaching only 1275 m during the convective day.

Figures 8a, b, c show the spatial pattern of surface temperature, SH and LH fluxes, respectively, at 1400 LT over the study region. On convective day we don't clearly see the difference between the forest and the deforested regions spatial plots as seen on a dry day (Figure 7).

7.3 Circulation patterns

Figure 8 shows the development of circulation patterns along the line AB in Figure 7 from early morning to late afternoon on 23rd March, the convective day. In the morning hours, there is early convective activity over the pastures, with cloud tops reaching about 9 km. Note that convection is found only over pastures and not over the forests in the early morning. By 1000 LT strong updraft regions are developing in the convective regions and cloud tops reach about 12 km in height. At this time convective activity is initiated over the forests near the forest-pasture boundary. By local noon convective activity is found over the entire region, both over pastures and forests, with cloud tops reaching 13-14 km. Note the regions of strong updraft. By 1400LT the forest regions have ceased convection along this AB line, although very strong convection with very large updrafts is found over the pastures. By late afternoon, convective activity is ceasing, but with isolated cells over both pastures and forests.

Note that the cloud top heights are approximately the same over both pastures and forests. To examine the realism of these results, GOES infrared imagery was obtained for 23 March 2003. Cloud top heights were examined over both forested and deforested regions, and there were no significant differences in the results (not shown). Once generated the convective clouds in all situations modeled have sufficient CAPE to reach the tropopause level.

8. Discussion and conclusions

High surface temperatures together with the troughs associated with the cold fronts coming from the north tend to destabilize the air. This together with the local sea-breeze increases the potential energy required for increasing convection. And as we see from the results, convective activity with precipitation starts from the mid of March 2003. This together with the prevailing high surface temperature destabilizes the air leading to more convective storms during the latter days of March 2003.

During dry conditions LH fluxes are very low over both pastures and forests, and SH fluxes are a factor of three to ten times larger. PBL heights reach 2000 m during the heat of the day and there is minimal cloud cover on the order of 13% and virtually no precipitation for these dry conditions.

A very different scenario occurs during wet convective conditions. Under these conditions convection is initiated early in the morning hours over the pastures and not over the forests. By mid-day convection is found over both pastures and forests, and by late afternoon convection decreases over both regions, but much more so over the forested regions. LH fluxes become very large (~700 Wm-2) and are five to ten times larger than the SH fluxes. Cloud cover over the forests increases to about 22% during mid-day but up to about 38% over the pastures. Overall, substantial precipitation rates of 61 mm were found over the pastures compared to less than half this amount (~25 mm) for the forests. PBL heights are much lower over pastures than over forests.

These results are consistent with *Negri et al.* [2004] who utilized GOES data for their Amazonia study, a wet region. Higher cloud covers and precipitation rates are found over deforested regions. *Manoharan et al.* [2010] also used GOES data and found higher cloud cover over deforested regions in Guatemala. However, note that such conditions of higher cloud cover and precipitation are found only under wet, convective conditions and are not found under dry conditions. Furthermore, the results are consistent with *Wang et al.* [2000] who reported deforestation can create mesoscale circulations with rising motions that trigger dry season moist convection. The present results clearly demonstrate that much strong convective circulations are created over pastures (deforested) regions than over forests.

In terms of the sustainability of the lowland corridor regions in the proposed MBC, the results strongly suggest that forested corridors will experience warmer conditions due to higher temperatures in surrounding deforested areas. Also from the observational study by *Manoharan et al.* [2010], we see severe dryness and drought prevailing in the region during 2003. However, by far the most important factor is precipitation. During the first half of the month, there is little or no rainfall, whereas, during the latter days of the month (March 2003) we see significant convective activity over the region. This is the result of the increase in energy that initiates convection by increased SH release during the initial days of March and associated local-sea breeze. Thus, the forested corridors will receive higher than normal precipitation rates due to the fact that surrounding warmer deforested regions generate higher convective activity. The above scenario implies a "climate tipping point" will not occur in the proposed corridor regions within the lowland regions of Guatemala in the study area which would threaten their stability and sustainability.

9. References

Avissar, R., R. R. Da Silva, and D. Werth (2004), Implications of tropical deforestation for regional and global hydroclimate. In: *Ecosystems and land use change*, R. S. DeFries, G. P. Asner, and R. Houghton (eds.), AGU, Geophysical Mongraph 153, 73-83.

Avissar, R., P. L. Silva Dias, M. A. Silva Dias and C. A. Nobre (2002), The Large Scale Biosphere Atmosphere Experiment in Amazonia (LBA). Insights and future needs, *J. Geophys. Res.*, 107, doi:10.129/2002JD002704.

Avissar, R., and Y. Liu (1996), A three-dimensional numerical study of shallow convective clouds and precipitation induced by land-surface forcings, *J. Geophys. Res.*, 101, 7499– 7518.

Bastable, H.G., W.J. Shuttleworth, R.L.G. Dallarosa, G. Fisch and C.A. Nobre (1993), Observations of climate albedo, and surface radiation over cleared and undisturbed Amazonian forest, *Int. J. Climat*, 13, 783-796.

Beltrán-Przekurat, A., C. H. Marshall, and R. A. Pielke Sr. (2008), Ensemble reforecast of recent warm-season weather: impact of a dynamic vegetation parameterization, *J. Geophys. Res.*, 113, D24116, doi:10.1029/2007JD009480.

Beltrán, A. B. (2005), Using a coupled atmospheric-biospheric modeling system (GEMRAMS) to model the effects of land-use/land-cover changes on the near-surface atmosphere, Ph.D. dissertation, 186 pp., Colorado State University, Fort Collins, CO.

Bennett, A. F. (1999), Linkages in the landscape: The role of corridors and connectivity in wildlife conservation, IUCN, Gland, Switzerland.

Bergen, J. D. (1985), Some estimates of dissipation from the turbulent velocity component gradients over a forest canopy. In: *The forest-atmosphere interaction*, B. A. Hutchinson and B. B. Hicks (eds.), 613-630, Dordrecht, The Netherlands, D. Reidel.

Bonan, G. B. (2004), Biogeophysical feedbacks between land cover and climate. In: *Ecosystems and land use change*, R. S. DeFries, G. P. Asner, and R. Houghton (eds.), AGU, Geophysical Monograph 153, 61-72.

Carr, M.H., Lambert, J.D., Zwick, P.D. (1994), Mapping of Continuous Biological Corridor Potential in Central America, Final Report, Paseo Pantera, University of Florida.

Chen, D. X., and M. B. Coughenour (1994), GEMTM: A general model for energy and mass transfer of land surfaces and its application at the FIFE sites, *Agric. For. Meteorol.*, 68, 145-171.

Chen, D. X., and J. H. Lieth (1993), A two-dimensional, dynamic model for root growth distribution of potted plants, *J. Am. Soc. Hortic. Sci.*, 118, 181–187.

Cochrance, M. A. (2003), Fire science for rainforests, *Nature*, 421, 913-919.

Costa, M. H., S. N. M. Yanagi, P. J. O. P. Souza, A. Ribeiro and E. J. P. Rocha (2007), Climate change in Amazonia caused by soybean cropland expansion, as compared to caused by pastureland expansion, *Geophys. Res. Lett.*, 34, L07706, doi:10.1029/2007/GL029271.

Cotton, W. R., et al., (2003), RAMS 2001: Current status and future directions, *Meteor. Atmos. Phys*, 82:5–29

Coughlan, M, D. Jones, N. Plummer, A. Watkins, B. Trewin and S. Dawkins (2003), Impact of 2002-03 El Niño on Australian climate, Drought.com Workshop.

Cutrim, E.D., W. Martin, and R. Rabin (1995), Enhancement of cumulus clouds over deforested lands in Amazonia, *Bull. Am. Meteorol. Soc.*, 76, 1801-1805.

D'Amato, G., G. Liccardi, M. D'Amato and M. Cazzalo (2001), The role of outdoor air pollution and climatic changes on the rising trends in respiratory allergy, *Respir. Med.*, 95, 606-611.

Dale, V. H., S. M. Pearson, H. L. Otterman, and R. V. O'Neill (1994), Relating patterns of land-use change to faunal biodiversity in the Central Amazon, *Conservation Biology*, 8, 1027-1036.

DeFries, R. S., G. P. Asner, R. Houghton (2004), Trade-offs in land-use decisions: Towards a framework for assessing multiple ecosystem responses to land-use, In: *Ecosystems and land use changes*, R. S. DeFries, G. P. Asner, and R. Houghton (eds.), AGU, Geophysical Monograph 153, 1-9.

DeFries, R.S., L. Bounoua and G.J. Collatz (2002), Human modification of the landscape and surface climate in the next fifty years, *Glob. Chang. Biol.*, 8 (5), 438-458, doi:10.1046/j.1365-2486.2002.00483.x.

De Souza, A. C., K. E. Petersont, E. Cufino, M. I. Do Amaral, and J. Gardner (2001), Underlying and proximate determinants of diarrhoea-specific infant mortality rates among municipalities in the state of Ceara, north-east Brazil: And ecological study, *J. Biosocial Sciences*, 33, 227-244.

Didham, R. K., and J. J. Lawton (1999), Edge structure determines the magnitude of changes in microclimate and vegetation structure in tropical forest fragments, *Biotropica*, 31, 17-30.

Durieux, L., L.A.T. Machado and H. Laurent (2003), The impact of deforestation on cloud cover over the Amazon arc of deforestation, *Remote Sens. Environ.*, *86*(1), 132-140(9), doi:10.1016/S0034-4257(03)00095-6.

Eltahir, E. A. B. (1996), Role of vegetation in sustaining large-scale atmospheric circulations in the Tropics, *J. Geophys. Res.*, 101, 4255–4268.

Eltahir, E. A. B., and R. L. Bras (1994), Sensitivity of regional climate to deforestation in the Amazon basin, *Adv. Water Resour.*, 17, 101–115.

Eastman J. L., M. B. Coughenour, and R. A. Pielke Sr. (2001), The regional effects of CO_2 and landscape change using a coupled plant and meteorological model, *Global Change Biol.* 7: 797-815.

Eshleman, K. N. (2004), Hydrological consequences of land use change: A review of the state-of-science. In: *Ecosystems and land use changes*, R. S. DeFries, G. P. Asner, and R. Houghton (eds.), AGU, Geophysical Monograph 153, 13-29.

Ferreira, L. V., and W. F. Laurence (1997), Effects of forest fragmentation on mortality and damage of selected trees in central Amazonia, *Conserv. Bio.*, 11, 797-801.

Food and Agriculture Organization of the United Nations (1999), State of the World's Forests 1999, FAO Forestry Program.

Ford, M. J. (1982), *The changing climate*, London, George Allen and Unwin.

Gash, J. H. C., and C. A. Nobre (1997), Climatic effect of Amazonian deforestation: Some results from ABRACOS, *Bull. Am. Meteorol. Soc.*, 78, 823– 830.

Goudriaan, J (1977), Crop micrometeorology: A simulation study, Simulation Monographs, Pudoc, Wageningen, 249pp.

Herrera (2003), Implementation of the Convention on Biological Diversity in Mesoamerica: environmental and developmental perspectives.

Hansen, A. J. and J. J. Rotella (2002), Biophysical factors, land use and species viability in and around nature reserves, *Conserv. Biol.* 16, 1-12.

Hansen, A. J., R. P. Neilson, V. Dale, C. Flather, L. Iverson, D. J. Currie, S. Shafer, R. Cook, and P. Bartlein (2001), Global change in forests: interactions among biodiversity, climate and land use, *BioScience* 51, 765-779.

Holton, J. R. (2004), An Introduction to dynamic Meteorology, 4th Edition, Elsevier Academic press.

IPCC (2001), *Climate change 2001: Synthesis report*, Cambridge University Press, Cambridge, UK.

IPCC (2007), *Climate Change 2007: The Physical Science Basis. Contribution of Working Group I to the Fourth Assessment Report of the Intergovernmental Panel on Climate Change* [Solomon, S., D. Qin, M. Manning, Z. Chen, M. Marquis, K.B. Averyt, M.Tignor and H.L. Miller (eds.)]. Cambridge University Press, Cambridge, United Kingdom and New York, NY, USA

Kain, J. S., and J. M. Fritsch (1993), Convective parameterization for mesoscale models: the Kain-Fritsch scheme, The representation of cumulus convection in numerical models, *Meteor. Monogr.*, 46, 165-177.

Kapos, V. (1989), Effects of isolation on the water status of forest patches in the Brazilian Amazon, *J. Trop. Ecol.*, 5, 173-185.

Laurence, W. F. (2004), Forest-climate interactions in fragmented tropical landscapes, *Phil. Trans. R. Soc. Lond. B*, 359, 345-352, doi 10.1098/rstb.2003.1430.

Laurence, W. F., A. K. M. Albernaz, G. Schroth, P. M. Fearnside, E. Ventincinque, and C. Da Costa (2002), Predictors of deforestation in the Brazilian Amazon, *J. Biogeogr.* 29, 737-748.

Laurance, W.F., Delamonica, P., Laurance, S.G., Vasconcelos, H.L., Lovejoy, T.E., (2000), Conservation – rainforest fragmentation kills big trees, *Nature*, 404, 836.

Laurence, W. F., L. V. Ferreira, J. Rankin-de Merona, and S. G. Laurence (1998), Rain forest fragmentation and the dynamics of Amazonian tree communities, *Ecology*, 79, 2032-2040.

Lawton, R. O., U. S. Nair, R. A. Pielke and R. M. Welch (2001), Climate impact of tropical lowland deforestation on nearby montane cloud forests, *Science*, 294, 584-587.

Lovejoy, T. E., R. O. Bierregaard, Jr., A. B. Rylands, J. R. Malcolm, C. E. Quintela, L. H. Harper, K. S. Brown, Jr., A. H. Powell, G. V. N. Powell, H. O. R. Schubart, and M. B. Hays (1986), Edge and other effects of isolation on Amazon forest fragments, 257-325 In: M. E. Soule (Editor), Conservation Biology: The science scarcity and diversity, Sinauer Associates, Mass.

Lyons, T. J., P. Schwerdtfeger, J. M. Hacker, I. J. Foster, R. C. G. Smith, and H. Xinmei (1993), Land-atmosphere interaction in a semiarid region: The bunny fence experiment, *Bull. Am. Meteorol. Soc.*, 74, 1327–1334.

Manoharan, V. S., R. M. Welch and R. O. Lawton (2009), Impact of deforestation on regional surface temperatures and moisture in the Maya lowlands of Guatemala, *Geophys. Res. Lett.*, 36, L21701, doi: 10.1029/2009GL040818.

Miller, K., E. Chang, N. Johnson (2001), Defining Common Ground for the Mesoamerican Biological Corridor. World Resources Institute, Wash, DC. 45 pp.

Mittermeier, R.A., C. G. Mittermeier, N. Myers, G. A. B. Da Fonseca, J. Kent (2000), Biodiversity hotspots for conservation priorities, *Nature*, 403, 853.

Nair, U.S., R.O. Lawton, R.M. Welch and R.A. Pielke Sr. (2003), Impact of land use on Costa Rican tropical montane cloud forests: sensitivity of cumulus cloud field characteristics to lowland deforestation, *J. Geophys. Res.*, *108*(D7), 4206, doi:10.1029/2001JD001135.

Negri, A. J., R. F. Adler, L. Xu, J. Surratt (2004), The impact of Amazonian deforestation on dry season rainfall, *J. Clim.*, 17, 1306-1319.

Nepstad, D. C., et al. (2002), The effects of partial throughfall exclusion on canopy processes, aboveground production, and biogeochemistry of an Amazon forest, *J. Geophys. Res.*, 107(D20), 8085, doi:10.1029/2001JD000360.

Otterman, J., A. Manes, S. Rubin, P. Alpert, and D. O'C. Starr (1990), An increase of early rains in southern Israel following land-use change? *Bound. Layer Meteor.*, 53, 333-351.

Patz, J. A., P. Daszak, G. M. Tabor, A. A. Agruirre, M. Pearl, J. Epstein, N. D. Wolfe, A. M. Kilpatrick, J. Foufopoulos, D. Molyneux, and D. J. Bradley (2004), Unhealthy landscapes: Policy recommendations on land use change and infectious disease emergence, *Environ. Health Perspect.*, 112(10), 1092-1098.

Patz, J. A., T. K. Graczyk, N. Geller, and A. Y. Vittor (2000), Effects of environmental change on emerging parasitic diseases, *Int. J. Parasitol.*, 30 (12-13), 1395-1405.

Peters, R. L., and J. D. Darling (1985), The greenhouse effect and nature reserves, *Biosciences*, 35, 707.

Pielke, R.A. Sr. (2001), Influence of the spatial distribution of vegetation and soils on the prediction of cumulus convective rainfall, *Reviews of Geophysics, 39*, 151-177.

Pielke, R. A. (1984), *Mesoscale Meteorological Modeling*, 612 pp., Academic, San Diego, California.

Pielke R. A. Sr. (1992), A comprehensive meteorological modeling system RAMS, *Meteorol. Atmos. Phys.*, 49, 69-91.

Pimm, S. L. (2001), *The world according to Pimm: A scientist audits the Earth*, McGraw-Hill, NY

Rabin, R. M., and D. W. Martin (1996), Satellite observations of shallow cumulus coverage over central United States: An exploration of land use impact on cloud cover, *J. Geophys. Res.*, 101, 7149-7155.

Rabin, R. M., S. Stadler, P. Wetzel, D. J. Stensrud, and M. Gregory (1990), Observed effects of landscape variability on convective clouds, *Bull. Amer. Meteor. Soc.*, 71, 272-280.

Ramakrishnan, P. S. (2001), *Ecology and sustainable development*, National Book Trust, New Delhi, India.

Ray, D.K., R.M. Welch, R.O. Lawton and U.S. Nair (2006), Dry season clouds and rainfall in northern Central America: Implications for the Mesoamerican Biological Corridor, *Global Plant. Chang.*, 54, doi:10.1016/j.gloplacha.2005.09.004.

Ray, D.K., U.S. Nair, R.M. Welch, Q. Han, J. Zeng, W. Su and T.J. Lyons (2003), Effects of land use in Southwest Australia: Observations of cumulus cloudiness and energy fluxes, *J. Geophys. Res.*, 108(D14), 4414, doi:10.1029/2002JD002654.

Reichle, R. H., R. D. Koster, J. Dong, and A. A. Berg (2004), Global soil moisture from satellite observation, land surface models, and ground data: implications for data assimilation, *J. of Hydrometeor.*, 5, 430-442.

Sampaio, G., C. Nobre, M. H. Costa, P. Satyamurty, B. S. Soares-Filho, and M. Cardoso (2007), Regional climate change over eastern Amazonia caused by pasture and soybean cropland expansion, *Geophys. Res. Lett.*, 34, L17709, doi:10.1029/2007GL030612.

Saunders, D. A., H. J. Hobbs, and C. R. Margules (1991), Biological consequences of ecosystem fragmentation: A review, *Conservation Biology*, 5, 18-27.

Segal, M., R. Avissar, M. C. McCumber, and R. A. Pielke (1988), Evaluation of vegetation effects on the generation and modification of mesoscale circulations, *J. Atmos. Sci.*, 45, 2268- 2292.

Shukla, J., C. Nobre, and P. Sellers (1990), Amazon deforestation and climate change, *Science*, 247, 1322-1325.

Sizer, N., and E. V. J. Tanner (1999), Responses of woody plant seedlings to edge formation in a lowland tropical rainforest, Amazonia, *Biol, Conserv.*, 91, 135-142.

Sud, Y. C., W. C. Chao, and G. K. Walker (1993), Dependence of rainfall on vegetation: Theoretical considerations, simulation experiments, observations, and inferences from simulated atmospheric soundings, *J. Arid Environ.*, 25, 5 – 18.

Swift, L. W., W. T. Swank, J. B. Mankin, R. J. Luxmoore, and R. A. Goldstein (1975), Simulation of evapotranspiration from mature and clear-cut deciduous forest and young pine plantations, *Water Resources Research*, 11, 667-673.

Tauil, P. L. (2001), Urbanization and dengue ecology, *Cadernos de Saude Publica.*, 17 (Suppl.), 99-102.

Viana, V. M., A. A. Tabanez, and J. Batista (1997), Dynamics and restoration of forest fragments in the Brazilian Atlantic moist forest, In: *Tropical forest remnants: ecoogy, management, and conservation of fragmented communities*, W. F. Laurence and R. O. Bierregaard (eds.), 352-365, Universityof Chicago Press.

Walko, R. L., et al., (2000), Coupled atmosphere-biophysics-hydrology models for environmental modeling. *J. Appl. Meteorol.*, 39, 931-944.

Wallace and Hobbs (2006), Atmospheric Science: An Introduction Survey, Academic Press.

Wang, J., R. L. Bras and E. A. B. Eltahir (2000), The impact of the observed deforestation on the mesoscale distribution of rainfall and clouds in Amazonia, *J. Hydrometeor.*, 1, 267-286.

Weaver, C. P., and R. Avissar (2001), Atmospheric disturbances caused by human modification of landscape, *Bull. Am. Meteorol. Soc.*, 82, 269–282.

Wetzel, P. J., S. Argentini, and A. Boone (1996), Role of land surface in controlling daytime cloud amount: Two case studies in GCIP-SW area, *J. Geophys. Res.*, 101, 7359-7370.

Whitehead, P. G., and M. Robinson (1993), Experimental basin studies: An international and historical perspective of forest impacts, *J. Hydrology*, 145, 217-230.

3

Impacts of Deforestation on Climate and Water Resources in Western Amazon

Ranyére Silva Nóbrega
University Federal of Pernambuco
Brazil

1. Introduction

The Amazon importance in several areas of research demonstrates how the region affects the balance of South America and, depending on the scale used, on the planet. The biodiversity, mineral wealth, water resources wealth, carbon sequestration, transport of energy in the atmosphere are examples of important aspects of the region. Another important phenomenon that occurs in the Amazon are the energy flows between soil-vegetation-atmosphere dynamics that affect the climate, water resources and the advection of moisture to the surrounding parts.

Deforestation is the major environmental problem in the Amazon River basin nowadays, and its impacts affect both the local and global scale. In fact, this region is responsible for approximately 13% of all global runoff into the oceans (Foley et al., 2002) and its abundant vegetation releases large amounts of water vapor through evapotranspiration leading to a recycling in precipitation of about 25-35% (Brubaker et al, 1993; Eltahir and Bras, 1994; Trenberth, 1999).

Rondonia State, located in Western Amazon, already has a large area of vegetation changed by deforestation. Historically, there were tax incentives and government so that there was an expansion of development. Today the concern with changes in environmental balance of the Amazon basin, Rondônia, is justified by the increasing pressure on various forms of exploitation of the region, for example, timber extraction and agricultural expansion, the construction of hydropower, exploitation of biological and mineral riches.

Krusche et al. (2005) suggest the following reasons for this progress: between 1970 and 1990 there was a surge in state occupancy with settlers coming from other regions, extensive cattle ranching became the main economic activity and the state ground most is old and weathered, with the exception of some basins, promoting agriculture in an appropriate area. According to the authors, the pattern of occupancy was observed of the "fishbone" associated with the opening of roads.

The highway BR-364construction, responsible for turning the region with the rest of the country, was one of the factors that triggered large projects of colonization / occupation. Earlier this deforestation process was seen as boon, as a prerequisite for applying for tenure and subsequent legalization of land (Santos, 2001). Fearnside (2007) assert that the main aspect of change in land use / land cover in region is deforestation, and that it has grown over the years.

Deforestation rates of the State during the period 1988 to 2007 followed, in general, the same degradation that deforestation in the Amazon, since it has to be estimated. The most relevant peaks occurred in 1994 and 2004, showing a slight decrease for the years 2005 to 2007, however, leaving the state responsible for a higher percentage than the last 10 years earlier, reflecting a greater intensity on change in coverage plant that closed in the rest of the Amazon.

Year	Deforestation (km2)		%
	Rondönia	Amazon	
1988	2340	21050	11.10
1989	1430	17770	8.00
1990	1670	13730	12.20
1991	1110	11030	10.10
1992	2265	13786	16.40
1993	2595	14896	17.40
1994	2595	14896	17.40
1995	4730	29059	16.30
1996	2432	18161	13.40
1997	1986	13227	15.00
1998	2041	17383	11.70
1999	2358	17259	13.70
2000	2465	18226	13.50
2001	2673	18165	14.70
2002	3067	21651	14.50
2003	3620	25396	14.40
2004	3834	27772	14.00
2005	3233	19014	17.00
2006	2062	14286	14.70
2007	1611	11651	13.82
2008	1136	12911	10.00
2009	482	7464	6.45%
2010	427	6451	6.6%

Table 1. Annual deforestation rates - Amazon and Rondônia (Source data: Prodes, Inpe)

Public policies has been working to combat deforestation across the Amazon, we can observe the decrease in the rate since 2007, however, has been observed that in conservation areas this rate is increasing.

The proposal chapter is investigating the impacts that deforestation and climate change can lead on hydrological cycle in the region, as well as the feedback system of climate and hydrological cycle.

2. Metodology

The study is centered in Rondônia state, whose area of about 234.000 km2. The state's network runoff is represented by Madeira river (an important tributary of the Amazonian river basin) and its streams that form eight important sub river basins, among them, it is the Jamari sub river basin (Fig. 1). About 28% of the Rondônia state have already been deforested, because of this, is used as the test catchment study.

The Jamari river basin has suffered a substantial deforestation due to the advance of the agricultural frontier in the Rondônia state. The basin is crossed by two important rivers namely Jamari and Candeias. Jamari river has its nascent in the southwest part of "Serra do Pacaás Novos", in Rondônia, and streams northward flowing into the right bank of Madeira river, whose river basin is defined by the geographical coordinates 08° 28'S to 11° 07'S of latitude and 62° 36'W to 64° 20'W of longitude with about 29.066.68 km² of area.

The semi-distributed hydrological model SLURP with more detailed input parametric information will be used in this research in order to investigate the impacts caused by deforestation as well as climate changes on hydrological processes in Jamari River basin. Realistic and extremes scenarios of deforestation will be analyzed, and also scenarios of temperature rise and precipitation increase/decrease.

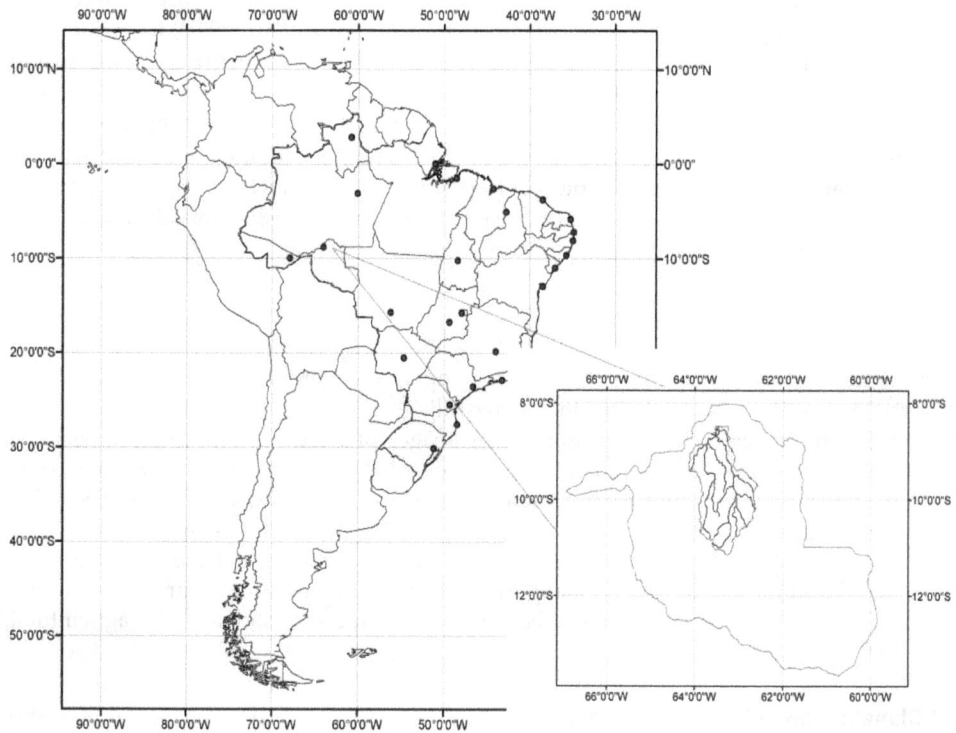

Fig. 1. Localization and drainage network Jamari sub river basin.

2.1 Hydrological model

Semi-distributed Land Use-based Runoff Processes - SLURP is a basin model that simulates the hydrological cycle from precipitation to runoff including the effects of reservoirs, regulators and water extractions (Kite, 2005). First divides a basin into sub-basins using topography from a digital elevation map. These sub-basins are further divided into areas of different land covers using data from a digital land cover classification. Each land cover class has a distinct set of parameters for the model.

This model uses basically three types of data: i) digital elevation data (DEM); ii) land cover data; and iii) climatic data. The matrix data of both DEM and land cover data must have the same dimension. Climatic data should contain: precipitation, air temperature, dew-point temperature (or relative humidity), solar radiation and wind intensity. Firstly, it divides a hydrological basin into sub-river basins and then divides each sub-river basin into land cover components using the public-domain topographic analysis software TOPAZ (and Martz and Garbreht, 1999). These homogeneous areas are based on the hydrological response unit (HRU) concept described by Kite (2005). SLURP defines these areas as Aggregated Simulation Areas (ASA).

The model has been applied in many countries for small hectares basins (Su et al., 2000) to large basins such as Mackenzie (Kite et al., 1994) and it was developed to make maximum use of remote sensing data. Applications of the model includes studies of climate change (Kite, 1993), hydropower (Kite et al., 1998), water productivity (Kite, 2005), irrigation (Kite and Droogers, 1999) and wildlife refuges (de Voogt et al., 1999), contribution of snowmelt to runoff (Laurente and Valeo, 2003; Thorne and Woo, 2006), and large mountainous catchment (Thorne and Woo, 2006). However, the SLURP model was not used in the Amazon, and its conceptual approach allows its use in regions with little data, as well as the possibility and direct use of remote sensing data which allows to retrieve physical parameters with good accuracy, even in basins with small slopes, as found in some sub-basins in the Amazon River.

2.2 Data

Digital Elevation Model - DEM from the Shuttle Radar Topography Mission (SRTM) with 90-m resolution horizontal was used to obtain topography. In order to correct failures, it was used the technique of space filtering, interactive filling.

For actual land cover data it was used seven images of Landsat 7 scenes 2007 over the Jamari sub-river basin, resolution of 30m, provided by Amazonian Protection System (SIPAM). Firstly, the scenes were georeferenced and then a mosaic was composed. Secondly, NDVI performed a supervised classification to obtain the land-cover image. Then, the data was sampled again to 90m resolution, since the SLURP requires that the matrix of land cover has the same size of DEM. Finally, the data was classified into four classes: water, forest, non-forest and man-modified (urbanized). The non-forest class includes agricultural areas and the savannah.

2.3 Climatic, rainfall and runoff data

Climatic, rainfall and runoff data are some of the main difficulties in hydrometeorological modeling in the Amazon. The time series available is short and has many flaws. Was used data set from four stations with information about precipitation, air temperature and dew point of the Agency for Environmental Development in Rondônia (SEDAM). We also use data from five rainfall gauge of the National Water Agency (ANA). The data sets are from the period between 1 January 1999 and December 31, 2007.

2.4 Model performance evaluation criteria

Model performance was evaluated by using four different error measures: Nash and Sutcliffe (NS), Percent BIAS (PBIAS), Daily Root Mean Square (DRMS) error criteria (Zhi et al., 2009; Moriasi et al., 2007), and Deviation Volume (D%) (Kite, 2005). The equations were given as showed below:

$$NS = 1 - \frac{\sum\limits_{i=1}^{n} (Q_{obs} - Q_{mod})^2}{\sum\limits_{i=1}^{n} (Q_{obs} - \overline{Q_{obsd}})^2} \tag{1}$$

where Qobs and Qmod are the measured and modeled data, respectively; . $\overline{Q_{obsd}}$. is average modeled data; and n is the total number of data records. The coefficient can range from -∞ to 1 and represents the amount of data oscillation that is explained by the model. The model is considered optimal if NS = 1, appropriate and good if NS > 0.75, acceptable if 0.36 <NS <0.75, and unacceptable if NS <0.35. If NS < 0, the predictor is worse than the average (Nóbrega, 2008).

$$PBIAS = \frac{\sum\limits_{i=1}^{n} (Q_{obs} - Q_{mod})}{\sum\limits_{i=1}^{n} Q_{obs}} \tag{2}$$

where Qobs and Qmod are the measured and modeled data, respectively. The optimal value of PBIAS is 0. Low magnitude values indicate accurate model simulation; PBIAS > 0 indicate model underestimation bias; and PBIAS < 0 indicate model overestimation bias (Zhi, 2009).

$$PBIAS = \frac{\sum\limits_{i=1}^{n} (Q_{obs} - Q_{mod})}{\sum\limits_{i=1}^{n} Q_{obs}} \tag{3}$$

RSR varies from optimal value of 0, which indicates zero RSME or residual variability and therefore perfect model simulation, to large positive values; the smaller RSR the better the model simulation performs (Zhi, 2009; Kannan et al., 2007).

$$D_V(\%) = 100. \frac{Q_{obs} - Q_{mod}}{Q_{obs}} \tag{4}$$

This criterion is simply a change in the pattern of calculated and observed average in the simulated period, which is a statistical test comparing the simulated discharge volumes to measures during the event, generating information about performance of the water balance total modeled. A value of zero indicates optimal modeling, or no difference between the volumes measured and simulated. A positive value indicates underestimation of the simulated volumes (losses in origin). A negative value indicates that the calculated average flow is high (losses in sinks) (Kite, 2005).

3. Simulations

Based on the percentage of deforestation in the basin obtained from the PRODES data (Table 2) it was defined two trends scenarios: i) DEFOR+20, 20% more deforestation area, and ii) DEFOR+30, 30% deforestation area. It was defined three extreme scenarios of land cover for investigating the relationship between soil-cover change and runoff within SLURP model. The experiments are: i) 100%FOR, one hundred per cent with forest and water; ii) 100%NOFOR, one hundred per cent with savannah plus pasture and water; and iii) 100%MANMODIF, one hundred per cent man-modified area and water. For climatic impacts analysis has been used two scenarios. The scenarios were based in climate futures projections up to 2050 A2 HadCM3 model, discussed in Marengo (2006b). In both scenarios is assumed that the temperature rise 2oC, and rainfall varies 20%, decreasing and increasing. The climate scenarios are: i) P+20, meaning 20% increase in rainfall, with an increase in temperature of 2 degrees, and ii) P-20, meaning 20% reduction in rainfall, with an increase in temperature of 2 degrees.

SCENARIO	DESCRIPTION
DEFOR+20	20% more deforestation area
DEFOR+30	30% more deforestation area
100%FOR	100% forest area
100%NOFOR	100% savannah plus pasture
100%MANMODIF	100% man-modified area
P+20	20% increase in rainfall, and 2oC temperature increase
P-20	20% decrease in rainfall, and 2oC temperature increase

Table 2. Simulations scenarios

3.1 Problems identified – implementation of necessary remedial measures
SLURP model needs data from weather stations which contains precipitation, temperature, humidity and wind. The average is calculated using the Thiessen polygons method for each ASA. If there is no such data, the model does not perform the simulation (for example, when there is rainfall data, but there is not temperature).

In some countries, such as Brazil, it is common to have only rainfall station, instead of weather station, making the network rainfall much denser than the weather. But the compilation of the model does not allow the use of this data. Aiming to overcome this limitation, we tried to develop a methodology that would use the climatic stations without changing the source code of the model. Adopted method is based on the concept that the spatial variability of precipitation is less than the other data, such as temperature. Then, without the model, it was calculated the mean rainfall for each ASA also using the Thiessen method, but including the data from rainfall stations. After that, the files of average precipitation for each ASA were replaced.

4. Results and discussion

Jamari sub-river basin was automatically divided by SLURP into five aggregate similar areas (ASAs) according to the DEM and land cover data (Fig. 2). For each ASA it was obtained the percentile area of land cover occupied for each of the four classes: i) water; ii) forest; iii) non-forest; and iv) man-modified (Table 3). The total basin area obtained by the model is 28.847 km^2 (~99% of the total area, according to Government State official data).

The model was calibrated and checked by the two different types of data: i) weather stations data (OBS1); and ii) weather station data added rainfall gauge station (OBS2). Obviously it is expected that the use of a denser network of precipitation within a basin simulation results in improvements, since the data quality is consistent, but it was not clear if the model would accept the manual modification.

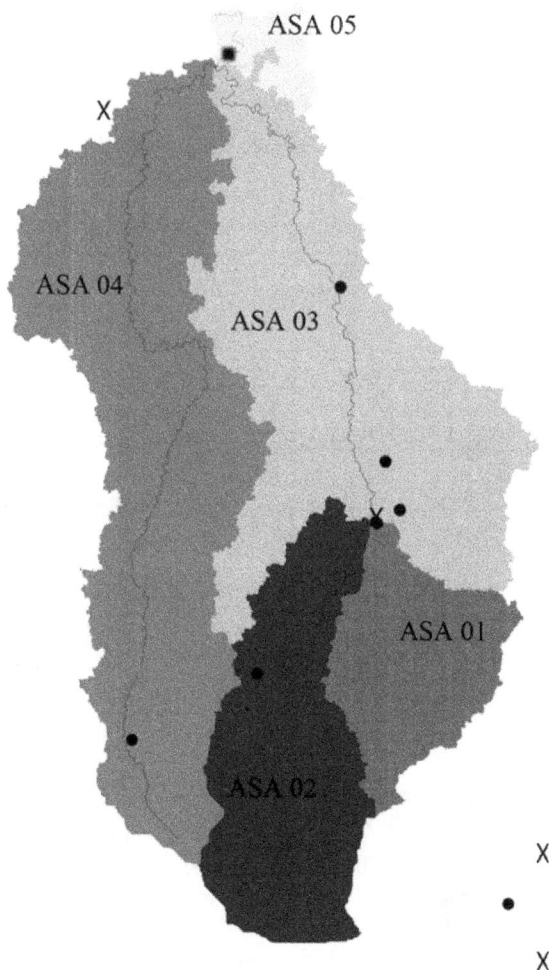

Fig. 2. ASAs for Jamari sub river basin; X – weather stations; ● – rainfall station

ASA Name	Water	Forest	Non forest	Man modified	Total (km²)
ASA 01	9.0	63.0	7.8	20.2	3025.81
ASA 02	3.6	67.3	11.7	17.4	5007.65
ASA 03	9.3	61.3	7.0	22.4	9239.68
ASA 04	3.9	62.2	11.4	22.5	10999.55
ASA 05	4.1	93.8	2.0	0.1	573.94

Table 3. Land coverage and total area for each ASA (%)

The NS, RSR, PBIAS and D(%) for OBS2(OBS1) calibration period was 0.88(0.74), 0.31(0.45), -7%(-12%) and -0.94%(-10.1%), respectively. The NS, RSR, PBIAS and D(%) for OBS2(OBS1) validation period was 0.84(0.71), 0.34(0.48), -8%(-15%) and -13.4(-10.3). The verification of the model efficiency criteria indicates that the values are acceptable during both the calibration and validation period. (Table 3), but it is clear that the model did better with the inclusion of climatic stations. From this point, we used OBS2 data with SLURP climate input.

4.1 Calibration and verification

Model was calibrated and verified by the two different types of data: i) weather stations data (OBS1); and ii) weather station data added rainfall gauge station (OBS2). Obviously it is expected that the use of a denser network of precipitation within a basin simulation results in improvements, since the data quality is consistent, but it was not clear if the model would accept the manual modification.

The NS, RSR, PBIAS and D(%) for OBS2(OBS1) calibration period was 0.88(0.74), 0.31(0.45), -7%(-12%) and -0.94%(-10.1%), respectively. The NS, RSR, PBIAS and D(%) for OBS2(OBS1) validation period was 0.84(0.71), 0.34(0.48), -8%(-15%) and -13.4(-10.3). Model efficiency criteria verification indicates that the values are acceptable during both the calibration and validation period (Table 4), but it is clear that the model did better with the inclusion of climatic stations, therefore, used OBS2 data with climate input.

OBS2 (OBS1)	NS	RSR	PBIAS	D(%)
Calibration	0.88 (0.74)	0.31(0.45)	-7%(-12%)	-0.94 (-10.1)
Verification	0.84 (0.71)	0.34(0.48)	-8% (-15%)	-13.4 (-10.3)

Table 4. Model performance

4.2 Deforestation impacts

Taking into account the current deforestation rate in the area which is being studied, the trend scenarios can be designed by 2013 and 2016, respectively. The results for DEFOR+20% and DEFOR+30% indicated increased runoff compared to the average from 1999-2007, 825.3 $m^3.s^{-1}$, to 1048.1 $m^3.s^{-1}$ and 1163.7 $m^3.s^{-1}$, resulting in an increase of 27% and 41%, respectively. During the dry season (characterized by a weak runoff), the flow trends to increase remarkably, what can be a concern for local population who use these rivers for human supply, navigation (in some places, the only kind of transportation), and also for power generation. If these scenarios become real, the rivers of the basin will be subjected to a different runoff pattern that might cause some socioeconomic impact. Although the extreme scenarios are not realistic, the results are instructive because they clarify the non-linear response of the hydrological cycle to the progressive changes in land cover.

When modifying the land cover to 100%FOR, the annual calculated runoff average decreased from 825.3m^3.s^{-1} to 329.1 m^3.s^{-1}, i.e., a decrease of about 60%. On the other hand, for the scenarios with 100%NOFOR and 100%MANMODIF, runoff increased to 2313.1m^3.s^{-1}, and 1729.4m^3.s^{-1}, an increase of 181%, and 109% of the observed annual runoff average, respectively.

The parameters that most influenced the results in these scenarios were related to the amount of available soil water for evapotranspiration and canopy interception, which were modified according to the soil cover. The high interception in scenario 100%FOR leads to a reduction of precipitation that reaches the soil and thus reduces runoff. It also reduces the amount of water available for evaporation. Furthermore, the increase of flowing in scenarios 100%NOFOREST and 100%MANMODIF is due to the substantial decrease in evapotranspiration and rainfall interception by the canopy. It is worth mentioning that this study used the same series of precipitation for all scenarios, but the precipitation in the region is a variable that has its intensity, largely influenced by local evaporation. Hence, decrease in evapotranspiration tends to reduce rainfall. The use of SLURP coupled to an atmospheric model can reveal more about this feedback mechanism in further studies.

The elements of water balance are shown in Fig. 3. It may be noticed that evapotranspiration varies slightly between the scenarios, except the 100%NOFOR, where evapotranspiration is approximately 90% the value of the other scenarios. However, when the exchange of water between the surface and the atmosphere is divided into evaporation and transpiration, the peculiarities of each scenario are quite evident. In 100%FOREST, transpiration contributed to the increase of evapotranspiration. This is quite obvious, since in this scenario, the interaction between the surface and free atmosphere is dominated by the exchange of processes in the vegetation canopy. In 100% NOFOREST, the land cover formed by the typical savanna and pasture vegetation makes the transpiration to be twice as much as evaporation. In 100% MANMODIF, the deforested soil with urban characteristic implies a greater contribution in evaporation than in transpiration.

It can be observed that the evapotranspiration and groundwater reduces slightly with the decrease of forest areas when we compare with the trend scenarios. In terms of numbers the evapotranspiration decreased from 20% to 30%, respectively; groundwater flow decreased 20% and 8%, respectively. As long as deforestation leads to less water interception by vegetation, the contribution of evaporation increases with the expansion of the deforested area.

Several studies suggest the local contribution of evapotranspiration as responsible for about 50% of the precipitation that occurs in western Amazonia (Nóbrega, 2005; Marengo, 2006a; Nóbrega, 2008). Therefore, a decrease in vegetation cover over a region can alter the precipitation regime is this region (and neighborhood), decreasing the amount of water vapor originated there, because the evapotranspiration decreased in these simulated scenarios, and this trend, combined with the increase in flowing during dry periods, may worsen social and environmental problems during more critical periods.

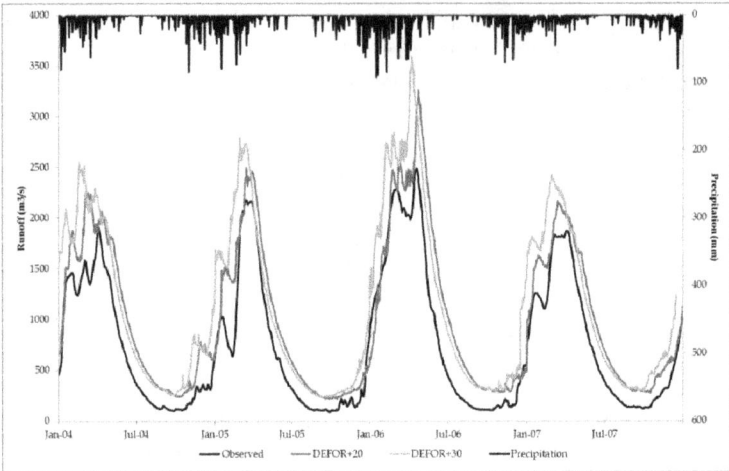

Fig. 3. Observed and Simulated water balance components for scenarios DEFOR+20% and DEFOR+30

4.3 Climate change impacts

Daily changes in runoff due to temperature and precipitation variations are shown in Fig. 4. The effect when the rainfall increases or decreases in 20% was as expected. An increase in rainfalls tends to increase runoffs, and a decrease in rainfalls tends to decrease runoffs, which is more noticeable during the rainy season, when rains are more significant. For the scenario P +20, the runoff increased 31%, and setting P-20 decreased 13%, indicating that a increase rainfall will respond more significantly than the decrease in this specific region.

These results allow an analysis that a decrease in rainfall may have a critical effect especially during the rainy season, affecting navigation, agricultural production, human consumption and power generation. Fig. 5 shows the seasonal average flow for the investigation period, confirming that the impacts are more significant during the rainy season.

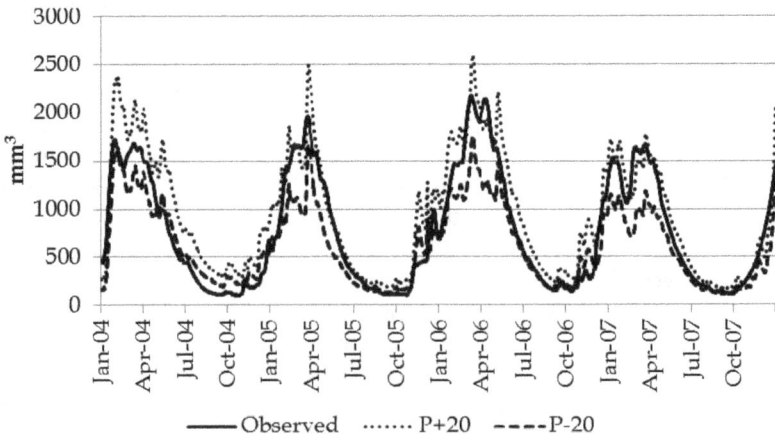

Fig. 4. Observed and Simulated runoff for scenarios P+20 and P-20

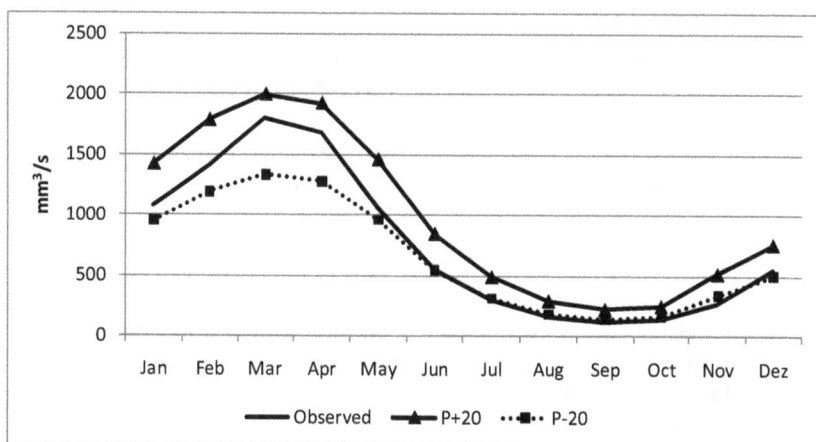

Fig. 5. Superficial runoff monthly average Observed and simulated for scenarios P+20 and P-20

Water balance changes due to changes in temperature and precipitation, it was observed that an increase in temperature tends to increase transpiration in both scenarios, and a decrease in rainfall tends to reduce evaporation (Fig. 6). The evapotranspiration increases 30% and 54%, evaporation decreases 21% and increases 3%, transpiration increases 37% and 41%, and groundwater decreases 35% and 33% for the sets of P-20 and P +20 respectively.

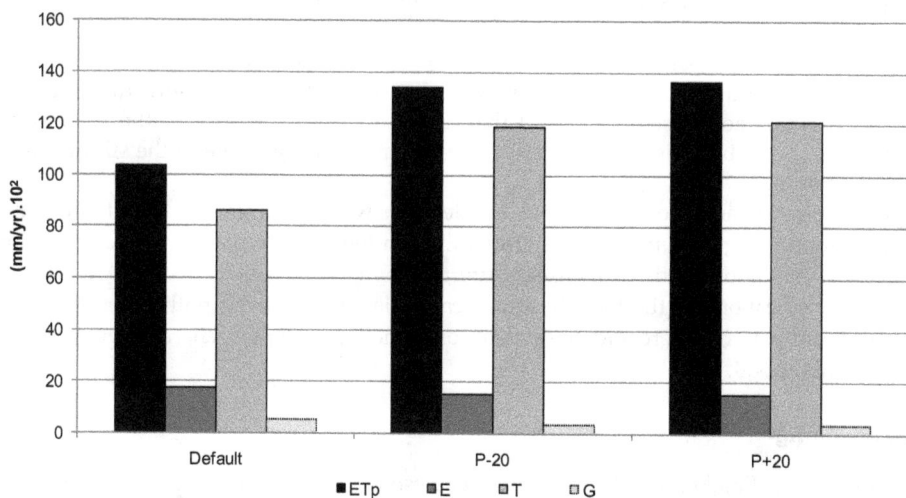

Fig. 6. Observed (Default) and simulated water balance components for scenarios P+20 and P-20

It is important to mention that the simulation did not consider the impacts that deforestation would cause on climatic variables. Although with the results obtained from the simulation, it is possible to conclude that deforestation modifies substantially physical processes of the hydrological cycle.

Impacts are easier to be seen during the rainy season. Furthermore, with less water available for evapotranspiration, the vegetation may suffer from water stress. Although some authors found a precipitation increase in deforested areas (Li et al., 2007), if the deforestation keeps on increasing, the changes in the hydrological cycle might become unsustainable. The change in cover / land use can lead the current system to a new dry equilibrium, and vegetation should be modified to adapt itself to climate changes.

5. Conclusions

Firstly, it was needed to make sure if the model could be used in that region, due to the lack of some meteorological data and the small slope of the region. Based on the NASH, RSR, PBIAS and D% criteria, the results indicate acceptable values. Furthermore, since it is a semi-distributed model, it requires less startup parameters than the distributed models, and also is able to calculate results faster.

Deforestation in Amazonia has been occurring for some decades and the rate of annual growth is noticeable. In addition, this might be influenced by climatic and social-economics factors. Land cover/use changes simulations indicated that the runoff can be changed. The results suggest that there is an increase in runoff when deforestation occurs in the extreme and trend scenarios, associated with less interception of water by the canopy. If the average rate of deforestation continues to be about 3.45% per year in the basin, our simulations predict that the annual runoff will increase about 27% by 2013 and 41% by 2016. Samuel Hydropower, located on the Jamari river, began to be built in 1982. Between 2004 and 2006, the hydropower floodgates had to be opened because the river level reached its maximum level.

The results make us believe that the ongoing deforestation could be responsible for opening these floodgates, since the observed data do not indicate more rain than the average. In used model, sediment load that affects the level increase of the river is not taken into account, but it is likely to result in a sediment increase due to the silt produced by deforestation.

Evapotranspiration and groundwater tend to decrease with deforestation. Results show that the main impact might occur on transpiration, which tends to decrease with deforestation, while the evaporation tends to increase. Alterations in water balance in the Amazon can result in modifications in the local hydrological cycle, and agreeing with other studies, it will affect rain patterns there and close areas, once the water vapor generated goes straight to the neighborhood.

6. References

Brubaker, L.K., Entekhabi, D., and P.S. Eagleson. (1993). Estimation of continental precipitation recycling. J. Climate, Vol. 6, pp. 1077-1089.

de Voogt, K., Kite, G.W., Droogers, P., and H. Murray-Rust. (1999). Modeling water allocation between a wetland and irrigated agriculture in the Gediz Basin, Turkey. Research Report, International Water Management Institute, Colombo, Sri Lanka

Eltahir, E.A.B., and R.L. Bras. (1994). Precipitation recycling in the Amazon basin. Quart. J. R. Met. Soc., Vol. 120, pp. 861-880.

Fearnside, P.M. (2007). Deforestation in Amazonia. Encyclopedia of Earth. Eds. C.J. Cleveland (General Editor) & M. Hall-Beyer (Topic Editor). Environmental Information Coalition, National Council for Science and the Environment, Washington, D.C., U.S.A.

Foley, I.A., Botta, A., Coe, M.T., and M.H. Costa. (2002). The El Niño-Southern Oscillation and the climate ecosystem and river of Amazonia. G. Biogeochemical Cycles, Vol.16, pp.1132-1144.

Kannan, N. et al., (2007). Hydrological modelling of a small catchment using SWAT-2000 – Ensuring correct flow partitioning for contaminant modeling. J. Hydrology, Vol. 3(34) pp. 64-72.

Kite, G.W. (2005). Manual for the SLURP hydrological model. 236 p.

Kite, G.W., and P. Droogers. (1999). Irrigation modeling in the context of basin water resources. J. Water Resources Development, Vol.15, pp. 43-54.

Kite, G.W., Danard, M., and B. LI. (1998). Simulating long series of streamflow using data from an atmospheric model. Hydrological Sciences, Vol.43(3).

Kite, G.W., and P. Droogers. (1999). Irrigation modeling in the context of basin water resources. J. Water Resources Development, Vol.15, pp. 43-54.

Krusche, A.V., Ballester, M.V.R., and R.L. Victoria. (2005). Effects of land use changes in the biogeochemistry of fluvial systems of the Ji-Paraná river basin, Rondônia. Acta Amazônica, Vol. 35(2), pp. 192-205.

Laurent, MESt, C. Valeo. (2003). Modeling runoff in the northern boreal forest using SLURP with snow ripening and frozen ground. Geophysical Research Abstracts, Vol. 5, pp. 06-30.

Li, K.Y., Coe, M.T., Ramankutty, N., and R. Jong. (2007). Modeling the hydrological impact of land-use change in West Africa. J. Hydrology, Vol. 337, pp. 258-268.

Marengo, J.A., (2006a). On the hydrological cycle of the Amazon basin: a historical review and current state-of-the-art. Rev. Brasil. Meteorologia, Vol. 21(3), pp. 1-19.

Marengo, J.A., (2006b). Global Climate Changesand biodiversity efects. 201 p.

Martz, W., and J. Garbrecht. (1999). An outlet breaching algorithm for the treatment of closed depressions in a raster DEM. Computers & Geosciences, Vol. 25, pp. 835-844.

Moriasi, D.N. et al., (2007). Model evaluation guidelines for systematic quantification of accuracy in watershed simulations. Vol. 50(3), pp. 885-900.

Nóbrega R.S., (2008). Modeling impacts of deforestation in water resources of river basin Jamari (RO) using data surface and TRMM. D.Sc. Thesis. Federal University of Campina Grande, Paraíba, Brasil. 212p.

Santos, C. A. (2001). Fronteira do Guaporé. Porto Velho/RO: EDUFRO.

Su, M, Stolte, W.J., and G. van der Kamp. (2000). Modeling Canadian prairie wetland hydrology using a semi-distributed streamflow model. Hydrological Processes, Vol.14(14), pp. 2405-2422.

Thorne R., and M. Woo. (2006). Efficacy of a hydrologic model in simulating discharge from a large mountainous catchment. J. of Hydrol, Vol. 30(1-2), pp.301-312.

Trenberth, K.E. (1999). Atmospheric Moisture Recycling: Role of Advection and Local Evaporation. J. Climate, Vol. 12, pp.1368-1381.

Zhi, L., et al., (2009). Impacts of land use change and climate variability on hydrology in an agricultural catchment on the Loess Plateau of China. J. Hydrology (doi: 10.116/j.jhydrol.2009.08.007)

4

Deforestation Dynamics:
A Review and Evaluation of Theoretical
Approaches and Evidence from Greece

Serafeim Polyzos and Dionysios Minetos
University of Thessaly, Department of Planning and Regional Development,
Pedion Areos, Volos,
Greece

1. Introduction

Amongst others, forest land use changes occur for multiple reasons and from interacting processes and mechanisms. Human-driven changes, at an array of scales, are affecting forest ecosystems accelerating changes such as global warming with adverse consequences on human well being. Research has demonstrated that, in the long term, there is to not a single factor or set of factors that can explain the emerging patterns of land uses and their associated changes (Chomitz and Gray 1996; Lambin, Turner et al. 2001; Aspinall 2004). Yet, the importance of forests on global environmental issues such as biodiversity loss and global warming is apparent.

Deforestation processes have different characteristics across space and time (Verburg, Schulp et al. 2006). A particular combination of factors that may explain deforestation pattern somewhere, might not be applicable for justifying change in any other location or time period (Mahapatra and Kant 2005). Therefore, there is a need for conducting empirical investigations in order to analyse and understand the geographical and historical context of land use changes. In addition, forest fragmentation, conversion and modification have significant economic, social and environmental implications (Elands and Wiersum 2001; Walker 2001; Platt 2004; Verburg, Overmars et al. 2006) such as disruption in continuity of the natural landscape, forest and open-land constriction between agricultural and urban land uses, deterioration of vital habitats that sustain valuable biodiversity as well as broader issues such as air pollution.

In this paper, we concentrate on reviewing and evaluating deforestation-related theoretical schemata. We introduce a representative collection of theories dealing, directly or indirectly with deforestation processes in order to provide guidance to an appreciation of the past and the future land use patterns. In addition, we give evidence of recent deforestation dynamics in Greece.

2. Framework of review and evaluation

Theories are presented in a timeline context. There is also a set of criteria used for the evaluation. The criteria are presented in table 1.

EVALUATION AXES	SPECIAL CRITERIA OF EVALUATION
A. SCALE	A.1 Human decision level A.2 Spatial Level A.3 Temporal Level (time step)
B. FOCUS OF THE THEORY	B.1 Raking of the phenomenon B.2 Descriptive B.3 Causal B.4 Predictive
C. MAIN MECHANISM OF LAND USE CHANGE	C.1 Economic Mechanisms C.2 Social Mechanisms C.3 Administrative-Political Mechanisms C.4 Natural Resources
D. INTERDISCIPLINARY	D.1 Transectoral focus D.2 Sectoral focus
E. NATURE OF THEORY	E.1 Descriptive E.2 Geometrical E.3 Mathematical
F. LEVEL OF DEVELOPMENT	F.1 Developed countries F.2 Developing countries

Table 1. Framework for evaluating of deforestation-related theoretical schemata.

a.   The scale in which the theory can be applied for describing and interpreting land use changes.

The significance of scale is critical and concerns:

- The human decision-making process (single person, household etc).
- The *geographical unit of analysis* (spatial analysis) in which the theory can be effectively applied (city, region, country).
- The *time unit of analysis,* (temporal analysis) which is connected with the ability of the theory approaching short-term, medium-term or long-term changes.

b.   The central aim of the theory.

Theories, mainly attempt to classify, describe, explain or even forecast spatial tranformation phenomena. Classification is a way of categorizing observed land use changes, so that any likely differences or similarities in the way that spatial phenomena transform space, are better understood. Descriptive theoretical approaches, in addition to classifying observed complex geographical formations, also attempt to determine concrete functional relationships and processes that, in turn, reveal likely associations between different uses of land. Explanatory theoretical perspectives concentrate on the determination of factors and dynamics that produce, wear out or eliminate certain land uses. Finally, predictive theoretical perspectives focus on the projection of spatial phenomena into the future aiming at forecasting any future composition of land use system.

c.   Main mechanism of land use change.

The comprehension of underlying causes associated with land use change processes as well as The understanding of the major mechanisms of land use allocation is a critical matter. The existing theoretical pool in the field of land use change includes diverse approaches that either focus on economic processes and factors or on social and administrative processes, or support that the spatial distribution of different land uses is determined by the existing distribution of natural resources. Therefore, it is extremely significant to understand the particular mechanism or set of mechanisms that each theoretical perspective puts forward in order to approach successfully complex spatial phenomena.

d.   Sectoral or multi-sectoral approach.

The choice of how to approach a particular land use change process is of critical importance. Some theoretical perspectives follow a sectoral approach whereas some others employ an inter-sectoral logic. In the first case the theory focuses in only one category or subcategories of land use (rural use, urban use, forest use, or residence, industry, tourism, etc.). In the second case, the focus is shifted towards interpretation of intersectoral phenomena connected with observed land use transformations (e.g. urban sprawl etc).

e.   Nature of theory.

The way that each theory has been stated or presented, relates to the level of formalism and scientific severity. Generally speaking, most theories adopt a verbal, geometrical, or mathematic approach or a combination of them, as a basic platform in putting forward their key statements.

f.   Level of development

Remarkable differences in the levels of economic growth between countries have caused the emergence of at least two general categories of theories. There are theories suitable for satisfactorily interpreting spatial phenomena in developed societies, and also theories that mainly apply best in the case of less developed countries.

Below, an attempt is made to critically review some of the most important theoretical approaches that directly or indirectly refer to processes in forest land use change.

3. Review of theoretical approaches

This section provides a brief review and evaluation of theoretical approaches on the phenomenon of forest land use change. The purpose of the review is to locate the major proximate and underlying causes of deforestation proposed by the literature. Recent attempts to theorise forest land use changes have yielded some noticeable contributions on the field of deforestation.

One such contribution referred to as *"forest transition theory"* has been put forward by Mather, Grainger and Needle since the early '90 (Grainger 1995; Reid, Tomich et al. 2006). According to this theoretical perspective, an overview of forest land use changes in the long run, provides firm evidence that while initially forest land areas retreat at a high speed, at same point, depletion starts slowing down. There is even a critical point over which the process of depletion reverses and forest land recovers by expanding into new areas. *Prosperity level* seems to have a key role in the whole process. The main land use change mechanism suggested by this theory is of economic nature and also has some common places with Kuznets' Environmental Curve (Koop and Tole 1999; Ehrhardt-Martinez, Crenshaw et al. 2002) that links national or regional environmental quality with the state of economic development. The spatial level of analysis that the theory best applies is to nationwide or higher.

Based on the aforementioned perspective, Mather (2006) and Grainger (1995) state that most of the developed countries have been in a state of forest transition as forest land has been expanding for several decades. Explanations concerning forest transition processes are sought both in *development theory* and *modernization theory*. These theoretical perspectives focus on the importance of economic and social changes that spring from the adjustment of the economic, social and political structure to technological advances. In this respect, during the course of development increased pressure is put on forest land due to higher demand for land and forest related products. As a spatial unit moves to higher stages of development, pressure on forest land retreats because technological innovations allow for increased productivity in the primary sector, limiting the needs for expansion on forest land. At the same time, the increased rates of urbanization drive large waves of population away from the countryside into cities and towns lowering the pressure on forests by human activity in exurban remote areas.

A sizable body of empirical research on forest transition theory has generated considerable evidence in favour of some of the theory's reasoning (Koop and Tole 1999; Ehrhardt-Martinez, Crenshaw et al. 2002; Geist, McConnell et al. 2006; Reid, Tomich et al. 2006). Empirical evidence systematically suggests a negative relationship between the rate of deforestation and the rate of urbanization or the level of new technology adoption in the primary sector (Perz 2007).

In a recent attempt of improving the context of the theory, Angelsen (2001) suggests that the stages of forest transition theory (low deforestation, intense deforestation, containment of deforestation, afforestation) can be better understood by focusing on the fundamental characteristics of the long-term relationship between *agricultural land rent and forest land rent*. Mathematically, this means that forest land use changes [DU]$_{fr}$ are a function of agricultural land rent [LR]$_{agr}$ and forest land rent [LR]$_{frs}$ of the type

$$DU_{fst} = f\left(\frac{LR_{agr}}{LR_{frs}}\right).$$

In this case, the most important part of the analysis is to identify the way and magnitude of the influence of applied policies on agricultural and forest land rent. However, Angelsen points out that diagnosis and measurement of the actual influence of policies on land rent is, at least difficult to achieve, since on top of the obvious and direct impacts, economic phenomena are usually involved in numerous feedbacks and interactions that produce new waves of influences on land rent. For instance, the adoption of a new technology in agriculture may initially make agricultural activities more profitable fueling the expansion of agricultural land on forest land. In the long run, however, balancing effect will emerge due to changes both in agricultural goods prices and wages in agriculture leading to the containment of the expansion or even a reverse process of forest advancement.

The aforementioned analysis by Angelsen focuses on tropical regions, where one of the most important proximate causes of deforestation is believed to be agricultural activity. However, in cases where urban land uses control the process of deforestation, the relationship between forest and urban land uses in terms of land rent might constantly fuel urban expansion. An additional consideration regarding the Agelesen's land rent approach rests on the property regime of forest resources. In several parts of the world, the great share of forest land belongs to the state and not to private owners. In those cases, where the state cannot assure its one rights on land there is more scope for forest land encroachment and exploitation through logging and cultivation. In these cases, forest land use change processes maybe better understood on the basis of Hardin's theory of *the tragedy of the common* (1968). Individuals when act independently might have no motives for considering the need for a sustainable use of shared natural resources (Herschel 1997). The negative external economies by the depletion of forest resources are spread to society whereas the economic profits from wood exploitation and agricultural production in formerly wooded land are capitalized on by the intruders.

Approaches for unsustainable forest resource exploitation are not scarce. Several researchers (Roberts and Greimes 2002; Shandra, London et al. 2003), drawing from the *"theory of dependence"* by Baran, Frank and Amin as well as from *"world systems theory"* by Wallerstein, claim that the developed regions have establish a particular system of economic exchange that imposes certain land use patterns to the less developed regions. This particular economic exchange process between the developed and the less developed peripheries takes place on unequal terms resulting in the unsustainable use of natural resources in the less developed areas. Power, wealth and prosperity are, therefore, related to the depletion of forest resources in the developing regions and to the sustainable use and possible expansion in prosperous regions.

Both the importance and the adaptive nature of strategies employed by agents in order to maintain their prosperity level were firmly established in the context of *"multi-phasic response theory"* proposed by Davis in 1963. It is now believed that applied more broadly, *multi-phasic response theory* can help to understand how agents decisions impact land use changes (Lambin, Geist et al. 2006).

A similar body of approaches attempts to apply concepts from *"game theory"* in order to capture agents' behaviour in forest land use change process. (Fredj, Martvn-Herran et al. 2004). According to some of these perspectives, the major decisive force of the way forest resources are utilized is state policies. State policies are far from static paying particular importance to economic growth during periods of economic difficulties where environmental concerns are ranked low in social agenda. Therefore, regardless of the level of economic development, the changing conditions of economy over a period of time could result in the adoption of a sustainable or a less sustainable behaviour towards forest resource. However, in some cases the unsustainable behaviour adopted by agents may not lay on the difficulties brought about by unfavourable economic conditions. It might be, in fact, an act of land speculation based on either high tolerance shown by the political system or insufficient administrative and environmental monitoring mechanisms. Yet, in the context of game theory it is proposed that it is possible to arrive to sufficient and sustainable solutions to the issue of deforestation through cooperation and coordination of the involved parties and individuals (Fredj, Martvn-Herran et al. 2004; Stern 2006).

On the other hand, there is a quite different view concerning the current social behaviour and state intervention towards forest resources. It is widely believed that current afforestation policies are not merely an opportunistic reaction to the undisputed acute depletion of forest resources and its associated impacts. Instead, they also revile a much deeper transformation of social attitudes and ethics towards the environment. Mather et. al (2006) argue that at least in the case of developed countries, a great part of society is driven by the principles of *post-productivism philosophy*. Profit maximisation is not the only as well as the central axis of individual behaviour formation. Economic growth is possible to coexist with protection and restoration of the environment. Several aspects of this new philosophy of *post-productivism* can especially be traced in the countryside (Shucksmith 1993) in the form of certain environmentally sensitive policies that are voluntarily embraced by farmers. Amongst other, afforestation policies, organic farming measures and codes of good agricultural practice aim to establish alternative agricultural land management as well as sustainable ways of agricultural production. Current agricultural land uses are increasingly characterized by the aforementioned *post-productivist* features (Reid, Tomich et al. 2006) as the demand for environmental services has risen sharply in developed regions during the last decades. Therefore, land use changes in the countryside could possibly be better understood in the light of society's priorities in the context set by *post-productivism*.

In this respect, *Maslow's theory of the hierarchy of needs* maintains considerable potential in explaining certain emerging land use conversions as well as land use qualitative modifications in exurban land use systems. The importance of the new *post-consumption social motives and ethics* stresses Inglehart (1990) pointing out that current economic, technological, and sociopolitical changes have resulted in an apparent transformation of the fundamental cultural characteristics of developed societies. This transformation is ceaseless and has a greater impact on new generations. As new generations replace the older ones in the system of political, social and economic organization, the characteristics of adopted economic development strategies change, as well incorporating more environmental considerations. The implied redirection in priorities, behaviours and ethics is mostly reflected on agricultural policies and thus on rural land uses. Nowadays farmers receive more subsidies and other benefits in order to sustain and improve the quality of soil than to increase agricultural goods production.

Recently, the concept of *post-productivism* has been enriched with a spatial dimension grounded on the observation that some regions have developed stronger *post-ponductivism* structures than others (Agarwal, Green et al. 2002; Reid, Tomich et al. 2006). *Post-ponductivism* is thought of as a spatial phenomenon too, that is strongly connected to the rural patterns of land uses. Amongst others, Marsden (1998), Groot et. al. (2007) and van der Ploeg et. al. (2000) underline the fact that there exist considerable spatial differences within developed countries at the regional level in relation to the intensity one can observe evidence of *post-productivism*. Nevertheless, Mather argues that in spite of the observed spatial differences in the strength of *post-productivism* characteristics amongst developed countries and regions, the phenomenon is present and lies on deep social and cultural transformations in progress in the developed countries. Such transformations are capable of inducing constructive institutional interference which is both a cause and a consequence of new policy planning and application.

The emergence of *post-productivism* concept as a mean of interpreting the observed rural land use changes has received considerable criticism. Among others, Evans believes (2002) that productivism and post-productivism carry a dualistic meaning just like fordism and post-fordism, and therefore, cannot contribute to an in-depth understanding of complex spatial phenomena. Alternatively, more solid theoretical approaches should be considered such as the *regulation theory*. According to this perspective, new economic patterns and social structures are the result of capitalist economic crises as well as efforts to overcome these crises. Accordingly, the observed spatial inequalities and the associated land uses and characteristics of agricultural sector are analogous to the established production relationships, to spatial division of labour and to changing power allocation in the field of governance.

Recently, *ecological modernisation theory* has been proposed as an alternative to *post-productivism* argument in terms of rural land use change theorisation. According to the advocates of this view (Andersen and Massa 2000; Buttel 2000; Marsden 2004), it is possible and therefore it should be pursued, economic growth and social prosperity to be in line with environmental protection. Technological advances are at the core of this perspective as they are thought of being able of substantially contributing to the effective utilization of natural resource and to the decrease in the volume of both utilized raw material and produced waste.

Another recent view on managing deforestation proposes the concept of *"compensated reduction"* (Santilli, Moutinho et al. 2005). According to this view, countries that are chosen to lower their national level of deforestation should receive post facto compensation if they commit to stabilize or even reduce deforestation in the future. In other words, designing and offering large scale incentives might be a strategy capable of managing high deforestation rates as it occurs in tropical forest regions. This proposal works similarly to the Certified Emissions Reductions (CERs) or the Clean Development Mechanism (CDM). It also relates to the notion of valuation of the "unpriced" services of forests in order to reduce deforestation through economic mechanisms.

In several instances, there exist perspectives linking deforestation directly to the expansion of agricultural activities. Deforestation due to agriculture expansion is threatening several critical aspects of the environment such as biodiversity. Regarding the driving factors, it is thought that technological development and international prices are the basic drivers of crop expansion. Nevertheless, local actors can develop ways to apply sustainable changes in the economic activities and reduce negative impacts managing deforestation rates through conservation policies. It seems, that the powerful position of some actors in the predominant production and distribution chains of agricultural goods is influenced immensely where generated prosperity accumulates (van der Ploeg 2000; Allen, FitzSimmons et al. 2003). In several instances, the greatest share of added value deriving from the ongoing restructuring of agricultural sector is not yielded by local producers and as a result the future course of sustainable agricultural systems appears uncertain (Smith and Marsden 2004). It is suggested that a reasonable response to the aforementioned issue could be the placing of restrictions on the size of food distribution enterprises as well as localising their characteristics (Raynolds 2000; Allen, FitzSimmons et al. 2003; Seyfang 2006; Feagan 2007). However, such arguments are directly opposite to the prevailing process of globalization in food production and distribution. It might be more realistic as well as effective to accent and emphasize the fundamental qualitative differences between globalization and localization approaches.

Globalization of food production and distribution systems abstracts and alienates economic transactions from their social and environmental context (Raynolds 2000; Seyfang 2006). Heterogeneity of rural social and natural forms and processes is neglected. In globalisation era, it is almost impossible for a consumer of a particular agricultural commodity to identify the particular social relations and environmental circumstances under which the commodity was produced (Allen, FitzSimmons et al. 2003). Contrarily, the suggestions of *green economy* favour the identification and designation of the social and environmental framework of production allowing the emergence of *economies of place* (Seyfang 2006).

The success of economies of place, however, presupposes reconnecting effectually consumers and producers and establishing trustfulness and reliance amongst all involved actors (Seyfang 2006). Hence, it might be worth focusing on raising *social capital in agriculture* as well as building sufficient stocks of trust, communication and cooperation (Rahman and Yamao 2007). Even though, some critical views (Hinrichs 2003; Winter 2003; Winter 2005) point out that economies of place form a defensive strategy towards the ongoing globalization-driven transformations in agricultural production and distribution.

This strategy is even likely to result in certain spatial disparities in the near future. At the regional and national scale the shift of society towards quality local products may apply high pressure on the developing regions and nations which primarily rely on exports of agricultural products. Sustainable consumption, therefore, has a worth investigating spatial dimension needing more thorough consideration. In this respect, the notion of *global ecological citizenship* suggests that duties and obligations should be perceived in a wider scale and that consumption of agricultural goods should be guided by global sustainable thinking (Dobson 2003; Smith 2005).

The theoretical schema of desakota by McGee (2007; 2008), is a relatively recent perspective trying to put urban expansion and land use change in the broader context of globalization. The perspective attempts to integrate new economic developments, technological change and other higher level forces with lower level factors such as distance, availability of infrastructure and new business opportunities. Although the model has mostly been tested in Asian regions, it is believed to hold significant potential for western Europe as well (Xie, Batty et al. 2007).

Finally, the theoretical schemata concerning the structure and evolution of urban space include several perspectives based on urban geography and political economy. Among others the theoretical steam of expanding city addresses the importance of current technological progress in the field of information technologies, the massive increase in the volume, flow speed and spatial extent of goods and services exchanged as well as the new social values and ways of living (Munoz 2003; Zhang and Sasaki 2005). According to Ingram (1998), the contemporary city is characterized by a strong tendency of sprawl for both people and employment (Thurston and Yezer 1994).

Despite their diverse origins and spatiotemporal scales of employment, the theoretical perspectives presented share some common features. It seems that both distance and accessibility have a significant influence on deforestation and land use patterns. Moreover, technological changes in transportation play a key role in urban evolution. The new social ethics, behaviours, preferences and ways of living also influence considerably the structure of space. Population and demographics which are traditional forces of change need also be taken into account in the context of regional and urban development.

Summing up the discussion of the above perspectives, it can be argued that the common ground between the aforementioned theoretical schemata lies on the ascertainment that exurban land use conversions and modifications are advancing at a high rate (McCarthy 2005; McCarthy 2008). To some theorists, those changes are circumstantial or adventitious reactions of invested capital on rural areas in order to protect and insure its reproduction (Evans, Morris et al. 2002). Therefore, the current trajectories of rural land use changes could rapidly shift to new directions due to changes in the global economic environment. On the other hand, a growing number of arguments suggest that emerging land use patterns rely on deeper and stable changes in fundamental characteristics of society in terms of the people's attitude towards natural environment (Inglehart 1990; Buttel 2000). However, it is still evident that the rural land use systems is still a matter of concern and an issue for study and evaluation. In the following sections, we attempt to produce empirical evidences on some of the theoretical arguments described above concerning the case of recent forest land use changes in Greece.

4. Synthesis

Following the literature review on deforestation, It is evident that a variety of economic and social forces might work competitively or additively towards the configuration of land use patterns across the regions or the prefectures of a country (Verburg, Soepboer et al. 2002;

Verburg 2006). Following, we make an attempt to connect all the important aspects of deforestation in a coherent conceptual framework. Based on the conceptual framework, an empirical forest land use change model is proposed and the possible effects of all explanatory variables are discussed hypothesizing that the effect of a particular variable may differ between geographical areas.

The underlying causes of forest land use change vary from locality to locality as well as amongst countries (Wood and Skole 1998; Lambin, Turner et al. 2001; GLP 2005). These economic, socio-cultural and political forces are capable of impacting negatively or positively the extent and distribution of forest land. Their influence on forests usually results in quantitative changes as well as qualitative changes. Both conversions and structural habitat modifications are of great importance to policy makers. Reality is even more complex because of the usual linkages and interactions between positive and negative factors of change (Mahapatra and Kant 2005). The likely aggregated outcome of the combination of negative and positive factors is difficult to understand and predict. However, it may be more feasible to identify some dynamic processes through which socioeconomic and political factors operate resulting in distinct patterns of land uses. In this respect we propose urban sprawl, location decisions of economic activities, agricultural expansion and agricultural abandonment as the main factors of forest land use change.

In most cases, quantification of these forces is a difficult task, as is also difficult finding the appropriate methodology capable of giving reliable estimations of the magnitude and importance of the relationships involved. A wide variety of approaches and techniques have emerged for this reason, with the intention to rationalise decision-making about land use change issues (Upadhyay, Solberg et al. 2006). How and to what extent existing methodologies have satisfactorily reached this target is also a matter of research. Among applied methodologies, statistical techniques concerned with land use change dynamics are the most widely used. These models primarily focus on the causes of deforestation rather than its sources. Observed land use patterns are tightly connected to urban and regional development policies and to the enlargement of the regional economy. Their ceaseless transformation is fuelled by the need for serving the rapidly changing economic and social requirements as well as for fulfilling newly arising demands as a result of economic liberalization, privatization and transformation of lifestyle.

The morphology and evolution of land use patterns have been extensively studied and theorised by scientists of different disciplines (Wood and Skole 1998; Irwin and Geoghegan 2001; Walker 2001; Verburg, Schot et al. 2004; Walker 2004). Thus, a plethora of theoretical and modelling approaches have been developed so far in order to provide possible explanations of land allocation processes. Two general categories of land cover /use changes are described in the literature: conversions and modifications (Baulies and Szejwach 1998; Briassoulis 2000; Lesschen, Verburg et al. 2005). Land cover conversion refers to a change from one cover type to another whereas land cover modification implies structural or functional transitions in cover without loss in initial determinative characteristics. Similarly, land use conversion refers to a complete change from one use type to another whereas land use modification implies structural or functional alterations in use without loss of initial attributive characteristics. Finally, the driving forces (causes) of LUC change can be divided into two categories: Proximate causes and underlying causes. Proximate causes of land use change are associated with coarse anthropogenic operations that directly influence spatial patterns as, for instance, urbanisation, agricultural expansion and forest exploitation (Geist and Lambin 2001; Lesschen, Verburg et al. 2005). Underlying causes of land use change are associated with generative agents that weave proximate causes, such as economic, socio-demographic and technological factors (Geist and Lambin 2001; Lesschen, Verburg et al. 2005).

Proximate causes	Land use change forces at the Macroscopic level	
	Functioning and expansion of urban land forms and activities	Transportation and other kinds of infrastructure
	Functioning and expansion or shrinkage of agricultural activities	Livestock breading systems and forest resources exploitation
Underlying causes	**Economic Factors**	
	Sector arrangement of the regional economy	Sectoral Employment
	Sectoral structure	Investments and business location decisions
	Size and synthesis of imports and exports	Consumption patterns
	Productivity	Diffusion of technology and adoption of innovations
	Competitiveness of the economy	Mean size of businesses
	Technological level of the economy	Scale and agglomeration economies
	Taxation	Investment incentives and development policies
	Income distribution	Added value
	Social factors	
	Population skills level	Housing policy
	Education level	Institutions
	Social infrastructure	Population quality in the public sector
	Social security	Life style
	Demographic factors	
	Population changes	Indirect population potential
	Urban and rural population	Direct population potential
	Age of the population	Population mobility
	Environmental factors	
	Soil fertility	Biodiversity
	Topography	Ecosystem productivity
	Climatic conditions	Water resources
	Stretch of the coastline	Insular or mainland area

Table 2. Proximate and underlying causes of deforestation

Another useful distinction regarding land use change driving forces. could also be taken into account. They can be categorised into "endogenously changed or shifting or metamorphotic forces" that usually change very quickly over time (e.g. employment patterns of the new economy, location and relocation decisions of certain types of firms, supply and demand of certain products and services) "slow-shifting forces" (e.g. population size and other demographical characteristics) and "conditioning forces" which usually exhibit a temporal stability (e.g. soil types, geomorphology). The last categorisation of driving forces, in a way, implies that a steady state of land use patterns should almost never be expected. This endogenous, ever-changing nature of certain forces has been pointed out by theoretical approaches such as game theory and has also certain modelling implication in land use change studies. As Arthur (2005) states, out-of-equilibrium situations or the emergence of equilibria and the general unfolding of patterns in the economy calls for an algorithmic approach. Land use patterns may represent temporarily fulfilled or unfulfilled complex expectations not necessarily rationally formed as in the case of El Farol Bar problem (Arthur 1994). Table 2 provides a summary of some deforestation driving forces under the aforementioned categorisation framework.

Regarding economic factors, their main effect on land use changes depends mostly on the changes of the sectoral structure of economy. For instance, there might be labour transfer from an economic sector to another as in case of agricultural expansion or large-scale tourism development that may follow a decrease in industrial employment. The aforementioned structural changes are likely to influence decisively the land use system and also result in changes in the allocation of labour and land.

Population movements between regions are also important influential factors of land system. Such movements may have a direction from rural areas to large urban concentrations or there may be a reverse process of rural rebound where people move away from cities towards rural regions. In the first case, there might be intensification of the use of land in peri-urban space. In the second case, land use change happen in the countryside.

Finally, energy policy and taxation in resources such as natural gas and petrol are possible to result in an increase in the use of fuelwood and thus in deforestation or they may result in the development of alternative energy sources such as wind energy, solar energy etc. The likely results on land use system and on forests are complex and difficult to predict. They depend on the applied economic policy and also on the level of environmental awareness of people and authorities.

The impacts of the ongoing economic crisis on forests and on land use system in general, are difficult to forecast as there is no information on the likely duration of economic crisis, and on the particular counties, regions and economic sectors that will be affected most.

5. Evidence from Greece

Land use changes involve several positive and negative impacts on economic, social and environmental aspects. These impacts could have limited territorial scope or wider territorial implications, causing changes in the use of land in a greater geographical scale. (Chhabra, Geist et al. 2006). Impacts can also, have short-term, medium-term or even long-term action, be additive, synergistic, reversible or irreversible. Overall implications depend on recipients' degree of sensitivity, the ability to absorb or cope with pressure, as well as the type, intensity, extent and duration of pressure. Consequently, recognition, estimation and evaluation of likely economic, social and environmental impacts connected to forest land use changes, is a difficult process (Chhabra, Geist et al. 2006). Fig. 1, presents the geographical distribution and magnitude of some key phenomena associated with forest land use changes in Greece. Based on the information of the Fig. 1, it is possible put forward some comments concerning land use change in Greece:

- High deforestation constitutes a significant process of land use change in several insular as well as mountainous regions of the country (Minetos and Polyzos 2010). A limited number of mainland coastal regions as well as some regions adjacent to large metropolitan areas, present high deforestation rates. Examining the information on the maps, it is obvious that deforestation, very often, coexists with the urban sprawl and illegal housing activity. Beyond the obvious impacts on the biodiversity of these regions, there also emerge several questions concerning erosion processes, flooding and loss of ground via rain water washings. Taking into account that edaphogenic processes follow the geological time scale, such changes should be considered as being irreversible. At the same time, the cost of protecting human activities from flooding events increases, the available fresh water resources lower and microclimate and living conditions at the local level change. In the long term, reduction of biodiversity is expected to affect negatively development opportunities and to also influence the regional level of prosperity (Minetos and Polyzos 2010). Finally, it is worth mentioning that deforestation processes at numerous localities accumulate affecting wider areas at the regional scale and also fuelling global environmental issues such as global warming and climate change.

- Increased conversion and modification of agricultural land present high rates in the case of regions with large urban concentrations, in regions adjacent to the aforementioned ones as well as in several insular regions (Minetos and Polyzos 2009). Processes that contribute to the configuration of this pattern are: (a) Urbanisation of agricultural land and, (b) abandonment of marginal agricultural land. Therefore, in a great number of the aforementioned regions, the loss of agricultural land is connected to pressures deriving from urban sprawl and illegal housing activity (Minetos and Polyzos 2009). In the rest of the regions, loss of agricultural land is connected to the low competitiveness of agricultural sector and the problematic environmental and demographic characteristics within which agricultural activity takes place.

- It is apparent that economic forces such as land-rent, lead to structural changes to the economic base of regional areas in question (Polyzos 2009). However, if we take into consideration the way in which this economic transformation (illegal housing, urban sprawl) is happening, then it is likely that several negative economic, social and environmental impacts will emerge having long lasting action. More specifically, the shrinkage of the economic base of the regional spatial units and, in certain cases, the observed orientation of local economic base to a single activity generates phenomena of "monoculture" (e.g. tourism) in the economy.

Urban sprawl, concerns most regions of the country but it presents particular intensity in the western and southern areas as well as in most of the islands. High sprawl of urban activities in ex-urban location is observed in also relatively remote areas (Polyzos and Minetos 2009). They are also frequent commercial linear developments following the major interregional transportation routes as well as extensive low density areas of urban forms (residential units, tourism infrastructure, etc). A development pattern like this needs to be supported by a large amount of infrastructure (road axes, networks of water supply, networks and installations of waste water treatment) the size of which might be disproportionate to the size of served population.

REGIONS	DEFORESTA-TION	AGRICULTURAL LAND LOSS	URBAN SPRAWL	ILLEGAL HOUSING
Attiki	L	VH	VH	L
Aitoloakarnania	H	H	H	M
Viotia	H	M	M	H
Evia	L	VH	M	H
Evritania	L	VH	VH	L
Fthiotida	M	L	M	L
Fokida	M	H	VH	L
Argolida	L	L	H	H
Arkadia	H	VH	VH	M
Achaia	M	VH	H	M
Ilia	H	H	VH	L
Korinthia	L	VH	M	M
Lakonia	H	VH	VH	M
Messinia	L	VH	VH	VH
Zakinthos	VH	VH	M	L
Kerkyra	VH	M	L	M
Kefallinia	VH	VH	H	L
Lefkada	H	H	H	L
Arta	L	H	M	L
Thesprotia	H	VH	H	M
Ioannina	M	VH	VH	M
Preveza	VH	H	H	L
Karditsa	L	L	M	L
Larisa	VH	L	M	M
Magnisia	L	M	H	H
Trikala	H	M	H	M
Grevena	L	L	M	L
Drama	L	M	L	L
Imathia	M	M	M	VH
Thessaloniki	H	H	L	L
Kavala	M	M	L	H
Kastoria	H	L	M	L
Kilkis	VH	H	L	H
Kozani	H	L	M	L
Pella	M	L	L	M
Pieria	L	L	L	H
Serres	H	VH	L	L
Florina	H	L	L	L
Chalkidiki	L	L	VH	VH
Evros	H	M	L	H
Xanthi	L	L	L	H
Rodopi	M	L	L	H
Dodekanisos	H	H	M	L
Kyklades	H	M	H	M
Lesvos	VH	H	VH	L
Samos	L	VH	VH	L
Chios	VH	VH	L	VH
Irakleio	VH	M	M	L
Lasithi	VH	VH	VH	L
Rethymno	VH	H	H	L
Chania	M	M	M	M

DEFORESTATION

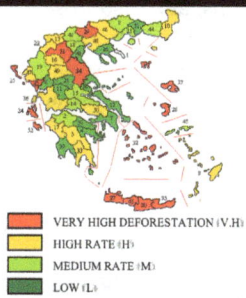

- VERY HIGH DEFORESTATION (V.H)
- HIGH RATE (H)
- MEDIUM RATE (M)
- LOW (L)

AGRICULTURAL LAND LOSS

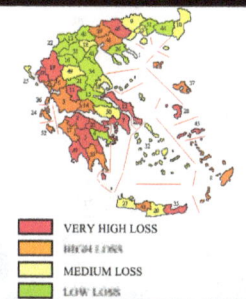

- VERY HIGH LOSS
- HIGH LOSS
- MEDIUM LOSS
- LOW LOSS

URBAN SPRAWL

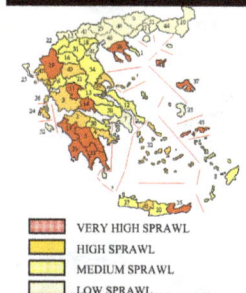

- VERY HIGH SPRAWL
- HIGH SPRAWL
- MEDIUM SPRAWL
- LOW SPRAWL

ILLEGAL HOUSING

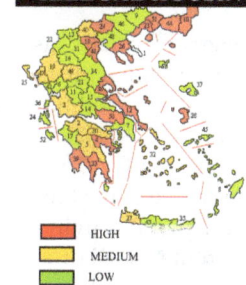

- HIGH
- MEDIUM
- LOW

Fig. 1. Matrix of regional spatial units in Greece and the magnitude of significant land use change phenomena (NSSG 1994; NSSG 1995; NSSG 1999; NSSG 2004; NSSG 2004; NSSG 2006).

Illegal housing seems to be high in almost all neighbouring regions to large urban concentrations, and also in several costal locations and remote areas (Polyzos and Minetos 2009). While in the past, illegal housing activity as a phenomenon was concentrated in urban and suburban space, resulting in significant negative consequences to the formation and functionality of cities, nowadays it appears that illegal housing phenomenon influences wider spatial units.

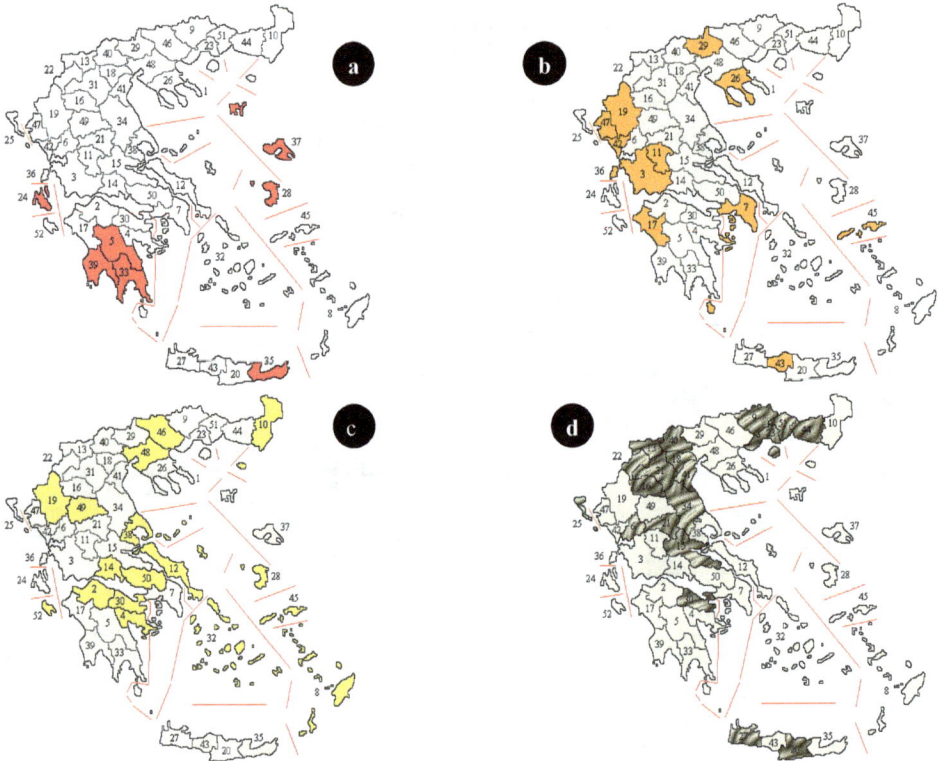

Fig. 2. Hot-spots of land use changes: a) Very significant land use changes b) significant land use changes c) moderate land use changes d) low land use changes

The phenomenon of land use change happens for multiple reasons and also presents important spatial differentiations. In order to acquire an overall understanding of land use transformations and modifications, we attempt to locate "hot-spots" (Reid, Tomich et al. 2006) or "regions of very high activity" across the country (Figures 1 and 2). In the first column of matrix in Fig. 1 we have coloured with red regions that present very high (VH) rates in at least two of the four spatial phenomena that mainly drive land use changes. They also present high (H) activity in at least one of the four aforementioned phenomena. These regions constitute the "hot spots" or "first level areas" of land use change. The spatial "distribution of "hot spots" is presented in the Fig. 2a. In these areas, land use changes are rapid and extensive and they perhaps jeopardize the fundamental regional characteristics as well as economic, social and environmental equilibrium of these areas.

We have also created a second category of hot spots (regions with orange shading in matrix of Fig. 1). These areas include regions with high or very high intensity in three out the four phenomena or regions with very high intensity in two out of four phenomena, excluded the prefectures of first category. These regions constitute constitute "second level hot spots. Their spatial distribution is presented in the Fig. 2b. They are regions in which land use changes are significant in magnitude and their future course may become particularly problematic.
A third category of regions consist of areas with high or very high rate in two out of four phenomena. In these areas, land use changes are of a moderate magnitude either because these regions are in a kind of recession compared to their past size of activity (eg Attica, Thessalonica, Dodekanisa etc), or it is expected that they are going to accelerated in the near future (Trikalas, Ioanninas etc). In most of the rest regions land use change phenomena present low intensity.
Summarising the above discussion, it could be supported that the applied spatial policy in Greece attracts a relatively low interest compared to other sectoral policies. Consequently, objectives regarding regulation of space are not always explicit and compatible while sustainable management of space, protection of environment and relaxation of regional inequalities still remain issues that need to be managed and placed into a proper policy context.

6. Conclusions

This paper has dealt with theoretical perspectives concerning forest land use change in general, as well as the factors of land use changes in Greece. Making informed land policy decisions is central to achieving sustainability at a regional level. Prior to formulating certain sustainable policy objectives and targets, the baseline information needed is the identification kind of the driving forces that influence current forest land use patterns. Generally speaking, these driving forces are closely associated with the economic, social and environmental context within which the regions exist and function. The effects on forest land of the predictor variables that where employed by this study, while significant in most regions are still characterized by many uncertainties. Some theoretically interesting explanatory variables have indicated that the effects of certain processes on land use changes may be important but not always straightforward.
A synthesis and evaluation of the results brings up some important issues relevant to the theoretical framework of the field. In particular, a noticeable argument relates to the course of development of deforestation rate in the long term, when the major competing uses to forests are urban and not agricultural land uses as it is assumed by Angelesen's (2007) model in the case of tropical deforestation. It seems that when the antagonism involves urban and forest land uses and when also the types of forest ecosystems fall into the category of not-productive forests (as it is the case for most Mediterranean forests) then land-rent generated by forest uses is unlikely to compensate for the one coming from urban development of the land. Urban land uses through market mechanisms will tend to outrage forests. As long as the Greek law is strongly opposed to the conversion of forest, it seems that, at least in the sort term, the only way of confronting this market-induced process might be the restructuring of the law enforcement mechanisms.
On the other hand, it seems that in most urban prosperous regions as well as in their vicinity, deforestation has slowed down, although additional data need to be analysed.

Deforestation moves to more remote regions as accessibility improves. There are interactions amongst indirect population potential and illegal housing activity that influence the spatial distribution of deforestation rate. This complex relationship implies a geographical transfer and dispersion of illegal housing phenomenon from urbanized regions to remote, less-urbanized ones. It is likely that urban populations remain the major source of illegal housing activity; however they now tend to exercise this activity in longer distances due to accessibility improvements. If this is the case then deforestation controllers are still in urban areas even though the impacts of their acts are being systematically "exported" to other areas. The geographical transfer of deforestation relieves forests in the places of origin but at the same time, it escalates pressure on the host areas' forests.

An additional issue, relevant to the abovementioned argument, relates to the geographical characteristics of the areas being deforested. In insular prefectures, land surface is a scare resource and areas covered with forests are limited. If such spatial units were involved in a lengthy urban-forest land use antagonism, it would probably be difficult for all stages of forest transition theory to take place. Due to geographical remoteness, small-sized insular regions cannot easily base part of their development on exploiting the land of adjacent regions. Thus, it is difficult for islands to *"export"* deforestation processes. Bearing in mind the scarcity of developable land, high rents associated with urban land uses, the lack of forest cadastral maps and the insufficient forest law enforcement mechanisms, it is more likely for insular forests to move to the direction proposed by the theory of the tragedy of commons rather than the one proposed by forest transition theory. Even if the country as a whole managed to increase its forests, there would probably be winners and losers at the regional scale.

Generally speaking, the spread of urban uses in the countryside has negatively affected forests either directly or indirectly. The improvement of transportation infrastructure, the expansion of urban plans, urban sprawl, illegal housing activity and legal building construction activity create a complex negative background for forest land uses. On the other hand, changes in Gross Domestic Product in agriculture and change in tourism infrastructure either have limited or zero adverse impacts on forests. It seems that the observed improvement in the performance of agricultural sector in the relevant regions has not occurred in the expense of forests through some expansion of cultivated land. In addition, new tourism accommodation infrastructure, possibly due to the mandatory environmental impact assessment introduced in the early '90s, had limited negative effects on forests. However, regions with high growth in their tourism accommodation in ex-urban areas show significant signs of deforestation.

In the context of planning a sustainable forest policy, accessibility issues as well as the spatial patterns generated by urban phenomena such as urban sprawl and illegal housing are of crucial importance. At first glance, the improvement in the regional level of prosperity might be associated with lower deforestation rate. However, it is not certain whether this additional prosperity has been achieved in a sustainable manner. It is likely that improvement in prosperity at some location has been achieved in the expense of forests at some other location. These results could guide further research into improving the understanding of spatial processes such as forest land use changes and into rationalizing forest-related decision making. Strategic project monitoring and appraisal as well as evaluation of project impacts on land uses can help spatial planners and land use decision-makers to introduce specific environmental protection objectives into land development and planning processes.

7. References

Agarwal, C., C. Green, et al. (2002). A Review and Assessment of Land-Use Change Models. Dynamics of Space, Time, and Human Choice. CIPEC Collaborative Report Series No. 1, Center for the Study of Institutions Population, and Environmental Change Indiana University.

Allen, P., M. FitzSimmons, et al. (2003). "Shifting plates in the agrifood landscape: the tectonics of alternative agrifood initiatives in California." Journal of Rural Studies 19(1): 61-75.

Amin, S. (1976). Unequal development Sassex, Harvester press.

Andersen, M. S. and I. Massa (2000). "Ecological modernization - origins, dilemmas and future directions." Journal of Environmental Policy and Planning 2(4): 337-345.

Angelsen, A. (2007). Forest Cover Change in Space and Time: Combining the von Thünen and Forest Transition Theories. Policy Research Working Paper 4117, The World Bank: 1-43.

Arthur, W. B. (1994). Inductive Reasoning and Bounded Rationality: The El Farol Problem. Given at the American Economic Association Annual Meetings, 1994, Session: Complexity in Economic Theory, chaired by Paul Krugman. Published in Amer. Econ. Review (Papers and Proceedings), 84, 406, 1994. Los Alamos, USA, Santa Fe Institute.

Arthur, W. B. (2005). Out-of-Equilibrium Economics and Agent-Based Modeling. "Paper prepared for Handbook of Computational Economics, Vol. 2: Agent-Based Computational Economics, K. Judd and L. Tesfatsion, eds, ELSEVIER/North-Holland, forthcoming 2005.". Los Alamos, USA, Santa Fe Intitute.

Aspinall, R. (2004). "Modelling land use change with generalized linear models--a multi-model analysis of change between 1860 and 2000 in Gallatin Valley, Montana." Journal of Environmental Management 72(1-2): 91-103.

Baulies, X. and G. Szejwach (1998). LUCC Data Requirements Workshop: Survey of needs, gaps and priorities on data for land-use/land-cover change research; in LUCC Report Series; no 3. Barcelona,, Institut Cartogràfic de Catalunya,.

Briassoulis, H. (2000). "Analysis of land use change: Theoretical and modelling approaches in: The web book of regional science, http://www.rri.wvu.edu/WebBook/Briassoulis/contents.htm."

Buttel, F. H. (2000). "Ecological modernization as social theory." Geoforum 31(1): 57-65.

Chhabra, A., H. Geist, et al. (2006). Multiple impacts of land-use /cover change. Land-use and Land-cover change: Local processes and global implication. E. F. Lambin and H. J. Geist. Wurzburg, Springer: 71-116.

Chomitz, K. M. and D. A. Gray (1996). "Roads, land use, and deforestation: a spatial model applied to Belize." World Bank Economic Review 10(3): 487-512.

Dobson, A. (2003). Citizenship and the Environment. Oxford Oxford University Press.

Ehrhardt-Martinez, K., E. M. Crenshaw, et al. (2002). "Deforestation and the Environmental Kuznets Curve: A Cross-National Investigation of Intervening Mechanisms." Social Science Quarterly 83(1): 226-243.

Elands, B. and F. Wiersum (2001). "Forestry and rural development in Europe: an exploration of socio-political discourses " Forest Policy and Economics 3(1-2): 5-16.

Evans, N., C. Morris, et al. (2002). "Conceptualizing agriculture: a critique of post-productivism as the new orthodoxy." Progress in Human Geography 26(3): 313-332.

Feagan, R. (2007). "The place of food: mapping out the 'local' in local food systems." Progress in Human Geography 31(1): 23-42.

Fredj, K., G. Martvn-Herran, et al. (2004). "Slowing deforestation pace through subsidies: a differential game." Automatica 40(2): 301-309.

Geist, H. and E. Lambin (2001). What drives tropical deforestation? A meta-analysis of proximate and underlying causes of deforestation based on subnational case study evidence; in LUCC Report Series; no 4.

Geist, H., W. McConnell, et al. (2006). Causes and trajectories of Land-use /cover change. Land-use and Land-cover change: Local processes and global implication. H. Geist and E. Lambin. Wurzburg, Springer: 41-70.

GLP (2005). Global Land Project, Science Plan and Implementation Strategy, IGBP Report No. 53 /IHDP Report No. 19. IGBP Secretariat. Stockholm: 1-64.

Grainger, A. (1995). "The Forest Transition: An Alternative Approach." Area 27(3): 242-251.

Groot, J. C. J., W. A. H. Rossing, et al. (2007). "Exploring multi-scale trade-offs between nature conservation, agricultural profits and landscape quality--A methodology to support discussions on land-use perspectives." Agriculture, Ecosystems & Environment 120(1): 58-69.

Hardin, G. (1968). "The Tragedy of the Commons." Science 162(3859): 1243-1248.

Herschel, E. (1997). "A General Statement of the Tragedy of the Commons." Population and Environment: A Journal of Interdisciplinary Studies 18(6): 515-531.

Hinrichs, C. C. (2003). "The practice and politics of food system localization." Journal of Rural Studies 19(1): 33-45.

Inglehart, R. (1990). Culture Shift in Advanced Industrial Society. Princeton, New Jersey, Princeton University Press.

Ingram, G. K. (1998). "Patterns of Metropolitan Development: What Have We Learned?" Urban Studies 35(7): 1019-1035.

Irwin, E. G. and J. Geoghegan (2001). "Theory, data, methods: developing spatially explicit economic models of land use change." Agriculture, Ecosystems & Environment 85(1-3): 7-24.

Koop, G. and L. Tole (1999). "Is there an environmental Kuznets curve for deforestation?" Journal of Development Economics 58(1): 231-244.

Lambin, E., H. Geist, et al. (2006). Introduction: Local processes with global impacts. Land-use and Land-cover change: Local processes and global implication. H. Geist and E. Lambin. Wurzburg, Springer: 1-8.

Lambin, E. F., B. L. Turner, et al. (2001). "The causes of land-use and land-cover change: moving beyond the myths." Global Environmental Change 11(4): 261.

Lesschen, J. P., P. H. Verburg, et al. (2005). Statistical methods for analysing the spatial dimension of changes in land use and farming systems - LUCC Report Series 7, The International Livestock Research Institute, Nairobi, Kenya and LUCC Focus 3 Office, Wageningen University, the Netherlands.

Mahapatra, K. and S. Kant (2005). "Tropical deforestation: a multinomial logistic model and some country-specific policy prescriptions." Forest Policy and Economics 7(1): 1-24.

Marsden, T. (1998). "Agriculture beyond the treadmill? Issues for policy, theory and research practice." Progress in Human Geography 22(2): 265-275.

Marsden, T. (2004). "The Quest for Ecological Modernisation: Re-Spacing Rural Development and Agri-Food Studies." Sociologia Ruralis 44(2): 129-146.

McCarthy, J. (2005). "Rural geography: multifunctional rural geographies - reactionary or radical?" Progress in Human Geography 29(6): 773-782.

McCarthy, J. (2008). "Rural geography: globalizing the countryside." Progress in Human Geography 32(1): 129-137.

McGee, T. (2007). "Many knowledge(s) of Southeast Asia: Rethinking Southeast Asia in real time." Asia Pacific Viewpoint 48(2): 270-280.

McGee, T. (2008). "Managing the rural–urban transformation in East Asia in the 21st century." Sustainability Science 3(1): 155-167.

Minetos, D. and S. Polyzos (2009). "Analysis of Agricultural land uses transformation in Greece: A multinomial regression model at the regional level." International Journal of Sustainable Planning and Development 43: 189-209.

Minetos, D. and S. Polyzos (2010). "Deforestation processes in Greece: A spatial analysis by using an ordinal regression model." Forest Policy and Economics 12(6): 457-472.

Munoz, F. (2003). "Lock living: Urban sprawl in Mediterranean cities." Cities 20(6): 381-385.

NSSG (1994). Census of buildings on 1st December 1990. Pireas, Hellenic Republic.

NSSG (1995). Pre-census data of the Agriculture and Livestock Census of the year 1990/1991. Pireas, Hellenic Republic.

NSSG (1999). Tourist statistics 1994-1996. Pireas, Hellinic Republic.

NSSG (2004). Census of buildings on 1st December 2000. Athens, Hellenic Republic.

NSSG (2004). Statistical Yearbook of Greece. Athens, Hellenic Republic.

NSSG (2006). Building Activity Statistics for the Years 1997-2006. Pireas, Hellinic Republic.

Perz, S. G. (2007). "Grand Theory and Context-Specificity in the Study of Forest Dynamics: Forest Transition Theory and Other Directions." The Professional Geographer 59(1): 105-114.

Platt, R. V. (2004). "Global and local analysis of fragmentation in a mountain region of Colorado." Agriculture, Ecosystems & Environment 101(2-3): 207-218.

Polyzos, S. (2009). "Regional Inequalities and Spatial Economic Interdependence: Learning from the Greek Prefectures." International Journal of Sustainable Development and Planning 4(2): 123-142.

Polyzos, S. and D. Minetos (2009). "Informal housing in Greece: A quantitative spatial analysis." Theoretical and Empirical Researches in Urban Management Journal 2(11): 7-33.

Rahman, H. and M. Yamao (2007). "Community based organic farming and social capital in different network structures: Studies in two farming communities in Bangladesh." American Journal of Agricultural and Biological Science 2(2): 62-68.

Raynolds, L. (2000). "Re-embedding global agriculture: The international organic and fair trade movements." Agriculture and Human Values 17(3): 297-309.

Reid, R., T. Tomich, et al. (2006). Linking land-change science and policy: Current lessons and future integration. Land-use and Land-cover change: Local processes and global implication. E. Lambin and H. Geist. Wurzburg, Springer: 157-176.

Roberts, T. and P. Greimes (2002). World-System theory and the environment: Towards a new synthesis. Sociological theory and the environment: Classical foundations, contemporary insights. R. Dunlap, F. Buttel, P. Dickens and A. Gijswijt. New York, Rowman & Littlefield Publishers: 167-196.

Santilli, M., P. Moutinho, et al. (2005). "Tropical Deforestation and the Kyoto Protocol." Climatic Change 71(3): 267-276.

Seyfang, G. (2006). "Ecological citizenship and sustainable consumption: Examining local organic food networks." Journal of Rural Studies 22(4): 383-395.

Shandra, J. M., B. London, et al. (2003). "Environmental degradation, environmental ssustainability, and overurbanisation in the developing world: A quantitative, cross-national analysis." Sociological Perspectives 46(3): 309-329.

Shucksmith, M. (1993). "Farm household behaviour and the transition to post-productivism." Journal of Agricultural Economics 44: 466–478.

Smith, E. and T. Marsden (2004). "Exploring the 'limits to growth' in UK organics:beyond the statistical image." Journal of Rural Studies 20: 345–357.

Smith, G. (2005). "Green Citizenship and the Social Economy." Environmental Politics 14(2): 273-289.

Stern, N. (2006). "What is the Economicsof Climate Change?" WORLD ECONOMICS • Vol. 7 No. 2 April–June 7(2): 1-10.

Thurston, L. and A. M. J. Yezer (1994). "Causality in the Suburbanization of Population and Employment." Journal of Urban Economics 35(1): 105-118.

Upadhyay, T. P., B. Solberg, et al. (2006). "Use of models to analyse land-use changes, forest/soil degradation and carbon sequestration with special reference to Himalayan region: A review and analysis." Forest Policy and Economics 9(4): 349-371.

van der Ploeg, J. D. (2000). "Revitalizing Agriculture: Farming Economically as Starting Ground for Rural Development." Sociologia Ruralis 40(4): 497-511.

van der Ploeg, J. D., H. Renting, et al. (2000). "Rural Development: From Practices and Policies towards Theory." Sociologia Ruralis 40(4): 391-408.

Verburg, P. (2006). "Simulating feedbacks in land use and land cover change models." Landscape Ecology 21(8): 1171-1183.

Verburg, P., K. Overmars, et al. (2006). "Analysis of the effects of land use change on protected areas in the Philippines." Applied Geography 26(2): 153-173.

Verburg, P., P. Schot, et al. (2004). "Land use change modelling: Current practice and research priorities." GeoJournal 61(4): 309-324.

Verburg, P., C. Schulp, et al. (2006). "Downscaling of land use change scenarios to assess the dynamics of European landscapes." Agriculture, Ecosystems & Environment 114(1): 39-56.

Verburg, P., W. Soepboer, et al. (2002). "Modeling the Spatial Dynamics of Regional Land Use: The CLUE-S Model." Environmental Management V30(3): 391-405.

Walker, R. (2001). "Urban sprawl and natural areas encroachment: Linking land cover change and economic development in the Florida Everglades." Ecological Economics 37(3): 357-369.

Walker, R. (2004). "Theorizing land-cover and land-use change: The case of tropical deforestation." International Regional Science Review 27(3): 247-270.

Winter, M. (2003). "Embeddedness, the new food economy and defensive localism." Journal of Rural Studies 19(1): 23-32.

Winter, M. (2005). "Geographies of food: agro-food geographies, food, nature, farmers and agency." Progress in Human Geography 29(5): 609-617.

Wood, C. H. and D. Skole (1998). "Linking satellite, census, and survey data to study deforestation in the Brazilian Amazon." People and Pixels: Linking Remote Sensing and Social Science: 70-93.

Xie, Y., M. Batty, et al. (2007). "Simulating Emergent Urban Form Using Agent-Based Modeling: Desakota in the Suzhou-Wuxian Region in China." Annals of the Association of American Geographers 97(3): 477-495.

Zhang, Y. and K. Sasaki (2005). "Edge city formation and the resulting vacated business district." The Annals of Regional Science 39(3): 523-540.

5

Deforestation and Water Borne Parasitic Zoonoses

Maria Anete Lallo
Universidade Paulista (UNIP), São Paulo
Brazil

1. Introduction

Disease emergence or re-emergence is often the consequence of the societal and technological change and manifests frequently in an unpredictable manner. It has been estimated that among emerging diseases, 75% has zoonotic characteristics. Many factors influencing the emergence of zoonoses, such as environmental change and land use, changes in demographics, changes in technology and industry, increasing international travel and commerce, breakdown of public health measures, and microbial adaptation and change (Broglia & Kapel, 2011).

Deforestation is one of the most disruptive changes affecting parasitic and vector populations. When the forest is cleared and erosion of the soil strips away the former state, if indeed, it is permitted and able to regenerate. The response of tropical forests to perturbation is affected by soil type, elevation, mean precipitation, and latitude. Cleared tropical forests are typically converted into grazing land for cattle, small-scale agricultural plots, human settlements or, left as open areas. Expansion of existing human settlements and movement of human populations create a need for increased food supply, leading to changes in the types and amounts of vegetation, thereby providing changed ecological niches and conditions for proliferation of newly arriving and/ or adaptive existing vectors and their parasites (Slifko et al., 2000). Deforestation is one of the changes that most affect the ecological niches of the disease, favoring the transmission of them (Patz et al., 2000; Slifko et al., 2000).

The waterborne or food is the main route of transmission of parasitic diseases. Zoonoses such as giardiasis, cryptosporidiosis and microsporidiosis are waterborne diseases that include the participation of domestic and wild animals and man. Environmental changes and ecological disturbances, due to both natural phenomena and human intervention, have exerted and can be expected to continue to exert a marked influence on the emergence and proliferation of zoonotic parasitic diseases. They change the ecological balance the ecological balance and context within which vectors and their parasites breed, develop, and transmit disease (Patz et al, 2000).

Interest in the contamination of drinking water by enteric pathogenic protozoa has increased considerably during the past three decades and a number of protozoan parasitic infections of humans are transmitted by the waterborne route (Patz et al., 2000; Slifko et al., 2000).

Waterborne transmission is one of the main risk factors for intestinal diseases causing an important morbidity and mortality worldwide. Over 50% of the waterborne infections are produced by unknown agents. In the last years, for economic and environmental reasons, spreading sewage sludge on agricultural lands has increased. This might affect not only the circulation of recognized pathogens such as *Cryptosporidium* and *Giardia*, but also emerging pathogens, such as microsporidia. The general impression is that treatment of water has demonstrated a high efficacy of pathogen removal, however, as viable pathogen have been detected in water. It is important understand that the presence of human pathogens in surface water may suggest the presence of living environmental reservoirs, such as domestic and wild animals. Aquatic birds may play an important role in the transmission of different pathogens (Izquierdo et al., 2011).

The parasites *Cryptosporidium*, *Giardia* e microsporidia are major of diarrheal disease in human, worldwide and have also been recognized as the predominat causes of waterborne diseases. Cryptosporidium, *Giardia* and microsporidia have life cycle wich are suited to waterborne and foodborne transmission. Their life cycle are completed within and individual host, with transmission by fecal-oral route. The transmissible stages, *Cryptosporidium* oocysts or *Giardia* cysts or microsporidia spores, are produced in a large numbers and are infectious when excreted, a marked resistence to environmental and water treatment stresses, wich assists their dissemination, and have the potential to be transmitted from non-human to human hosts (zoonoses) and vice-versa, enhancing the reservoir of (oo)cysts or spores markedly (Smith et al, 2007). The purpose of this chapter is to show the biological and epidemiological aspects most relevant of the parasites *Cryptosporidium*, *Giardia* and microsporidia responsible for waterborne parasitic diseases most important.

2. Cryptosporidiosis

It is a parasitic disease caused by protozoa of the genus *Cryptosporidium*, which affects amphibians, birds, mammals, reptiles and fish and is characterized by impairment of the digestive system. Since its first description in the 1970s, cryptosporidiosis has been considered an opportunistic infection in immunodeficient individuals, it is known today, however, that it is a prevalent disease in immunocompetent individuals also important. Many studies have also revealed the prevalence of this protozoan infection in animals, although the zoonotic transmission of the disease is not yet fully understood (Fayer et al., 2010).

2.1 About the agent and the disease

Cryptosporidium is a protozoan that has about 22 identified species (Table 1). Although many species have been described to date, *C. parvum* is the most widespread species of mammals, including man. It is known that *C. parvum* is not a homogeneous species since the isoenzyme analysis and DNA sequencing revealed differences between oocysts isolated from various animal species. 7 genotypes have been identified - of cattle, humans, mice, pigs, opossuns, dogs and ferrets (Fayer 2010; Plutzer Karanis, 2009).

As in other coccidian monoxenic its life cycle is not employing intermediate host. *Cryptospodirium* oocysts are small and contain four sporozoites inside free. When they are ingested by the host, the oocysts release of sporozoites in the small intestine, which invade intestinal cells. The sporozoites begin multiplying to form asexual meront type I and type II with 4 and 8 merozoites, respectively. The type II merozoites give rise to the sexual phase of the cycle or gametogony with differentiation stages in male (microgametes) and female (macrogametes). The microgametes penetrates macrogamete leading to the formation of a zygote that develops into oocyst. Are two types of oocysts produced - a kind of thin-walled autoinfectante able to release within the host, starting a new cycle and a thick-walled, highly resistant to environmental conditions, which is eliminated in feces. The cycle time is variable and may occur in up to 48 hours or 14 days depending on the host species. Unlike other coccidia, which eliminate the non-sporulated oocysts, the oocysts of *C. parvum* undergo sporulation within the host, eliminating the already infective for the environment (Carey, 2004; Chalmers & Davies, 2009).

Species	Hosts
	Fishes
Psicicryptosporidium cichlidis	*Oreochromis miloticus* e *Tilapia zilli*
Psicicryptosporidium reichenbachklinkei	*Trichogaster leeri*
Cryptosporidium molnari	*Sparus auratusDicentrarchus labrax* (*Gilthead*)
C.scophthalmi	*Scophthalmus maximus*(*Turbot*)
	Amphibians and reptiles
C. serpentis	*Elaphe guttata* (*Corn snake*)
C. varanii	*Varanus prasinus* (*Emerald monitor*)
C. fragile	*Duttaphrynus melanostictus* (*Black-spined toad*)
	Birds
C. meleagridis	*Meleagris gallopavo* (turkey)
C. baileyi	*Gallus gallus* (chicken)
C. galli	*Gallus gallus* (chicken)
	Mammals
C. muris	*Mus musculus* (mice)
C. parvum	*Mus musculus* (mice)
C. wrairi	*Cavia porcellus* (guinea pig)
C. felis	*Felis catis* (cat)
C. andersoni	*Bos taurus* (cattle)
C. canis	*Canis familiaris* (dog)
C. hominis	*Homo sapiens* (man)
C. suis	*Sus scrofa* (pig)
C. bovis	*Bos taurus* (cattle)
C. fayeri	*Macropus rufus* (kangaroo)
C. ryanae	*Bos taurus* (cattle)
C. macropodum	*Macropus giganteus* (kangaroo)

Table 1. Species of *Cryptosporidium* by Fayer (2010).

The pathogenesis and clinical picture of criptoporidiose are influenced by several factors, including animal species, age, immune response and association with other pathogens. The infection can range from subclinical to severe, And have more severe disease (Table 2). In humans, the incubation period is 20-10 days and the duration of the disease in immunocompetent individuals, up to 3 weeks. In immunodeficient or immunosuppressed individuals, infection is chronic, with symptoms and elimination of persistent oocysts. The clinical signs manifested in cryptosporidiosis include profuse watery diarrhea, vomiting, anorexia, weight loss, abdominal pain, fever and dehydration. In immunocompetent individuals, these symptoms are mild and transient. The spread of infection to the gallbladder and bile ducts, pancreas and respiratory system is common in AIDS patients (Barr, 1998; Chalmers & davies, 2009).

The diagnosis of cryptosporidiosis is based on the meeting of the parasite in feces, using methods of concentration of oocysts, such as formaldehyde or ether flotation with saturated sucrose solution, associated with staining techniques, such as Ziehl- Nielsen, Kinyoun, fuchsin or safranin. Different immunological techniques have been used for the diagnosis of human cryptosporidiosis. Among them, ELISA or immunofluorescence with monoclonal or polyclonal antibodies are used to detect oocysts in the feces. The PCR technique is an alternative to both conventional diagnosis of *Cryptosporidium* in fecal specimens and in environmental samples. Although PCR is rapid, sensitive and accurate, has limitations as the detection of nucleic acid of viable organisms, naked nucleic acid and the possibility of laboratory contamination. We recommend its use for oocysts in water samples (Fayer et al., 2000; Marquardt et al., 2000; Xiao, 2002).

Characteristics	Immunodeficient individuals	Immunocompetent individuals
Susceptible population	Immunocompromised persons of all ages, especially with AIDS	Children, first with less than 1 year of age and adults of all ages
Infection sites	Intestinal or extraintestinal	Intestinal usually
Enteric form	Asymptomatic or transient or chronic diarrhea or fulminant	Asymptomatic, acute and persistent
Clinic form	Diarrhea, fever, abdominal pain, weight loss e vômitos	Diarrhea, fever, abdominal cramps, weight loss, nausea and vomiting

Table 2. Characteristics of cryptosporidiosis in humans.

2.2 Epidemiology and prophylaxis
Cross-transmission studies revealed that oocysts obtained from humans are infective to other mammals, as oocysts from animals that are infectious to other species, it is a zoonotic disease potential. The epidemiology of cryptosporidiosis is influenced by the capacity of thick-walled oocysts survive in the environment (Table 3) (Xiao, 2002).

Temperature	Survival Time
25 e 30°C	3 months
20°C	6 months
15°C	7 months
-20 e -70°C	Feel hours

Table 3. Survivel condition of *Cryptosporidium* oocysts

The oocyst is the oral-fecal route and occurs until now, have been described various forms of transmission, can be highlighted - from person to person by direct or indirect contact, including sexual activities, from animal to animal, animal to man, by drinking water or recreation, from food and air. The number of oocysts required to establish an infection is small, it is estimated that the infectious dose varies from 9 to 1,000 oocysts (Fayer et al., 2010).

Feces containing oocysts contaminate soil, food and water. The movement of oocysts in the environment is favored by the winds, the rain water, the movement of animals and the actions of man himself. The major outbreaks of cryptosporidiosis reported in HIV-negative people are linked to the ingestion of water contaminated with oocysts derived from cattle or sheep. This water containing oocysts also contaminated foods, especially vegetables and fruits, and is another important way of transmission in outbreaks of cryptosporidiosis. Additionally, food can be contaminated by the hands of manipulators (Robinson et al., 2010).

A variety of risk factors are associated with infection by *Cryptosporidium*, among them stand out from the deficiency of the immune response, the presence of concomitant infections, ingestion of contaminated food and water, poor sanitary conditions and occupational exposure, is the contact with animals or infected humans (Plutzer et al., 2009).

From the earliest descriptions of cryptosporidiosis in humans, the number of cases has continued to grow, in the case of individual reports or outbreaks, such infection is attributed to *C. parvum*. The average prevalence in industrialized countries by 2.2% in immunocompetent individuals and 14% in HIV-positive. Already in developing countries, these numbers increase and may reach 8.5% in immunocompetent patients and 24% in HIV-positive. This means that proper sanitary conditions and a rigorous treatment of the water, as seen in developed countries, the spread of the disease can be decreased (Hajdušek et al., 2004).

In the United States, an estimated 50% of the animals from cattle herds eliminate oocysts of *C. parvum*, however, the disease is preferentially observed in calves that manifest from the 4th day of life until the fourth week. In other domestic animals, the prevalence of cryptosporidiosis is less valued, however it is known that predominates in neonates and young people (Thompson et al., 2009).

Human or animal cryptosporidiosis can only be controlled if the oocysts of the parasite is eliminated or destroyed. The oocysts can spread and persist in the environment for a long time. Moreover, it is known that this parasitic form also resists water to conventional treatments such as chlorination and filtration. *Giardia* is 14 to 30 times more susceptible to water treatment with chlorine or ozone. Either way, ozone is the most effective chemical agent in the inactivation of *Cryptosporidium* oocysts (Thompson et al., 2008).

To reduce the risk of infection of individuals more susceptible to cryptosporidiosis, such as HIV-positive, immunosuppressed individuals and children, it is recommended that drinking water be boiled for about 1 minute before ingestion. The same recommendation should be made for the young or immunodeficient animals (Fayer et al., 2010; Thompson et al., 2008).

Others include general health care and proper cleaning of the hands of food handlers, proper disposal of animal waste, sewage treatment, washing litter boxes with boiling water, among others (Chalmers & Davies, 2009).

One should keep susceptible individuals, HIV-positive or immunosuppressed, have contact with the feces of pets, especially if they have less than 6 months old. If this is not possible, you should recommend the appropriate use of gloves. Not recommended the removal of this person's contact with your pet because of the strong emotional bond that unites them. Immunosuppressed patients should ideally acquire an animal older than 6 months, you do not have diarrhea and it has been previously examined by a veterinarian (Fayer et al, 2000; Xiao, 2010).

Of the various species of *Cryptosporidium*, *C. parvum* has been observed in a larger number of human infections, and reaches a large number of animal species. Within this species, two genotypes are, most of the time, described the infection - the human (38% of infections) and cattle (62% of infections). Thus, for the human cryptosporidiosis occurring bovine genotype is necessary to contact with infected animals such as cattle, sheep and goats, or that there is environmental contamination, especially water and food. Several cases of cryptosporidiosis have been reported in veterinary students and animal handlers, resulting probably from contact with infected animals, especially cattle, which reinforces the possibility of such transmission. Recently, the dog genotype of *C. parvum* has also been identified in human infections(Fayer et al, 2000; Xiao, 2010).

All these evidences indicate that a large number of hosts and genotypes may be involved in cryptosporidiosis and molecular characterization of this parasite should facilitate understanding of the epidemiology of the disease. While these points are not completely understood, it is recommended the adoption of control measures in situations of potential risk (Xiao, 2010).

3. Giardiasis

It is a disease caused by the flagellate protozoan *Giardia*, intestinal parasite of a wide variety of animals, including man, constituting one of the most prevalent intestinal parasites, known and described throughout the world (Marquardt et al.,2000).

3.1 About the agent and the disease

The most accepted classification of the genus *Giardia* is based on its morphological characteristics, with six described species - *G. agilis*, *G. Muris*, *G. psittaci*, *G. ardeae*, *G. microti* and *G. duodenalis* (Table 4). *G. duodenalis* is also known as *G. intestinalis* or *G. lamblia*. This protozoan has two simple forms of life - the trophozoite and cyst. The trophozoite lives in the small intestine where they act by the scourges, but many are adhering to the intestinal mucosa. The cyst is ovoid and is surrounded by a proteinaceous fibrous wall that confers resistance to the environment conditions (Hopkins et al., 1997; Volotão et al, 2007).

Giardia has a direct life cycle and its transmission is oro-faecal route. Drinking water contaminated with cysts represents a major cause of giardiasis in humans and other animals, which is therefore considered a waterborne disease (Kulda & Nohýnková, 1995).

Inside the host, the cyst release two trophozoites that attach to the small intestine. The trophozoites start their asexual multiplication by binary fission and by action of bile salts and alkaline pH suffer encystment, but the mechanism and where this occurs remain unknown (Bogitsh & Cheng, 1998).

Giardiasis may present as an asymptomatic or symptomatic, Acute or chronic disease. In adult animals, the infection is usually asymptomatic and is rarely detected. Already in young animals aged less than one year, clinical signs and symptoms may be present and identification of the parasite is more easily obtained (Thompson, 2000).

In general, clinical signs and symptoms observed in giardiasis include diarrhea, acute or chronic, steatorrhea, abdominal pain, lethargy, anorexia, flatulence, fatigue, abdominal distension, nausea, mucus in stools, growth deficits and weight loss (Bogitsh & Cheng, 1998).

The elimination intermittent cysts and lack of specific clinical signs makes the diagnosis of *Giardia* more difficult and requires multiple fecal examinations are carried out within 4 to 5 days (Kulda & Nohýnková, 1995).

Species and genotypes	Hosts
G. microti	Muskrat *and voles*
G. agilis	Amphibians
G. muris	Rodents
G. psittaci	Birds
G. ardeae	Birds
G. duodenalis (*G. intestinalis* ou *G. lamblia*)	Mammals
Genotypes of *G. duodenalis*	
Genotype A	Humans, primates, dogs, cats, cattle, rodents, wild animals
Genotype B	Humans, primates, dogs, horse, cattle
Genotype C	Dogs
Genotype D	Dogs
Genotype E	Ungulates
Genotype F	Cats
Genotype G	Rodents

Table 4. Species and genotypes of *Giardia*.

Techniques that promote the fluctuation of the cysts, using saturated solutions of zinc sulphate and sugar, are methods that allow diagnosis of most cases. Another method of diagnosis is an ELISA assay, which detects *Giardia* antigen in faeces preserved in formalin or kept under refrigeration. In humans, the sensitivity and specificity of this test is high (100% and 96%, respectively), allowing quick diagnosis, however, its use in dogs and cats revealed similar results flotation techniques, which are preferred for their low cost and ease of implementation.The indirect immunofluorescence and polymerase chain reaction (PCR) have been used for epidemiological studies or as research tools (Kulda & Nohýnková, 1995).

3.2 Epidemiology and prophylaxis

Giardiasis is spread throughout the world being described more than 250 million symptomatic cases in humans for years and is now included as the list of neglected diseases made by the World Health Organization in humans, its prevalence depends on the level of hygiene and sanitary facilities, ranging from 2% to 43%. Children are more susceptible to infection because of its low immunity and their lack of hygiene. In adults, infection may provide a certain degree of resistance to subsequent infections, reducing its prevalence in this age group (Mohammed Mahdy et al, 2008; Monis &Thompson, 2003).

Although giardiasis is classified as a zoonosis by the World Health Organization, it is unclear the exact participation of animals in the epidemiological chain of transmission of this disease. However one must consider that the interaction of multiple hosts associated with environmental conditions are essential links to propitiate the occurrence of giardiasis, which just as cryptosporidiosis, are the waterborne disease (Mohammed Mahdy et al, 2008; Monis &Thompson, 2003).

Of all the species of *Giardia* only *G. duodenalis* has been observed in humans, livestock and pets. Many of these animals possess a particular genotype, but they all have infections with genotypes A and B also found in humans. The number of molecular studies involving the infection observed in man with concomitant infections in animals is very restricted in aboriginal Australians were detected 13 human cases and 9 dogs in the genotype A, as found cases of giardiasis by genotypes A and B simultaneously in dogs and humans in Bangkok. Some wild animals such as beavers and rats, which have a high prevalence of giardiasis by genotype B, has historically been considered important sources of water contamination (Caccíó & Ryan 2008).

The intake of only 10 cysts of *Giardia* is sufficient to determine the disease and an infected person can eliminate up to 300 million cysts per ml of feces (Inpankaew et al, 2007).

The infection is acquired by ingesting contaminated food or water. Cysts remain viable for weeks in water, which facilitates its transmission. The survival of cysts in the water the same temperature dependent and may remain viable for up to 2 months in water at 8 °C and for only 4 hours in water at 37 °C (Xiao & Fayer, 2008).

Water contamination is through human sewage or feces from infected animals. The occurrence of giardiasis in sparsely populated areas, such as the Arctic region of Canada, reinforces the hypothesis that wild animals such as beavers, constitute important sources of infection. The beavers have 44% prevalence of *Giardia* and therefore are reservoirs for human infection. As the transmission of this disease occurs by contamination of water, food and environment as described in Table 5, the prevention of giardiasis include environmental sanitation (Hunter &Thompson, 2005).

Lack of hygiene of food handlers
Use of contaminated feces, sewage and manure as fertilizer for agriculture
Grazing cattle close to agriculture
Defecation of infected wild hosts in plantations
Vector-borne contamination of food contaminated by sewage
Use of animal manure or soil contaminated with human feces in agriculture
Use of contaminated water for irrigation
Use of contaminated water to dilute insecticides or fungicides
Wash salads and raw foods in contaminated water
Use of contaminated water to make ice or frozen foods
Use of contaminated water to make food that they receive the least amount of heating or treatment with preservatives

Table 5. Possible sources of food contamination.

The water should be protected from possible contamination by proper disposal and treatment of sewage from humans or animals. Because *Giardia* is resistant to routine chlorination of water, it is recommended that the flocculation, sedimentation and filtration of water are carried out before use. When the water is not subjected to such treatment, there should be boiling water, which promotes the complete removal of *Giardia* cysts. The greatest risk of zoonotic transmission is related to the genotype of *G. duodenalis*, which has been described in humans, farm animals, dogs, cats, beavers and in rodents (Hunter & Thompson, 2005; Jerlström-Hultqvist, 2010).

Farm animals, especially cattle, are an important source of infection for man. Cattle with giardiasis can eliminate up to one million cysts per gram of feces, so few infected animals are a major risk to public health (Thompson et al, 2008).

Although the clinical consequences of infection with *Giardia* in dogs and cats seem to be minimal, there are grounds for believing that such animals are important sources of infection in urban areas. Molecular epidemiology studies have shown that some genotypes are closely related to the infection in humans and dogs (Smith et al., 2007).

4. Microsporidiosis

Microsporidia are small eukaryotic intracellular parasites considered to be true, due to the presence of nuclear envelope coated core of intracytoplasmic membrane systems and separation of chromosomes by mitotic spindle. However, these protozoa also possess characteristics of prokaryotes, as a small ribosomal RNA (rRNA) and the absence of mitocrôndrias, peroxisomes and Golgi cisterns. These indicators point to the fact that they are phylogenetically very primitive protozoa, however, molecular studies reveal a close proximity of this group of parasites with the fungi, a fact that still causes doubts in the classification of these parasites. Diarrhea is the most frequent health problem caused, mainly in immunocomopromised people. The transmission routes indicated are via airborne, person-to-person, zoonotic, and waterborne means (Didier, 2005; Theiller, Breton, 2008).

4.1 About the agent and the disease

The life cycle of microsporidia includes three distinct phases: the spores responsible for transmission of infection, the proliferation of vegetative forms of intracellular schizogony or merogony calls, and finally the formation of spores or sporogony (Mathis, 2000).

The spores are very resistant structures to environmental conditions in its interior is an extrusion apparatus (polar tubule), whose function is to inject the infective material (sporoplasm) into the host cell. In the presence of favorable conditions, the extruded polar tubule and enters the host cell to inoculate sporoplasm in the cytoplasm of the same. The sporoplasm injected into the host cell starts proliferation merogony stage. His rapid multiplication occurs by binary or multiple fission. Then sporogony begins when meront acquire a dense amorphous layer around the cell, being called sporonts. These again can grow and multiply by binary or multiple fission sporoblasts to form, which, in turn, develop distinct cytoplasmic organelles and a thick wall, making the spores mature. The spores spread through the tissues of the host by infecting new cells and continuing the cycle (Mathis, 2000; Weiser, 2005).

So far, no evidence of intermediate hosts or vectors of microsporidial infection in man. For some genera that infect mosquitoes, has been reported a complex sequence of development, involving alternation between different invertebrate hosts (Weber et al., 2002).

The development of some species of microsporidia can be confined to cells of a single organ. However, other species can cause systemic infection. The clinical manifestations of infection are dependent on the species infected and the competence of the host immune response. The imbalance in the parasite-host relationship results in the proliferation and spread, causing cell destruction (Didier, 2005).

So far have been described about 1000 species of microsporidia belonging to 100 genera, however, this number will increase as new hosts are being researched. In humans, the infection was first described by Matsubayashi et al. in 1959 and since then an increasing number of cases have been reported particularly in immunocompromised individuals. The clinical manifestations include ocular lesions (conjunctivitis, chorioretinitis), muscles (myositis), kidney (nephritis), neurological (encephalitis), liver (hepatitis), peritoneal (peritonitis) and others (Abreu-Costa, 2005; Lallo et al., 2002; Lallo & Bondan, 2005; Weber et al., 2002).

Some species of microsporidia infecting invertebrates and can cause serious economic losses. Two well-known examples are the *Bombyx mori*, which affects the creations of silkworm and *Nosema apis*, which infects bees and honey production decreases. On the other hand, some species are used in biological control of some pests, for example, *Nosema locustae* used to control locusts (Mathis, 2000; Méténier & Vivarès, 2001).

More than 60 species, 11 genera of microsporidia have been described in fish. The most important genera are: *Glugea, Pleistophora* and *Sprague*. In freshwater fish and saltwater, the infection is highly contagious and can be fatal, the parasite is found in the intestine, bile ducts, liver, mesenteric lymph nodes, nerve ganglia, in the subcutaneous tissue, testes and ovaries (Rodriguez-Tovar et al.,2011).

We found cysts of *Giardia* were found in faecal samples from 2 prehensile-tailed porcupines (*Coendou villosus*) and *Cryptoporidium* oocysts in 3 rodents - montane akodont (*Akodon montensis*), ebony akodont (*Thaptomyces nigrita*) and guainan squirrel (*Sciurus aestuans*). Microporidia spores were seen in the stools of small rodents, including 3 montane akodonts, 1 prehensile-tailed porcupine and 2 pigmy rice rats (*Oligoryzomys* sp.), as well as of 3 marsupials, including 1 gray slender mouse opossum (*Marmosops incanus*) and 1 big eared opossums (*Didelphis aurita*), and of 3 hairy-legged vampire bats (*Diphylla ecaudata*). This was the first description of microsporidiosis in wildlife animals in Brazil. The study emphasizes the importance of the animals, particularly small wild mammals, as potential sources of parasite infection to other animal populations, including man, in areas of deforestation (Lallo et al. 2009; Pereira et al. 2009).

Few cases of human microsporidiosis had been reported until the advent of AIDS. However, it is now considered emerging and cosmopolitan. Among the species that cause human microsporidiosis, *E. bieneusi* is responsible for most infections occurring in approximately 40% of AIDS patients who have chronic diarrhea. It is not clear, however, what the risk factors that may be related to the prevalence of infection. Most reports occur in male patients, HIV-positive and CD4+ lymphocyte count at or below 100 cells/mm3, with few cases observed in women and children (Malčekova, 2010).

4.2 Epidemiology and prophylaxis

There is much controversy about the mechanisms of transmission of microsporidia. It is believed that the ingestion of spores is an important route for the species that infect the gastrointestinal tract of man and that environmental contamination occurs by the spread of the spores contained in faeces, urine and other excretions. In foxes, domestic dogs and squirrels, it was observed that transplacental transmission is an important mechanism of spread of the disease (Bern, 2005; Weiss, 2001).

Although microsporidia spores are resistant to the environment, can be inactivated when exposed for 30 minutes at 70% alcohol, 1% formaldehyde or hydrogen peroxide 1% as well as when they are autoclaved for 10 minutes at 120°C (Méténier & Vivarès, 2001).

The phenotypic and genotypic differences among strains of *Encephalitozoon cuniculi* can be used to indicate the main sources of infection for man. The lack of a specific host, coupled with the fact that the primate is susceptible to encephalitozoonosis suggests that man can become infected when exposed to an infected animal (Bern, 2005).

The possibility that microsporidiosis is a zoonotic disease is still obscure. However, one must consider that these primitive protozoa have great capacity to adapt, since they are distributed among different groups of invertebrates and vertebrates (Didier, 2005).

Transmission of invertebrate microsporidia to humans has been considered impossible. This hypothesis is based on the temperature differences between the two classes of animals. However, Trammer et al. (1997) obtained infection of athymic mice inoculated with *Nosema algerae*, one of Microsporidian of the culicids. These results demonstrate for the first time it is possible the development of invertebrate microsporidia in mammalian hosts. Therefore, one should consider the possibility that invertebrates may be sources of infection for human microsporidiosis.

Diarrhea is the most frequent health problem caused by microsporidia, mainly in immunocompromised people. Waterborne transmission of microsporidia spores has not yet been appropriately addressed in epidemiological studies, due to the small size of spores. Their presence have been associated with waterborne outbreaks and also with recreational and river water (Fournier et al., 2000; Izquierdo et al., 2011).

Several drugs have been used to treat microsporidiosis. Fumagillin was the first drug to have effective results in the control of *Nosema apis* and was subsequently described as a broad-spectrum drugs to combat various species of microsporidia that parasitise insects. This drug has managed to control the multiplication of *E. cuniculi* in vitro and has been effective in the treatment of ocular infection by *E. hellen* in HIV-positive (Mathis, 2000).

Other active ingredients have been used to combat the microsporidia of invertebrates and small vertebrates. However, there are few drugs licensed for human use. Albendazol seems to be the most appropriate drug to combat infection by microsporidia (Bern, 2005; Weiss, 2001).

Measures aimed at preventing infection by microsporidia are not specific, since the mode of transmission and sources of infection for the disease remain uncertain. However, as the transmission of this parasite appears to be based on the ingestion of spores from feces and urine, preventive measures aimed at controlling the intake of them. In hospital settings precautions with body fluids, and good personal hygiene of infected individuals. These measures are particularly important for the prevention of eye infection, which results from hands and fingers contaminated by spores of respiratory fluids or urine (Didier, 2005).

The infective potential of the spores can only be evaluated in vitro for species that grow in cell culture. Experimental studies show that microsporidia spores can survive for months or years, depending on humidity and temperature, due to their chitin wall. Disinfectants usual, the simple boiling and autoclaving can kill the spores (Weiser,2005).

5. Relationship between deforestation and the perpetuation of the parasites and enviromental

Deforestation and ensuing changes in landuse, human settlement, commercial development, construction of roads, water control systems (dams, canals, irrigation systems, reservoirs), and climate singly and in combination have been accompanied by global increases in morbidity and mortality from a number of emergent parasitic diseases.

The nature and extent of change in the incidence of parasitic disease are affected by changes in landuse and settlement, the time interval from one landuse to another/others, changes in type of soil and its degree of water absorption, changes in vegetation characteristics, changes in the types and amounts of bodies of water, their size, shape, temperature, pH, flow, movement, sedimentation and proximity to vegetation and, changes in climate (Table 6).

Deforestation envariomental changes
Grazing land (cattle):
create supportive habitat for parasites (specially *Cryptosporidium* genotype cattle)
Small-scale agricultural plots:
promove food contamination by water
sugar cane or rice may be surrounded by a network of irrigation ditches and artificial bodies of farms and wells
Climate changes:
warmer temperatures cause increased precipitation intensity and more rainfall events, favoring the survival of cysts, oocysts and spores of the parasites in the environment
Open areas:
cleared lands have neutral pH and steep inclines and streams make large pools, both changes promote the creation of a microclimate favorable to the persistence of parasites in the environment
Humans settlements:
migrants become a new reservoir for transmission of parasitic disease (migrants, foreign or non-immune resident settlers, indigenous)
animals and humans are exposed to new contacts in new environments
roads facilitate acceleration of crop farming, ranching, logging, mining, commercial development, tourism and, building of dams and hydroelectric plants and new settlements, increase erosion and blocking the flow of streams

Table 6. Environmental changes caused by deforestation and its relationship with the emerging parasites.

6. Conclusion

Cryptosporidium, Giardia and microsporidia are single-celled, microscopic organisms and are disease-causing parasites that may infect people through contaminated drinking water and recreational waters. Symptoms resulting from infection include mild to severe diarrhea, abdominal cramps, weight loss, bloating, and vomiting. Chlorine, a commonly used disinfectant in water supplies, can not eliminate *Giardia, Cryptosporidium* or microsporidia from water sources. Rates of deforestation have grown explosively since the beginning of the twentieth century. Driven by local to global demand for agricultural and forest products and expanding human population centers, large swaths of species-rich tropical and temperate forest, and prairies, grasslands, and wetlands, have been converted to specie-poor agriculatural and ranching areas. The global rate of tropical deforestation is continuing at staggering levels well into this decade, with more than 2.3% of humid tropical forests cleared between 2000 and 2005 alone. Parallel to this habitat destruction is an exponential growth in human-wildlife interaction and conflict, which has resulted in exposure to new pathogens for humans, livestock, and wildelife. This way, we can conclude that deforestation can contribute to transmission of waterborne diseases.

7. Reference

Abreu-Acosta, N. et al. (2005). *Enterocytozoon bieneusi* (microsporidia) in clinical samples from immunocompetent individuals in Tenerife, Canary Islands, Spain. *Trans. Royal Soc. Trop. Med. Hyg.*, Vol. 99, pp. 848-855.

Barr, S.C. (1998). Cryptosporidiosis and cyclosporidiosis. In: *Infectious diseases of the dog and cat*. GREENE, C.E., pp.518-524, W B Saunders, Philadelphia.

Bern, C. et al. (2005). The epidemiology of intestinal microsporidiosis in patient with HIV-AIDS in Lima, Peru. *J. Infec. Dis.*, Vol.191, pp.1658-64.

Bogitsh, B.; Cheng, T.C. (1998). *Human parasitology*. pp. 210. Academic Press, San Diego.

Broglia, A.; Kapel, C. (2011) Changing dietary habits in a changing world: emerging drivers for the transmission of foodborne parasitic zoososes. Vet. Parasitol., Vol.182, pp.2-13.

Cacció, S.M.; Ryan, U. (2008). Molecular epidemiology of giardiasis. *Mol. Biochem. Parasitol*, Vol.160, pp.75-80.

Carey, C.M. et al. (2004). Biolgy, persistence and detection of *Cryptosporidium parvum* and *Cryptosporidium hominis* oocysts. *Water Research*, Vol.38, pp.818-862.

Chalmers, R.M.; Davies, A.P. (2009). Minireview: clinical cryptosporidiosis. Exp. Parasitol. Vol.76, pp.1976-1979.

Didier, E.S. (2005) Microsporidiosis: an emerging and opportunistc infection in humans and animals. *Acta Trop.*, Vol.94, pp.61-76.

Fayer, R. et al. (2000). Epidemiology of *Cryptoporidium*: transmission, detection and identification. *Intern. J. Parasitol.* Vol.30, p.1305-1322.

Fayer, R. (2010) Taxonomy and species delimitation in *Cryptosporidium*. *Exp. Parasitol.* Vol.124, pp.90-97.

Fayer, R. et. al. (2010) *Crypstosporidium ubiquitum* n. sp. in animals and humans. *Vet. Parasitol.* Vol.172, pp.23-32.

Fournier, S. et al. (2000). Detection of microsporidia in surface water: a one-year follow-up study. *FEMS Immunol. Med. Microbiol.*, Vol.29, pp.95-100.

Hajdušek, O. et al. (2004). Molecular identification of *Cryptosporidium* spp. in animal and human hosts from the Czech Republic. *Vet. Parasitol.* Vol.122, pp.183-192.

Hopkins, R.M. et al. (1997). Ribossomal RNA sequencing reveals differences between the genotypes of *Giardia* isolates recovered from humans and dogs living in the same locality. *J. Parasitol.* Vol.83, pp.44-51.

Hunter, P.R.; Thompson, R.C.A. (2005). The zoonotic transmission of *Giardia* and *Cryptosporidium*. *Int. J. Parasitol.* Vol.35, pp.1181-1190.

Inpankaew, T. et al. (2007). Canine parasitic zoonoses and temple communities in Thailand. *Southeast Asian J. Trop. Med. Public. Health.* Vol.38, pp.247-255.

Izquierdo, F. et al. (2011). Detection of microsporidia in drinking water, wastewater and recreational rivers. *Water Res.*, Vol. 45, pp. 4837-4843.

Jerlström-Hultqvist, J. et al. (2010). Is human giardiasis caused by different *Giardia* species? Gut Microbes. Vol.1, pp.379-382.

Kulda, J.; Nohýnková, E. (1995). Giardiasis in humans and animals. In: *Parasitic protozoa*. KREIER, J.P., Academic Press, pp.225-422, San Diego.

Malčeková, B. et al. (2010). Seroprevalecne of antibodies to *Encephalitozoon cuniculi* and *Encephalitozoon intestinalis* in humans and animals. *Res. Vet. Sci.*, Vol.89, pp.358-361.

Lallo, M.A. et al. (2002). Infecção experimental pelo *Encephalitozoon cuniculi* em camundongos imunossuprimidos com dexametasona. *Rev. Saúde Pública*, Vol.36, pp.621-626.

Lallo, M.A.; Bondan, E.F. (2005). Experimental meningoencephalomyelitis by *Encephalitozoon cuniculi* in cyclophosphamide-immunosuppressed mice. Arq. Neuro-Psiquiatr., Vol.63, pp.246-251.

Lallo M.A. et al. (2009). Ocorrência de *Giardia*, *Cryptosporidium* e microsporídios em animais silvestres em área de desmatamento no Estado de São Paulo, Brasil. Ciência Rural, Vol.39, pp.1465-1470.

Marquardt, W.C. et al. (2000). *Parasitology & vector biology*. Haucourt Academic Press, pp. 205, San Diego.

Mathis, A. (2000). Microsporidia emerging advances in understanding the basis biology of these unique organism. *Int. J. Parasitol.*, Vol.30, pp.795-804.

Méténier, G., Vivarès, P. (2001). Molecular characteristics and physiology of micrsporidia. *Microbes and Infection*, Vol. 3, pp.407-415.

Mohammed Mahdy, A.K. et al. (2008). Risk factors for endemic giardiasis: highlighting the possible association of contaminated water and food. *Trans R. Soc. Trop. Med. Hyg.*, Vol.102, pp.465-470.

Monis, P.T.; Thompson, R.C.A. (2003). *Cryptosporidium* and *Giardia*-zoonoses: fact or fiction? *Infect. Genet. Evol.*, Vol.3, pp.233-244.

Patz, J.A. et al. (2000). Effects of enviromental change on emerging parasitic diseases. *Int. J. Parasitol.*, Vol.30, pp.1394-1405.

Pereira, A. et al. (2009). Ocorrência de microsporídios em pequenos mamíferos silvestres no Estado de São Paulo. Arq. Bras. Med. Vet. Zootec., Vol.61, pp.1474-1477.

Plutzer, J.; Karanis, P. (2009). Genetic polymorphism in *Cryptosporidium* species: an update. *Vet. Parasitol.*, Vol.165, pp.187-199.

Robinson, G. et al. (2010). Re-description of *Cryptosporidium cuniculus* Inman and Takeuchi, 1979 (Apicomplexa: Cryptosporidiidae): morphology, biology and phylogeny. *Intern. J. Parasitol.*, Vol.40, pp.1539-1548.

Rodriguez-Tovar, L.E. (2011). Fish microsporidia: Immune response, immunomodulation and vaccination. *Fish & Shellfish Immunology*, Vol.30, pp.999-1006.

Smith, H.V. et al. (2007). *Cryptosporidium* and *Giardia* as foodborne zoonoses. *Vet. Parasitol.*, Vol.149, pp.29-40.

Slifko, T.R. et al. (2000). Emerging parasite zoonoses associated with water and food. *Int. J. Parasitol.*, Vol.30, pp.1379-1393.

Theiller, M.; Breton, J. (2008). *Enterocytozoon bieneusi* in human and animals, focus on laboratory identification and molecular epidemiology. *Parasite*, Vol.15, pp.349-58.

Thompson, R.C.A. (2000). Giardiasis as a re-emerging infectious disease and its zoonotic potential. *Int. J. Parasitol.*, Vol.30, pp.1259-1267.

Thompson, R.C.A. et al. (2008). The public health and clinical significance of *Giardia* and *Cryptosporidium* in domestic animals. *Vet. J.*, Vol.177, pp.18-25.

Thompson, R.C.A. et al. (2009). Parasite zoonoses and wildlife: emerging issues. *Int. J. Environ. Res. Public.*, Vol.6, pp.678-693.

Volotão, A.C. et al. (2007). Genotyping of *Giardia duodenalis* from human and animal samples from Brazil using beta-giardin gene: a phylogenetic analysis. *Acta Trop.*, Vol.102, pp.10-19.

Weber, R. et al. (2002). Human microsporidial infections. *Clin. Microbiol. Rev.*, Vol.7, pp.426-61.

Weiser, J. (2005). Microsporidia and the society for the invertebrate pathology: A personal point of view. *J. Invert. Pathol.*, Vol.89, pp.12-18

Weiss, L.M. (2001). Microsporidia: emerging pathogenic protist. *Acta Trop.*, Vol.78, pp.89-102.

Xiao, L. (2002). Host adaptation and host-parasite co-evolution in *Cryptosporidium*: implications for taxonomy and public health. *Int. J. Parasitol.*, Vol.32, pp.1773-1785.

Xiao, L.; Fayer, R. (2008). Molecular characterization of species and genotypes of *Cryptosporidium* and *Giardia* and assessment of zoonotic transmission. *Int. J. Parasitol.*, Vol.38, pp.1239-1255.

Xiao, L. (2010). Molecular epidemiology of cryptosporidiosis: an update. *Experimental Parasitol.*, Vol.124, pp.80-89.

6

Landslides Caused Deforestation

Diandong Ren[1], Lance M. Leslie[2] and Qingyun Duan[3]

[1]*Australian Sustainability Institute, Curtin University, Perth,*
[2]*School of Meteorology, University of Oklahoma, Norman,*
[3]*Beijing Normal University*
[1]*Australia*
[2]*USA*
[3]*People's Republic of China*

1. Introduction

This study investigates landslide caused disturbances to ecosystem-vegetation burial and overturning of soil horizons which can accelerate vegetation loss, or even lead to complete deforestation of the landscape. A summary is presented of the historical development and recent advances in understanding and prediction of landslides/debris flows, which are major global natural hazards that have caused great loss of life and damage to property and infrastructure. The focus here is on rainfall triggered landslides/debris flows, and their deforestation effects. Debris flows have large destructive power because the solid and fluid forces interact within the sliding mass (Iverson 1997): the embedded boulders can exert great impulsive loads on objects they encounter; while the high fluid contents make debris flows travel long distances even in channels with modest slopes and therefore can inundate vast areas. Thus the whole spectrum of debris flows has deforestation effects (even small scale debris flows can denude vegetation). The first two sections of this chapter are concerned, respectively, with understanding ecosystems in terms of their soil and vegetation types, especially in regions that have sloping terrain. The emphasis is on the impact of landslides on deforestation, and this aspect has not yet been studied extensively. The third section focuses on the dynamics of landslides and provides a comprehensive description of their lifecycles, and the fourth section discusses the development and application of empirical and descriptive landslide models. Such models have contributed to knowledge of storm-triggered shallow landslides/debris flows. However, the deficiencies of empirical models are well-known and an alternative modelling technique, pioneered by the first author, is described in detail. This modeling approach is to solve the fully three dimensional, Navier-Stokes, and to that end a multi-rheological scalable, extensible geo-fluid model modeling system, known as SEGMENT, was developed. When coupled with a landslide model the system is called SEGMENT-landslide. The SEGMENT-landslide model has been used extensively in a predictive mode for both relatively short lead decadal times, and for century long predictions of landslides, in several very different locations. Here, the performance of SEGMENT-landslide is assessed for the Yangjiashan creeping slope in China, in predictive mode for the decade 2010-2109. Investigation of the impact of possible future precipitation morphological changes over this region utilizes 21st century simulations from 17 Climate General Circulation Models. SEGMENT-landslide performed encouraging well in both hindcast and forecast mode, providing prior and future vulnerability of the Yangjiashan creeping slope to landslides and rainfall rate thresholds for sliding to occur. SEGMENT-landslide also has been used to estimate landslide potential in other vulnerable regions, including southern California. Future work will focus on the influence of a future warming on the occurrence of landslides. Knowledge of upcoming changes in precipitation morphology is critical for predicting storm-triggered landslides and desertification. Advanced dynamical models, such as SEGMENT-landslide, which have a physical basis, are needed to supplement documentation of landslide occurrence. As there are biogeochemical submodels coupled in the SEGMENT system, it can be used to investigate significant environmental consequences of landslides, notably deforestation.

2. Ecosystem of sloping terrain, soil and vegetation

A unique feature of terrestrial ecosystems is that vegetation acquires its resources from two very different environments; air (for CO_2) and soil (for inorganic minerals of nutrients). Landslides cause disturbances to ecosystem productivity by displacing the soil mantle. In addition to be the medium in which most decomposer organisms and many animals live, the physical soil matrix provides a source of water and nutrients to plants and microbes and is the physical support system in which terrestrial vegetation is rooted. For these reasons, landslide displacement of the soil mantle has severe ecosystem consequences.

Sloping terrain creates unique patterns of microclimates through surface energy budgets, hydrology and availability of nutrients. For example, slopes facing the equator receiving more solar radiation than opposing slopes and hence usually have warmer and drier conditions. In colder or moister climates, the warmer microclimate of the equator facing slopes provides conditions that enhance productivity, decomposition and other ecosystem processes (including the formation of soils), whereas in dry climates, the low moisture levels on these slopes limits such production. Further, microclimatic variations associated with slope and aspect allows stands of an ecosystem type to exist hundreds of kilometres beyond its major zone of distribution. These outlier populations are important sources of colonizing individuals during times of rapid climate change and are therefore important in understanding species migration and the long-term dynamics of ecosystems. Topography also influences climate through drainage of cold, dense air in the form of katabatic winds, forming strong near surface air temperature inversions. Inversions tend to be strongest at night and during winter, when there is less warming of the surface, either in the form of shortwave solar radiation or longwave thermal from clouds, and hence insufficient convective mixing to remove inversions. Inversions are climatologically important because they increase the seasonal and diurnal temperature extremes experienced by ecosystems in low-lying areas. In cool climates, inversions greatly reduce the length of the frost-free growing season. The third aspect of the impact of sloping surfaces on ecosystems is surface hydrology. The surface runoffs on sloped surfaces are much larger than over flat terrain. For dry climates, this places severe water-limitations on production. Before addressing the slope effects on nutrients, we will review the soil formation process.

Under a given climatic regime, soil properties are the major control over ecosystem processes. Soils are the locality where geological and biological processes intersect. Soils mediate many of the key reactions in the giant global reduction-oxidation cycles of carbon, nitrogen and sulphur, and provide essential resources to biological processes that drive these cycles. As the intersection of the bio-, geo- and chemistry in biogeochemistry, soils play such an integral role in ecosystem processes that it is impossible to separate the study of soils from that of ecosystem processes. Soils are formed from weathered metamorphic rocks. The presence of living organisms accelerates the soil formation processes. Water is a pathway for nutrients entering an ecosystem and also is crucial in determining whether the products of weathering accumulate or are lost from a soil, especially the soluble minerals. Top soils generally are more fertile because weathering rates generally are larger at the surface. Also, leaching processes tend to transfer soluble ions (e.g., chelated complexes of organic compounds, iron or aluminium ions from precipitation or released in the weathering of upper layers soil) downwards. During the downward movement, they can react with ions encountered at depth under new chemical environments (e.g., increased pH value), or may precipitate out the system when dehydration occurs (e.g., water is evaporated in semi-arid or arid climate zones). As a consequence the levels of silica and base cations in the secondary minerals usually increase with depth and result in a nutrient poor deeper soil horizon. As iron and aluminium ions soluted in the soil water move downwards, slight changes in ionic content and the microbial breakdown of the organic matter both can cause the metal ions to precipitate as oxides. The deeper soil horizon containing iron-rich minerals usually are hardened irreversibly. These layers can impede water drainage and root growth. This is the case for tropical iron-rich soils, and similar processes exist for calcium (or magnesium) soils of arid and semiarid temperate climate zones. The hard calcic horizon at depth is formed when calcium carbonate precipitation occurs under conditions of increased pH, or under saturation concentrations of carbonate with evaporation of soil moisture. If debris flows remove the fertile top soils, the deeper horizon is exposed and this layer has poor water retention ability and is nutrient poor. Moreover, roots cannot develop within it and thus cannot support re-growth. Even for the depositional alluvial fans, the exposed deep soil and deep buried top soil forms a nutrient sink for the ecosystem, as the majority of the roots can use nutrients only in the upper one or two meters.

3. Landslides are double-edged swords

The surface of the earth, both land and beneath the oceans, is continually being modified by mass movements operating in response to gravitational forces. In this sense, landslides reduce the hill slopes to stable angles. They can assume the form of rockfalls, slumps and slides, and debris flows (Cruden 1991). For this study, the term landslide includes downslope movement over a variety of scales and velocities and also includes those related to vegetation cover. Rainstorms always are the prime cause.

Topography influences soil through its effects on climate and the differential transport of fine soil particles (Amundson and Jenny 1997). Characteristics such as soil depth, texture, and mineral content vary with hillslope position. Erosion processes, such as landslides, preferentially move fine-grained materials downslope and deposit them at lower locations. Depositional areas at the base of slopes and in valley bottoms therefore tend to have deep fine-textured soils with a high soil organic content and high water-holding capacity. These areas also supply more soil resources (water and nutrients) to plant roots and microbes and provide greater physical stability than do higher slope positions. For these reasons alluvial fans at valley bottoms typically exhibit higher rates of most ecosystem processes than do ridges or shoulders of slopes. In brief, soils in lower slope positions have greater soil moisture, more soil organic matter content, and higher rates of nitrogen mineralization and gaseous losses than upslope soils (Matson et al. 1991). Therefore, on geological timescales, landslides help produce stable land suitable for agricultural and habitation (for tropical islands, also tourism) and provide materials that form fertile plains and valleys, beaches, and barrier islands. However, on a scale typical of a human life span, the benefits accruing from landslides are overshadowed by their destructive characteristics; they are hazards that should be understood and mitigated as much as possible.

While it is reasonable to generalize that landslide activity is important, landslide impact on deforestation, on ecosystems in general, have not been well addressed. The following comments are not an exhaustive review but are intended to indicate the relevance of landslides to deforestation.

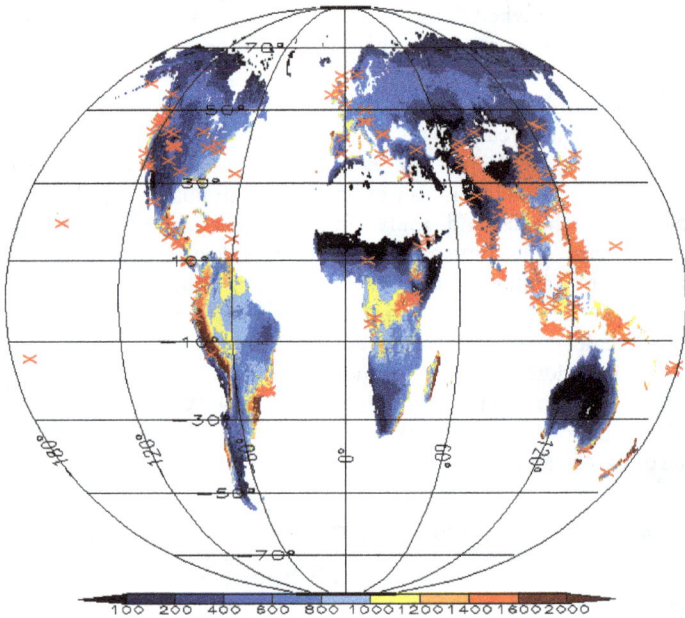

Fig. 1. Global distribution of Net Primary Production (NPP, color shading, in g/m²/yr) and occurrence (red crosses) of storm-triggered landslides (2003-2007).

The main geological hazards of volcanic activity, earthquakes and landslides are commonly, but not entirely, associated with processes occurring in areas near subduction zones. It is there that elevations are high enough to generate instability and sources of hazardous agents are abundant. This explains why the landslides belt coincides with the Earth's major earthquake belts. The Himalayas, for example is a hot spot (Figure 1). Figure 1 shows the primary landslides that are deeper than 2.5 m thick and contain $>10^5$ m³ of solid material, enough to cause vegetation damages. Plotted are the cases for years 2003-2007.

Landslides are one type of geological hazard. Geological hazards become so only where population, services or structures are at risk. In this sense, although northwestern China is a hot spot for landslides, the low population, by itself, reduces the hazard to a vanishingly small likelihood. In contrast, India and Indonesia both have large populations under the threat of storm-triggered landslides. It is reported (BNBP, 2009) that during the period 1998-

2007, 569 landslide events took place in Indonesia which caused 1326 fatilities and around 1500 people missing, around 170,000 people evacuated. In the humid tropics, soil loss after landslides are also very high as soil will be broken and exposed to rainfall wash-resulting in increased sediment load in the streams and causing flooding in low lying areas. As the population of vulnerable regions increases, previously unoccupied alluvial fans are used for habitation (e.g., the Zhouqu county in China) and landslide events therefore have a great potential to impact human settlements or activities.

The state of California in the USA is a locality where the combined effects of earthquakes and landslides have been documented extensively. The precipitation of this region, being regularly affected by ENSO and potentially also sensitive to climate change, has a large uncertainty in the future occurrence of storm-triggered landslides (Ren et al. 2011b).

From Figure 1, there are fewer landslide hazards for regions with NPP<200g/m^2/yr, indicating that soil, vegetation and landslides are closely tied together in the Earth's environmental system. The landslides free areas occur either because of their limited liquid precipitation (deserts and polar regions) or because of a lack of soil mantle. In both cases, the annual net primary productivity is low.

To evaluate the ecosystem consequence of landslides, it is necessary to know why landslides occur. The possibility of a landslide depends on the balance between the driving forces for downslope movement and the forces that resist this movement. Horizontal stress gradients caused by uneven topography are the drivers of sliding. At the stationary phase, the cohesion between soil particles is the resistive force. Once sliding is initiated, the resistive associated with flow shear is the primary stress that eventually stops the flow.

4. Landslide dynamics

Many factors influence the yield strength of a soil mass (Selby 1993). These include the sliding friction between the material and some well-defined plane (i.e., the movement of coarser soil over a saturated clay layer), but most commonly the internal friction caused by the friction among individual grains within the soil matrix. Cohesion among soil particles is sensitive to soil moisture/water content. For sandy soil, a small amount of water enhances cohesion among particles, whereas high water content reduces the frictional strength through increased pore pressure. The situation for clay soil is a simple reduction of strength as moisture content increases. From the perspective of dry internal frictional angle, fine-particle soils have lower slope thresholds of instability and are more likely to lead to slope failure than are coarse-textured soils. The situation is further complicated by hydrological processes and the presence of vegetation (Figure 2).

As illustrated in Figure 2, the presence of aboveground vegetation introduces the following effects: aboveground biomass loading (gravitational), growing season soil moisture extraction by live roots (hydrological), fortification/reinforcement of the soil within its extension range (mechanical), changing chemical environment of the soils and therefore the bonding strength among unit cells (chemical), and wind stress loading (meteorological). The overall effects are an interaction of the above factors and it is difficult to generalize before a detailed analysis is carried out that is specific to a certain situation. For example, the fortifying roots have yield strength larger than dry soils and the existence of roots is commonly thought to increase the resistance of soils. However, the presence of roots, especially when there is precipitation, also facilitates water channelling into deeper depths.

After the soil is moistened, the cohesion between soil and the root surface is reduced greatly (to negligible strengths <0.01Mpa) and the root strength cannot be effectively exerted. Also, the effect of roots is to 'unify' the soil particles within root distribution range. Once the entire rhyzosphere soil layer is saturated, the fortifying effects will be totally lost. The wind stress and the biomass loading will further deteriorate the situation (e.g., falling trees may join the sliding mass). Therefore, the effects of roots depend on precipitation rates in specific cases and in meteorological conditions in general. The hydrological and chemical effects are not expanded upon here but for interested readers details can be found in Ren et al. (2008, 2011a,b). Here, we provide some details that are highly relevant for 'progressive bulking' type of landslides, such as the pathway by which water leaves the landscape and the erosion ability of running surface water.

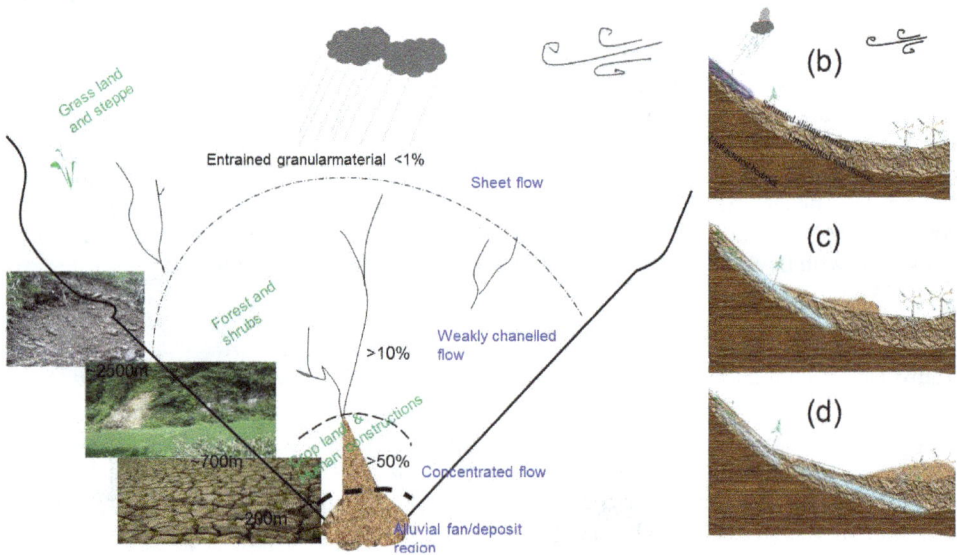

Fig. 2. A characteristic storm-triggered landslide (debris flow). Panel (a) is a plane view of the entire (solid material) collection basin. The elevation divisions are only for reference. The section with concentrated solid material creeping is only a small portion of the entire area. This means of mass redistribution is referred as "progressive bulking". Vegetation mortality is cause by that portion with >50% granular material concentration, especially when boulders are entrained. Panels (b), (c) and (d) illustrate a full life cycle of the debris flow: (b) initiation, (c) sliding, and (d) cessation. Precipitation generated surface runoff washes the fine-grained material downslope. Because the steepness of the slope is graded (steep at top and gentler at the toe), granular material (saturated soils) are thicker toward the toe of the slope. As the sliding material run downslope, it entrains the pebbles and small stones (granular material) at the bottom. This will significantly change its rheological properties as it becomes drier and more viscous. The sliding material eventually ceases at places with gentle slopes, or where the slope angle reverses. In the lowest inset panel, the cracked rice paddy is indicative of effects of roots on holding the soil particles together: almost all rice bundles are located at the centers of the cracked cells. Major cracks rarely run directly across a bundle of rice sprouts.

Water typically leaves a landscape by one of several pathways: groundwater flow, shallow subsurface flow, or overland flow (when precipitation rate exceeds infiltration rate). The relative importance of these pathways is strongly influenced by topography, vegetation, and material properties such as the hydraulic conductivity of soils. Drainage (ground water) and shallow subsurface flow dissolve and remove ions and small particles that cement large soil particles when being dry. Overland flow (runoff) causes erosion primarily by surface sheet wash, rills and rain splash. Runoff is strong for bared ground (e.g., arid soil-mantled landscapes) or disturbed ground (by soil animals or human construction). Runoff of 3 mm/s suffices to suspend clay and silt particles and move them downhill (Selby 1993). As water collects in gullies, its velocity, and therefore erosion potential, increases. For clear (not dense) debris flow, erosion potential can be approximated as $a = C_1 (gh)^{1/2} \alpha^{1/3} (D/D_0)^{0.27}$, where C_1 is a coefficient depending on vegetation condition, α is slope, D is particle size and D_0 is a reference particle size, h is runoff water depth, and g is gravity acceleration.

Vegetation and litter layers level/spread out the peak of runoff and increase infiltration and drainage through reducing velocity with which raindrop hit the soil, thereby preventing surface compaction, and through the inter-connected channels webbed by roots and soil animals. For progressive bulking debris flows, these are apparent preventative features. To evaluate the ecosystem consequence of landslides, we also need to know the size of a landslide. It is apparent that the disturbance of landslides to ecosystem is through displacement of the fertile top soil layer. Whether or not the existing vegetation can be destroyed depends on the severity of the landslide. The severity of a landslide depends on its size, velocity and material composition. The first two points are obvious. The composition of the sliding material is important for landslide destructive potential primarily because the size of granular material. If the debris flow entrained big boulders during its downslope movement, the destructive potential to trees and buildings will be much greater than similar debris flows containing only sands and silts. The drag of debris flows to obstructions can be expressed as $\eta_{eff} \dfrac{\partial U}{\partial Z}$, where U is flow speed, η_{eff} is effective dynamic viscosity (Pa•s). For a fast flow debris flows (3 m/s) and 50% solid material sludge of about 2 meter depth, the stress exerted on obstacles is on the order of 10^5 Pa. When a large boulder encounters an obstacle, the energy is transformed primarily in the form of longitudinal waves (e.g., similar to acoustic waves in the air). The stress (pressure) impact on the obstacle is $0.5 \rho_b U V$, with V the sound wave speed in solids (obstacle medium, at standard pressure of 1 atmosphere and temperature 25 °C, about 4500 m/s for bridges and concrete-steel buildings), and ρ_b is the boulder's density. In contrast to turbid sludges, a boulder of about 2 m^2 cross-sectional area (assuming this also is the area of simultaneous contact with an obstacle), of bulk density of 2.7×10^3 kg/m^3, can exert 10^8 pa pulse pressure, which is three orders of magnitude larger than the sludge.

Figure 2 is a conceptual sketch of storm-triggered landslides/debris flows. Clearly there is usually a large (one order of magnitude larger than the channelled, streaming, concentrated flows of the dense mud) collection region, providing solid material for the downstream area. Along the flow, the sliding material is denser and viscosity is larger. At the spread region, because of the gained kinetic energy, it will not stop even the slope angle is less than the repose angle. In actuality, it continues to spread until the speed is reduced to zero, usually at bed slopes much smaller than the stable repose angle. In Figure 2, panel (a) is a plane view of the mass collecting region, panels (b) to (d) illustrate the life cycle of a debris flow. We see that the solid material is collected by surface runoff from a much larger area than seen with naked eye as mudslide. The portion with direct ecosystem damage primarily is the region with concentrated stream flows, which typically constitutes only about 2% of the entire collecting area.

The August 8, 2010 Zhouqu landslide is a characteristic 'progressive bulking' type of debris flow. On August 7, Zhouqu's Beishan slope received over 80 cm of rain within two hours, leading to widespread shallow landslides and debris flow generation. The town of Zhouqu is built on the sloping surfaces of previous debris fans formed at the base of steep rocky hillslopes.

The debris fan at the mouth of the rocky creek, Sanyanyu, is usually stable because of the elaborated root system laced through the stony, loose soil. Prior to the intense cloudbursts, there was a long period of drought in the region and the ground surface was cracked, especially the mid-slope (1200-2500 m elevation range). In addition to causing rock falls (providing more solid sliding material), the Wenchuan earthquake, barely two years earlier, deepened the bedrock crevasses. Consequently, the drought stressed vegetation cover had little ability to intercept the rainfall and dampen the peak runoff. Much of the runoff water was channelled directly into the crevasses and deepened the shear zone. As the runoff water flowed down slope, it progressively increased the solid material contents (as a result of entrainment) and also its ability to further entrain. This positive feedback continues and at elevation of 2500 m, large boulders of 1 meter diameter can be picked up by the turbid mud. As water filled the crevasses, patches of soil layers up to 4 meters were made unstable and were scoured out and descended along rock creeks. The scoured was generally still dry (actually only the surface several centimetres are saturated). As it ran down the steep canyon, it further picked up debris as it travelled at ~5 m/s. By the time it reached lower, gentler slopes (~1200 m elevation), its mass had increased by one order of magnitude, but the overall water content was still low. Some parts (the finer granular components) simply came to a halt. The coarse granular material and boulders continued their motion and smashed into constructions at the mouth of the rocky creek. Subsequently, material disturbed by the slide, including wood and constructions, was washed by following slides and moved down the Sanyanyu creek, depositing debris all the way to the Bailongjiang River. This tragedy, and others in the previous rainy seasons following the Wenchuan earthquake, led to the awareness of the need to develop warning system, compile hazard maps, and adopt new legislation concerning forest practice and soil reservation, for regions on active faults.

5. A recently developed landslide dynamics model: SEGMENT-landslide

Because of their frequent occurrence, storm-triggered shallow landslides/debris flows have been actively studied. Empirical and descriptive landslide models have contributed much to the public awareness of landslide hazards and have led to valuable accumulated experience in identifying the key causal factors (Caine 1980; Cannon and Ellen 1985; Sirangelo and Versace 1996; Godt et al. 2006). Caine (1980) proposed the seminal rainfall intensity-duration threshold line, above which shallow landslides may occur (ID method hence forth), based on 73 landslides worldwide.

In intensity-duration thresholds, a dataset consisting of rainfall intensity (I, mm/day) and rainfall duration (D, hr) of landslide events is first made/prepared. A scatter graph is then generated with rainfall duration as x-axis and rainfall intensity as y-axis. The equation of rainfall threshold is a power-law curve that fit the points in the scatter plot (actually a lower envelope in that is a point lies to the right upper of the curve, landslides may occur), usually take the form $I = aD^{-b}$, where a and b are positive constants that vary with soil, vegetation and land use.

Godt et al. (2006) suggested that landslide-triggering rainfall must be considered in terms of its relationship with antecedent rainfall. For example, a heavy rainfall event within a dry period is not likely to trigger shallow landslides, while the opposite is true for lighter rainfall within a wet period. As it directly affects soil moisture conditions, Godt et al. (2006) correctly claim that antecedent rainfall must be included in an empirical model's assessing of a rainfall's potential in causing landslide. Godt et al. (2006) therefore is a significant improvement over Caine's (1980) seminal rainfall intensity-duration threshold line approach. The antecedent rainfall index is usually defined as a red noise of the accumulative rainfall amount 3 days (for tropics) and 7 days (temperate climate zone) prior the landslide event. Recent empirical methods also compile many soil hydrological parameters by using water-balance models with little physical basis but are convenient for estimating soil moisture conditions. For example, Godt et al. (2006) uses a detailed assessment of rainfall triggering conditions, hill slope hydrologic properties, soil mechanical properties, and slope stability analyses. The accumulation of sliding material is a slow process (either rockfalls or aeolian processes or damaging erosion processes from weathering) compared with the sliding. Previous sliding will reshape the sliding material profile and may even completely remove the sliding layer. These will increase the stability of the slope and a similar rainfall amount may cause sliding on a reduced scale, or not at all. Thus, the empirical parameters (a and b) in the ID approach vary not only spatially but also temporally. In this sense, all previous ID approaches still lack the important time varying features.

A synthetic consideration of preparatory and triggering factors, however, demands a more comprehensive modeling of the physical processes involved in landslides (Costa, 1984; Iverson, 1997). The overview by Iverson (1997) suggested several criteria for dynamic landslide models, including that a model should be capable of simulating the full start-movement-spread-cessation cycle of the detached material, and should cover a wide spectrum of debris flows. With continued growth and expansion of human population, rain-triggered shallow landslides increasingly result in loss of life and significant economic cost. From an ecological viewpoint, landslides are an important factor in desertification over mountainous regions because they are very effective in transferring biomass from live to dead respiring pools (Ren et al., 2009).

Along these lines of walking, there are physically-based slope stability models to simulate the transient dynamical response of pore pressure to spatiotemporal variability of rainfall (e.g., Transient Rainfall Infiltration and Grid-based Regional Slope-Stability Analysis – TRIGRS, Baum et al. 2008); commercially available numerical modeling codes for geotechnical analysis of soil, rock and structural support in three dimensions (e.g., FLAC-3D, www.itascacg.com/flac3d), and fully three dimensional, full Navier-Stokes and multi-rheological modeling systems such as the scalable, extensible geo-fluid model, known as SEGMENT (Ren et al., 2008; Ren et al., 2009, Ren et al. 2010; Ren et al. 2011a,b).

Slope stability models are based on the following reasoning: On a sloping surface, the gravitational force can be partitioned into a component normal to the slope (F_n), contributing to friction that resists sliding erosion, and a component parallel to the slope (F_p) that promotes sliding. A stability parameter, S, is defined as $S = F_n \times \eta / F_p$, where η is the internal friction coefficient. This is the general form of S, but there are many specific forms, based on the fact that, landslides, as the movement of a mass of rock, debris or earth downslope (Cruden 1991; Dai et al. 2002), occurs when shear stress (F_p) is higher than shear strength (i.e., when $S<1$). For example, the stability factor can be defined as

$$\left(S\times G-C_{eff}L\sin\alpha\right)/\left(\cos\alpha+tg\phi\sin\alpha\,/\,S\right)tg\alpha = C_{eff}L+\left(S\times G-C_{eff}L\sin\alpha\right)/\left(S\times\cos\alpha+tg\phi\sin\alpha\right)tg\phi$$

(derivations to be detailed), where G is the total weight of the sliding mass, L is the slope length, α is slope angle (of the sliding surface), C_{eff} is effective cohesion (mechanical property of soil and weathered rock), and ϕ is internal frictional angle (mechanical property of sliding material). If a slope is composed of several segments of sub-slopes with very different mechanical properties, this expression can be summed to obtain the gross stability of the entire slope. If the iteratively obtained S is 1, it is assumed that slope is in critical stable condition.

The stability parameter is intended to diagnose when the sliding initiate. Because only static mechanical properties are included, slope stability model does not include the sliding process and how the sliding material redistributes after sliding ceases. Two dimensional slope stability models have difficulty in implementing realistic lateral boundary conditions and cannot be applied to a regional area. Three dimensional (3-D) models allow all the known physical processes and can be applied to regional areas. The following is an outline of a recently established 3-D dynamics model for studying storm-triggered landslides.

To describe the physics involved in Figure 2, the following equations are used in SEGMENT-landslide. For the sliding material, a coupled system is solved for conservation of mass:

$$\nabla\cdot\vec{V}=0 \tag{1}$$

and momentum:

$$\rho\left(\frac{\partial\vec{V}}{\partial t}+\nabla\bullet(\vec{V}\otimes\vec{V})\right)=\nabla\bullet\sigma+F \tag{2}$$

under the multiphase rheological relationships, for water, condensed mud, and wet sliding granular materials, granular viscosity parameterized as:

$$v=\left(\mu_0+\frac{\mu_1-\mu_0}{I_0/I+1}\right)\frac{S}{\left|\overset{\bullet}{\varepsilon_e}\right|} \tag{3}$$

where ρ is bulk density, $\vec{V}$ is velocity vector, σ is internal stress tensor, and F the body force (e.g. gravity $\rho\vec{g}$). Here v is viscosity, $S=\left(R_{kk}-\rho g(h-z)\right)/3$ is the spherical part of the stress tensor σ, μ_0 and μ_1 are the limiting values for the friction coefficient μ, $\left|\overset{\bullet}{\varepsilon_e}\right|$ is the effective strain rate and $\left|\overset{\bullet}{\varepsilon_e}\right|=\left(0.5\overset{\bullet}{\varepsilon_{ij}}\overset{\bullet}{\varepsilon_{ij}}\right)^{0.5}$, I_0 is a constant depending on the local slope of the footing bed as well as the material properties, and I is inertial number defined as $I = \left| \dot{\varepsilon}_e \right| d / (S / \rho_s)^{0.5}$, where d is particle diameter and ρ_s is the particle density. Soil moisture enhancement factor on viscosity is assumed varying according a sigmoid curve formally as Eq. (9) of Sidle (1992) but with the time decay term replaced by relative saturation.

As derivatived from Eq. (1), the prognostic equation for surface elevation, $h(x,y)$, is

$$\frac{\partial h}{\partial t} + \left(\vec{V} \bullet \nabla_H\right)\Big|_{top} h - w\Big|_{top} = 0 \qquad (4)$$

Where $X\big|_{top}$ indicates evaluation at the free surface elevation. In the case with slope movements, Eq. (4) is solved regularly to update the sliding material geometry. The w terms may include sedimentation rate and entrainment rate. Once the material is entrained inside the sliding material, it changes the rheological property of the medium and is advected within the sliding material.

The viscous term in Eq. (2) implies an energy conversion from kinetic energy to heat. To make a full closure of energy, we need the following thermal equation:

$$\rho c \left(\frac{\partial T}{\partial t} + (\vec{V} \cdot \nabla)T\right) = k\Delta T + \frac{2}{v} \cdot \sigma_{eff}^2 \qquad (5)$$

Where c is heat capacity (J/kg/K), T is temperature (K), κ is thermal conductivity (W/K/m), and σ_{eff} is effective stress (Pa). The last term is 'strain heating', which is the converting of work done by gravity into heat affecting the sliding material by changing viscosity or causing a phase change. Above is the landslide component of SEGMENT-landslides. Other components are shown in Fig. 3.

To describe the full start-slide-stop cycle, we boundle internal stress tensor as

$$\sigma = \phi + C + S\mu + \delta_E \qquad (6)$$

, where $\phi = \rho g h$ is the gravitational potential, C is effective cohesion, and δ_E the pressure perturbation caused by earthquake or human-induced disturbances at that location. $\mu <$ $\mu_1 = tg\phi$, with ϕ granular repose angle. For conditions with ground water, hydrostatic pressure is usually included in S for convenience. The extreme values of the middle two terms on the right hand side are the yielding strength (shear strength) of the sliding material $\tau_f = C + S_f\mu$ (with subscript 'f' means failure). Note that, in addition to be soil moisture and soil chemical components dependent, C and μ are functions of shear stress (e.g., Schofield 2006). For unfractured bedrock, C is the dominant term, usually three orders of magnitudes larger than the remaining three terms combined together. For most of the soil (except pure sandy soil), cohesion and internal friction are both important in maintaining stability slopes. For fractured rocks and sandy soils, the internal friction becomes dominant term (not necessarily larger than gravitational potential, though. But, it is the horizontal gradient of the gravitational potential that caused motion, not the bulk term). When vegetation roots are involved, the mechanical effects are included in cohesion. Note that it is the interaction of distributed roots and the surrounding soil particles, not merely the

2

additive of root tensile strength and soil cohesion. In a certain way, it is like the iron web reinforcement inside concrete, except that soil moisture does not play a role in the concrete case for shear strength.

The system described by Eqs. (1-6) is most suitable for study either deep-seated rotational landslides or shallow or storm-triggered debris flows. It is also convenient to investigate the positive feedback between deforestation, land use changes, undercutting of slope for road construction and expansion of settlement areas and landslides. For example, the effects of vegetation can be fully considered in this modelling system. The weight loading is set as upper stress boundary. The hydrological effects manifest in the parameterization of viscosity (dissolve of certain chemical bonds in clay soils, effects on cohesion and internal friction for sandy soils), the pore pressure adjustments of the spherical part of the stress tensor, and (minor) changes to the loading corresponding to the soil water weight (Smith and Petley 2008). The mechanical properties of the roots are implemented in the effective cohesion. The water distribution effects of vegetation roots are parameterized in a land surface sub-model. This sub-mode provides the soil moisture conditions for the sliding sub-model. These explain how antecedent rainfall influence the saturation of soil and ground water level for a vegetated slope and its instability (Crosta 1998; van Asch et al. 1999) and provide a solid basis for discussing deforestation effect on landslide and the positive feedback that lead to further deforestation. Interestingly, ID empirical approach and the slope stability models are various forms of reduced form of the above equation set. For example, if the time dependence is neglected and three dimensional topography reduced to only including x-z plane, the governing equations can be written as (in component form):

$$\frac{\partial \sigma_{xx}}{\partial x} + \frac{\partial \sigma_{xz}}{\partial z} = 0 \tag{7}$$

$$\frac{\partial \sigma_{xz}}{\partial x} + \frac{\partial \sigma_{zz}}{\partial z} - \rho g = 0 \tag{8}$$

Further simplify the slope geometry to be of constant bedrock slope (α) and uniform sliding material thickness (thus, surface slope also is α), the volume integration of equations (7) and (8) yield

$$\left(G - C_{eff}L\sin\alpha / S + u \times tg\phi\sin\alpha / S - \Sigma_v\right) / \left(\cos\alpha + tg\phi\sin\alpha / S\right) \times \sin\alpha -$$
$$\left(C_{eff}L + \left\langle\left(G - C_{eff}L\sin\alpha / S + u \times tg\phi\sin\alpha / S - \Sigma_v\right) / \left(\cos\alpha + tg\phi\sin\alpha / S\right) - u\right) \times tg\phi\right\rangle / S \tag{9}$$
$$\times\cos\alpha + \delta E + \Sigma_h + u = 0$$

where u is the hydrostatic pressure from ground water, Σ_v is the net vertical support at top and toe (determined by boundary conditions), Σ_h is the horizontal support evaluated at top and toe, and δE is the horizontal force exerted during earthquake. Eq. (9) indicates that slope stability is influenced by various factors, such as slope gradient (α), soil properties (implicit in G, C_{eff} and α), ground water table and geomorphology (u and the boundary conditions). As for triggering mechanism, it not only include the storm trigger (excessive rainfall), but also the earthquake and volcanic activity (δE).

In the case without ground water and no effects of earthquake, taken the form

$$\left(S \times G - C_{eff}L\sin\alpha\right)/\left(\cos\alpha + tg\phi\sin\alpha \,/\, S\right)tg\alpha =$$
$$C_{eff}L + \left(S \times G - C_{eff}L\sin\alpha\right)/\left(S \times \cos\alpha + tg\phi\sin\alpha\right)tg\phi$$

, as discussed in the review of slope stability model in the previous subsection.

Similarly, rainfall intensity-duration methods (IDs) based on the soil moisture sensitivity of the resistive terms (the middle two terms on the right hand side of Eq. (6)). Sliding material are a structured mixture, with particular spacing patterns and arrangements, of solid particles, pore water and, in some cases, cementitious material accumulated at particle-particle contacts. In addition to factors such as the electrical charge of the particles, and the chemistry of pore water, the chemical bonds are the key to the soil yield strength. As an example, there are the cementation-particles connected through a solid substance, such as recrystallized calcium carbonate formed in dry climate when seeping water experience new environments of high pH value and (or) the solvent get evaporated, as mentioned in Section 1. Soil water, especially of acidic pH value (<6), can dissolve these chemical bonds. Increased pore pressure also tends to reduce the effective spherical pressure (S) and reduces the yield strength of the sliding material. Comparatively, the extra loading of the water mass is insignificant for increasing slope instability. Unfortunately, IDs do not follow a processed modelling of the relationship. Rather, they have empirical/statistical relationship constructed between precipitation and occurrence of sliding. In addition to shortcomings already discussed in the previous section, current IDs do not take rain water acidity and soil profile properties into account. The empirical parameters therefore are sensitive to changes in climate and soil/vegetation/land use.

By implementing the three dimensional dynamics and the known soil physics/chemistry, SEGMENT-landslide is a recently developed and extensively tested mechanistic, process-based modeling system for monitoring and predicting storm-triggered landslides and their ecosystem implications. It is a tool for investigating the roles of triggering factors in sliding, and offers a unique opportunity for regional scale assessment. In SEGMENT, landslides are simulated using a full three dimensional Navier-Stokes solver. This is necessary because, as a mechanism for releasing unevenly concentrated stress, landslide occurrence at a specific location affects the stability of adjacent areas. 'Legacy effects' (Casadei et al. 2003) is one example. While possibly missing landslides without historical precedent, many current empirical and theoretical procedures also tend to over-predict the area that may fail in a given rainstorm (Casadei et al. 2003; Wooten et al. 2008; Y. Hong and Z. Liao 2010, personal communication). Over-estimation occurs because many empirical models employ indices based on historical events. However, the occurrence of a slide, by releasing the stress build-up, makes the surrounding area more stable and less likely to experience future sliding.

There are situations when previous landslides do assist subsequent landslides. Individual landslides transform limited amounts of potential energy into heat (Eq. (5)). Sliding can cause intense heating inside the material, especially near the bottom. This can be a positive feedback mechanism which accelerates sliding through further reduction of the shear band of loose granular material strength. However, in the case of earthquakes, the heat released by thousands of simultaneous landslides is significant enough to interact with local convective processes. The intense storms that follow may cause widespread landslides in the same area. This is well illustrated by the Wenchuan earthquake, which induced over

5000 landslides and ~3600 rock falls. The massive amount of potential energy was transferred into heat and enhanced local convection. The following storms incurred ~358 debris flows, resulted in a direct economic loss of near \$US60 billion. SEGMENT thus simultaneously solves the thermal equation, the dynamic equations and the surface kinematic (continuity) equations.

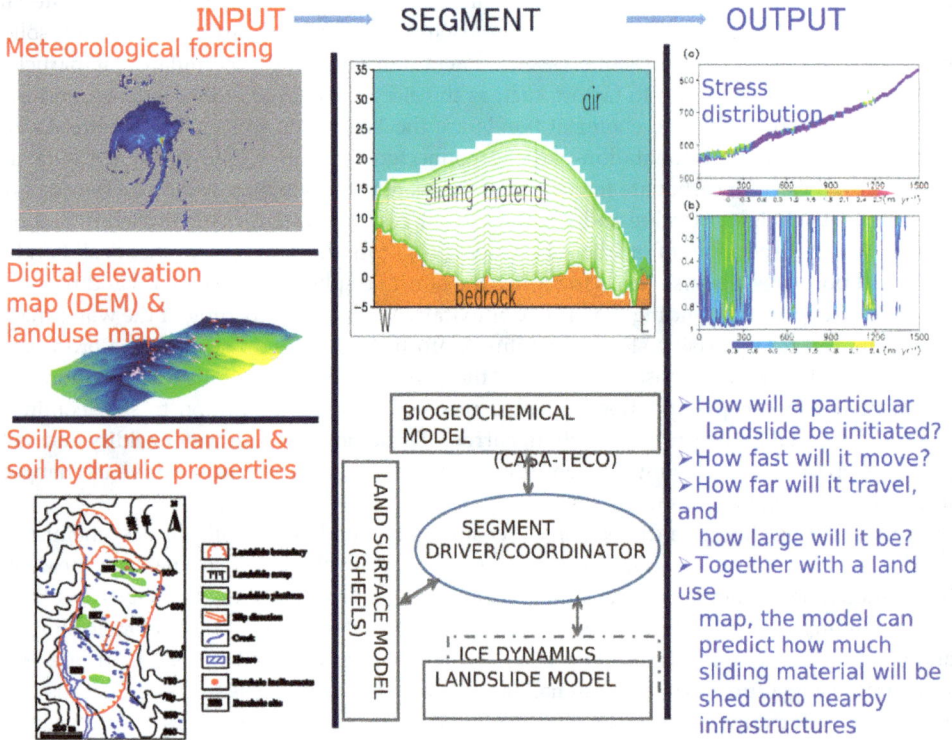

Fig. 3. Conceptual framework of SEGMENT-landslide for the projection of storm-induced landslides. The numerical techniques, model physics, input and output parameters are described in Ren et al. (2010b). The model has wide applications in many other geophysical flows including glaciers and ice-sheets, snow and mud avalanches, soil and coastal erosion, sea level change following ocean bottom tsunami, and pyroclastic flows such as magma from volcanic eruptions. The land surface model component considers hydrological processes of soils and vegetation. The mechanical properties of roots and the biomass loading also are implemented in the landslide model component.Adapted from Ren et al. 2011a.

SEGMENT, because it synthetically simulates landslides over a continuous regional area, can, in principle, minimize the false-alarm tendency of most empirical procedures. Against the 2007 wild fire-burn background, using the observed precipitation, SEGMENT simulated the landslide cases during year 2008 for a region with documented landslides (Ren et al. 2011b). Inserting the landslide model component into a scalable and extensible system (Fig. 3) also makes implementation of newly identified physical processes more convenient. This advantage is apparent in a recent simulation of Zhouqu landslides, where SEGMENT satisfactorily demonstrated why rainfall intensity is a critical factor affecting slope stability for cracked slopes.

For investigating environmental issues, the landslide model is inserted into the generalized scalable, extensible modelling system (SEGMENT). To illustrate the viability of SEGMENT-landslide, a case study is presented of a landslide in Yanjiashan of Hubei province, China.

6. Case study for the Yangjiashan creeping slope

SEGMENT-landslide uses the data typically required by a sophisticated land surface scheme. In addition to a surface elevation map and bedrock topography, the model also requires as input various geo-mechanical parameters such as cohesion, angle of repose (dry), density, porosity, field capacity, and saturated hydraulic conductivity. SEGMENT-landslide also requires a vegetation weight mask and a root distribution profile.

Detailed geological surveys are conducted only for areas with important infrastructure in danger of destruction by natural hazards. Fortunately, we have access to one such geological dataset for a small region along the Qingjiang River, a tributary of the Changjiang River. The Yangjiashan creeping slope (YC) in China, is located at 29°50′ N and 109°14′ E (see Fig. 4). In addition to a recently detailed geological map, we also performed an engineering study of the soil and rock profiles from bore-hole drilling, extracting soil samples for laboratory testing (Table 1), and continuous displacement measurements within the creeping zone (Fig. 5). The shear strengths of the various materials present in the region also have been obtained. Both drained and non-drained shearing tests are carried out on the soil specimens (Table 1). In addition to recovering creeping zone rock and fluid for laboratory analyses, intensive down-hole geophysical measurements and long-term monitoring provided the following information: the composition and geomechanical properties of active creeping zone rocks; the nature of the stresses responsible for sliding; and the role of pressurized water in controlling landslides recurrence, for field conditions with a wide dynamic range. Displacement surveys have been carried out continuously since July 2007, using the RST-IC3500 digital inclinometer. Displacement data from five boreholes (see Fig. 4, BH6-9) are analyzed in this study. The boreholes are labeled BH6, BH7, BH8, BH8-1, and BH9 (see Table 1).

The YC region has a sub-tropical moist climate, with a typical monsoonal precipitation pattern. The annual mean precipitation ranges between 1100 and 1900 mm. This section of the Yangjiashan Mountains, because of its proximity to a dam, is well-instrumented and thus is an ideal region to verify the numerical model, which calculates the roles of pore pressure, biomass loading and root distribution, and the intrinsic friction of the stress distribution. In this region the major slides are preceded by creeping movements. Since the 1960s, the YC slope has experienced repeated failures, as a result of exceptionally heavy rainfall periods in 1960, 1980 and 1997. During a storm in July 1997, very heavy rainfall fell over 2-3 days and caused substantial erosion in one small canyon. Field studies indicate that after a major rainstorm toe-slope failure occurs first, reducing the stability of the upper slope, and the failure then moves gradually to the upper slopes. The scenario therefore is that towards the end of a heavy rain storm, a block of material was undercut by the stream and moved into the canyon, its downhill movement left an unsupported upslope block which followed the movement. This, in turn, was followed by a third block as the movement retrogressed up the slope

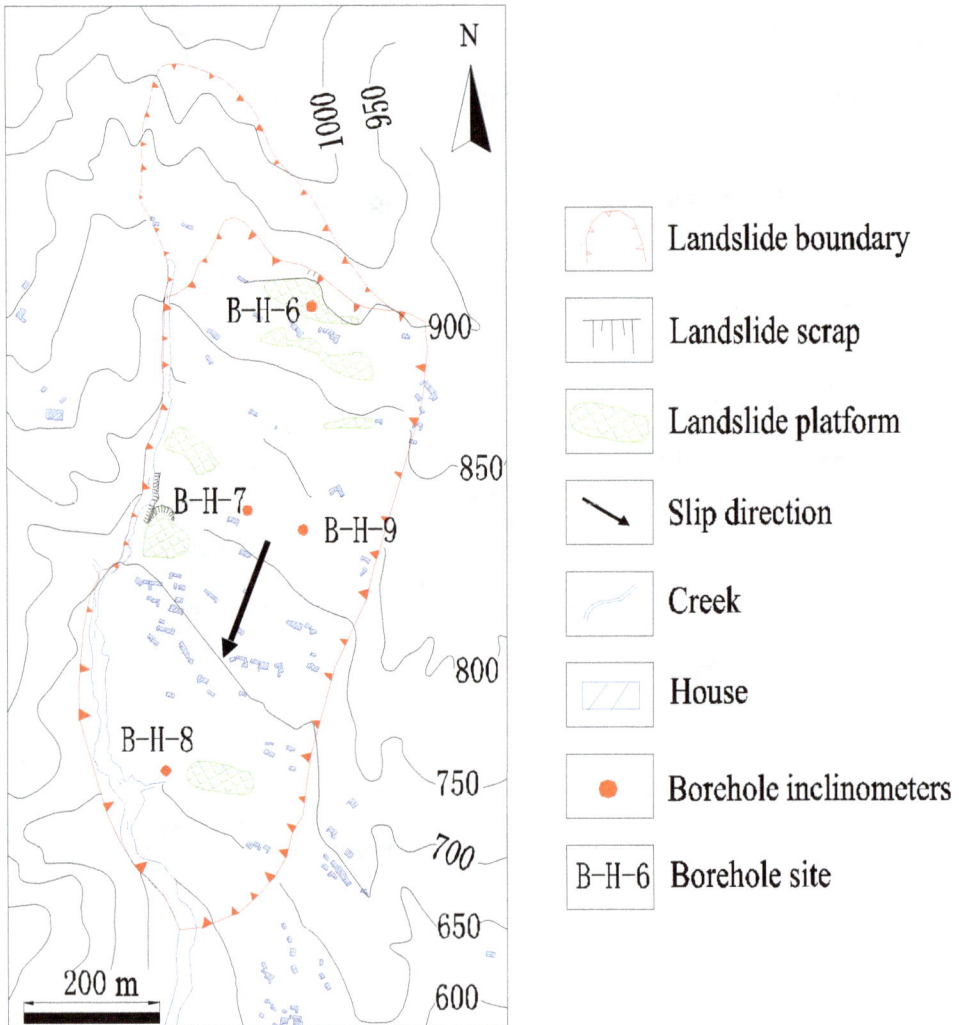

Fig. 4. Topographic map of Yangjiashan Creeping (YC) slope (the upper map shows its location in China). Contours are surface elevation (m). Borehole locations are labeled with a red circle (e.g., BH06-9). Color shading indicates surface maximum attainable creeping speeds larger than 1mm/s, after a 50-year recurrence storm event. The primary sliding direction (red arrow) is determined according to subsiding and swelling belts. The X- and Y-axis are distances from the SE corner (29.9N, 109.1 E).

The Yangjiashan community is situated on hills composed of inter-bedded siltstones and sandstones, occasionally interspersed with altered clay layers. The rocks range from highly to completely weathered at the ground surface. The weathered rocks date from the Paleozoic and Mesozoic eras, the 200-900 m thick yellowish interbedded sandstone and siltstone dates from the Silurian period, and the grey siltstone is from the Triassic period. The infiltration of rainfall through macro pores, which are well-developed in the soil and rock mass, plays a critical role in slope stability. The hills intersect with canyons in which increased erosion takes place during the spring and fall rainy seasons. Although many of the drainage patterns in this region have been altered by human activity, thereby increasing the slope stability, some remain unchanged, even in the inhabited areas (eastern part of the slope).

	Young's modulus E(GPa)	Poisson's raio μ	density ρ (kg/m³)	cohesion C(MPa)	internal friction angle φ (°)
Insitu rock and soils	0.16	0.24	2150	0.38	20
Slip-surface material	0.032	0.32	1950	0.04	16
Middle/lightly weathered mudstone	1.6	0.2	2600	1.6	28

Table 1. Geomechanical observations and parameters obtained by field surveys and laboratory tests of saturated specimens. Rock strength and deformation properties are obtained from tri-axial compression tests using INSTRON-1346. Four saturated specimens of each layer were tested. During tests, confining pressure was applied step-wise in 3 MPa increment (i.e., 3, 6, 9, and 12 MPa), and vertical load are applied at displacement rate at 0.1 mm/s.

6.1 A prediction for the Yangjiashan slope for the 2010-2019 decade

The Triassic siltstone of YC is especially sensitive to pore pressure changes. Because of the complicated stratification, the model is initialized with five borehole soil/rock profiles (18-50 m depths within the granular soil mantle), with a horizontal resolution of 10 m in delineating the 3000 m by 5500 m simulation domain. In order to reduce spurious numerical boundary effects, the simulation domain encompasses the entire region shown in Fig. 4. The sliding material forms a characteristic alluvial fan, that is with the down-slope section thicker than the upslope section. A thin plate splines (Burrough and McDonnell, 2004) interpolator is used to obtain a sliding mass depth distribution over the grid.

In the YC site, there are three model sub-layers to delineate the sliding mass, and one layer to represent the montmorillonite, within which sub-layers 10-12 are assumed to define the slip surface. This is further supported by the creep monitoring data (Fig. 5). This layer has the same chemical composition as the overlying layer but is physically fractured. We use 3 sub-layers to delineate the regolith layer because, although only ~1 m deep, it is the critical layer controlling water infiltration into the creeping slide mass. As it represents granular material under high confining pressure, the viscosity of this thin bed of finer-grained materials is smaller than that of the adjacent layers. The deep underlying rock layer is divided into 7 sub-layers, with mechanical properties specified from laboratory test results. As a link between strain and stress, viscosity Eq. (3) is the most important parameter in continuum modeling of slope movement. Under the same pressure gradient force, the smaller the viscosity, the faster the creeping rate. Non-fractured rocks have viscosity as high as 10^{19} Pa•s. The creeping rate thus is minimal for non-fractured rocks. Fractured rocks (granular material), can have viscosities 10 orders of magnitude smaller, depending on confining pressure and lubrication condition (usually water content under natural conditions). A creeping curve indicates that the vertical flow shear is very large within the granular layer. Above and below the granular layer, flow shear is much smaller. For a non-slip lower boundary condition, the creeping rate usually is imperceptible (below 53 m in Fig.5) before reaching the granular layer.

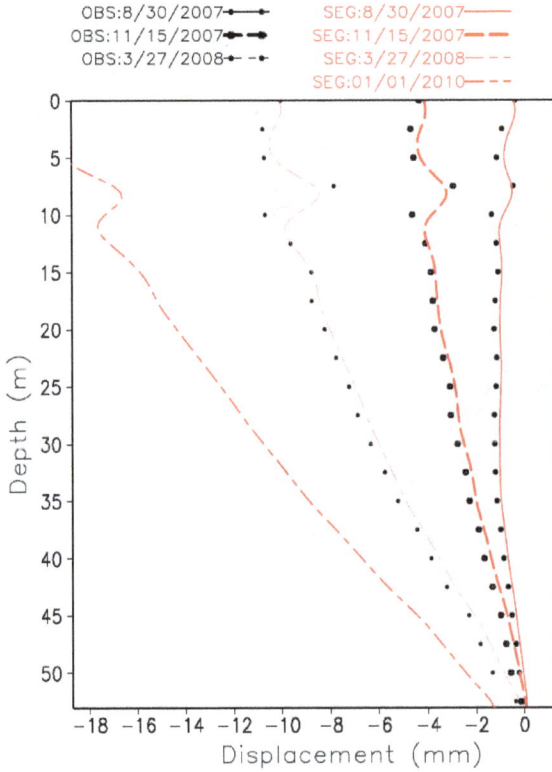

Fig. 5. The observed (markers) and modeled (lines with same style) displacement-time curves for BH8.

The absence of vegetative root strength binding soils increases susceptibility to sliding in loose soil on steep slopes during intense rainstorms (Dietrich and Perron 2006). Extra biomass loading also may contribute to slope instability during wet seasons. Ground surface biomass loading information for YC, because of its limited extent of only ~20 km², was collected by a geological survey team following a request by the first author.

Using mean annual soil moisture conditions (obtained from NCEP/NCAR reanalyses, http://www.cdc.noaa.gov/data/gridded/data.ncep.reanalysis.surfacefluxes.html),   a creeping rate is simulated using the SEGMENT model near BH8 of 17, 17.2, 16.2, 8.5, 5.2 and 3.1 mm/yr respectively at 1.5, 3, 10, 30, 40 and 45 m depths, agreeing well with the observed measurements. The root mean squared error, when compared with the 108 measurement data grids, is overestimated by only 0.42 mm/year, well within instrumental error range. The depth of the sliding surface (49 m at this location) is accurately delineated. Sliding will eventually accelerate along this plane of weakness, which is composed of highly fractured sandy shale. Interestingly, under natural conditions, the creeping speed curve is not a simple dilatant profile: there are local minima/maxima. The faster creeping locations near the surface in the curves are partially due to the low viscosity of this layer and partially due to the uneven surface loading in the down-slope direction of the borehole. At the time of this study, measurements are available only up to March 27, 2008.

The displacement curve is assessed for January 1, 2010 based on the climatological precipitation over the region (see Fig. 5). The basal sliding attributed to the Wenchuan earthquake of 2008 also was taken into account. Compared with the initial measurements, the shallow level displacements are as large as 18 mm. Based on our analysis of the effects of the Wenchuan earthquake on the YC creeping slope, the creep soon will accelerate. The Wenchuan earthquake reduced the natural creeping period by at least five years. For a crevasse near BH06, the surface crack enlarged from 10 to 18 cm. If the crack geometry is assumed to be constant, the crack depth almost doubles. Estimated depth changes were made of other cracks. The changes in its natural sliding cycle are obtained by comparing its creeping speed under current conditions with the pre-quake conditions. For example, after a major slope adjustment, say in 1998, it is assumed that there are no cracks due to strain. In 2008, before the earthquake, the cracks are already monitored. They all are located at model locations with large strain rates. The natural cycle is not difficult to estimate; 1mm/day is the critical value for next major sliding event. Even if there is little change in the precipitation morphology, over the next ten years there likely will be significant slope movements. However, if there is an intense rainstorm with over 150 mm/day at any time in the future period, then sliding becomes imminent.

Three historical landslides (1960, 1980 and 1997) reported in the YC study area are separated by ~20 years (Fig. 6). In 1960 and 1997, high annual precipitation values of 1819 and 1771 mm, respectively, were recorded. However, 1980 was relatively dry with 1200 mm total precipitation compared to 1600 and 1360 mm respectively for 1979 and 1981. Examination of the daily precipitation series from 1979 and 1980 indicates that in 1979 over 90% of the annual precipitation occurred in the latter half of the year, with no significant precipitation before June. Although the total precipitation for 1980 is small, the heavy precipitation in January 1980 immediately following the previous year's precipitation events formed an extended wet period. There was a significant precipitation event in January, reaching a rate of 58.3 mm/day that lasted 4.7 hours on January 15. As a result, the deep soil moisture (0.23 volume per total volume) remained relatively high for the remaining several months. What triggered the landslide was a very heavy precipitation event of a 100-year recurrence frequency, with a single day precipitation of 230 mm on June 11th.

For storm-triggered landslides, those precipitation events separated by less than two dry days can be considered as one single 'super' rain event. Thus, unlike many previous studies (e.g., O'Gorman and Schneider 2009), which count daily precipitation one day at a time (traditional precipitation analysis), we count those extended super-rain-events, defined as a somewhat continuous rainfall period nowhere separated by more than two consecutive dry days (rain-event analysis, Ren et al. 2011a). Figure 6 compares two methods of rainfall analyses. Using a traditional analysis, the 1997 slide does not correspond any significant daily rainfall event (>100 mm/day). However, following Ren et al. (2011), all major landslides (red arrows in Fig. 6) are triggered by rain events with higher rainfall totals. For the YC, heavy precipitation is both an enhancement factor, and a determining factor, in triggering landslides. However, current observational data are less than 2 years long, which is far too short to unambiguously resolve this hypothesis. SEGMENT-landslide simulations indicate that, under long term mean soil moisture conditions, the creeping will achieve a near-surface movement rate of 0.1 mm/day near the head of the slope, sometime within the next 40 years. Ironically, this creeping rate may never have been realized during the previous half-century. It has been interrupted by heavy rain storms and, consequently, slides have occurred before this movement rate ever was reached.

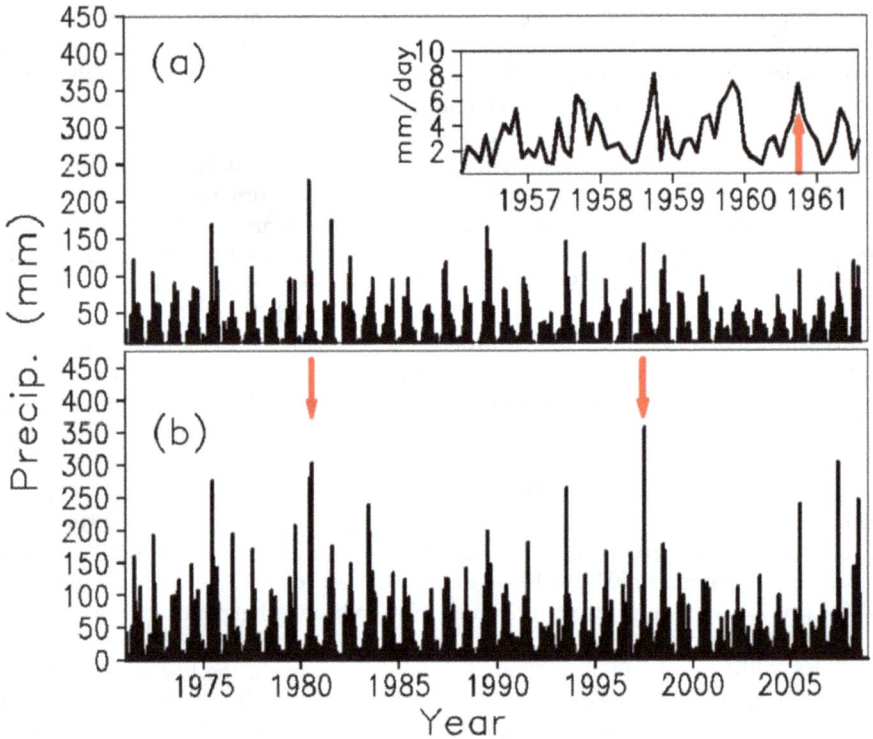

Fig. 6. (a) Daily precipitation and (b) rain event analyses, with the three major historical landslides indicated by red arrows. In panel (a) the daily rainfall time series reveals that that the 1997 event was not an intense rainfall event but was composed of two consecutive rainfall events over several days. In panel (b), rainfall totals are plotted for each rain event, by calculating the cumulative rainfall for each rain event and placing the total at the center of the start and end times. In panel (b), it is clear that both the 1980 and the 1997 events correspond to large total rainfall events. However, large rainfall totals alone do not necessarily trigger slope movements but, as discussed in the text, result from the combined effects of a number of factors. There is a lack of observed daily precipitation prior to 1970 to perform similar analyses. The inset of panel (a) is the NCEP/NCAR reanalysis monthly precipitation data. For example, the year 1960 not only had intense August precipitation, it also followed immediately after 1959, which was a very wet year.

For a storm with a 50-year recurrence frequency (~170 mm/day) more than 80% of the total water mass is channeled to the sliding surface through macro-pores, so movement rates can become dramatic toward the end of such a storm. The areas of significant deformation (i.e. maximum attainable surface sliding speed greater than 1 mm/s) after a 50-year storm are shown in Fig. 4 (with color shades). Thus far, the geological survey team has identified at least 10 landslide scars and slide debris deposits, all within the color shaded areas in Fig. 4. For example, there are obvious landslide platforms near BH8 and another major one near the position labeled point 'A'. A shear surface exists at point 'A' with a depth of about 70 m. When water drains down to this surface, the material strength at the shear surface is reduced to its residual value. The artesian pressures along the failure surface add to the instability of the sliding mass.

SEG:8/30/2008 × ×   SEG:8/30/2007 ① ①

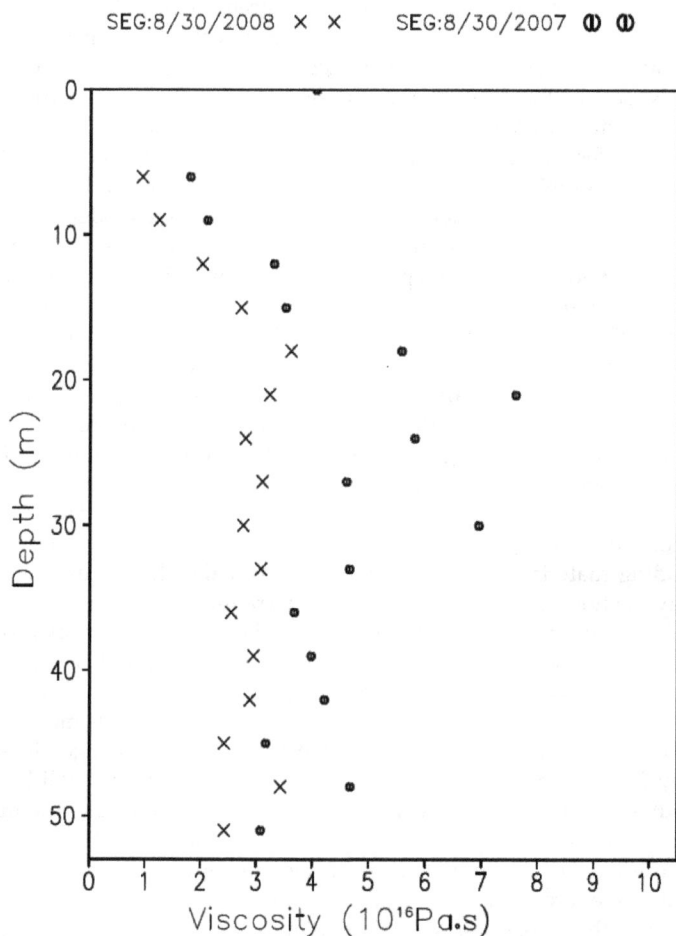

Fig. 7. Viscosity changes between August 30 2007 and August 30 2008, at borehole location BH8.

During heavy rainfall periods, water penetration reduces the strength of materials. In addition, hydrodynamic pressure along the slip surface further reduces stability of the sliding mass. The model simulations indicate that point 'C' is highly unstable under an extreme precipitation event. Failure will occur around this point first, triggering a failure of the upper portion. The sliding mass spreads about 50 m downslope and was brought to rest by the lateral stress from the walls of the V-shaped gully which parallels the 750 m elevation contour. The sliding material can become up to 20 m thick in the lower elevations. The accumulated material also can block the gullies and enhance infiltration of rainfall into deep layers and cause pore water pressure increases, which is a lubricating effect in the model parameterization. This is especially important for the land segment lying between the 700 m and 600 m elevation, which is where the primary human residential areas are located. To obtain additional information for verifying the model credibility for slope stability at Point 'C', several more bore-holes are required in slopes adjacent to the gully on the west bank.

In the above scenario, a volume of 6.3×10^7 m^3 of soil and rock is estimated to be creeping. In retrospect, it appears landslides have been occurring in this region at intervals through history; but only part of the total creeping mass is involved in any particular landslide. Specifically, one portion may slide, causing a reduction in the stability of an adjoining portion; then, years, decades or even centuries later, a subsequent landslide will occur. As a consequence, the topography of the area is hilly and highly uneven. What determines the creeping rate of a slope is the material viscosity. We examined the modeled viscosity change near BH8 (Fig. 7). Because the rocks are heavily weathered, the viscosity is on the order of 10^{16} Pa s, which is two orders of magnitude smaller than for the same material in an undisturbed state. So, when dealing with fractured rocks as a whole, they must be viewed as granular material. The viscosity of granular material changes as strain accumulates. In this case, the viscosity is reduced substantially with time at all levels (e.g., it is reduced by >40% from year 2007 to 2008). This explains why the creeping tends to accelerate with time.

The SEGMENT-landslide system is valuable for monitoring creeping because it can provide a dynamical representation of changes in the strain distribution inside the sliding material. Figure 8 shows the modeled creeping velocities for January 2012. A vertical cross-section is provided along the direction of the primary sliding direction, located near the demarcation line in Fig. 4, for the current geometry and a climatological mean soil moisture conditions. Because the sliding material depth is only one-tenth of the slope dimension, for a more effective display the flow field is transformed into terrain-following sigma coordinates. The surface corresponds to $\sigma=0$ and the bottom of the sliding mass corresponds to $\sigma=1$. For clarity, the portion with flow speeds less than 0.3 m/yr is filtered out. The formation of the local maximum speed 'core' (the band along the $\sigma=0.8$ level) near the bottom (Fig. 8b) is attributed to movement within the fractured layer. Varying the soil moisture conditions indicates that the movement of this layer is most sensitive to changes in soil moisture conditions. Any factors preventing surface water entering the ground will help reduce the acceleration of the sliding mass and delay future landslides. The landslides are sensitive to soil moisture conditions, but a qualitatively persistent feature is that the maximum strain area is located upslope, as shown by the warm color shading near ~750 m elevation in Fig. 8a. At present there are crevasses with openings wider than ~5 cm of horizontal displacement. That the maximum speed cores are relatively isolated indicates that the sliding surface is not fully connected. In the upcoming 10 years, the causal mechanism for major slides remains the same, namely, the storm charging of the artesian aquifer and the lubrication of the granular layer by drainage water.

To investigate the impact of possible future precipitation morphological changes over this region, 21st century simulations are analyzed from 17 Climate General Circulation Models (CGCMS) (see Ren and Karoly 2006) under the SRES A1B (moderate) emission scenario (Nakicenovic and Swart 2000), which assumes a balanced energy source in a future of rapid economic growth. Future precipitation rates under the SREA A1B scenario are expected to intensify in the upcoming decades, which is in accord with the consensus of the wider climate research community (e.g., Groisman et al. 2004, Karl and Trenberth 2003). In actuality, the total rainfall amount also increases, indicating that the primary mechanism may be the increase in atmospheric vapor concentration, as described by O'Gorman and Schneider (2009).

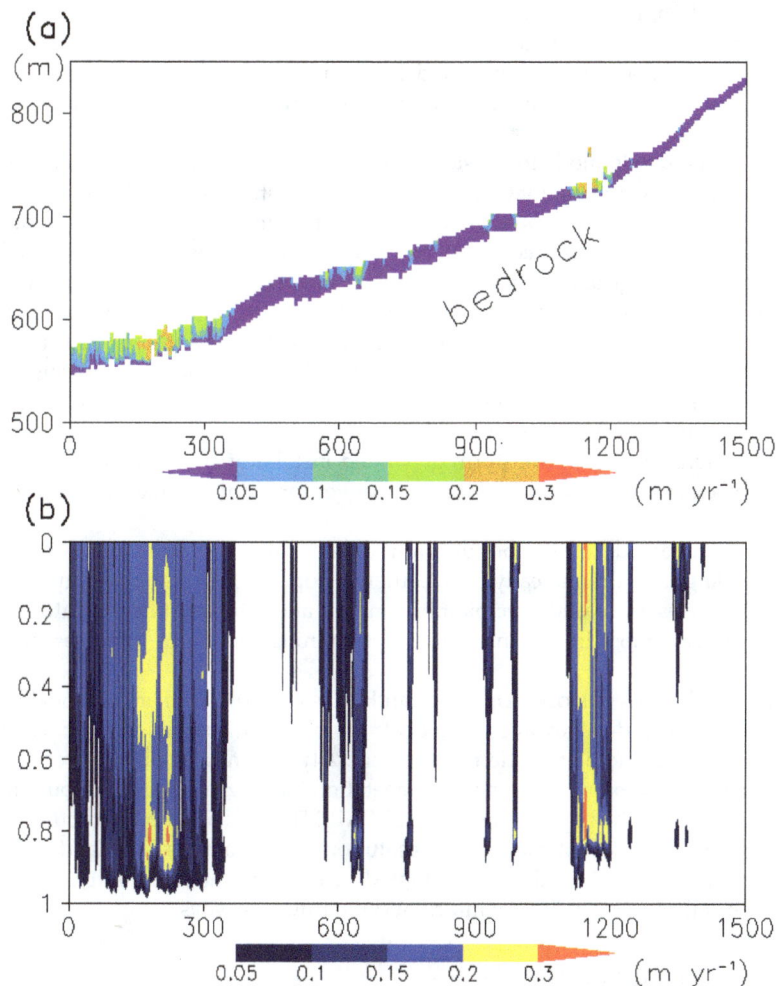

Fig. 8. Creeping speed for January 2012, within a vertical cross-section (along the red arrow in Fig. 4) under climate mean soil moisture conditions. The top panel is displayed in physical space (vertical axis is elevation; horizontal axis is distance from the origin). The bottom panel is displayed in the σ terrain-following coordinate system. So σ = 0 corresponds to the surface and σ = 1 corresponds to the bottom of the sliding mass. The color shading is the magnitude of the full 3-D velocity. In panel (b) maximum cores (roughly at the σ = 0.8 level) corresponding to the "creamy" basal sliding layer.

6.2 Discussion

A measure of our understanding of slope sliding processes is our ability to predict the future behaviour of slopes under a range of conditions. In this study, the SEGMENT-landslide model (Ren et al., 2009; Ren et al. 2010; Ren et al., 2011a, b) successfully reproduced three historical storm-triggered landslides that occurred during the past half-century, for the Yangjiashan creeping slope (YC) in China. SEGMENT also was used to make one further, long term prediction, for the forthcoming decade, 2010-2019, which quantified the stability of the YC region and showed that slope movements will occur during the next decade and that, even more significantly, a major landslide is imminent if an intense rainstorm with over 150 mm of rainfall occurs at any time in that period, even in the near or immediate future.

The SEGMENT landslide modeling system has demonstrated, from numerical experiments carried out over the YC region, that it can anticipate how strain accumulates. For example, it shows how load increases with precipitation and that there is an accompanying decrease in yield strength. SEGMENT can predict when sliding, or rapid slope failure, is probable, given the available meteorological parameters, soil properties and land cover conditions.

For the YC, its particular geological constituents are the main cause of its landslide susceptibility to triggering by storm events. Heavy storms are enhancement as well as triggering factors. Our study points out an aspect that requires close monitoring because it likely is responsible for the upslope cascading of storm-triggered landslides. Model sensitivity experiments established a stability feature not investigated by the survey team. The model demonstrated that rainstorm generated instability at a given location forms the first sliding block in a sequence, by acting as a trigger for a domino-like slide that moves up the slope.

We find that increased infiltration of groundwater into the sub-surface from storms increases creeping rates dramatically for weathered slopes. For slopes experiencing repeated failure-restore cycles, increased precipitation amount and intensity under a future warming climate are the two most important factors determining long-term increases in landslide frequency.

Quantitative predictions of storm triggered landslides require a numerical modeling system like SEGMENT-landslide. However, some of the requirements of SEGMENT-landslide, especially the input and verification data, generally are not available even in current geological maps. These parameters include vegetation loading and root distributions in soils and weathered rocks. The application of the SEGMENT-landslide model to other regions is limited primarily by a lack of these high resolution input datasets. The landslide features implemented in SEGMENT-landslide, if adopted by the relevant community, hopefully will encourage the collection of such vital information in future surveys.

7. Landslides in a future climate

There is a considerable and expanding body of opinion which suggests that earth may suffer marked temperature increases over the next 50-100 years (Rind 1984) due to heat retention by the atmosphere caused by increased levels of the greenhouse gases (GHG) such as CO_2, CH_4 and oxides of nitrogen. The levels in the atmosphere have increased quite dramatically in the last 70 years and are expected to continue to rise. One consequence is an intensified hydrological cycle. Estimates of the effects vary widely but all predict some increase in storm-triggered landslides.

The global hydrological cycle also is assumed to be intensified. However, different regions may respond very differently (Ren et al. 2011a). For example, both monsoonal regions (e.g., the Yanjiashan creeping slope located in the Asian monsoonal region) and Mediterranean regions (e.g., California) show significant increases in extreme precipitation, but the average annual precipitation changes are different. Precipitation over the Asian monsoon region increases significantly, consistent with increasing tropospheric specific humidity, as pointed out by Allen and Ingram (2002). In fact, the increase in annual mean precipitation over this region is due mainly to the shift toward heavier precipitation events. The CCSM simulated precipitation trend over southern California under the SRES A1B scenario, counter-intuitively indicates that total precipitation decreases by more than 0.1 mm/day on an annual basis. Storms become more intense but farther apart in time, favoring a drought-flood bipolar temporal pattern as suggested by Trenberth (1999). Importantly, current climate models show strong inter-model consistency for this finding.

In landslide terms, the phenomenon has been postulated to be a potential eventuality in two different ways. One proposes that the debris flows will occur more frequently on smaller scales. The other suggestion, considered more likely by some concerns the less frequent but more disastrous outburst of storm-triggered landslides. One thing for sure is that many of regions with minor risks could be magnified if the earth's climate undergoes significant changes over the next 50-100 years as a consequence of continued burning of large amount of fossil fuels.

8. Summary

Natural hazards are an ever-present threat to human lives and infrastructure. The need for greater predictive capability has been identified as one of 10 Grand Challenges in Earth Sciences (NRC, 2008). As an effort toward the goal of a reliable landslide mapping and warning system, we present a modeling system (SEGMENT) that systematically estimates the potential for landslides over a regional area, rather than for a single slope. The promising performance of the model is attributable to the use of a new, fully three-dimensional modeling framework based on a newly proposed granular rheology, and to the use of a land surface scheme that explicitly parameterizes the hydrological characteristics of macro-pores. Some requirements of the model, such as vegetation loading and root distribution in soils and weathered rocks, are not available even in present geological maps. Applications of SEGMENT to other regions are limited primarily by a lack of high resolution input data sets. However, the new concepts implemented in the model, if adopted by the community, may encourage the collection of such information in future surveys.

The anticipated future climate warming has influence on the occurrence of landslides caused by elevated water content in the ground. Changes in precipitation morphology are highly relevant for storm-triggered landslides and subsequent desertification, because the root system of vegetation has adapted to the current precipitation climatology and likely is not prepared for human-induced changes in climate. Under the current relatively stable astronomical boundary conditions there are natural "rhythms", whereas human induced changes are likely to transition significantly in one direction, leading to a climate state not experienced before by the existing terrestrial ecosystem. Microclimatic variations associated with slopes allow stands of an ecosystem type to exist far beyond their major zones of distribution (Chapin et al. 2002). These outliers act as important colonizing individuals during times of rapid climate change. Destroying these outlier species (transitional belts on

Holdridge's chart) by landslide burial, accompanying extreme precipitation, may slow down ecosystem migration in accord with climate change. Landslides are localized events. Advanced dynamical models with physical basis should be used as there is a need for prediction, rather than simply documenting the occurrence of landslides. SEGMENT-landslide is an effort in this direction. Because there are biogeochemical submodels coupled in the SEGMENT system, it also is an ideal tool for investigating the environmental consequences of landslides, including deforestation and an associated decrease in productivity.

9. References

Allen, M., and W. Ingram, 2002: Constraints on future changes in climate and the hydrologic cycle. Nature, 419, 224-232.

Amundson, R. V., and H. Jenny (1997), On a state factor model of ecosystem. BioScience 47, 536-543.

Baum, R. L., W.Z. Savage, and J.W. Godt (2008), TRIGRS-A Fortran program for transient rainfall infiltration and grid based regional slope-stability analysis, version 2.0: U.S. Geological Survey Open-File Report, 75pp.

BNPB, 2009. Indonesian Disaster Data and Information. Badan Penanggulangan Bencana Nasional (National Disaster Management Agency) Retrieved 21th May, 2009. http://dibi.bnpb.go.id/

Burrough, P.A., and R.A. McDonnell (2004), Principles of geographical information systems, Spatial Information Systems and Geostatistics, 333pp. Oxford University Press.

Caine, N. (1980), The rainfall intensity¬duration control of shallow landslides and debris flows. Geografisker Annaler, Series A, 62, 23¬27.

Cannon, S.H., and S. Ellen, 1985: Rainfall conditions for abundant debris avalanches. San Francisco Bay Region, California. California Geology, 38, 267-272.

Casadei, M., W. Dietrich, and N. Miller (2003), Testing a model for predicting the timing and location of shallow landslide initiation in soil-mantled landscapes, Earth Surf. Processes Landf., 28, 925-950.

Chapin III, F., P. Matson, and H. Mooney, 2002: Principles of terrestrial ecosystem ecology. Springer Science and Business Media, Inc., NY, 436p.

Costa, J. E. (1984), Physical geography of debris flows, in Costa, J. E., and Fleisher, P. J. (eds). Developments and Applications in Geomorphology: Springer Verlag, pp. 268-317.

Crosta, G. 1998. Regionalization of rainfall thresholds: an aid to landslide hazard evaluation. Environmental Geology 35 (2-3): 131-145.

Cruden, D. M., and D. J. Varnes (1996), Landslide types and processes; in, Landslides Investigation and Mitigation, A. K. Turner and R. L. Schuster, eds.: National Research Council, Transportation Research Board, Special Report 247, p. 36-75.

Cruden, DM. 1991. A Simple Definition of a Landslide. Bulletin International Association of Engineering Geology 43: 27-29.

Dai, FC, Lee, CF and Ngai, YY. (2002), Landslide risk assessment and management: an overview. Engineering Geology 64 (1): 65-87.

Dietrich, W. E., and J. T. Perron (2006), The search for a topographic signature of life. Nature, 439, 411-418.

Godt, J., R. Baum, and A. Chleborad (2006), Rainfall characteristics for shallow land-sliding in seattle, Washington, USA, Earth Surf.Processes Landforms, 31, 97-110.

Groisman, P., R. Knight, T. Karl, D. Easterling, B. Sun, and J. Lawrimore (2004), Contemporary changes of the hydrological cycle over the contiguous United States trends derived from in situ observations. J. Hydrometeorology, 5, 64-85.

Iverson, R. (1997), The physics of debris flows. Review of Geophysics, 35, 245-296.

Karl, T., and K. Trenberth (2003), Modern Global Climate Change. Science, 302, 1719-1723.

Matson, P.A., C. Volkmann, K. Coppinger, and W.A. Reiners (1991), Annual nitrous oxide flux and soil nitrogen characteristics in sagebrush steppe ecosystems. Biogeochemistry, 14, 1-12.

Nakicenovic, N., and R. Swart (eds.) (2000), Special report on emissions scenarios (SRES). Cambridge University Press, Cambridge and New York, 612pp.

NRC, 2008: Origin and Evolution of Earth: Research Questions for a Changing Planet. National Academy Press, 137 pp.

O'Gorman, P., and T. Schneider (2009), The physical basis for increases in precipitation extremes in simulations of 21st century climate change. PNAS, 106, 14773-14777.

Ren, D., R. Fu, L. M. Leslie, and R. Dickinson (2011a), Predicting storm-triggered landslides. BAMS. DOI: 10.1175/2010BAMS3017.1.

Ren, D., R. Fu, L. M. Leslie, and R. Dickinson (2011b), Modeling the mudslide aftermath of the 2007 southern California wildfires. J. Natural Hazards. DOI: 10.1007/s11069-010-9615-5.

Ren, D., R. Fu, L. M. Leslie, R. Dickinson, and X. Xin, 2010: A storm-triggered landslide monitoring and prediction system: Formulation and case study. Earth Interactions. Paper 12 of Volume 14.

Ren, D., J. Wang, R. Fu, D. Karoly, H. Yang, L. M. Leslie, C. Fu, and G. Huang (2009), Mudslide caused ecosystem degradation following Wenchuan earthquake 2008. GRL,36, doi:10.1029/2008GL036702.

Ren, D., L. M. Leslie, and D. Karoly (2008), Mudslide risk analysis using a new constitutive relationship for granular flow, Earth Interactions, 12, 1-16.

Ren, D., and D. Karoly (2006), Comparison of glacier-inferred temperatures with observations and climate model simulations. Geophysical Research Letters, 33, L23710.

Rind, D. (1984), The influence of vegetation on the hydrological cycle in a global climate model. *Climatic processes and climate sensitivity*. J. E. Hansen and T. Takahashi, Eds., Amer. Geophys. Union, Washington, DC, 73-91.

Schofield, A. N., and T. Telford, 2006. Disturbed soil properties and geotechnical design, ISBN 0-7277-2982-9.

Selby, M.J. (1993), Hillslope Materials and Processes. Oxford university press, Oxford, UK.

Sidle, R. C. (1992), A theoretical model of the effects of timber harvesting on slope stability. Water Resources Research, 28, 1897-1910.

Sirangelo, B., and P. Versace, 1996: A real time forecasting model for landslides triggered by rainfall. Meccanica, 31, 73-85.

Smith, K and D. Petley (2008), Environmental Hazards: Assessing Risk and Reducing Disaster, Fifth Edition. Routledge, London, 414p. ISBN 0-203-88480-9

Trenberth, K., 1999: Conceptual framework for changes of extremes of the hydrological cycle with climate change. Clim. Change, 42, 327-339.

Van Asch, T., and Van Beek, J. 1999. A View on Some Hydrological Triggering Systems in Landslides. Geomorphology 30 (Elsevier Science): 25-32.

Wooten RM, K. A. Gillon, A. Witt, R. Latham, T. Douglas, J. Bauer, S. Fuemmeler, and L. Lee, 2008: Geologic, geomorphic, and meteorological aspects of debris flows triggered by Hurricanes Frances and Ivan during September 2004 in the southern Appalachian Mountains of Macon county, North Carolina (southeastern USA). Landslides, 5,31–44.

7

Dinaric Karst – An Example of Deforestation and Desertification of Limestone Terrain

Andrej Kranjc
Slovenian Academy of Sciences and Arts
Slovenia

1. Introduction

Dinaric Mountains are one of the main mountain systems of the Balkans. The name was given by of the imposing Dinara Mountain (1913 m) at the border between Herzegovina (Bosnia and Herzegovina) and Dalmatia (Croatia). Under the name of Dinaric Alps it appeared already in the 18th century (Hacquet 1785). The part of Dinaric Mountains which is mostly built by carbonate rocks, limestone predominating, is called Dinaric Karst. The name Karst as well as the international term "karst" derived from the plateau Kras (Carso in Italian, Karst in German), the north westernmost plateau of the Dinaric Karst ridges (Kranjc, 2011). Dinaric Mountains are a mountain chain approximately 650 km long and up to 150 km large, covering an area of about 60 000 km², stretching between 42° and 46° of northern latitude (Fig. 1).

Fig. 1. Delimitation of the Dinaric Karst after Roglić and Gams (Mihevc & Prelovšek, 2010).

Geologically, Dinaric Mountains consist of two parts: Inner Dinarides at Northeast and External Dinarides at Southwest (Mihevc & Prelovšek, 2010; Zupan Hajna, 2010). While in the Internal Dinarides non-carbonate rocks prevail, in the External one the carbonate rocks are predominant – therefore there is karst. A. Penck's student of Vienna "geomorphological school", Jovan Cvijić was probably the most influential scholar to propagate karst and to substantiate the karst science. In his basic works of 1893 and 1895 (Cvijić, 2000) he stated: "All the forms on the bare limestone, made by water, we will call karst features". Cvijić's connotation of karst is "bare limestone landscape". The travellers who travelled from Vienna to Austrian Adriatic port of Trieste were the most impressed by a sudden change of landscape. After Postojna, they entered a bare rock land, without surface water and especially without any greenery. In 1689, Valvasor in his topography wrote about the Kras (Karst) plateau: "Somewhere it is possible to see for some miles, but everything is only grey, nothing green, because all the country is covered by stones." Illustration from the same work shows the cultivated land at the bottom of dolines only (Valvasor 1689). In many parts of the Dinaric Karst it is true for the actual situation (Fig. 2). On 18th century military maps the entire Kras surface is shown as ""Steinigte Terrain" (rocky terrain) (Fig. 3). Description of individual settlements added as a comment to the maps often stated: "There are no forests or trees, just some bushes one hour away from the village" as shown by the example of the village Gabrovitza (actual Gabrovica pri Komnu) (Rajšp, 1997).

Fig. 2. On Dinaric Karst cultivated land is mainly in the bottom of dolines only (photo A. Kranjc).

Fig. 3. Military map of Kras plateau from the second half of the 18th century: great majority of the surface is "Steinigter Terrain" (Stony terrain) (Rajšp, 1997).

Fig. 4. About 1850 the nowadays woody hill Sovič above Postojna was bare (Schmidl, 1854).

Impressions of travellers across karst terrain between Postojna and Trieste are all depressing (Fig. 4). Count Karl von Zinzendorf wrote in 1771:"The country is *affreux*. All these terrible rocks and in the midst of them some small cultivated parts of land encircled by stones...; " and B. F. J. Hermann in 1780: "Anywhere you look, it is only desert..." (Panjek, 2006). To the end of the 19th century and even later the impression of karst got from the published works of scholars as well as of laymen was one of bare rock and dry landscape. But it was not always such. On the Dinaric Karst, nowadays there are completely bare landscapes, mostly on the Mediterranean side, but there are also extensive forests covering slopes of high mountains and the tops of karst plateaus in the interior. Good examples of preserved forests are Rajhenau primeval forest (Kočevski Rog plateau above Kočevje in Slovenia) (Rajhenavski pragozd, 2011) and the forest Lom (*Piceo-Abieti-Fagetum illyricum*) on the Klekovača Mountain in Bosnia (Prašuma Lom, 2011). The first one occupies about 50 ha of *Abieti-Fagetum dinaricum*. The forest of Kočevje is a part of the biggest uninterrupted forest complex in the Western and Central Europe, stretching from the Kočevje region (Slovenia) to Gorski Kotar (Croatia).

The aim of the case study of this chapter is to show that man is the main factor both at destroying his natural environment and at restoring it. The man is capable of both. In our case that means a complete deforestation, the changing of a heavy wooded landscape to a bare rocky desert and back again to a dense, although, to be true, not a "natural" or optimal, forest, as it shows the further text. Of course the time scale is different as well as the attitude towards these processes triggered in both cases by the human itself. The first dwellers millennia ago have not seen, they did not know and they could not imagine what a process they have started by cleaning land for pastures and fields. And the process of deforestation and finally desertification lasted thousands of years. If reforestation was premeditated by well planned actions the actors knew well their aim and purpose. Comparing the lasting of reforestation with deforestation this was a short but an intensive action. The karst terrain, Dinaric Karst especially, is such a terrain where the human activities leading towards desertification have shown their most disastrous consequences and where the opposite action, reforestation, demanded extremely great efforts and financial input. This case study is not meant to be just a history of a forest but also a warning what can happen, not only in the mist of history, but also nowadays.

2. Deforestation

Deforestation started in prehistoric times already, by the arrival of people with Neolithic culture, leading Neolithic way of life, the transition from gathering and hunting to stockbreeding and farming, the so called Neolithic revolution. The Balkan Peninsula is a sort of bridge between Near East, across Asia Minor towards Central and Western Europe and the Neolithic culture reached it between 6 500 and 6 000 BC. Neolithic farmers did not enter far into the Dinaric Mountains. Instead they advanced across the fertile plains along the Danube River on the North and along the Adriatic coastal strip in the South, so avoiding the mountainous regions (Velušček, 1999). So their impact on the forest of Dinaric karst had to be negligible, with some exceptions - the Butmir locality for example. Pollen analyses show that the intensive deforestation phase occurred due to grazing during prehistory on the plateau of Kras (Slapšak, 1995). To confirm the prehistoric deforestation Gams' (1991) research on Rillenkarren is very illustrative; he found out that they started to be formed on the plateau Kras 3 000 – 3 500 years BP, which is when the forest was destroyed and bare rock started to appear on the surface (Fig. 5).

Fig. 5. Rillenkarren on a limestone rock: they start to form about 3 000 – 3 500 years ago while the smooth lower lying surface was covered by soil much longer (photo K. Kranjc).

During the Bronze Age, the situation changed drammatically. The population increased and due to their economy (farming, stockbreeding, and ore mining) entered deeply into the Dinaric Karst where they had to clear and cut the timber, which was used for buildings and defence installations as well as for ore smelting. In Dinaric karst the settlements were concentrated mainly in two border zones: on the Adriatic coastal plains (Low Periadriatic Karst) and on the karst plains and hills along the Pannonian plain (Low Peripannonian Karst). The innermost parts of Dinaric Karst seem to remain quite untouched. Thanks to archaeological research we know that during this population expansion the climax *Abieti-Fagetum* forest was already being replaced by lower association of *Quercus* type in some parts of Dinaric Karst (Turk et al., 1993).

During the Early Iron Age (Hallstatt culture) practically all the Dinaric Karst was settled and during the Late Iron Age (La Tène culture) Dinaric region entered into history: the native (Illyrian) tribes are known by names, in the southern Dinaric coast and islands Greek colonies were founded, from the North came a Celtic invasion and from the West the Roman one. From that period, the names of peoples living in the region are known: Illyrian tribes, Celtic tribes, Greeks and Romans. Regarding the use of timber and wood and the economy in general, the cause of deforestation, the choice of motives is very large: Illyrians deep in the interior of Dinaric Karst tended big flocks of sheep and goats using the transhumance system, and the farmers not abreast with the time used slash-and-burn system. To increase pasture areas shepherds also used the fire. On the Dinaric Karst, considering its climate, a forest fire does not only destroy trees and surface vegetation. Its consequence is much more important because it increases rain and wind erosion processes and rock aridity. In such cases woodlands can be really transformed into "rock deserts". Metallurgy using so-called "bean ore" (Bohnerz) which is very frequent in karst soil and clay, had to consume big quantities of wood, as shown by modern experiments (Kranjc, 2002). Near Straža village by the town of Novo mesto (NW part of Dinaric Karst, Slovenian Low Peripannonian Karst), 24 smelting furnaces were burning simultaneously as is shown by the archaeological research (Dular & Božič, 1999). While the Romans (the army at the beginning and colonists later) and Greeks used wood for construction, industry (including metallurgy, charcoal and lime production, pottery), heating (thermae with big basins of hot water were necessity of townspeople everyday life) and many other needs connected with high developed culture.

Along the Dinaric area - that is the Adriatic coast, shipbuilding was important too, for both the Illyrians as well as for Greek colonies. From historical sources it is known that the fleets consisted of a big number of smaller boats (lemba and liburna), but also big boats were constructed. Ancient authors, Polybius for example, often mentioned numbers of ships of Illyrian fleets:"Scerdilaedas ... provided 40 lembas and Demetrios of Pharos 50". To build the liburna, 33 m long and 5 m large ship with 36 oars, quite a lot of wood was needed. Polybius reported that Macedonian king Philippos the 5th ordered to construct 100 liburnas during the winter 217/216 BC. For oars alone they needed 3 600 adequate straight young trees, not accounting for spare oars. That this was an important question proved the report of Andokidos who came to help to the army of Samos: "For the beginning, I prepared wood for oars for your army of Samos." When Brazida conquered the town of Amphipolis this fact "... provoked great fear among the Athenians ... because from there they got the wood for shipbuilding...", both quotations taken from Thucydides (Cabanes, 2002).

From the opposite part of the Europe, from Scandinavia it is reported that the Viking's shipbuilding came to a serious deadlock because there were no more suitable trees in their homeland (Atkinson, 1979). In the Dinaric Karst such a problem is not known from historical times but aroused seriously in the 19th century, as seen from the text below. The Ljubljana Moor lies at the foot of high karst plateaus of NW Dinaric Karst. Between the towns of Nauportus at the foot of the plateau with very important karst spring, and Emona, the Roman legion had to build a road across the Moor in the first years AD. For the base of the road structure they placed thinner round tree trunks on the marshy soil. Illyrians (and other native peoples) to defend their oppida used palisades while Romans used them to defend their legion camps and also to strengthen walls of stone and wood combined to protect towns.

Of course there is little direct evidence of larger deforested surfaces existing during Prehistoric and first Historic times. In the Smederevo polje of Lika (Croatia), the position of skeletons in burial mounds shows that the landscape was open with a thin soil cover, that means there were no (more) forest at the time of a funeral (Horvat, 1957). Many parts of the Dinaric Karst that are nowadays forest-covered again were dramatically different during the Iron Age, when this was an open country with small fields and pastures, and fortified hilltop settlements. In Bosnia and Herzegovina the majority of actual barren landscape is found around the former Illyrian hill forts and settlements (Djikić, 1957). Nevertheless, at the beginning of the historical times there must still have been much of forest-covered land as proved by the topographical names. The Island of Korčula (Dalmatia) was called Korkyra Melaina by Greeks and Korkyra Nigra by Romans, both names meaning Black Korkyra because of its dense pine forest cover. On the plateau Kras cultivated land was in form of islands around the settlements and quite a lot of forest remained in between (Slapšak, 1995).

Knowing people and their history it is sure that the wars and clashes of arms existed in Prehistoric times, too. From the beginning of history on there are records of them from all periods. These violent activities had and has (just to remember the Vietnam War) a very great impact on forest. "Plunder and burn" was the most common motto: to get the booty and to devastate enemy's country, that means to ruin it economically. Burning a forest was not just an economical measure; it was used as a war or raid tactics, too. For example, the Turkish regular army cut down and burned woods to make easier way for the cavalry and heavy artillery, and to destroy Hajduks (Balkan guerrillas) hidouts and dwelling places. The army was ordered to destroy forest on both sides of roads to get clear zones to thwart ambushes by the Hajduks. From the 17th century there are picturesque descriptions of such activities in the diary of the Turkish traveller Evliya Çelebi (Evlija, 1957). Reciprocally, forest was sometimes burned too by the defending population in attempts to prevent the enemy's attack.

During the late Antiquity another reason joined "traditional" deforestation – invasions of Barbarian peoples. Between 3rd – 8th centuries they crossed and often settled the territories of the Roman Empire, in our case the Roman provinces of Illyricum and Dalmatia. Especially frequented was the direction of the Roman road Aquileia - Carnuntum, leading across Dinaric Karst on the section between Tergeste (Trieste) and Emona (Ljubljana). The consequence of their approaching and settling was the movement that is flight, of Romanised inhabitants towards the coast and especially on the normally overgrown islands which are stretching all along the Adriatic coast. Today these islands are effectively uninhabited but traces proved that they were relatively densely populated during the Late Antiquity. Maybe the population increase was not the worst. Such troubled times favoured stockbreeding over farming and overgrazing together with burning to increase pasture lands left many of the previous mentioned islands completely bare. Such conditions remained to modern times and now these islands are used as meagre pasture for sheep only, not to mention the summer tourism (Fig. 6).

Fig. 6. Kornati islands, practically nothing but dry pastures for sheep (photo A. Kranjc).

The next troubled times after the barbarian invasions regarding human pressure and impact on the forest was the Turkish occupation of the majority of the Balkans and Turkish raids into the neighbouring countries, which seriously began in the 14th century. The interior of the Balkans was Turkish Empire, much of the Adriatic coast belonged to the Venetian and Dubrovnik (Ragusa) Republics, and small part of the NW part belonged to Austria, either in the frame of so-called Military Zone (nowadays Croatia) or the Duchy of Carniola. Before the 14th century the forests which were not commune were divided into forests for hunting, oak forests and small forests, in the frame of the Austrian lands. Animal grazing was forbidden in them and for cutting or burning wood severe fines were foreseen. An Act from 1550 allowed all the inhabitants of Trieste, mule drovers and butchers to cut wood and grass in all of the commune's forests. In 1689 two revisers reported: "In the town, there is a shortage of fire-wood, it is impossible to make a stock, because all the forests are destroyed". In 1719 the port of Trieste was proclaimed a "free port". To the Austrian annexation of Bosnia in 1878 in outline the situation remains the same. A significant part of the Christian population fled before the Turks and re-settled in Austrian, Hungarian, Venetian or Dubrovnik lands not far from the borders. These migrations included some long distance displacements. For instance, entire "tribal families" from the inner parts of Serbia moved to the Istria Peninsula in the most north westerly corner of the Adriatic Sea, and even to Carniola – nowadays Slovenia (Cvijić, 1966). These groups moved with their flocks and commonly they continued primarily as owners and herders of grazing animals. Pressure on grazing land led to another increase in the rate of deforestation. The emigrants brought their slash-and-burn techniques, too. Transhumance together with burning, later also cutting of the forest, was preserved locally until the 20th century, when they were observed by the first researchers, J. Cvijić among them. According to the eyewitness Gušić, the main reason of deforestation was clearing the land for new pastures or meadows and sowing of grain in "novine" (new fields), used once only (Gams & Gabrovec, 1999). That cutting and burning of a forest could locally trigger accelerated soil erosion is proved by the practice in the near past, when farmers in remote mountains have burned forest in order to create such "novina" (Kranjc, 1979). This process is not connected with the Dinaric Karst only but largely with the Mediterranean (Fig. 7). On Baleares Islands on the land not suitable for agriculture grazing was practised at least from the Catalan conquest (1229). Traditional economy was based on the repetitive burning of herbaceous brushwood of *Ampelodesmos mauritanica*. This released active soil removal as well as the progressive degradation of scrub formation. At the end of the chain the bare rock results (Ginés, 1999). For some of authors, the main reason of not only deforestation but of the degradation to the bare rock country was grazing, grazing by goats especially (Wessely, 1876).

Fig. 7. In some parts of Dinaric Karst the burning of shrubs is still practised (photo K. Kranjc).

In parts of the Dinaric Karst under Austrian and Venetian influence, the agrarian pressure was not the only economic reason for deforestation. In the Venetian territory construction work and shipbuilding demanded large quantities of timber. It is reported that 1 200 000 tree trunks were needed for use as piles to support the church of Santa Maria della Salute in Venice (Horvat, 1957). It is difficult to imagine the whole number of piles, using for the churches, palaces and other buildings in Venice. Without doubt a great part of them came from the Dinaric Karst. There is a popular saying that the Venice "stands on oaks from Karst". In the time of the French Revolution (1792), the duty of the French consul in Albania was to take care of "cutting down construction timber for navy base at Toulon". Marshal Marmont for example, the Governor of the Illyrian provinces under Napoleon, ordered to cut off the tops of all the trees in one of the still remaining oak forest in the vicinity of Trieste, called Frned, to use them as a timber for ships. As a consequence the forest decayed completely to 1820. In the Austrian part of the Dinaric Karst the farmers (villeins and serfs) did not have the right to cut timber for trade before the so-called "Land Release" issued by the Empress Maria Theresa in 1848. In NW part of the Dinaric Karst the Austrian Navy had forest reserves, mainly oak. In these forests it was forbidden to fell timber for other purposes. The Navy's demand was great: the navy's forester (by the way the inventor of the vessel screw, too) J. Ressel reckoned up that to construct and maintain during its 150-year life a wooden battle ship 120 000 tree trunks were needed. In that time Austrian merchant navy had 523 big ships. To maintain the number 6.244 m³ of wood would be needed, while the production of the Istrian forests was 7.030 m³ (Piškorić, 1993). Good husbandry would thus not cause the deforestation by itself. But In 1819 the marine forest reserve was cancelled and massive felling programme started. Timber was sold to Venice, France, and especially to England (Murko, 1957). It is not surprise that the emperor Maximillian when visiting Trieste in 1850, described the plateau Kras as a rock desert with a curse hanging over it (Anonym, 2001) (Fig. 8).

Fig. 8. The view of the plateau Kras above Trieste in 1901 (Anonym, 2001).

The consequence of the mentioned "Land-Release" was disastrous for the forests in the Austrian parts of the Dinaric Karst. According to this act farmers became owners of the land which included right to cut down trees and to sell them. And they used new rights to a full extent, not thinking of replanting young trees. Parallel to this process, industrial development, especially mining and metallurgy contributed significantly to deforestation, even on remote karst plateaus in Slovenia and Bosnia and Herzegovina. During the second half of the 19th century, special narrow-gauge railroads were laid down to facilitate exploitation of the Dinaric Karst forests. The changes or regression in some branches of industry, metallurgy especially, can show indirectly the changes in forest structure or deforestation even. An example is the decrease of iron industry of the well known industrial Ž. Zois of Kranjska (Carniola) at the end of the 18th century. Some of his ironworks went short of fuel that is of charcoal. The so called "Slovene furnace" needed 50-60 % more charcoal than ore. Zois tried to use charcoal made of soft trees (spruce) instead of hardwood (beech). This is also one of the reasons of the change of the forest structure: for the shipbuilding the oak was over exploited and for the iron industry the beech (Kranjc, 2002) (Fig. 9).

Fig. 9. "Cooking" of charcoal on the Dinaric Karst at the beginning of the 20th century (Anonym, 2001).

As indicated in the text above, there were different factors causing deforestation of the Dinaric Karst and there are regions affected by different steps of deforestation. In any case the factor was man, either through his economy as stockbreeding and transhumance, slash-and-burn agriculture, fire wood gathering, construction and different branches of industry, shipbuilding and metallurgy emphasized or other, hostile activities, as "plunder and burn", army movements, attacks and protection of them, and last but not least the pressure on agricultural land. Some parts, relatively small and rare, of the Dinaric karst are practically unaffected by the process of deforestation and still nowadays covered by a dense forest; some others have still forest cover but heavily changed, and the last stage is "šikara", shrubberies and thickets. Big surfaces are pastures without any trees and some parts of the Dinaric Karst are bare rock landscape. Generally speaking the bareness of the Dinaric Karst is lesser in the central parts, and going towards the Adriatic coast, it increased reaching real rocky desert on Adriatic slopes and on the islands.

Fig. 10. Bare high ridge of the Velebit Mountain (photo A. Kranjc).

Fig. 11. Stony pasture on the Pag Island, sea side slopes of the Velebit Mountain in the background (photo A. Kranjc).

While the number of inhabitants increased, economic facilities did not follow the population growth. Data from the Karlovac district of Croatia, which has an extremely great proportion of karst landscape, can be shown as an example. In the middle of the 18th century, there were 940 inhabitants per square mile, while hundred years later, in 1850, there were 1824. This means that the population nearly doubled in a hundred years (Wessely, 1876). It was not a specific of the Karlovac district, in many parts or even in majority of Dinaric Karst the greatest population pressure on karst land was during the 19th century. Regarding the available data the example of Dinaric Karst in Slovenia can be taken into consideration. In

Slovenia as a whole there was a minimum of forest cover around 1875. Forest then covered 37 % of the surface, while in the district of Postojna, which included a great part of the Dinaric Karst in Slovenia, the forest covered 26 % in the year 1880 (Azarov, 1994). According to Gams (1991) the plateau Kras had only 20 % of forest surface in 1900. In 1989 the share of forest increased to 51 % and nowadays the rate of forest still increases, its share being estimated to be over 60 % of Slovenia and over 50 % of the Kras plateau. On Kras, it is mostly the monoculture black pine tree forest. The course of reforestation is now going on by itself; pastures are becoming overgrown by shrubs and being slowly transformed into forests. The surface is slowly overgrown first by tall herbs, then with shrub species, and finally by trees and forest ground flora. The front of pine forest, if not interfered by human, progresses at a maximum of 17 metres per year; computer modelling forecasts that the whole region of Kras will be overgrown by shrubs and trees till 2013 (Pertot, 1989).

On the karst of Croatia, forest cover decreased to an alarming state during the 19th century, too. In this time there was hardly any forest on the coastal side of the Dinaric Mountains. But the records of "Trieste Commercial Commissariat" for 1756 showed millions of trunks (Wessely, 1876). In the second half of the 19th century on the "Mountain Karst" (Fig. 10) of Croatia 39 % of surface was categorized as bare (non-productive) land, the "karst" in narrow sense of meaning, while on the "Sea Side Karst" such category includes 93 %, as shown by the same author (Fig. 11). It is difficult to imagine that nearly the entire country was "unproductive" bare rock landscape. In that time, the meaning of karst was just a "bare limestone landscape". And what was the prospect for the future: "General-Domänen-Inspektor und Forstakademie-Direktor a.D." reckoned that the surface of forest diminishes every year for 1 % in the frame of the Austrian Littoral and Dalmatia.

3. Natural vegetation of the Dinaric Karst

Available data suggest that originally about 90 % of Dinaric Karst were covered by a forest, in some areas even more. Many temperate taxa appear to have survived in the region during the last glacial in low but persistent population. A greater diversity of taxa existed in the mid to high altitude sites probably where the climate was more humid. At the lateglacial/Holocene transition many tree taxa expanded simultaneously. Changes in the composition of the early Holocene woodland included a change in the forest dominants between 8 000 and 7 000 BP, and the appearance and increase of *Carpinus orient./Ostrya, Abies, Carpinus betulus* and *Fagus* between 7 500 and 5 000 BP (Willis, 1994). In the north-western part of Dinaric Karst mixed oak forest (*Quercetum mixtum*) prevailed during the first postglacial warm (Boreal) period. During the Atlantic period fir and beech (*Abieti-fagetum*) had already developed as a climax forest. Dinaric Mountains' flora belongs to both the Mediterranean and the Euro-Siberian-North American phytoregions. In the Mediterranean region the main forest type includes Mediterranean (evergreen) oak (*Quercus ilex*) and Black Dalmatian pine (*Pinus nigra dalmatica*) while Dinaric fir and beech forest (*Abieti-fagetum dinaricum*) prevails in the interior. Black pine (*Pinus nigra*) is indigenous to some small karst areas of Slovenia (Culiberg et al., 1997; Šercelj, 1996) (Fig. 12). Development of the present day landscape started at approximately 4 500 BP with the onset of anthropogenic disturbance. Clearance resulted in increase of open ground herbaceous types with grasses (Willis, 1994).

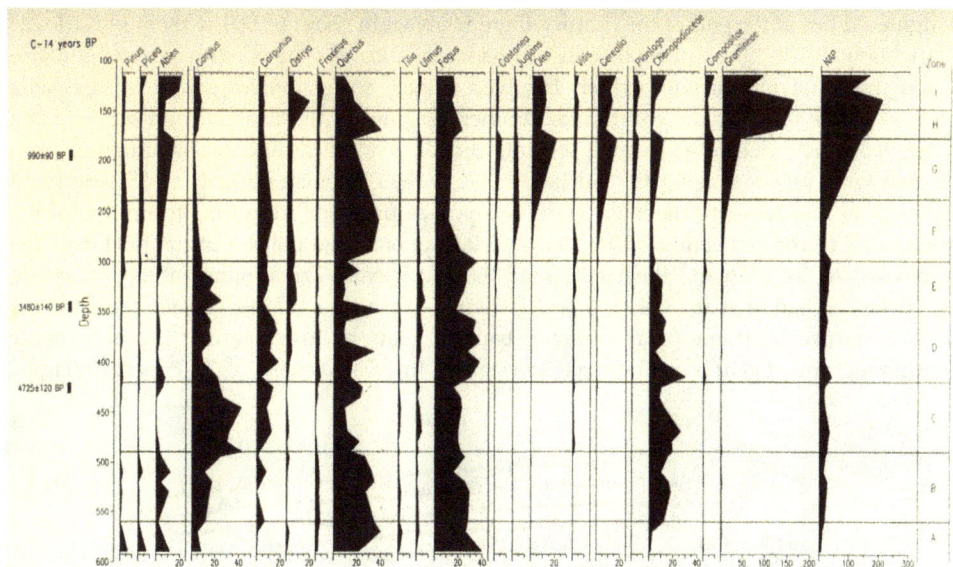

Fig. 12. Pollen diagram, showing great change around the year 1 000 BC as well as around 1 000 AD (Culiberg et al., 1997).

4. Forest protection and reforestation of the Dinaric Karst

As it was seen the main cause of the deforestation of the Dinaric Karst was man. But it does not mean that nobody cared for forests and did not see their importance. Officials, administrators, town councils realized at an early stage already, that deforestation could be a great threat and even an economic catastrophe, and a disaster for the everyday life of people. Many attempts of protecting and safeguarding forests by administrative, economic, penal and other measures are known from history, sometimes very strict. In spite of them deforestation reached alarming proportions.

From the 12th century on various Acts and documents are known, attempting to regulate tree cutting, protection of forest and reforestation. The town of Trieste, an important port at the foot of the plateau Kras introduced such an act in 1150 already. Similar to Trieste, who has no (more) forest in his hinterland, other towns of the Adriatic coast and islands, such as Korčula, Trogir, Dubrovnik, Skradin, Hvar, and Poljica, regulated the exploitation and protection of forest by town statutes enacted between 1214 and 1444. Venice edited a forest act in 1452 (Fig. 13) while for Istria Peninsula, Friuli and Karst the "Waldordnung" (the Forest Act) from 1541 is well known (Gašperšič & Winkler, 1986).

Despite the concern for forests these acts show that at that time the deforestation by cutting, clearing and burning had already seriously started. However in some presently barren places, forests still existed (Horvat, 1957). That the matter was taken seriously shows the example of Trieste, where an armed guard had to be organised by the town to protect the local forest in 1583 (Guttenberg, 1881) (Fig. 14). The said Acts commonly included a ban on goats, or a complete ban on grazing in the forest. In 1764 the edict was issued banning the free pasture of goats in the Military Croatia. Thousands of goats have been sold or slaughtered but at the end of the century there were again about 64 000 of them in Karlovac district only. In 1771 the Court office in Vienna issued a ban on the practice of transhumance across the karst of Carniola, which involved winter sheep grazing in Istria Peninsula lowland along the Adriatic coast, and summer grazing on the Kras and on higher karst plateaus (Nanos, Snežnik). Despite the interdiction a long distance transhumance, from Bosnian karst plateaus along the Adriatic coast to the Slovenian part of the Dinaric Karst and back to Bosnia along the Sava River valley, was practised from time to time until the second half of the last century. The last time it happened the police put the flock and the shepherds on the train and they were send back to Bosnia. The administration repeatedly issued acts on ban of goats: 1844 in Istria, 1870 in Gorizia, and 1874 in Trieste. The Republic of Slovenia (in the frame of Yugoslavia) banned goats by a decree in 1952, and finally Yugoslavia banned goat breeding, except in stables, in 1954 (Wessely, 1976; Papež, 1991).

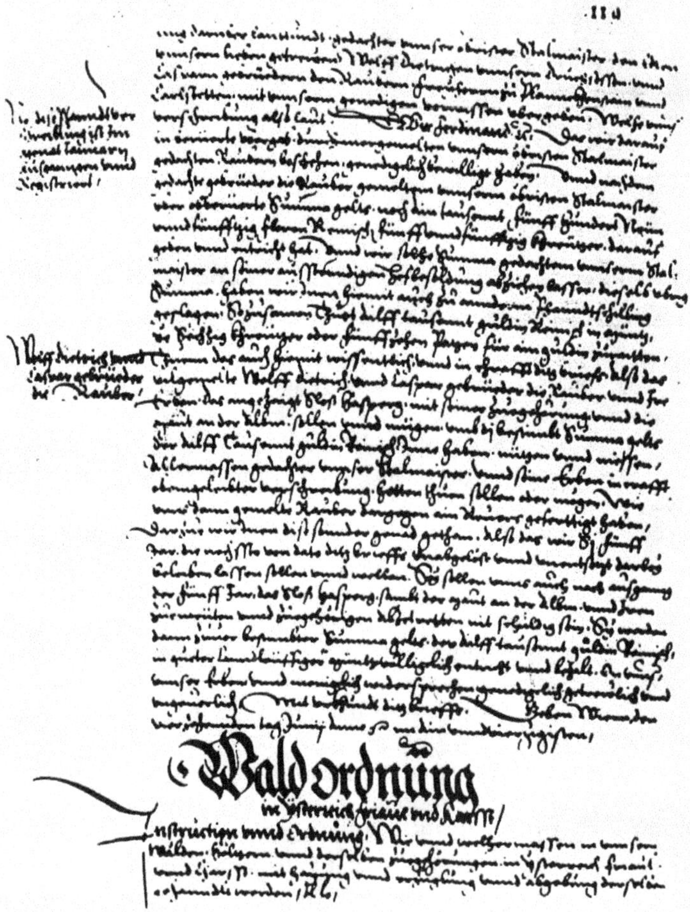

Fig. 13. Venice "Waldordnung" (Forest Act) of 1452 (Gašperšič & Winkler, 1986).

Fig. 14. The first black pine plantations of Kras plateau needed an armed guard (Anonym, 2001).

Even the Turkish threat sometimes had results that were favourable for forest conservation. To maintain the protection zone of dense forest against the advance and passage of irregular Turkish raiding bands, it was forbidden to touch any tree or bush in the forest within a few hours ride of the south-eastern border of Carniola.

But all the administrative measures were of little help and there was less and less forest. Early already some specialists suggested the reforestation. In his book "Hydrographical letters from Carniola" (Kranjska, the Austrian hereditary land) which is in fact a description of Carniolian karst, T. Gruber proposed reforestation as the most effective measure against the wind "burja", which caused quite an important damage and transport obstacle, especially in winter (Gruber, 1781). On the Istria Peninsula and in other parts of Dinaric coast under the Venetian Republic, all the oak forests and all oak trees everywhere were reserved for its shipyards, the Arsenal. The owner of the forest has the right to use only those forest products, trees, which the Arsenal did not need. Sentences were severe, including capital punishment. Of special value were naturally curved trunks of *Quercus pubescens*. In 1815, after the Vienna Congress, former Venetian territories of Istria and northern Adriatic Sea belonged to Austria. Soon after, in 1819, Austria had cancelled the navy oak reserve. The main forest keeper and navy manager of forest in Carniola, J. Ressel, realised that soon there would not be enough timber for navy needs. He proposed another type of navy forest reserve, different from the Venetian one, based upon constraint and punishment. He named his system "buying reserve" – the owner would get paid in advance, for each 10-years addition through growth, before cutting down a tree. Later, in 1842 Ressel proposed "Die Wiederbewalderung der Gemainde Gründe Istriens" (The new reforestation of the commons of Istria) project. He tried to achieve reforestation by planting acorns, stating that the oak to be the best, but it proved unsuccessful (Piškorić, 1993). Later, in 1852 he proposed a similar plan for the part of Kras belonging to the towns of Trieste and Gorizia (Fig. 15).

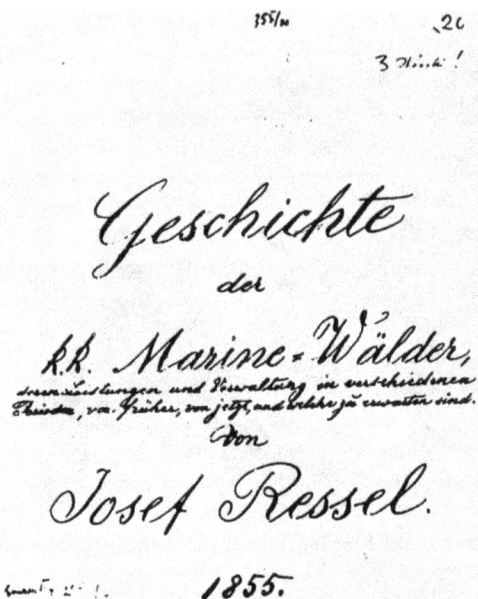

Fig. 15. Josef Ressel (1793-1857) and the first page of his "History of E.&K. Navy Forests" of 1855 (Gašperšič & Winkler, 1986).

Meanwhile, the Ministry of Agriculture began to stimulate and finance reforestation, in response to complaints from karst communities and professionals, from the forestry bodies and J. Ressel especially, about the rigours of life on a barren rock landscape. A detailed study of such a life was made by Wessely (1876). In 1857, the first railway which crossed the Karst was completed, that is the connection of Vienna to Trieste. There were great problems protecting the railroad against the strong northeast wind called Burja in Slovene, Bura in Croatian, and Bora in Italian language, probably meaning or coming from "Borealis". The major obstacle were snowdrifts formed by this wind. The Southern Railways Company had to build high drystone walls along the route of the railway; but they found it cheaper and better to plant trees along it; so they support and even join the reforestation programme. In 1859 the first successful reforestation occurred, using young black pine (*Pinus nigra* var. *austriaca*) seedlings on the plateau of Kras at Bazovica in the vicinity of Trieste, by J. Koller. The first successes boosted confidence and the activity spread to other parts of the karst within and outside of the Austrian lands (Fig. 16).

In 1885 finally the "Reforestation Act for Carniola (=Kranjska)" was issued, regulating the entire process, which included major work and investment as tree nurseries, wall construction, land preparation, and seedlings planting. On the karst terrains there were entire villages where reforestation provided the main or perhaps the only income. Everybody, men, women and children, was involved in these works: men were digging holes; women were planting seedlings, while children were bringing water and watering seedlings (Goll, 1898) (Fig. 17). Publications showing their success were published on different occasions (Goll, 1898). From all over Europe specialists came to Kras to see this successful reforestation, the senator Marquis de Campo and forestry engineer don Carlos de Mazeredo from Spain as an example (Rubbia, 1912).

Fig. 16. The first black pine plantations along the railway towards Trieste (Goll, 1898).

Fig. 17. Managing of black pine plantation above the Trieste in 1905 (Anonym, 2001).

Different activities were taking place in other parts of the Dinaric Karst too, the difference being due to local political and administrative particularities. There are also differences in the accessibility of documents and of publications. For some regions, in the frame of Austria for example, there is quite a lot of published works, technical documentation even, while for the parts which were in the frame of the Ottoman Empire, documentation is maybe more scarce and in any case more difficult to find and more difficult to understand, if written in Turkish language and in Arabic script. Under Hungarian administration Croatia was divided into Military Zone or Military Croatia (along the border with Turkish Empire) and

Provincial (Civil) Croatia. Due to Napoleon's ultimatum, Doge Ludovico Manin surrendered unconditionally on May 12, 1797 and abdicated, while the Major Council declared the end of the Venetian Republic. The Illyrian Provinces were created by the Treaty of Schönbrunn in 1809, and embraced a big part of the Dinaric Karst, beside Carniola the former Venetian territories along the Adriatic coast. Bosnia and Herzegovina remained in the Ottoman Empire until 1878 when it was annexed by Austria. In 1918, when the Kingdom of Serbs, Croats and Slovenes, later the Kingdom of Yugoslavia, was founded, Bosnia and Herzegovina as well as the great majority of the Dinaric Karst were included in it.

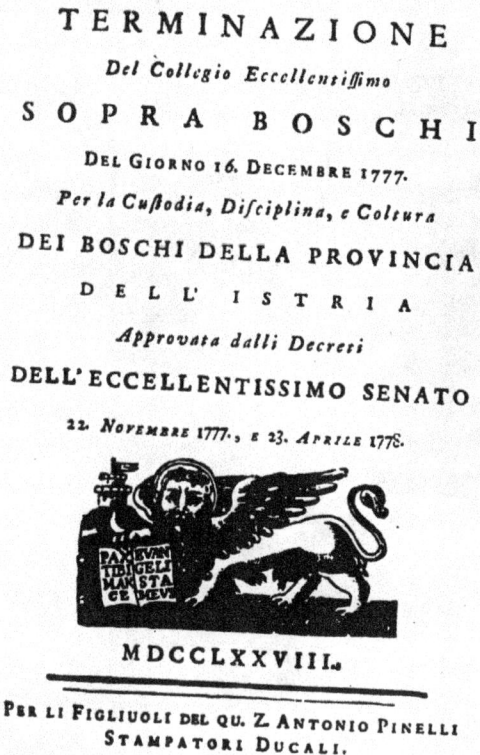

Fig. 18. The title page of the Venetian Forest Act for Istria of 1777 (Gašperšič & Winkler, 1986).

In 1871 when the military organisation of Croatia ended, the Austro-Hungarian Emperor wrote: "… the income of the sale of wood from the country's state's forests has to be used for investing, especially for the reforestation of karst." (Wessely, 1876). This was followed by the 1864 "Waldordnung" (The Forest Act) of the Military Croatia aimed at planting the barren land with a beech and a fir. Interesting illustration is the example of so-called "Laudonov gaj" (General Laudon's Wood) at Krbavsko polje in the Lika region. The western part of the polje was covered by moving sands. The later famous Austrian "Generalfeldwachtmeister" (major-general) G. E. Laudon served there, at Bunić in Karlovac region in 1740s. After becoming the major-general, he ordered to plant the forest there to fix the sands and to prevent damage to agriculture. Under the Military administration about

700 ha was planted with black pine and oak, in the form of a military formation as said. In 1965, it was proclaimed a forest reserve (Jaić, 2011).

Reforestation of the karst pastures in the hinterland of the port of Rijeka began in 1857 (Horvat, 1957). In Dalmatia, they tried to start reforestation under the Venetian regime already, by the so-called Grimani Act of 1756, but due to bad administration and corrupt officials the work did not even start (Fig. 18). During the time of the French Illyrian Provinces (1809-1813) each commune had to plant a "sacred wood" (bois sacré), but the provinces' period was too short to achieve the desired results. Successful reforestation began in the 1880s following the example of Trieste. Yet in the countries of the Ottoman Empire, there are no Acts and no activities for protection of the forest or for reforestation known. From 1918 practically all of the Dinaric Karst was within the borders of the Kingdom of Yugoslavia. Not earlier than in 1929 the State's Act on forest was issued with the essential prescription: "…all deforested lands especially on karst has to be set apart in the period of 10 years with the aim to be afforested as soon as possible. Reforestation has to be realized in 50 years…". Even the state did not last 50 years and such a decree was impossible to implement, so it survived on paper only (Djikić, 1957). Systematic reforestation slowly spread over the entire of the Dinaric Karst and continued into the 1950s. In the socialist Yugoslavia, immediately after the end of the 2nd World War, massive actions of reforestation in the form of "Mladinske delovne brigade" (Youth work brigade), a form of a voluntary youth work were organized in the form of summer camps. On the karst of Slovenia one of the last actions was the reforestation work in 1950s on the Vremščica Mountain between Postojna and Trieste. From these times on, reforestation is mainly the duty of the forestry organization and of the owners of the forests.

Fig. 19. On a karst plateau, a pasture started to be overgrown by trees (photo K. Kranjc).

5. Conclusions

Defforestation, degradation and in some cases desertification even of Dinaric Karst started early in prehistoric times. They reached the peak in the second half of the 18th and in the first half of the 19th century. In any case, the reason was human factor, the economy with no attitude to sustainability at all. During the last thousand years it is possible to see the attempts to prevent the forest or even to meliorate, to reforest degraded lands. By the middle of the 19th century such attempts were mainly unsuccessful, but from that time on the situation started to change rapidly. At the beginning, reforestation was a sort of mass-activity while nowadays other factors join it. The general perception of the importance of a forest and of the sustainability helped a lot, but also the change of economy and activity of the population of the Dinaric Karst, the decline of the agriculture emphasized. Maybe the Dinaric Karst is turning to the other extreme – to be overgrown (Fig. 19). In Slovenia, on the Kras particularly specialists as well as laymen started to ask: how to prevent the Kras from becoming overgrown? "How to reasonably stop the overgrowing of Kras" is the title of a round table organized by the review "Kras" at Nova Gorica in 1997. The discussion also showed that foresters suggested replacing slowly black pine with oak (Mlinšek 1993).

The foresters also suggested that Kras should be a field experimental laboratory of international importance to study the revitalization of a completely degraded landscape. Especially important should be the study of the revitalization of thermophile associations, which are the most affected and at the same time the most suppressed and neglected by the World's public (Mlinšek, 1993).

6. References

Anonym (2001). *Pogozdovanje krasa* (reprint). Avtonomna dežela Furlanija Julijska krajina, Deželno ravnateljstvo za gozdove in parke, Trst

Atkinson, I. (1979). *The Viking Ships.* Cambridge University Press, ISBN 0 521 21951 5, Cambridge – London etc.

Azarov, E. (1994). Črni bor na Krasu. *Kras,* No. 4, (1994), pp. 18-21, ISSN 1318-3257

Cabanes, P. (2002). *Iliri od Bardileja do Gencija.* Svitava, ISBN 420228055, Zagreb

Culiberg, M., Kaligarič, M., Lovrenčak, F., Seliškar, A., Zupančič, M. (1997). Soil & Vegetation. In: *Slovene Classical Karst Kras.* Kranjc, A., pp. 103-129, ZRC SAZU, ISBN 961-6182-42-0, Ljubljana

Cvijić, J. (1966). *Balkansko poluostrvo i južnoslovenske zemlje,* Zavod za izdavanje učbenika SR Srbije, Beograd

Cvijić, J. (2000). Karst, geografska monografija. In: *J. Cvijić Sabrana dela.* Knjiga I, Stevanović, Maletić, Ranković, Kulenović-Grujić, Jovanović, pp. 203-323, SANU & Zavod za izdavanje udžbenika SR Srbije, ISBN 86-17-08301-8, Beograd

Djikić, S. (1957). Historiski razvoj devastacije i degradacije krša u Bosni i Hercegovini. *Proceedings of Savezno savjetovanje o kršu,* 3, Split, June-July 1957

Dular, J., & Božič, D. (1999). Železna doba. In: *Zakladi tisočletij,* Aubelj, B., pp. 98-183, Modrijan, ISBN 961-6183-68-0, Ljubljana

Evliya, Ç. (1957). *Putopis, odlomci o Jugoslovenskim zemljama.* Svjetlost, Sarajevo

Gams, I. (1991). The origin of the term Karst in the time of transition of Karst (Kras) from deforestation to forestation. *Quaderni del Dipartimento di Geografia,* No. 13, (1991), pp. 1-8, ISSN 1120-9682

Gams, I., & Gabrovec, M. (1999). Land use and human impact in the Dinaric Karst. *Int. J. Speleol.*, Vol. 28 B, No. 1/4, 55-70, ISSN 0392-6672

Gašperšič, F., & Winkler, I. (1986). Ponovna ozelenitev in gozdnogospodarsko aktiviranje slovenskega krasa. *Gozdarski vestnik*, No. 5, pp. 169-183, ISSN 0017-2723

Ginés, Á. (1999). Agriculture, grazing and land use changes at the Serra de Tramuntana karstic mountains. *Int. J. Speleol.*, Vol. 28 B, No. 1/4, pp. 5-14, ISSN 0392-6672

Goll, W. (1898). *Die Karstaufforstung in Krain*. Aufforstungs-Commission für Karstgebiet des Herzogthums Krain, Laibach

Gruber, T. (1781). *Briefe hydrographischen und physikalischen Inhalts aus Krain an Ignaz Edler von Born*. Krauß, Wien

Guttenberg, H. v. (1881). Der Karst und seine forstlichen Verhältnisse. *Zeitschr. des Deutschen und oesterreichischen Alpenvereins*, Vol. B XII, pp. 24-62

Hacquet, B. (1785). *Physikalisch-politische Reise aus den Dinarischen durch die Julischen, Carnischen, Rhätitschen in die Norischen Alpen im Jahre 1781 und 1783 unternommen von Hacquet*. I-II, Adam Friedrich Böhme, Leipzig

Horvat, A. (1957). Historiski razvoj devastacije i degradacije krša. *Proceedings of Savezno savjetovanje o kršu*, 2, Split, June-July 1957

Jaić, M. (2010). Od Udbine preko Krbavskog polja do Laudonovog gaja. 30.08.2011, Available from:
http://www.pustolovina.com/index.php?Itemid=55&catid=34:putopisi&id=46:od
-udbine-preko-krbavskog-polja-do-laudonovog-gaja-
&option=com_content&view=article

Kranjc, A. (1979). Kras v povirju Ljubije. *Geografski vestnik*, Vol. 51, (1979), pp. 31-42. ISSN 0350-3895

Kranjc, A. (1999). Reafforestation of *Kras* – improvement or degradation. *Proceedings of the International Seminar on Land Degradation and Desertification*, ISBN 972-98331-0-9, Aveiro, August-September 1998

Kranjc, A. (2002). The History of karst resources exploitation: an example of iron industry in Kranjska (Slovenia). *Theoretical and Applied Karstology*, Vol. 15, (2002), pp. 117-123, ISSN 1012-9308

Kranjc, A. (2011). The Origin and evolution of the term „Karst". *Procedia Social and Behavioral Sciences*, No. 19, (2011), 567-570 (www.sciencedirect.com)

Liburna. (September 2011). In: *Wikipedia*. 01.10.2011, Available from:
http://en.wikipedia.org/wiki/Liburna

Liburnians. (August 2011). In: *Wikipedia*, 22.08.2011, Available from:
http://en.wikipedia.org/wiki/Liburnians#cite_ref-25

Mihevc, A. & Prelovšek, M. (2010). Geographical Position and General Overview, In: *Introduction to the Dinaric Karst*, Mihevc, Prelovšek, Zupan Hajna, pp. 6-8, Inštitut za raziskovanje krasa ZRC SAZU, ISBN 978-961-254-198-9, Postojna

Mlinšek, D. (1993). Življenjski prostor "nizki kras", primer človekove destruktivnosti, energije življenja, upanje v človeka in trajen raziskovalni laboratorij. *Gozdarski vestnik*, Vol. 51, No. 5-6, pp. 280-293, ISSN 0017-2723

Murko, V. (1957). *Josip Ressel: življenje in delo*. Tehniški muzej Slovenije, Ljubljana

Panjek, A. (2006). *Človek, zemlja, kamen in burja*. Založba Annales, ISBN 10961-6033-76-x, Koper

Papež, J. (1991). Kozjereja, In: *Enciklopedija Slovenije*, Ivanič, M., pp. 354, Mladinska knjiga, ISBN-86-11-14288-8, Ljubljana

Pertot, M. (1989). Kraška gmajna se bo kmalu zarasla. *Proteus*, Vol. 52, No.2 (October 1989)., pp. 59-61, ISSN 0033-1805

Piškorić, O. (1993). Josip Ressel u hrvatskom šumarstvu. *Šumarski list*, Vol. CXVII, pp. 489-506, ISSN 0373-1332

Prašuma Lom (August 2011). In: Zvanična prezentacija opštine Petrovac, 22.08.2011, Available from http://www.drinic.rs.ba/lom.htm

Rajšp. V. (Edit.). (1997). *Slovenija na vojaških zemljevidih 1763 – 1787 (1804)*, 3, ZRC SAZU & ARS, ISBN 961-6182-013, Ljubljana

Rajhenavski pragozd (August 2011). 22.08.2011, Available from http://www.zrsvn.si/sl/informacija.asp?id_meta_type=63&id_informacija=491

Rjazancev, A. (1963). Vesti o bobovcih iz Julijskih Alp. *Železar Tehnična priloga*, Vol. 2, (1963), pp. 67-70

Rubbia, K. (1912). *Petindvajset let pogozdovanja na Kranjskem*. Pogozdovalna komisija, Ljubljana

Schmidl, A. (1854). *Die Grotten und Höhlen von Adelsberg, Lueg, Planina und Laas*. W. Braumüller, Wien

Slapšak, B. (1995). Možnosti študija poselitve v arheologiji. *Arheo*, No. 17, (1995), pp. 2-90, ISSN 0351-5958

Šercelj, A. (1996). Začetki in razvoj gozdov v Sloveniji (The Origins and Development of Forests in Slovenia). *Dela-Opera Classis IV. SAZU*, No. 35, pp. 1-142, ISSN 86-7131-102-3

Turk, I., Modrijan, Z., Prus, T., Culiberg, M., Šercelj, A., Perko, V., Dirjec, J., & Pavlin, P. (1993). Podmol pri Kastelcu, Novo večplastno arheološko nahajališče na Krasu, Slovenija. *Arheološki vestnik*, Vol. 44, (1993), pp. 45-96, ISSN 0570-8966

Velušček, A. (1999). Mlajša kamena in bakrena doba. In: *Zakladi tisočletij*, Aubelj, B., pp. 52-75, Modrijan, ISBN 961-6183-68-0, Ljubljana

Valvasor, J. W. (1689). *Die Ehre deß Herzogthums Crain*. I. Th., W. M. Endter, pp. 696. Laibach – Nürnberg

Wessely, J. (1876). *Das Karstgebiet Militär-Kroatiens und seine Rettung, dann die Karstfrage überhaupt – Kras hrvatske krajine i kako da se spasi, za tiem kraško pitanje uploške*. k.k. General-Commando in Agram, Agram

Willis, K.J. (1994). The vegetational history of the Balkans. In: *Science Direct - Quaternary Science Reviews*, Vol. 13, No. 8, (1994), pp. 769-788, 18.08.2011, Available from: http://www.sciencedirect.com/science/article/pii/027737919490104X

Zupan Hajna, N. (2010). Geology, In: *Introduction to the Dinaric Karst*, Mihevc, Prelovšek, Zupan Hajna, pp. 14-19, Inštitut za raziskovanje krasa ZRC SAZU, ISBN 978-961-254-198-9, Postojna

Part 2

Mapping Deforestation

8

Sustainable Forest Management Techniques

K.P. Chethan, Jayaraman Srinivasan, Kumar Kriti and Kaki Sivaji

TCS Innovation Labs Bangalore, Tata Consultancy Services,
India

1. Introduction

Forests being an indispensable resource play an important role in maintaining the earth's ecological balance. The major contributors of deforestation are logging off of trees (legal or illegal), tree theft, forest fire etc. Large scale deforestation has negative impact on the atmosphere resulting in global warming, flash floods, landslides, drought etc. Due to these adverse effects, forest management department all over the countries have taken steps for monitoring the forest to prevent deforestation. Several surveillance techniques have been employed for monitoring and prevention; they are broadly classified as Ground-based sensing techniques and Remote sensing techniques.

Surveillance plays an important role in forest management. It had been used in the past and is still being used for monitoring and information collection. Ground-based techniques generally include surveillance by on-site security staff and mobile patrols monitoring the property by water, land and air (Magrath et al., 2007). Some complementary systems such as Fixed Earth System are also used with observation towers located at strategic points with specialized personnel for observing and detecting the presence of fire. All these systems are expensive and time consuming, requiring a lot of resources.

Nowadays, remote sensing technologies are also used like, aerial photographs, automatic video surveillance, wireless surveillance systems and satellite imagery. Satellite imagery is very costly for use in monitoring any illegal activity like trespassers, tree theft and deforestation (if they are able to detect at all). On the other hand, with the technological advancements in wireless communication, various low power, and low cost, small-sized sensors nodes are available which can be readily deployed to monitor environment over vast areas. Wireless Sensor Networks (WSNs) technology is being used widely for monitoring and controlling applications. Currently three main wireless standards are used namely: WiFi, Bluetooth and ZigBee. Amongst them, ZigBee is the most promising standard owing to its low power consumption and simple networking configuration. Wireless sensor network based surveillance systems for remote deployment and control are more cost effective and are easy to deploy at location of interest. They can even reach those areas where satellite signals are not available. Moreover, they can be configured to monitor large areas and they have secure mode of data transmission.

Environmentally, WSNs finds immense application in land management, agriculture management, lake water quality management, forest fire detection, tree theft prevention and also in the prevention of deforestation. In addition, WSN system has also been used for strain monitoring in railway bridges. (Bischoff et al., 2009), developed an event based strain monitoring WSN system for railway bridge. They used low power MEMS acceleration sensors for detecting an approaching train. Whenever this event was detected, strain gauges were operated and measured data was used for cycle counting based fatigue assessment of steel bridges. This event based detection was developed to manage the power consumption and make the system more energy efficient. Moreover, solar rechargeable battery powered base station was used to increase the system lifetime.

WSN system has also been used for landslide monitoring and prevention. (Rosi et al., 2011) discussed the implementation and deployment details of a WSN system for landslide monitoring in Northern Italy Apennines. Six Micaz nodes having a 2 axis accelerometer for sampling vibrations were used. These vibrations were a result of slope movements caused by landslides. The data measured by the sensor nodes were routed to the brides and finally sent to the base station following a predefined static routing table. These examples give a fairly good idea of the amount of work and research going on in WSN area making them the most promising technology to use for monitoring and control purpose in diversified fields.

A lot of research has been done using WSN for forest monitoring either for fire prevention or for monitoring the illegal logging activity. Some researchers have proposed algorithms for detection and prevention and have simulation results verifying their control. On the other hand, some have come up with the design, implementation and deployment of the system. The work on forest monitoring is not limited to fire and deforestation detection and prevention but also includes preserving and conserving the flora and fauna of the forests. A brief summary of the work done in this domain is given below.

(Awang & Suhaimi, 2007), developed a WSN based forest monitoring system called RIMBAMON. This system consisted of sensor nodes deployed in the forest at specific distances for capturing temperature, light intensity, acoustic, acceleration and magnetic readings. MICA2 Mote was used for implementation for its long range in ISM band. These sensors were either mounted on the trunk of the tree at the ground level or kept along the roadside. The sensors used helped in monitoring any illegal logging activity in addition to detection of fire in the forest. Temperature and light intensity sensors gave an indication of both logging activity as well as presence of fire. On the other hand, acoustic sensors gave more information on logging activity alone owing to the abnormal sound associated with the usage of machinery, tractor or chainsaw. The system was simulated and tested well to capture and transmit data to the base station. It displayed the acquired data in form of graphs, tables and maps to help in taking prompt action. However, the system lacked remote monitoring through the web, which could be useful in monitoring hostile areas.

(Harvanova et. al., 2011) proposed a Zigbee based WSN system for detection of wood logging using real time analysis of sounds from surroundings. The WSN system periodically acquires sound samples, processes it and transmits it to the central server. Tools which are vastly used for deforestation are chainsaw. There is a characteristic sound associated with a logging activity. Whenever, the sound samples acquired from the sensors matches the sound samples of logging tools, a logging activity is detected and the responsible personnel is notified through an e-mail or a SMS alert.

(Soisoonthorn & Rujipattanapong, 2007), also studied the unique acoustic characteristics of the chainsaw and used it for detecting the activity of chainsawing leading to deforestation. The algorithm was based on a limited energy sensor node and combined three techniques which included adaptive energy threshold, delta pitch detection and energy band ratio in high frequency range. Since the energy characteristic of chainsaw is quite constant, state machine used was further simplified for detection purpose. They could achieve the detection accuracy of 90.8% with this method.

(Figueiredo et. al., 2009), studied the communication performance of WSN for preserving and conserving the flora and fauna of rainforests. A set of experiments were carried out to assess how data communication is affected by environmental parameters like, forest density, humidity and extreme temperature variations. It was concluded that communication range of a WSN system deployed in a dense forest gets reduced by 78% as compared to deployment in any other environment.

A lot of study has been done on early fire detection and a number of techniques and sensor combinations have been investigated. Techniques include remote sensing techniques as well as event detection for wireless sensor networks. (Bahrepour et. al., 2008) presented a survey on automatic fire detection from a wireless sensor network perspective. The survey included fire detection techniques for residential areas; for forests and contribution of wireless sensor networks in early fire detection. Since the sensors used for detection were prone to noise, multiple sensors were used to reduce the false alarms generated in case of single sensor usage. Usually temperature sensors are combined with gas concentration sensors for better fire detection. In this study, it was concluded that in residential areas, ION detectors are more beneficial for flaming fire detection. On the other hand, photo detectors are more beneficial for non-flaming fire detection. Fire Weather Index (FWI), which resulted from several years of forestry research, can be used as promising factor for forest fire detection.

(Lozano & Rodriguez, 2010), designed a WSN based system which monitored temperature and humidity for early detection of forest fires. Weather conditions especially temperature, humidity and rainfall determines the degree and speed by which fire spreads in the forest. The correlation between the various weather elements and flammability of the waste of branches and trees helps in predicting the risk of fire at any given location. Mesh topology was used to configure the communication network and temperature and humidity sensors were used to gather the data from the remote location. Through simulation it was shown that the system was capable of detecting fire at an early stage thereby, protecting the nature reserves.

(Zoltan Kovacs et. al., 2010), presented a case study of a simple, low cost WSN system implementation for forest fire monitoring. Smoke detectors and temperature sensors were used to detect forest fire. A simple star topology was used to cut down on the computation and power consumption. (Zhang et. al., 2009), proposed a Zigbee based WSN system for forest fire detection in real-time so that decision to prevent or extinguish fire can be taken in real-time. The sensor nodes comprised of humidity, temperature, wind speed, wind direction, smoke, pressure and other fire monitoring sensors. The data collected by them was sent to the cluster head which was responsible for data aggregation and transmission. Network co-ordinators were responsible for network building, access control and other network management functions. The data was transmitted to the routers which established local databases and sent the data to the host for monitoring purpose over the internet. Some important factors related to ad hoc network technology, forest-fire forecasting model and determination of effective communication distance was discussed.

(Wang et al., 2010), proposed a new wireless network implementation for forest fire monitoring based on Zigbee and GPRS technology. This work is quite similar to the one adopted by us in monitoring illegal logging of trees in the forest. This system was capable of transmitting the data collected by the wireless network to FTP server through GPRS so that real-time data can be made available to the experts to help in decision making. The hardware schemes and program flows of the system were given. (Fonte et. al., 2007), designed a low cost system-on-chip microwave radiometer on silicon for remote sensing of temperature to find application in fire prevention. A detailed system analysis was carried out by means of simulations to study its feasibility in civil and environment safeguard applications.

(Gil et. al., 2010), came up with a fire monitoring device which provides visualization services after gathering GPS and sensor data from the micro-system (quad rotor). The sensors used were CO2, humidity, fume and temperature. The data was wirelessly transmitted and displayed on the map using open map API to give information where the fire broke out. (Hefeeda & Bagheri, 2007), proposed a WSN system for forest fire modelling and early detection. Forest fire was modelled according to Fire Weather Index (FWI) system which is considered as one of the most comprehensive forest danger rating systems in North America. A k-node coverage problem in WSN for forest fire detection was studied and a approximation algorithm was proposed which had better convergence, promised optimal number of sensor usage and doubled the network lifetime than other existing algorithms.

Some problems on optimization related to sensor nodes deployment were also explored. (Al-Turjman et al., 2009), studied the various design factors important for WSN system deployment especially in harsh environment like coverage, connectivity and lifetime. They explored the problem of placement of the relay nodes in 3D forest space. They formulated an optimization problem which focuses on maximizing the network connectivity with a limit on the number of relay nodes used. They came up with a threshold on a minimum number of relay nodes used for desired connectivity in harsh environment.

Apart from using WSN for this application, some researchers have explored other technology also. (Luming et al., 2008), studied and came up with a new technology which combined the advantages of video monitor and GIS systems for fire prevention. These two techniques complemented each other well and helped in increasing the accuracy of fire detection and hence prevention, reducing the false alarm. Also, synchronous tracking of video monitored areas in GIS of forest resources helped in getting more accurate information of the land form of the affected area.

Monitoring deforestation is a very complicated process. It becomes even more complicated with the increasing need of resources. Our work addresses the issue of deforestation detection and prevention using an event based WSN system. The design and implementation details of the sensor nodes are given. Mesh routing algorithm is used here for routing data packets to the sink whenever an event is detected.

Following the brief introduction to the problem being addressed in this chapter, the other sections are organized as follows: Section 2 discusses the design concept of a WSN setup for monitoring large space like forests. This includes the advantages and challenges encountered in deployment of WSN for such an application. Followed by this, Section 3 gives a detailed description of the WSN prototype developed which finds application in the detection of tree theft, forest fire and deforestation. Section 4 discusses the challenges faced in the deployment of the proposed prototype. The power requirement of the sensor nodes is handled by the power management unit which has a provision of harvesting energy from the surrounding to increase the network deployment lifetime. The various 'energy harvesting' techniques which can be used for recharging the sensor nodes are discussed in

Section 5. Finally, we have Section 6 giving the summary and important conclusions of the work discussed.

2. Design concept of WSN system for forest monitoring

Designing, deploying, and evaluating a novel wireless systems at a large scale requires substantial effort. One of the major applications of wireless sensor networks is in event detection. Here, a sensor network is monitoring certain phenomenon and the respective communication node needs to get triggered on occurrence of a certain event. Subsequently the event needs to be reported to the sink node as quickly as possible. The communication nodes can be sleeping for most of the time to conserve power since most of the events are rare in nature. But there must be a mechanism to wake them up for quick event transmission through appropriate synchronization. Some of the prominent applications of this category are detection of fire, intrusion, earth quake, landslide, theft of assets and other surveillance applications. However, it is still a great challenge to design a wireless sensor network system for rare event detection; where network lifetime and robustness is a major concern. Some of the recent developments include campus-wide and community-wide wireless mesh networks (Bicket *et al.*, 2005; UCSD Active Campus; Camp *et al.*, 2006), and real-world sensor network deployments in environments as diverse as forests, active volcanoes, and bridges. WSN system design for forest monitoring involves:

- Sensors
- Design of low power wireless communication module
- Simulation and implementation of energy efficient protocol
- Deployment strategies
- Middleware

2.1 Sensors

One of the main goals of sensor network is to provide accurate information about a sensing field for an extended period of time. This requires collecting measurements from as many sensors as possible to have a better view of the sensor surroundings. However, due to energy limitations and to prolong the network lifetime, the number of active sensors should be kept to a minimum. To resolve this conflict of interest, sensor selection schemes are used. The sensor selection problem can be defined as follows: Given a set of sensors $S = \{S_1 \ldots S_n\}$, we need to determine the "best subset" $S\_$ of k sensors to satisfy the requirements of one or multiple missions. The "best subset" is one which achieves the required accuracy of information with respect to a task while meeting the energy constraints of the sensors. So, we have two conflicting goals: (1) to collect information of high accuracy and (2) to lower the cost of operation. This trade-off is usually modelled using the notions of utility and cost:

Utility: accuracy of the gathered information and its usefulness to a mission.

Cost: These consist mainly of energy expended activating and operating the sensors which is directly proportional to number of selected sensors k. Another cost factor that can be considered is the risk of detecting active sensors especially if wireless communication is used. Table 1 shows the selection of sensors for different hazards

Hazard	Application	Sensor Application
Earthquake/wind	Observation	Acceleration
	Experiment	Acceleration strain
	Structural control	Acceleration
	Health monitoring	Acceleration/strain/Displacement
	Damage detection	Acceleration/strain/Displacement
Fire	Fire detection	Temperature/Smoke/Acoustic
	Gas leak detection	Olfactory
	Alarm ,warning	Sounder
	Evacuation control	Temperature/Smoke/Acoustic/Light
Crime	Surveillance	Acceleration/Light/Acoustic/Camera
	Security alert	Sounder

Table 1. Different Sensors used for Sensing Different Hazards

2.2 Design of low power wireless communication module

Power Management is the major challenge in wireless sensor network design. Sensor nodes of the WSNs are battery powered due to their nature of application and deployment requirement. However, batteries life time is limited life which affects the performance of the WSN and it needs replacement from time to time. To overcome this issue, lifetime of the battery can be extended by adopting the following approaches:

1. Design of low power sensor nodes
2. Energy harvesting

Many applications require periodic monitoring rather than continuous monitoring of elements of interest. For such applications, the system need not be in awake state (high power consumption) all the time; instead it can be in sleep state (low power consumption) till it is required to monitor the elements of interest. This can lead to considerable reduction in power consumption. Many low power chipsets are now available which can be configured for such an application.

Additionally, harvesting energy from the surrounding can play a significant role in improving the self sustainability of the WSN system. Once WSN is deployed it is expected to work continuously and autonomously with minimum or no human intervention. Therefore there is a need for sensor nodes to be self sufficient in terms of energy consumption. Energy harvesting can be performed from sources like solar, vibration, RF etc for recharging the sensor nodes batteries, thereby increasing their lifetime. The different energy harvesting techniques which can be employed depends on the location of WSN deployment. Further, the different techniques and their implementation is discussed in Section 5.

2.3 Protocol section and simulation

In WSN, most sensor networks are application specific and have different requirements. On the other hand the sensor nodes have a limited transmission range, processing and storage capabilities, energy resources as well. The routing protocols for wireless sensor networks are responsible for maintaining the routes in the network and have to ensure reliable multi-hop communication under these conditions. In consequence, all or part of the above mentioned design objectives need to be considered in the design of sensor network protocol. (Singh et al., 2010) provided a survey on challenges involved in the design of protocols for WSN. Below is the list of requirements to be considered in order to design and develop a good quality application protocol for WSN.

- Small node size
- Low node cost
- Low power consumption
- Scalability
- Reliability
- Self-configurability
- Adaptability
- Channel utilization
- Fault tolerance
- Security
- QoS support
- Sensor locations
- Limited hardware resources
- Massive and random node deployment
- Network characteristics and unreliable environment
- Data aggregation
- Diverse sensing application requirements

2.4 Middle ware

As WSN vision evolves, multiple paradigms co-exist as single, multiple and internet-scale sensor networks. A diversity of approaches has been proposed to deal with the multitude of WSN application requirements. Current systems do not address most of these requirements adequately; especially the aspects like support for security, trust, transparency, mobility, and heterogeneity.

Middleware for sensor networks is an emerging and very promising research area. Most of the reported works on sensor middleware are at an early stage, focusing on developing algorithms and components for data aggregation, self organization, , network service discovery, routing, synchronization, optimization etc to build higher level of service structures. They often lack attention for integrating these algorithms and components into a generic middleware architecture, and for helping application developers to compose a system that exactly matches their requirements. There are still few widely accepted software standards for middleware. SensorML (Sensor Model Language) for service discovery and Global Sensor Networks (GSN) (Middleware for Sensor Networks) or integrating virtual sensors have the potential for adoption. We also see the relevance of context aware computing technologies (Ontologies and expert systems) in creating a semantic layer for WSN applications. However, in this chapter the implementation was performed in the hardware without any separate OS.

Microsoft has a web based visualization service called Sense Map (SensorMap) which can be used to host and share sensor data. Google Maps, Google Earth, Virtual Earth, are also provided as interfaces to the final user. In addition to computers, PDA (Personal Digital Assistant) and mobile phones can be used to monitor the data and subscribe to alert services. Also it will be interesting to see how virtual reality environments can provide effective visualization environment for WSN applications. WSN characteristics require a specific approach for middleware development that goes beyond dealing with resource constraints. It involves an end-to-end approach that handles the WSN as a whole rather than a group of individual nodes. This implies considerable consequences for typical middleware services such as mobility, coordination, service discovery, security, data aggregation, quality of service, handling hardware heterogeneity, handling communication errors, scalability, and network organization. Good architectures are needed to integrate new sensor services to integrate safely with legacy systems; enhanced programming models, event propagation models, and data models to accommodate the requirements of sensor applications and services; and inventive design. There is no single middleware existing till date which addresses all these requirements.

2.5 Deployment

In general, deployment establishes an association of sensor nodes with objects, creatures, or places in order to augment them with information-processing capabilities. Deployment can be as diverse as establishing one-to-one relationships by attaching sensor nodes to specific items to be monitored (Przydatek et al., 2003), covering an area with locomotive sensor nodes (Bulusu et al., 2004), or throwing nodes from an aircraft into an area of interest (Karlof & Wagner, 2003). Due to their large number, nodes have to operate unattended after deployment. Once a sufficient number of nodes have been deployed, the sensor network can be used to fulfil its task. This task can be issued by an external entity connected to the sensor network.

3. Proof of Concept (PoC) for forest monitoring

The main objective of our system is to monitor tree theft/fire in the forest and alert using an event based wireless sensor network. In forest monitoring application, events like tree theft, fire etc. occur rarely, so in our implementation, the communication nodes are kept in sleep mode (until any event is detected) to cut down on the power consumption. Whenever an event is detected by sensor, it triggers the communication nodes. Subsequently, the event is reported to the sink node as quickly as possible and an alert is generated. In addition to event based monitoring, our system incorporates energy harvesting technique to power the sensor nodes and for carrying out other power management related tasks. The detailed description of the design and implementation of the proposed system is given below.

3.1 Communication node design

This section describes the sensor node architecture which has been designed and developed for the low power application stated above. Fig. 1 shows the functional block diagram of the sensor node developed for forest monitoring. The sensor node architecture mainly consists of a SOC (system-on-chip) for data collection, processing, networking and controlling; sensors for event detection; power supply for meeting the power requirements of the sensor node and RF energy harvester system for harvesting energy from RF. The hardware details of the various components of the sensor node are described below.

3.1.1 SOC (System-on-Chip)

MC13213 system on chip (SOC) from Freescale Semiconductor has been used for this implementation. The interrupt which is generated by the Key Board Interrupt (KBI) has been used to wake up the controller from the sleep mode. Whenever SOC receives an interrupt to KBI from the tilt/temperature sensor, it sends low signal on the corresponding KBI and wakes up the controller which starts its service from the corresponding service routine.

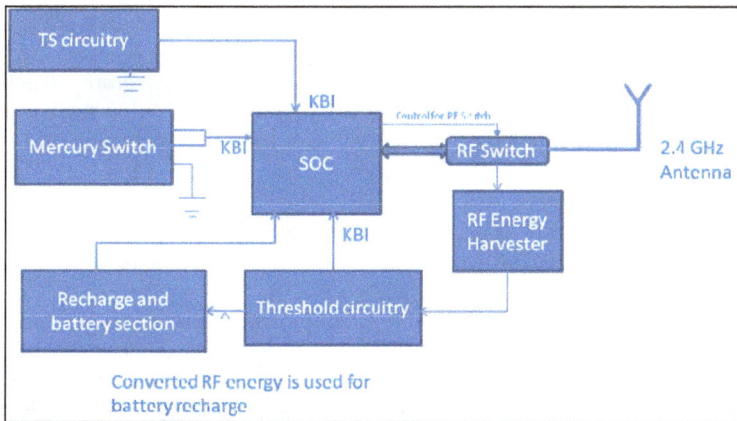

Fig. 1. Functional Block Diagram of the Sensor Node developed for Forest Monitoring

On the other hand, when the neighbor nodes want to send the data, it first sends the RF signal and then sends the data. Received RF signal will interrupt the node and energy storage will occur based on signal threshold. When RF signal is within the predefined thresholds, threshold circuit sends an active low signal to KBI. KBI wake up the controller and start the corresponding service routine to listen its neighbour. If the received RF energy is out of these two thresholds, it activates the energy harvest mode.

Energy of RF signal $=V_f$

if $TH_1 < V_f < TH_2$ –KBI interface

Else enable the RF harvesting

3.1.2 Sensor section

In our system, two sensors have been incorporated for detecting tree theft/fire. Mercury switch sensor (also known as a mercury tilt switch) has been used to detect tree fall/cut and simple thermistor based temperature sensor has been used to detect sudden high ambient temperature which is a probable cause of fire. The tree fall/cut is detected when the mercury switch is tilted in the appropriate direction, which causes mercury to flow and touch the contacts, thereby completing the electrical circuit. Tilting the switch in the opposite direction causes the mercury to move away from the set of contacts, thereby resulting in open circuit. This switch is interfaced to the micro-controller (SOC) to generate the interrupt whenever a tilt event occurs. When the tree bends or tilts more than 50°, a contact is made between the device ground and the KBI pin, resulting in fall event detection which wakes up the node for alert transmission. Fire is detected with the help of a simple thermistor sensor with circuitry to trigger the micro-controller (SOC) via KBI pin. Whenever the sensed ambient temperature goes beyond a prefixed threshold value (due fire catch etc.), this circuitry raises the voltage at the KBI pin of micro-controller which in turn takes care of sending an alert to the destination.

3.1.3 Power section

This section deals with the power supply requirements of the various components of the sensor node namely, sensors, SOC and threshold circuitry. The sensor node is powered using rechargeable alkaline battery (two 1.5V cells in series). Since our system is an event based system, the system spends most of the time in sleep mode (consuming uAmps of current), so the current consumption of this system is very low. Therefore, once the battery is fully charged the system runs for minimum six months. In addition to that, RF energy harvesting technique is implemented in this system to recharge battery, which further increases the sensor network lifetime. RF energy is extracted from RF transmitters/senders specially designed for this purpose. Typically, RF senders are required for every 50 nodes, whose job is to transmit the RF signal with high power so that the nodes which receive the signal with high power will recharge their battery. Such RF sender is required to run for one or two days once in six months.

RF harvesting technology can be used for multiple frequencies and can generate standard or custom output voltages. Batteries or any other energy storage devices can be recharged easily either in close proximity or remotely. In addition, some low power devices can be directly driven from the received RF power. In our application, harvester system for RF consists of Power harvester P1100 module from Powercast (Power harvester P1100 Module Datasheet), which converts received RF energy into DC power with high efficiency. With the help of a threshold circuit, this chip is used for two purpose here, one for waking up the sensor node and other for recharging the battery. When any neighbour node wishes to transmit data, it first sends the RF energy for approx.10 seconds. This is received by surrounding nodes and goes through P1100 module to get converted into DC power. The converted DC power is fed to the threshold circuit which decides whether the RF signal received is to wake up the controller or to recharge the battery. The decision depends entirely on the power available at the output of P1100 module and the threshold limits adopted. Depending on the decision, a KBI interrupt is generated if it is required to wake up the neighbouring nodes for data transmission.

3.2 Network design architecture

One of the important aspects of the network design is the communication of data among the sensor nodes. Therefore, it is highly necessary to design efficient routing algorithm considering multiple constraints that is inevitable in wireless environment. There are two types of routing possible based on the functionality of the nodes in the network namely, flat routing and hierarchical routing. In hierarchical routing, the whole network is divided into multiple hierarchies. Each node has different functionality with respect to the level of hierarchy. Zigbee routing is one such hierarchical routing protocol where the nodes are organized in a hierarchical manner. However, in flat routing protocol, also known as mesh routing protocol, all the nodes in the network are organized in the same hierarchy i.e. all nodes in the network have the same hardware and functional properties. Directed diffusion is one example of this type of routing. In this study, we propose a mesh routing protocol system suitable for monitoring the environment, wherein the nodes in the network are in the same hierarchy. The proposed protocol for this study is an event based protocol, where the nodes generate data corresponding to the event occurred and communicates it to the sink. The protocol has two phases: Configuration Phase and Routing Phase.

Configuration Phase: In this phase, the whole network is in wake up state. The sink node sends the CONFIG packet which traverses through the network in multiple hops. All the nodes receive the CONFIG packet and construct the routing table. Routing table is used to route the data packets towards the sink node. Once the network is configured, the sink broadcasts the sleep packet throughout the network and all the nodes go to sleep mode.

Routing Phase: In this phase, an event detected in the vicinity of the sensor, wakes up the sensor node and records the sensed data at the node. This node then transmits a beacon signal for a pre-defined period of time to wake up the sensor nodes within its range (neighbours). The node selects the upstream node (towards the sink) using routing table to forward the data packet. A similar procedure is carried out on the selected upstream nodes to route the data packets to the sink node. The nodes enter back into sleep mode after a specified period of time.

3.3 Visualization and monitoring

A tool for monitoring and visualization of the forest related events was developed. It supported major features required for middleware for many practical applications involving sensor data collection, visualization and monitoring. The developed framework had two flavours. The first one is a PC based standalone tool, named as Wi-SenseScape developed in Java. It has a graphical user interface (GUI) supporting commands for network visualization including topology, node parameters, and sensor data. The GUI supports the specification of background and foreground image files to be displayed to adapt to various application scenarios. Events can be defined by specifying a mathematical expression based on the sensor parameters. This expression will be evaluated in real-time whenever the dependent parameter changes. Once an event is detected, there is a facility in this tool to link an action to a specific event based on user's discretion.

The second one is a web based visualization system, named as Web-SenseScape (shown in Fig. 2), which integrates Google maps and other map sources as geographical reference layers. Markers are used for identifying the sensor node deployment which displays current details on mouse click. The WSN is represented in XML in a hierarchical structure. The structure is Network -> Clusters-> Nodes-> Sensors. A text based structure view is incorporated to enable the expandable tree visualization. The desired Sensor-ID could be selected to view the sensor data in text format or as a time series plot.

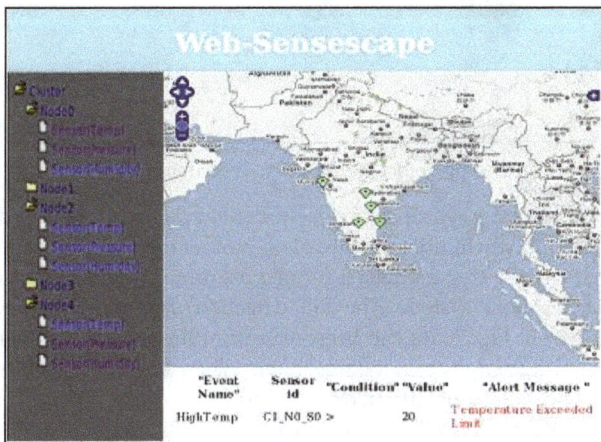

Fig. 2. Web based WSN Monitoring and Visualization

Specific to forest monitoring, this can be loaded with pre-defined blue-print of the forest which is to be monitored. An alert window provides real-time display of events that got triggered based on the events defined. Appropriate audio-visual messaging or alarms can be invoked on the occurrence of specific events. For example, whenever tree falls it shows which tree has fallen by highlighting the tree and proper audio message to take necessary action. This framework is being extended to incorporate visualization of multiple data types and sources including cameras, microphones, GPS (Global Positioning System), medical sensors etc. represented with SensorML. Network management functionality is also being incorporated to enable interactions with the sensor nodes deployed in the field. The middleware architecture has been developed keeping in view of the flexibility, scalability and portability required for supporting multiple networking standards, applications and platforms. The next section demonstrates the application in the deployment scenario.

3.4 Deployment of WSN for forest monitoring

The typical setup of the forest monitoring application is shown in Fig. 3. Few wireless nodes (say N1 to N10) which are responsible for sensing the desired physical entities and communicating this information are deployed in the forest. The information gathered by the nodes is transferred to their upstream routers. One Base Station (BS) is used which gathers the information from the routers. This BS is in turn connected to host system through wireless connection which finally processes the information received from the BS and takes appropriate decision.

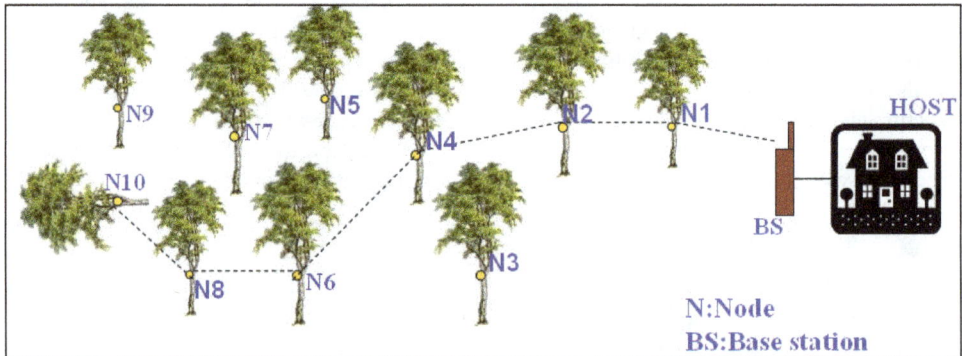

Fig. 3. Demonstration Scenario of detecting a Fallen Tree using WSN

In our application, the wireless sensor nodes are mounted on each tree in order to protect the tree from critical events like theft, fire etc. Wireless T-mote is the name given to each node. Few T-motes were attached to the tree models. It formed a simple dynamic tree network topology to route packets to the host system. The fall detection approach adopted by us can be explained with the help of Fig. 3. As can been seen from the figure, fall event is detected on N10 which may be generated in case of tree theft. When the tree is falling the tilt sensor generates the event and wakes up the node. This node transmits a beacon signal for a pre-defined period of time to wake up the sensor nodes within its range. A dynamic routing path created in the routing phase is used to send this message to the base station which is connected to the host system. In this example, routing path created via Node 8, 6, 4, 2, 1 is used for alert sending. Host system running the Wi-Sensescape application, provides services like: sensor data visualization and analysis, map of the network topology, alarm services, network analysis and filtration of sensor data. This helps in studying and analyzing the activities and behaviour of the WSN deployed for monitoring purpose.

3.5 Results and discussions

This section discusses the important results and observations of the proposed system. In this work we have considered three scenario to calculate the power consumption of the sensor node, they are: (a)when there is no event, (b)there is event from the sensor (tilt sensor/temperature) and (c) there is event from the neighbor to forward the data. After the event occurs, it wakes up the SOC and starts the operation. The SOC's current consumption is typically around 50mA, it continues till the communication is completed (generally it takes 3 to 4 seconds to complete its operation) and then it goes to sleep mode. Therefore, once the event occurs, the battery life is reduced by ~6 hours from the overall life of the battery. Table 2 shows the power consumed by the mote when there is no event.

Components	Mode of Operation	Average Current Drawn
Threshold circuit	Active	$100\mu A$
SOC	Sleep	$10\mu A$
Leakage	-	$15\mu A$
Total current required by the sensor node		$125\mu A$
Power required = 3V * Total current		$\sim 375\mu W$

Table 2. Power Calculation with No Event

$$Total\ hours = \frac{Battery\ Capacitor(BC)}{Battery\ Drain\ of\ Usage} = \frac{2000mA}{125\mu A} = 16000$$

$$Total\ days = \frac{Total\ hours(TH)}{Hours\ of\ Usage\ per\ day} = \frac{16000}{24} \cong 2 years$$

Routing Protocol: The simulation of the routing protocol has been performed in NS-2 ver 2.32 platform. We choose the amount of delay to analyze the performance of the protocol. Average Delay measures the average one-way latency observed between transmitting information from the source and being received by the sink. We study this metric as the function of the sensor network size. We generated variety of sensor network scenarios with different network sizes to study the performance of the routing protocol as a function of network size. In every experiment we study the performance with 10 different sensor network scenarios with the network size ranging from 50 to 150 nodes with the increment of 25 nodes. We perform the simulation by keeping the network density constant of about 100 nodes/m² throughout the experiment. We chose the transmission range of the nodes to be 10m. The other sensor network fields are generated randomly thereby keeping the node density constant. We used the default 802.15.4 MAC and physical layer stack provided by the NS-2. We carried out the experiment by making a single sink and single source node participate in the event generation. Finally, we averaged the results over 10 different generated sensor network scenarios. Fig. 4 shows the graph of average delay for the simulation scenarios discussed above. The simulation was carried out for two different cases where in case 1, the network is in wake-up mode throughout the simulation period. In this case, it is not necessary to use the RF signal to wake up the node. In case 2, the network is in sleep mode, where an event wakes up the sensor node and eventually intermediate nodes are woken up by the RF signal. We simulated this case by transmitting the RF signal continuously for a period of 1sec to wake up the nodes. As observed in case 1, the data packet reaches the sink node within negligible period as there is no overhead of waking up the nodes. But in case 2, as the number of hops to reach the sink increases, it takes a longer duration for a data packet to reach to sink as it includes the delay of wakening the nodes. Therefore, we concluded that the average delay is purely proportional to the number of hops, which in turn is dependent on the distance between the sink and event generating source node. The major advantage of using this protocol in relaxed latency constrained application is the amount of energy saved since the network is in sleep mode till an event occurs.

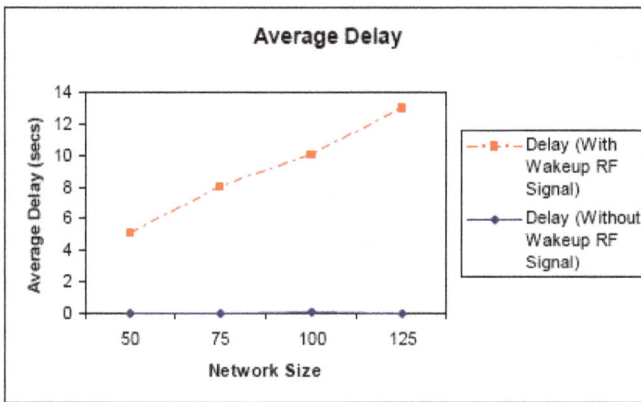

Fig. 4. Latency Time v/s Network Size

4. Deployment challenges

Wireless Sensor Network is a very promising technology for monitoring and controlling large, remote and hostile areas. They are finding extensive application in the fields of home, industry, healthcare, agriculture and environment. Some of the reasons of its popularity are discussed below. The sensor nodes can acquire and analyze the measured data collected over vast distributed areas using multi-hop communication at lower wiring cost. Moreover, they can be deployed very quickly and easily without requiring any pre-existing infrastructure. They can even be integrated with existing external instruments in hostile areas to help collect, analyze and transmit data to the base station for control action. Lot of research is also going on in making the sensor nodes more energy efficient thereby increasing the lifetime of network deployment.

Although WSN is very useful technology for precise monitoring of large, remote and hostile areas, it suffers from some disadvantages. Researchers have now found the difference between the predicted and observed behaviour of the wireless sensor networks after deployment in the field. Some of the constraints and challenges involved in designing a WSN based system for any application are:

- Optimizing the size, power, cost and their associated tradeoffs
- Selection of network protocols that account for key realities in wireless communication
- Selection of real-time routing protocol (e.g RAP protocol, SPEED protocol)
- Selection of sensor network with limited processor bandwidth and less memory
- Improvement of communication range (Wark *et al.*, 2008)
- Design of sensor network with low power consumption for long term deployment. Much of the current research focuses on how to provide full or partial sensing coverage in the context of energy conservation. In such an approach, nodes are put into a dormant state as long as their neighbours can provide sensing coverage for them. In addition, attention has been drawn towards event based systems which account for power consumption.
- Selection of optimum OS for middleware is a critical aspect of WSN.
- Design for security: Sensor nodes are often deployed in accessible areas, presenting a risk of physical attacks. The key challenges are establishment, secrecy and authentication, privacy, robustness to denial-of-service attacks, secure routing, and node capture.

5. Energy harvesting techniques

WSN nodes being battery powered are designed to be energy efficient so as to maximize the lifetime of deployment. Apart from the hardware design, substantial research has been done on designing energy-efficient networking protocols to maximize the lifetime of WSNs (Seah et. al., 2009, 2010). One of the major problems faced by WSNs is the lack of a reliable energy source. Most of the WSNs which are deployed are (primary) battery powered and they need replacement from time to time. Battery replacement might not be practically possible in many situations especially where the sensor nodes are embedded in structures and need to be installed for long duration so, there is a need of using rechargeable batteries which can be charged from time-to-time thus, continuously delivering power to the sensor nodes. The batteries can be charged from the mains but that would incur additional cost of wiring and cabling for each sensor node which might not be practically feasible. Therefore, different energy harvesting techniques are adopted to charge the batteries and make the nodes self sustainable.

Since the WSN nodes require low power, micro-scale energy harvesting techniques are used to extract power at low levels from the surroundings. Micro-scale energy harvesting systems are now coming up for producing self sustaining low power electronics which no longer depend on battery for their operation. These systems are capable of extracting milli-watts of power from sources like light energy, vibration energy, RF and thermal energy. Harnessing energy from these sources has always been a challenge as these sources tends to be intermittent and unregulated although being abundantly available. The energy which is extracted from these sources can be stored in a capacitor, super capacitor, or battery. (Kompis & Aliwell, 2008), gave a review of the different energy harvesting technologies that could be used for remote and wireless sensing along with the limitations associated with the energy sources. Table 3 summarizes the various sources used for micro-scale energy harvesting along with the power estimate which can be extracted from them (Raju, 2008).

Energy Source	Harvested Power Estimate
Light (photovoltaic)	10uW-10mW/cm2
Vibration/Motion	4uW-100uW/cm2
Temperature difference	25uW-10mW/cm2
RF	0.1uW-1uW/cm2

Table 3. Energy Sources and their Harvested Power Estimate

As seen from the table, light (photovoltaic), thermal and vibration/motion energy seem to be more promising sources for low power applications. On the other hand, energy harvested from RF is very small and is still under the development stage but it finds considerable application in wireless power transmission esp. for powering low power Wireless Sensor Networks (WSNs). We have used RF energy harvesting technique in our prototype. The choice of the energy source for a particular application largely depends on the location of deployment. This section describes two energy harvesting techniques which could be employed for powering WSN nodes, these are: solar energy harvesting and RF energy harvesting.

5.1 Photovoltaic harvesting system
In this method, ambient light energy is harnessed and converted into electrical energy with the help of solar cells. Solar cells are essentially semiconductor junctions. Solar cells work on the photovoltaic effect in which, the light incident on the solar cell generates electron-hole pairs on both sides of the junction, the generated electrons and hole then diffuse in the junction and are swept away by the electric field thereby, generating current. The conversion efficiency of the solar cells is quite low ~10-20%. Fig. 5 shows the general structure of a solar cell. The power output of the solar cell is DC and so can be directly used to power DC loads or can be used for battery charging applications. They can even be used to power AC loads with the help of an inverter. In addition to solar energy, with recent advancements, a large range of low power solar cells are now available which are capable of working in indoor environment i.e. under a fluorescent source.

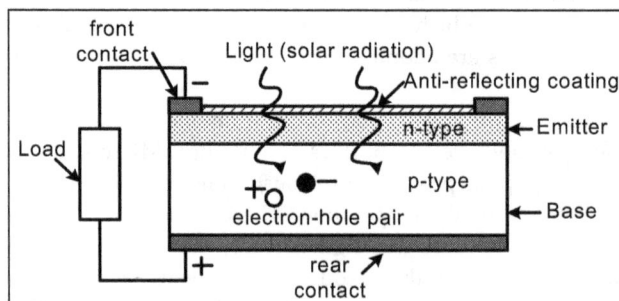

Fig. 5. General Structure of a Solar cell

Advantages: clean source, requires less maintenance, noise-less operation, flexible in configuration i.e. it can be easily connected in series or in parallel combination depending on the power requirement.
Disadvantages: power extracted is expensive due to high initial investment and low conversion efficiency, toxic chemicals like, cadmium and arsenic are used in the production

of solar cells which adversely impacts the environment, intermittent nature of the source in terms of power output.

Applications: low power solar cells find extensive application in calculators, portable lamps, watches, battery chargers, remote telemetry and communication.

5.2 RF Energy harvesting system

RF energy harvesting is the process by which energy is derived from external RF sources, (e.g., FM, TV Towers, Wi-Fi, Cell towers and Mobile phones etc.) captured and stored for different applications like in Wireless Sensor Networks(Seah et al., 2009, 2010) and Wearable electronics. In this method, RF energy harvesting receivers convert ambient energy (i.e. surrounding RF energy) into electricity. Receiver circuit consist of a rectenna i.e. a special type of antenna which directly converts RF energy into useful DC electricity (Mohmmed et al., 2010) as shown in Fig. 6. In this case also, the power output is DC so it can be used to power DC loads or for battery charging applications.

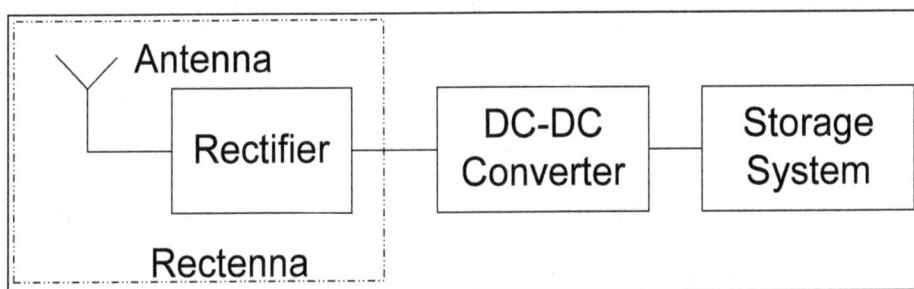

Fig. 6. General Block Diagram of RF Energy Harvesting Receiver System

Advantages: Power can be transferred to remote locations where wired power is not possible, controllable, and predictable. Power can even be transferred over long distances by proper system design.

Disadvantages: The amount of ambient energy captured is very small and irregular, so it can be used for powering very low power devices.

Applications: wireless sensor network, consumer electronics, industrial and transportation.

5.3 Photovoltaic harvesting system and its application in WSN

This section describes the solar photovoltaic harvesting system for low power application with the main focus being their application in WSNs. Nowadays, ultra low power solar cells are extensively finding applications in Wireless Sensor Networks (WSNs) deployed outdoors as well as indoors(Hande et al., 2007). The next sub-section describes the commercially available low power solar cells which can be used for such an application.

5.3.1 Solar cells (photovoltaic) for WSN

Solar photovoltaics are more popular in high power applications due to the fact that the high cost of the PV panel can be supported by such applications. However, with technological advancements, several low cost solar cells are now commercially available which can be used for WSN application esp. in industrial and hospital environment where indoor lights are operational continuously. Solar cells are available in crystalline silicon, thin film and many other varieties with a trade off between cost and efficiency. Solar cells are available for illumination levels starting from 200 lux (under a fluorescent source) to 1000W/m² (under 1.5 solar spectrum). Typical voltage and current at maximum power point range between 1.2V-16V and 4uA-85mA respectively under different illumination conditions. The voltage per cell is low, so a number of cells are connected in series (stack) depending on the voltage requirement of the sensor node. The current requirement is met by connecting several such stacks in parallel. The overall cell configuration, acting as a single energy source can then be directly connected to the electronic device or through a charge storage device (like super capacitor, NiCad, NiMH, or Li rechargeable batteries) with charge controller system which limits overcharging of the batteries.

5.3.2 MPPT for maximum utilization of the source

The solar cell output is significantly affected by changes in the irradiation and temperature levels. Fig. 7 shows the current-voltage and power-voltage characteristic of the solar cell at particular irradiation and temperature level (Kumar, 2010). Since the current-voltage characteristic of a PV cell is non-linear, for a particular irradiation and temperature, there is a unique point on the power-voltage characteristic at which the photovoltaic power is maximum. This point is termed as the Maximum Power Point (MPP). The power, voltage and current corresponding to this point are referred to as P_{MPP} (power at maximum power point), V_{MPP} (voltage at maximum power point) and I_{MPP} (power at maximum power point) respectively. As the irradiation level changes, the power output of the PV system changes, which in turn, changes the MPP. Fig. 8 shows how the MPP points changes under different irradiation levels.

It is desirable to make the solar cell operate at MPP so that the source is utilized efficiently at all the times. This is made possible interfacing the solar cell with power electronic converter working as a Maximum Power Point Tracker (MPPT) incorporating one of the MPPT schemes. Various MPPT schemes for solar photovoltaic systems have been reported in literature (Faranda & Leva, 2008; Hohm & Ropp, 2000) with respect to their tracking speed and accuracy.

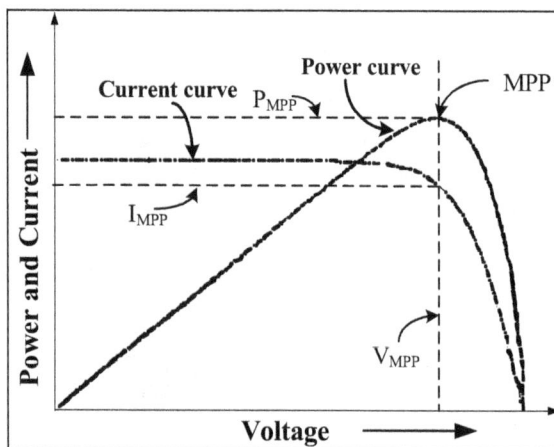

Fig. 7. Current-Voltage and Power-Voltage Characteristics of the Solar Cell

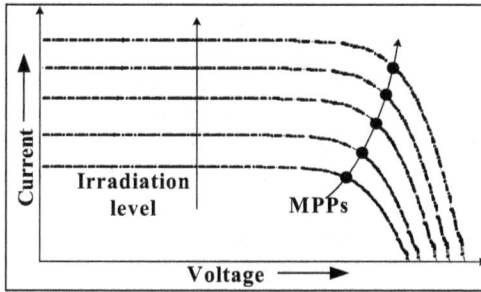

Fig. 8. Variation of MPP with the Variation in the Irradiation

5.3.3 Solar photovoltaic system for WSN application

Designing a power supply that involves, generation, storage and conversion is very challenging. A lot of problems are encountered both at the design level as well as at the implementation level and various solutions have been given by people to solve this problem (Mahlknecht & Roetzer, 2005). The solar photovoltaic harvester requirements change depending on the application at hand. A number of factors govern the power being delivered by the solar cell. Since they are employed to power WSNs deployed outdoors, the placement of sensors is one of the factors which play a major role. Best compromise of the placement position should be identified between the best illumination and location requirement. It should be ensured that the mounting place is not shadowed by the other objects and the place is sufficiently illuminated. Apart from the placement aspect, the radiation pattern of the location needs to be studied as it helps in the sizing of solar cell arrangement.

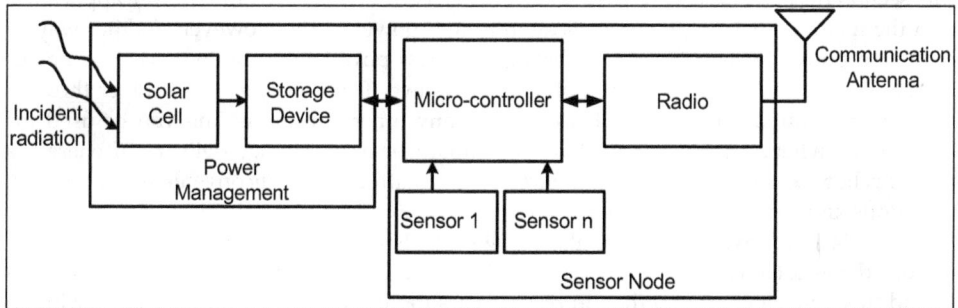

Fig. 9. Basic Block Diagram of a Solar Powered Wireless Sensor Node

The basic block diagram of solar powered wireless sensor node is given in Fig. 9. It typically consists of a power management unit, controller, sensors and radio transceiver. The power management unit comprises of the solar cell configuration (series/ parallel combination) and a storage element (capacitor/battery). The sizing of the solar cell configuration is done based on the power requirement of the wireless sensor node (RF Monolithics Application Note M1002, 2010). The storage element can be a super capacitor or battery or a combination of both with super capacitor providing the peak current and battery acting as the main back-up reservoir. Super capacitors are large capacitors available in the range of 50-100 Farad with operating voltage of 2.6V. Several of these can be stacked up in series/parallel combination to supply the peak power demand. There are four types of rechargeable batteries which can be for this application, these are: Lead Acid, Nickel Cadmium, Nickel Metal Hydride and Lithium Ion. Of all these, Lithium Ion battery is most preferred to power WSNs because of its operating voltage of 3.6V. Lead Acid and Nickel Cadmium battery although being cheap and easily available are not popular due to increasing environment concerns about the lead and cadmium content.

The next block is the controller unit which processes the data obtained from the sensors connected to it and transmits the information using the radio transceiver. Some configurations might include MPPT device to effectively utilize the PV source. MPPT is popular in high power applications of solar photovoltaic where huge amount of energy is extracted. However, in low power applications, MPPT might add to power consumption and fail to be useful. Therefore, for low power applications demanding a constant output, the solar cell is interfaced with a power electronic converter. Power electronic converter works as a regulator/charge controller thereby helping the solar cell in meeting the constant load demand/charging the batteries. To conserve energy, WSNs operate in sleep mode, waking up at regular intervals to acquire process and transmit the sensor data to its neighboring devices/master system for decision making. The amount of current consumed by them in the sleep mode and wake up mode is of the order of 5uA and 200mA respectively. The current consumption and the time duration in each mode help in sizing the solar cell configuration for powering the sensor nodes. In addition to that, choice of battery and the depth of discharge also have an important role in determining the power requirement of the sensor node.

5.4 RF energy harvesting system and Its application in WSN

This section describes the RF energy harvesting receiver system for low power application with the main focus being their application in low power WSN. However, another way of supplying power to sensor nodes is through wireless power transmission. Wireless power Transmission (WPT) is transmission of electrical power from one point to another through vaccum or an atmosphere without use of wire any other substance. This can be used for applications where either an instantaneous amount or a continuous delivery of energy is needed, but where conventional wires are unaffordable, inconvenient, expensive, hazardous, unwanted or impossible.

Nicola Tesla who invented Radio, also known as "Father of Wireless" was the first who conceived the idea Wireless Power Transmission and demonstrated it in 1899. The idea behind this investigation was that, in recent years the use of wireless devices is growing in many applications like mobile phones, medical implants (Huang et al., 2008) or sensor networks. This increase in portable wireless applications has generated an increasing use of batteries. Many researchers are working on energy alternatives to reduce down their dependence on batteries and come up with low power counterparts so as to increase device lifetime. The charging of wired applications is still easy because the user can do it easily, like for mobile phones. But for other applications, like medical implants or wireless sensor nodes located in difficult access environments, the charging of the batteries remains a challenge. This requirement still increases when the number of devices is large and are distributed over a wide area or located in inaccessible places. Wireless Power Transmission (WPT) can be used as one solution to overcome the above mentioned limitations or challenges.

Different methods exist by which electrical energy can be transferred from the source to a load without the use of wire. These are: Electromagnetic Induction, Magnetic Resonance and Electromagnetic Radiation. Out of these, electromagnetic radiation is most popular for powering WSN nodes. The various RF sources and RF harvester design for the WSN nodes are discussed in the subsequent sections.

5.4.1 RF energy sources

RF energy harvest method can be used to remotely charge the battery operated WSN device. The different categories of available RF sources (Kumar) which can be used for conversion are listed in Table 4. Along with the frequency range, the transmitted power for the various RF sources (transmitters) is also mentioned which can be captured and used for different low power applications.

RF Sources	Frequency Range	Tx Power
FM Tower	88-108MHz	10KW
TV Tower	180-220MHz	40KW
AM Tower	540-1600KHz	100KW
Wi-Fi	2.4-2.5GHz	10-100mW
Cell Tower	800,900,1800MHz	20W
Mobile Phones	GSM-900	2W
	GSM-1800	1W

Table 4. RF Energy Sources and their Harvested Power Estimate

When more power or continuous energy is required than what is available from ambient RF sources, RF energy can be broad casted in unlicensed frequency bands such as 868MHz, 915MHz, 2.4GHz and 5.8GHz. Each region has limitations on transmitted power like for e.g. its 4W in North America and 2W in Europe region.

5.4.2 RF harvesting receiver system for WSN application

The basic block diagram of RF powered wireless sensor node is shown in Fig. 10. It typically consists of a energy harvesting unit, controller, sensors and radio transceiver. Apart from the energy harvesting unit, rest of the blocks remain the same for designing any WSN node as described in section 5.3.3. In any energy harvesting unit, mainly three components are required, these are: energy conversion, harvesting and conditioning and energy storage. Here, the energy harvesting unit (Hagerty et al., 2005) comprises of the rectenna (antenna + rectifier) and a storage element (capacitor/battery). The design of the rectenna is done based on the power requirement of the wireless sensor node.

Since the amount of useful energy obtained from RF is very less, reduction in energy consumption of system makes RF energy harvesting more practical. Researchers have come up with new solutions to improve the amount of energy being harvested from RF sources. In particular, circular polarized antennas are being implemented in the rectenna design because they avoid the directionality of other antenna designs (Strassner & Chang, 2003; Ali et al., 2005; Ren & Chang, 2006). An array of rectennas is now increasingly being used to improve the power output (Kim et al., 2006). Several new rectenna design schemes (Park et al., 2004; Chin et al., 2005) have been proposed by researchers. (Harrist, 2004), discussed wireless battery charging using RF energy harvesting. A charging time of 4mV/sec was observed when mobile phone batteries were charged by capturing RF energy at 915MHz.

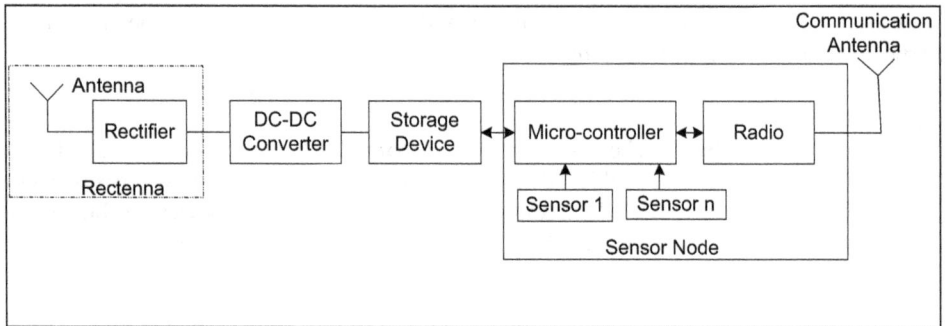

Fig. 10. Basic Block Diagram of RF Energy Harvested Sensor Node

6. Conclusion

This chapter discussed the aspects of using wireless sensor networks for forest tree monitoring and alerting using rare event detection with ultra low power consumption. In this prototype, two sensors (mercury sensor & temperature sensor) which work well for the detection of fire and tree theft were selected and mesh protocol was used for alert routing and event detection. Network lifetime and latency estimation for the deployment scenario showed the implementation feasibility of such a monitoring system for deforestation application. However, as the sensor nodes are battery powered, issues related to battery life and ease of battery replacement are major concerns for WSN applications that involve long term monitoring of vast area especially hostile areas. It is therefore necessary to have some means of recharging the batteries of the sensor node to increase the network lifetime. For this, one of the most common ways is to extract the energy from the surrounding environment. Life time of network is increased by adopting RF energy harvesting technique for recharging the sensor nodes. In addition to RF technique various other energy harvesting techniques are available that can be used for this purpose. The various energy sources which can be used for this prototype implementation have been explored here. A detailed description of solar and RF energy harvesting is given which can be used to charge the batteries and hence increase the lifetime of the deployed WSN system.

In future, the efforts can be taken to increase the robustness of the WSN setup in case of (a) self organization network, (b) failure of the sensor node (auto healing of sensor node) and (c) false alarms generated by sensor nodes.

7. References

Ali, M., Yang, G., & Dougal, R. (2005). A New Circularly Polarized Rectenna for Wireless Power Transmission and Data Communication. IEEE Propagation Letters on Antennas and Wireless, ISSN: 1536-1225, Vol. 4, pp. 205–208, August 2005.

Al-Turjman, F. M., Hassanein, H. S., & Ibnkahla, M. A. (2009). Connectivity Optimization for Wireless Sensor Networks Applied to Forest Monitoring. Proceedings of IEEE International Conference on Communications (ICC), ISBN: 978-1-4244-3435-0, pp. 1-6, Dresden, June 2009.

Awang, A., & Suhaimi, M. H. (2007). RIMBAMON©: A Forest Monitoring System Using Wireless Sensor Networks. Proceedings of IEEE International Conference on Intelligent and Advanced Systems(ICIAS), ISBN: 978-1-4244-1355-3, pp. 1101-1106, Kuala Lumpur, November 2007.

Bahrepour, M., Meratnia, N., & Havinga, P. (2008). Automatic Fire Detection: A Survey From Wireless Sensor Network Perspective. Centre for Telematics and Information Technology (CTIT), Technical Report TR-CTIT-08-73, ISSN 1381-3625, 13 Pages, December 2008.

Bicket, J., Aguayo, D., Biswas, S., & Morris, R. (2005). Architecture and Evaluation of an Unplanned 802.11b Mesh Network. Proceedings of 11th annual International Conference on Mobile Computing and Networking (MobiCom),ISBN:1-59593-020-5, ACM Press, Germany, August 2005.

Bischoff, R., Meyer, J., Enochsson, O., Feltrin, G., & Elfgren, L. (2009). Event-based strain monitoring on a railway bridge with a wireless sensor network. Proceedings of 4th International Conference on Structural Health Monitoring of Intelligent Infrastructure (SHMII-4), ISBN 978-3-905594-52-2, Zurich, Switzerland, July 2009.

Bulusu, N., Heidemann, J., Estrin, D., & Tran, T. (2004). Self-configuring Localization Systems: Design and Experimental Evaluation. Journal of ACM Transactions on Embedded Computing Systems (TECS), Vol. 3, Issue 1, February 2004.

Camp, J., Robinson, J., Steger, C. & Knightly, E. (2006). Measurement Driven Deployment of a Two-tier Urban Mesh Access Network. Proceedings of 4th International Conference on Mobile Systems, Applications and Services (MobiSys), ACM Press, Sweden, June 2006.

Chin, C. H., Xue, Q., & Chan, C.H.(2005). Design of a 5.8 GHz Rectenna Incorporating a New patch Antenna. IEEE Propagation Letters on Antennas and Wireless, ISSN: 1536-1225, Vol. 4, pp. 175–178, June 2005.

Eu, Z. A., Tan, H. P., & Seah, W. K. G. (2010). Wireless Sensor Networks Powered by Ambient Energy Harvesting: An Empirical Characterization. Proceedings of IEEE International Conference on Communications (ICC), ISBN: 978-1-4244-6402-9, pp. 1-5, May 2010.

Faranda, R., & Leva, S. (2008). A Comparative Study of MPPT Techniques for PV Systems. Proceedings of 7th WSEAS International Conference on Application of Electrical Engineering, ISBN: 978-960-6766-80-0, 2008.

Figueiredo, C. M. S., Nakamura, E. F., & Ribas, A. D. (2009). Assessing the Communication Performance of Wireless Sensor Networks in Rainforests', IEEE 2nd IFIP Wireless Days (WD), ISBN: 978-1-4244-5660-4, pp. 1-6, Paris, December 2009.

Fonte, A., Zito, D., Neri, B., & Alimenti, F. (2007). Feasibility Study and Design of a Low-Cost System-on-a-Chip Microwave Radiometer on Silicon. Proceedings of IEEE International Conference on Waveform Diversity and Design, ISBN: 978-1-4244-1276-1, pp. 37-41, Pisa, June 2007.

Gil, K. J., Prasetiyo, R. B., Park, H. J., Lim, S. B., & Eo, Y. D. (2010). Fire Monitoring System based on Open Map API. Proceedings of 6th IEEE International Conference on

Networked Computing and Advanced Information Management (NCM), ISBN: 978-1-4244-7671-8 , pp. 600-605, Seoul, August 2010.

Hagerty, J. A., Zhao, T., Zane, R., & Popovic, R. (2005). Efficient Broadband RF Energy Harvesting for Wireless Sensors. Proceedings of Government Microcircuit Applications & Critical Technology Conference (GOMACTech), Las Vegas, NV, pp. 1-4, April 2005.

Hande, A., Polk, T., Walker, W., & Bhatia, D. (2007). Indoor Solar Energy Harvesting for Sensor Network Router Nodes. Journal of Microprocessors and Microsystems, Elsevier Science Publishers, Vol. 31, No. 6, pp. 420-432, September 2007.

Harrist, D. W. (2004). Wireless Battery Charging System using Radio Frequency Energy Harvesting. M.S. thesis, Department of Electrical Engineering, University of Pittsburgh, PA, 2004.

Harvanová, V., Vojtko, M., Babiš, M., Ďuríček, M., Pohronská, M. (2011). Detection of Wood Logging Based on Sound Recognition Using Zigbee Sensor Network. Proceedings of International Conference on Design and Architectures for Signal and Image Processing, November 2011.

Hefeeda, M., & Bagheri, M. (2007). Wireless Sensor Networks for Early Detection of Forest Fires. Proceedings of IEEE International Conference on Mobile Adhoc and Sensor Systems (MASS), ISBN: 978-1-4244-1455-0, pp. 1–6, Pisa, October 2007.

Hohm, D. P., & Ropp, M. E. (2000). Comparative Study of Maximum Power Point Tracking Algorithms using an Experimental, Programmable, Maximum Power Point Tracking Test Bed. Proceedings of IEEE Photovoltaics Specialist Conference, ISBN: 0-7803-5772-8, pp. 1699-1702, Anchorage, AK, 2000.

Huang, C-C., Yen, S-F., & Wang, C-C. (2008). A Li-ion Battery charging Design for Biomedical Implants. Proceedings of IEEE Asia Pacific Conference on Circuits and Systems (APCCAS), ISBN: 978-1-4244-2341-5, pp. 400-403, Macao, November 2008.

Karlof, C. & Wagner, D. (2003). Secure Routing in Wireless Sensor Networks: Attacks and Countermeasures. Proceedings of the 1st IEEE International Workshop on Sensor Network Protocols and Applications, ISBN: 0-7803-7879-2, pp. 113-127, May 2003.

Kim, J., Yang, S.Y., Song,K. D., Jones, S., & Choi, S.H. (2006). Performance Characterization of Flexible Dipole Rectennas for Smart Actuator Use. Smart Materials and Structures, Vol. 15, No. 3, pp. 809–815, 2006.

Kompis, C., & Aliwell, S.(Editors). (2008). Energy Harvesting Technologies to Enable Remote and Wireless Sensing. Sensors and Instrumentation Knowledge Transfer Network Report, June 2008.

Kovács, Z. G., Marosy, G. E., & Horváth, G. (2010). Case Study of a Simple, Low Power WSN Implementation for Forest Monitoring. Proceedings of 12th Biennial Baltic Electronics Conference (BEC), ISBN: 978-1-4244-7356-4, pp. 161-164, Tallinn, October 2010.

Kumar, G. Radiation Hazards from Cell phones/Cell Towers. IIT Bombay.

Kumar, K. (2010). Modelling, Design and Fabrication of Grid-connected Photovoltaic Systems. M. Tech. Thesis, Department of Reliability Engineering, IIT Bombay, India, July 2010.

Lozano, C., & Rodriguez, O. (2010). Design of Forest Fire Early Detection System Using Wireless Sensor Networks. The Online Journal on Electronics and Electrical Engineering (OJEEE), Vol. 3, No. 2, Reference Number: W10-0097.

Luming, F., Aijun, X., & Lihua, T. (2008). A Study of the Key Technology of Forest Fire Prevention Based on a Cooperation of Video Monitor and GIS. Proceedings of 4th IEEE International Conference on Natural Computation, ISBN: 978-0-7695-3304-9, pp. 391-396, Jinan, October 2008.

Magrath, W. B., Grandalski, R. L., Stuckey, G. L., Vikanes, G. B., Wilkinson, G R. (2007). Timber Theft Prevention: Introduction to Security for Forest Managers. Sustainable Development-East Asia and Pacific Region, Discussion Papers, The World Bank Publication, August 2007.

Mahlknecht, S., & Roetzer, M. (2005). Energy Supply Considerations for Self-sustaining Wireless Sensor Networks. Proceedings of IEEE Second European Workshop on Wireless Sensor Networks, ISBN: 0-7803-8801-1, pp. 397-399, January 2005.

Middleware for Sensor Networks, available at http://lsir-swissex.epfl.ch/index.php/GSN:Home

Mohmmed, S. S., Ramasamy, K., & Shanmuganantham, T. (2010). Wireless Power Transmission – A Next Generation Power Transmission System. International Journal of Computer Applications, ISBN: 978-93-80746-12-8, Vol 1, No.13, 2010.

Park, J. Y., Han, S. M., & Itoh, T. (2004). A Rectenna Design with Harmonic-Rejecting Circular-Sector Antenna. IEEE Propagation Letters on Antennas and Wireless, ISSN: 1536-1225, Vol. 3, No. 1, pp. 52–54, December 2004.

Powerharvester P1100 Module from Powercast, datasheet available at http://powercastco.com/PDF/HarvesterDataSheetv2.pdf

Przydatek, B., Chan, H., Song D. & Perrig, A. (2007). SIA: Secure Information Aggregation in Sensor Networks. Journal on Computer Security-Special Issue on Security of Ad-hoc and Sensor Networks, Vol. 15, No. 1, ACM Press, January 2007.

Raju, M., (2008). Energy Harvesting ULP meets energy harvesting: A game-changing combination for design engineers. Texas Instruments White Paper, 2008.

Ren, Y. J., & Chang, K. (2006). 5.8 GHz Circularly Polarized Dual-Diode Rectenna and Rectenna Array for Microwave Power Transmission. IEEE Transactions on Microwave Theory and Techniques, ISSN: 0018-9480, Vol. 54, No. 4, pp. 1495–1502, April 2006.

Rosi, A., Bicocchi, N., Castelli, G., Mamei, M., & Zambonelli, F. (2011). Landslide Monitoring with Sensor Networks: Experiences and Lessons Learnt from a Real-World Deployment. International Journal of Sensor Networks, Vol. 10, No. 3, pp.111-122.

Seah, W. K. G., Eu Z. A., & Tan, H. P. (2009). Wireless Sensor Networks Powered by Ambient Energy Harvesting (WSN-HEAP) – Survey and Challenges. Proceedings of 1st IEEE International Conference on Wireless Communication, Vehicular Technology, Information Theory, Aerospace & Electronics Systems Technology (VITAE), ISBN: 978-1-4244-4066-5, pp. 1-5, Aalborg, May 2009.

SensorMap, available at http://atom.research.microsoft.com/ sensewebv3/sensormap

SensorML, Sensor Model Language, available at http://www.opengeospatial.org/standards/sensorml

Singh, S. K., Singh, M. P., & Singh, D. K. (2010). Routing Protocols in Wireless Sensor Networks – A Survey. International Journal of Computer Science & Engineering Survey (IJCSES), Vol.1, No.2, November 2010.

Sizing, Solar Energy Harvesters for Wireless Sensor Networks. Application Note M1002, RF Monolithics, Inc., 2010.

Soisoonthorn, T., & Rujipattanapong, S. (2007). Deforestation Detection Algorithm for Wireless Sensor Networks. Proceedings of IEEE International Symposium on Communication and Information Technologies (ISCIT), ISBN: 978-1-4244-0976-1, pp. 1413-1416, Sydney, December 2007.

Strassner, B.,& Chang, K. (2003). 5.8-GHz Circularly polarized dual-rhombic-loop traveling-wave rectifying antenna for low power-density wireless power transmission applications. IEEE Transactions on Microwave Theory and Techniques, ISSN: 0018-9480, Vol. 51, No. 5, pp. 1548–1553, May 2003.

UCSD Active Campus available at http://activecampus.ucsd. Edu/.

Wang, G., Zhang, J., Li, W., Cui, D., & Jing, Y. (2010). A Forest Fire Monitoring System Based on GPRS and ZigBee Wireless Sensor Network. Proceedings of 5th IEEE Conference on Industrial Electronics and Applications (ICIEA), ISBN: 978-1-4244-5045-9, pp. 1859-1862, Taichung, June 2010.

Wark, T., Hu, W., Corke, P., Hodge, J., Keto, A., Mackey, B., Foley, G., Sikka, P., & Br¨unig, M. (2008). Springbrook: Challenges in Developing a Long-Term, Rainforest Wireless Sensor Network. Proceedings of IEEE International Conference on Intelligent Sensors Sensor Networks and Information Processing (ISSNIP), ISBN: 978-1-4244-3822-8, pp. 599-604, Sydney, December 2008.

Zhang, J., Li, W., Yin, Z., Liu, S., & Guo, X. (2009). Forest Fire Detection System based on Wireless Sensor Network. Proceedings of 4th IEEE Conference on Industrial Electronics and Applications (ICIEA), ISBN: 978-1-4244-2799-4, pp. 520-523, Xian, May 2009.

9

Remnant Vegetation Analysis of Guanabara Bay Basin, Rio de Janeiro, Brazil, Using Geographical Information System

Luzia Alice Ferreira de Moraes

Federal University of the State of Rio de Janeiro UNIRIO

Brasil

1. Introduction

The importance of tropical forests and the surrounding environment has been increasing, as specific threats and problems (e.g. deforestation, timber logging, infrastructure development, and mining) are generating increased atmospheric carbon dioxide concentration, with severe present and future consequences in climate (Schulze in Carreiras et al., 2006).

Rio de Janeiro is a Brazilian state that presents the greatest diversity of ecosystems, including major portions of Atlantic Forest (Mata Atlântica in Portuguese), considered a hotspot for its importance and relevance in terms of natural resources and biodiversity. The high biodiversity in this biome is a function of the extreme variations in environmental conditions, and great differences in altitude, ranging from sea level to over 1800 meters.

The Atlantic Forest domain has the following delimitations established by the Brazilian Vegetation Map of IBGE (Veloso *et al.*, 1991): ombrophilous dense Atlantic Forest; mixed ombrophilous forests; open ombrophilous forests; semidecidual stational forests; decidual stational forests; the countryside swamps, the northeastern forest enclaves (regionally called "brejos") and the associated ecosystems - mangroves and restingas.

According to publication of SOS Atlantic Forest Foundation and the Brazilian National Institute for Space Research (2009) between 2005 and 2009, there was a loss of 1,039 hectares of Atlantic Forest in the state of Rio de Janeiro. The Forest is now fragmented in isolated remnants scattered throughout a landscape dominated by agricultural uses. According to Grimm *et al.* (2008), ecosystem responses to land changes are complex and integrated, occurring on all spatial and temporal scales as a consequence of connectivity of resources, energy, and information among social, physical, and biological systems. As terrestrial landscapes become increasingly fragmented, so do hydrologic connections between landscape elements (Pringle, 2001).

The extensive deforestation at the Basin took place over a historic period. The colonization process of the area, specially of dense ombrophilous forest, was initially with sugar cane, coffee and orange plantations, followed by cattle raising and annual crops (CIBG, 2010), which favored the formation of erosive processes as well as silting of water bodies. The only area with significant forest remnants is found in high slopes, inappropriate for agriculture, especially within the limits of Três Picos State Park. According to Freitas et al (2010), roads and topography can determine patterns of land use and distribution of forest cover, particularly in tropical regions.

Many factors have affected negatively and contribute to degradation of Baia de Guanabara Basin such as the growth of unplanned cities, mining areas, exotic monocultures planted without planning, industries, oil refinery, among others. According to Bidone & Lacerda (2003), there are 12 municipal districts, 7.8 million inhabitants and around 12,500 industries distributed unevenly over the drainage basin area (4,000 km2).

With the increasing deforestation, there were created several protected areas to conserve natural resources and biodiversity existing within it. The National System of Conservation Units (SNUC) was created in Brazil by the Federal Law No. 9985/2000, establishing criteria and standards for the creation, deployment and management of protected areas. The Brazil's conservation units such as parks, reserves and APAs (Environmental Protected Areas) are grouped into two categories: "sustainable use" and "integral protection."

Integral protection conservation units are protected areas which main purpose is the conservation of biodiversity. Only the indirect use of natural resources is permitted and natural processes shall take place without human interference. The following conservation units according to International Union for Conservation of Nature (IUCN) - management category I - III) fulfill this purpose: biological and ecological reserves/stations (I), national and state parks (II), natural monuments and wildlife refuges (III).

Sustainable use conservation units were created with the idea of combining the conservation of biodiversity compatible with the rational use of the natural resources, while respecting the legislation that applies to such resources. The following conservation units (IUCN - management category IV- VI) fulfill this purpose: areas of relevant ecological interest (IV), environmental protection areas (V), extractive/fauna/sustainable development reserves (VI).

Bidone et al. 1999 as cited in Bidone & Lacerda (2004), in accordance with land use criteria, classified the watersheds of the Guanabara Bay region into three types: (1) the pristine type, without anthropogenic activities, which generally belongs to legal environmental protection areas, with Mata Atlântica (i.e., a mountainous tropical rainforest type) and/or similar abundant vegetation on the slopes and natural coastal vegetation in the lowlands (grasses, savannas, "restingas"); (2) the weakly impacted type with well-preserved Mata Atlântica and/or other remnant vegetation on the slopes, and lowland sectors with human activities (small-scale farming, tourist-urban activities); and (3) the highly impacted watersheds, densely populated and/or industrialized.

Geographic Information System is a tool for evaluating and monitoring of environmental impacts (IBAMA, 2002) and has been widely used in watershed management, environmental zoning, support for studies of biogeography, monitoring animals, management of protected areas, evaluate deforestation, among others. GIS applications are tools that allow spatial analysis of landscape patterns and the consequences of human activities on these patterns (Tuominem et al. 2009). Though imagery resources can provide a reliable basis for measuring the amount and spatial configuration of forest clearing and exploitation.

2. Objectives

This chapter aims to provide information on land-cover in the Guanabara Bay Basin for monitoring its changing through time; to integrate remote sensing data aiming to provide a diagnose of the distribution of vegetation remnants at Baia de Guanabara Basin among three periods ; to calculate the remaining areas of Atlantic Forest at the basin and correlate them with the altitude and slope; to use map algebra to combine raster maps of vegetation class and some other maps as the different conservation units at the basin to predict the remnant vegetation amount; to provide useful information for decision-making purposes.

3. Study area

The Guanabara Bay Basin is located in southeastern Brazil in the state of Rio de Janeiro, and its geographical coordinates are Latitude -22°20'S and - 22°59'S and Longitude - 42°32'W and - 43°34'W (Figure 1). This Basin is located in the tropical zone, with a typical hot and dry climate (Amador 1997). The annual average temperature reaches 24°C in the coastal plain and 20°C in the mountainous regions. The precipitation annual averaged 2,000mm in the Serra do Mar and oscillated between 1,000 and 1,500mm in the Baixada Fluminense (Amador op cit).

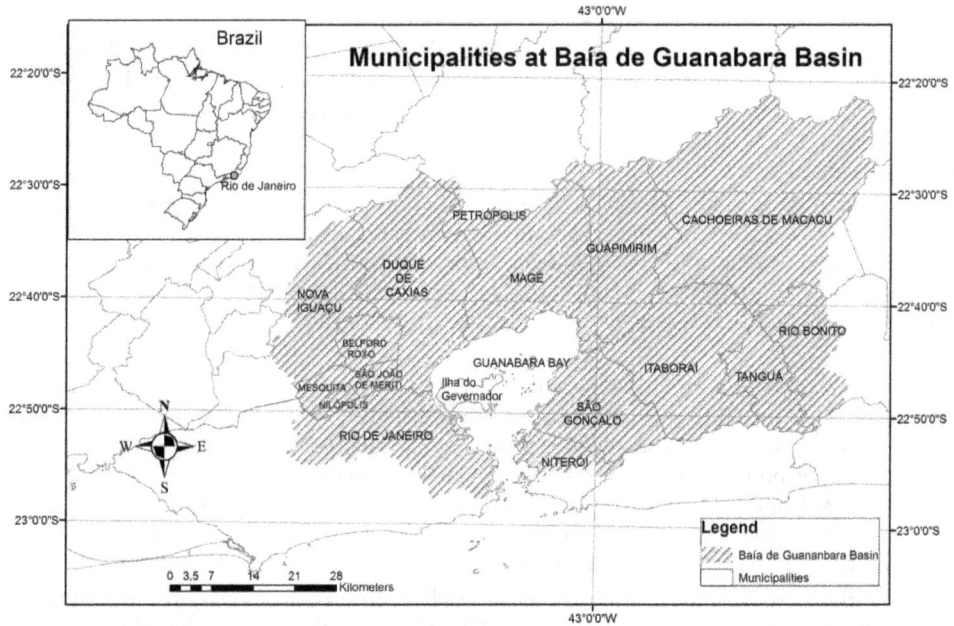

Fig. 1. Study Area

It covers an area of 4,198 km² and includes 16 municipalities that constitute part of the Metropolitan Region of Rio de Janeiro (IBG, 2010). Part of the Basin is located in the mountain range "Serra do Mar", mainly mountainous region and of rough relief, with steep slopes and small valleys. The western part of the Guanabara Bay is called "Baixada Fluminense", located in plain relief that belongs to the urban region of Rio de Janeiro. The Baixada encompasses especially the municipalities of Duque de Caxias, Nova Iguaçu, São João de Meriti, Nilópolis, Belford Roxo, Queimados and Mesquita.

4. Methodology

4.1 Data acquired
4.1.1 Landsat images
Image mosaics from Landsat 5 (Thematic Mapper) and Landsat 7 ETM+ (Enhanced Thematic Mapper plus) were obtained at INPE (Brazilian Institute of Spatial Research) and at USGS (United States Geological Survey's Earth Resources Observation and Science - EROS) websites.

The two images of three periods (1985, 2001 and 2010) were mosaiced to cover the area of Baia de Guanabara Basin, as below:

1985- Landsat 7 ETM+ scenes 217/75 and 217/76 from July 04 and August 05, respectively, obtained at websites of INPE and USGS;

2001- Landsat 7 ETM+ scenes 217/75 from September 04 and 217/76 from October 28, obtained at USGS website;

2010-Landsat 5 TM scenes 217/75 from May 06 and 217/76 from February 15, obtained at INPE website.

4.1.2 SRTM
Images of the Shuttle Radar Topography Mission V. 4.1 (SRTM) in 1-degree digital elevation model (DEM) were obtained at the site of CGIAR- Consortium for Spatial Information (Jarvis *et al.*, 2008), for the elaborations of maps of altitude and slope classes.

4.1.3 Shapefile data
Vector format at a scale of 1:50,000 relating to municipalities, hydrography and the conservation units, were obtained from government agencies like the Brazilian Institute of Geography and Statistics (IBGE); National Environment Institute (INEA), Guanabara Bay Remediation Program (PDBG), *Mata Atlântica* Biosphere Reserve (RBMA), Brazilian Institute of Environment and Renewable Natural Resources (IBAMA) and from municipalities.

4.2 Images processing and classification
This step is based on the application of techniques from digital image processing and visual interpretation of images to the acquisition of cartographic features. An image registration requires control points, a point whose coordinates reference is known. The three spectral bands of ETM+ and TM sensor with 30 meters spatial resolution (bands 3, 4 and 5) were registered through planimetrically correct maps. It was used the Universal Transverse Mercator (UTM) projection with longitude origin at 45°00'00"W and datum SAD69. All image pre-processing procedures were done in SPRING 5.1.7 (Georeferenced Information Processing System), a state-of-the-art GIS developed by Brazil's National Institute for Space Research (Camara, 1996) and available for free on the web. It was also made a datum transformation for integrating the different data.

The intent of the classification process is to categorize all pixels in a digital image into one of several land cover classes, or "themes". The software SPRING 5.1.7 was used to develop a statistical characterization of the reflectance for each information class for producing thematic maps of the land cover present in an image. It was made supervised classification (Atkinson, 2004; Foody, 2002; Richards, 1993), using Maximum Likelihood algorithm to extract the information and allow the mapping of land use and vegetation remaining (Waleed and Grealish, 2004). The area was classified into five major thematic categories as following: 1)vegetation - tropical rainforest (ombrophilous forest), forested wetlands include mangrove swamps and pioneer formations and reforestation; 2- fields – including deforested areas, fields of altitude called "campos rupestres"; agriculture and pasture; 3) anthropogenic (Urban or Built-up Land)- including urban and industrial areas and 4) exposed areas (Transitional Areas) including nonforest, temporarily bare areas as construction is planned for such future uses as residences, shopping centers, industrial complexes. and 5)water –representing Guanabara Bay.

4.3 Digital Elevation Model (DEM) from the shuttle radar topography mission
It was used the digital elevation model (DEM) from the Shuttle Radar Topography Mission V. 4.1 (SRTM) to create raster maps of altimetry and slope. The altimetry data were sliced into five class intervals: 0-8m 8-50m, 50-500m, 500-1500m and >1500m for generation a raster map. The slope classes was defined into six intervals as followed: 0-3% (plain terrain), 3-8% (gently sloping), 8-20% (sloping), 20-45% (moderately steep to steep), 45-75% (very steep - mountain slope) and >75% (scarped).
The vegetation classes were defined according to the altitude in: mangrove (0-7m), lowland (0-40m), lower montane forest (40-500m), montane forest (500-1500) and upper montane forest (1500-2200).

4.4 Map algebra application
Map algebra uses math expressions to combine raster layers using operators such as arithmetic, relational and boolean logic (Wang & Pulard, 2005). It was used the algebraic language as a tool to estimate the deforestation in the Guanabara Bay Basin using SPRING 5.1.7 software through Spatial Language for Algebraic Geoprocessing (LEGAL). Map algebra creates new features and attribute relations by overlaying the features from two or more input map layers. Features from each input layer are combined to create new output features. The thematic maps of classified images and some other maps of altimetry, slope and conservation units had been manipulated using Boolean algebraic expressions describing the rules and conditions involved in the evaluation and evolution of the deforestation process. Some Conservation Units were cut at the limit of the Guanabara Basin, since the target of this work is to verify the remnant vegetation belonging to the Basin. The integral and sustainable conservation units were overlayed with the maps of land cover classification to create a new map of the remnants vegetation areas in the three study periods (1985, 2001, 2010).

4.5 Maps elaboration
Finally, thematic maps of land cover classification, vegetation remnants according to altimetry and slope classes, vegetation remnants in the conservation units and vegetation fragments were prepared using the softwares Spring (INPE) and ArcGis (ESRI)

5. Results

5.1 Supervised classification of the three Images (1985-2001-2010)
The supervised Classification (Figures 2-4 and Table 1) shows a decrease of vegetated extent in 24.99 percent between 1985 and 2001. The removal of vegetation cover and riparian forest is directly linked to increase in pasture lands and agricultural lands over the three periods, as showed in land cover classification.

In the first time period (1985-2001) the vegetation clearance occurred in 321.989 square kilometers with an increase of agriculture and pasture lands. According to the image classification of 2010 period (Table 1) areas under or pasture use represent the major land-cover type in Guanabara Basin, with 44.91 percent of land-cover classified. Although in the same period, some areas previously occupied by fields became urban and peri-urban areas. The increase of anthropogenic class was probably due to unsustainable land management and city expansion especially in informal settlements ("favelas") with an increase of 1,035.973 square kilometers of total occupation. As geographers and urban sociologists have long observed, topography is a key-element contributing to the heterogeneity of residential segregation (Medeiros, 2009). Rio de Janeiro offers a particularly interesting case, with favelas populating the hills and mountains right next to the high income areas (Medeiros, op.cit). According to Freitas et al (2010) roads and topography are not the current drivers of deforestation, but they act as attractors of land-use change and deforestation. In Guanabara Bay basin the observed linearity is due to the high rates of population growth and to unplanned occupation of watersheds, without the proper infrastructure to cope with their effluents (Bidone & Lacerda, 2004). According to Moraes (2009), escalating drought, deforestation, capitation, irresponsible land use, and pollution are direct consequences that demand an integrated management scheme.

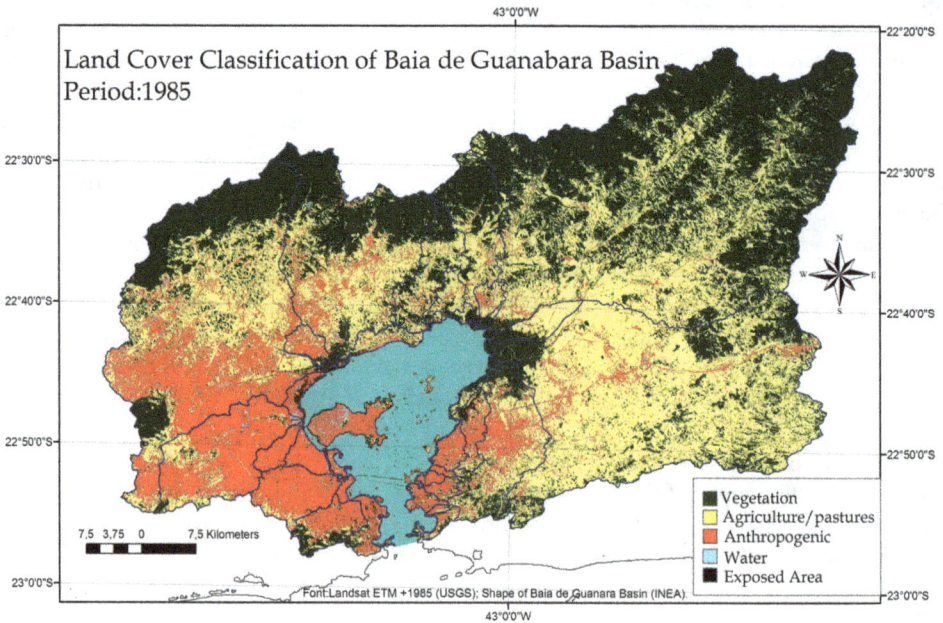

Fig. 2. Land Cover Classification map from 1985 period

Figure 4 of the 2010 period also shows a large exposed area over 10 square kilometers in Guapi-Macacu Basin in the municipality of Itaboraí. The area was exposed due to excavation and earthmoving activities for the implementation of Petrobras Industrial Complex. According to the Environmental Impact Report (EIA), the basic petrochemical unit of COMPERJ will process 150,000 bbl/day of domestic heavy oil to produce thermoplastic resins and fuels (Hernández, 2010). The establishment of petrochemical complex with the magnitude of COMPERJ can lead to an untenable situation due to the increase of the population rate in the Municipality of Itaboraí, which can cause serious damage to riparian vegetation and wetlands remaining in the eastern bay. Attention should be directed to potential social costs and impacts of large-scale projects in the Basin. According to Members of the Committee of the Guanabara Bay Basin (Hernández, 2010) water availability in Metropolitan Rio de Janeiro, considering the water imported from surrounded sub-basins, is no longer sufficient to meet the additional demand generated by the installation of the Petrochemical complex (Pedreira et al, 2009 as cited in Hernández, 2010). Given the tendency toward continued population and Industrial growth, water availability will decline over time, though water availability per se tends to remain fairly constant (in terms of flow, but not in terms of quality) (Hespanhol, 2008).

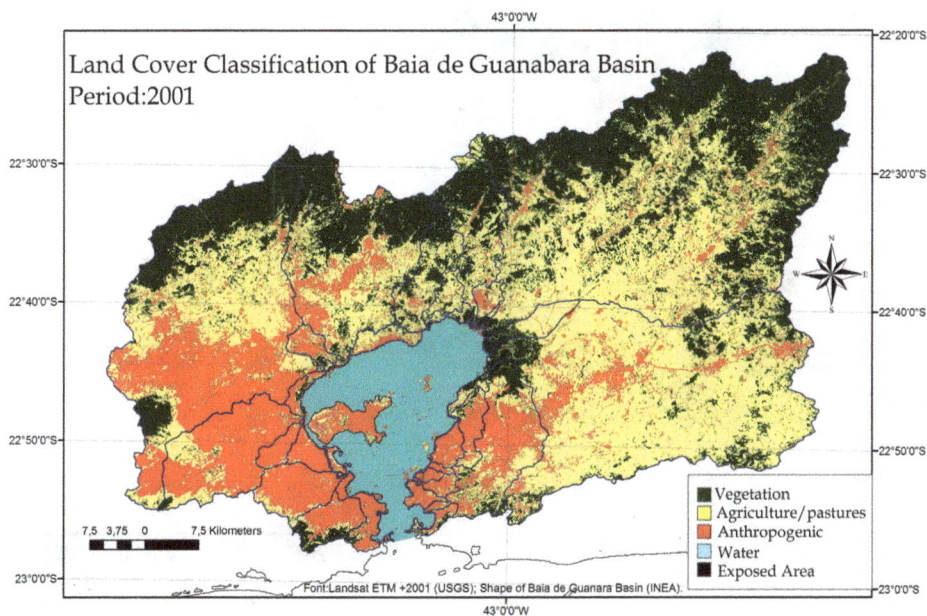

Fig. 3. Land Cover Classification map from 2001 period

According to Hernández op cit. adequate water quality management is necessary for water resource management in a river basin, specifically having a sound water quality monitoring system to indicate the status of water body. Two other large-scale infrastructure projects are undergoing in the Basin: The Metropolitan Arch which will connect Itaguaí municipality to three other major highways: the BR-040 to Belo Horizonte (Minas Gerais) and Brasília (Federal District), the BR-116 to Bahia, and the BR-101 to Espírito Santo and the Gasduct Camboinhas Reduc III with 179 kilometers of extension. The Gasduct was been made in an area of Environmental Protection in Cachoeiras de Macacu Municipality. According to Hespanhol, op cit water conservation in the form of demand management should also be encouraged in industry, pressing for the adoption of modern industrial processes and washing systems with lower water requirements, as well as water treatment stations for public supply through the adequate recuperation and reuse of water used to wash filters and decanters. The Niteroi Municipality Act (Law N°. 2.856/2011) establishes mechanisms to encourage the installation of collection system and wastewater reuse in public and private buildings. Under the Act, new public or private buildings, with an area over 500m² and water consumption greater than or equal to 20m³ per day are obliged to encourage and promote gray water reuse.

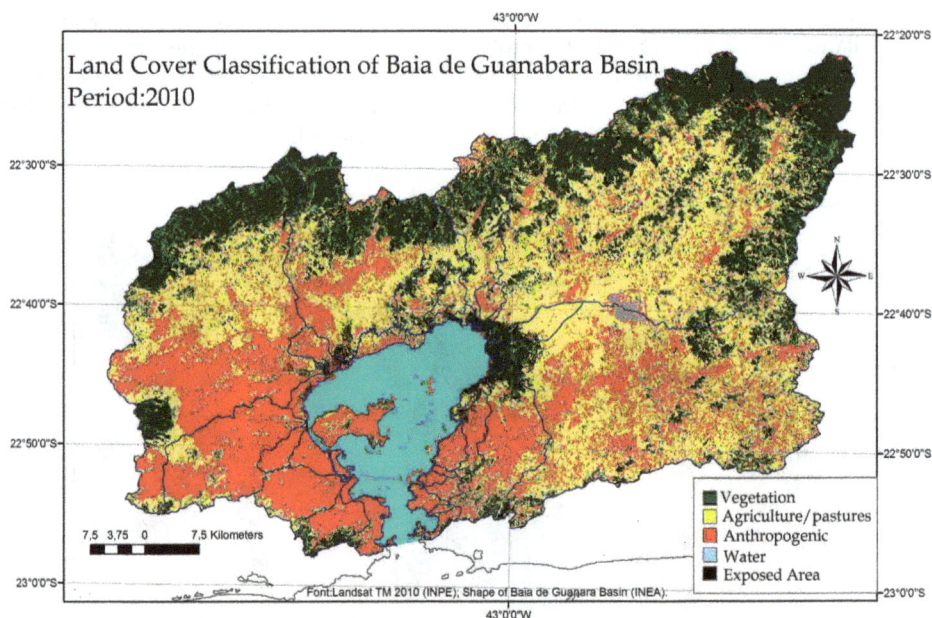

Fig. 4. Land Cover Classification map from 2010 period

Land Cover Classses	1985	2001	2010	Variation (%) 1985-2010
Vegetation	1,590.738	1,268.749	1,193.087	-24.99
Fields	1,791.635	1,985.152	1,833.160	+2.32
Anthropogenic	661.874	829.596	1,035.973	+56.52
Exposed	48.72	1.44	19.16	-60.67

Table 1. Land Cover areas (km²) in the three study periods.

5.2 Conservation units in the Guanabara Basin

The Figure 5, Table 2, shows the delimitation of the major conservation units in the Guanabara Bay Basin according to their uses as strict protection or sustainable use. The units are managed by Brazilian Institute of Environment and Renewable Natural Resources (IBAMA), Chico Mendes Institute for Biodiversity Conservation (ICMBio),

National Environment Institute (INEA) and municipalities, among which seventeen
Environmental Protected Areas (APAS), five Parks, a Biological Reserve, two Ecological
Stations, and an Ecological Reserve. The Atlantic Rainforest Central Mosaic which
includes 22 conservation units and the Sambê Santa Fé Corridor which encompasses the
mountains regions of Sambê, Santa Fé and Barbosão with well preserved forest stretches.

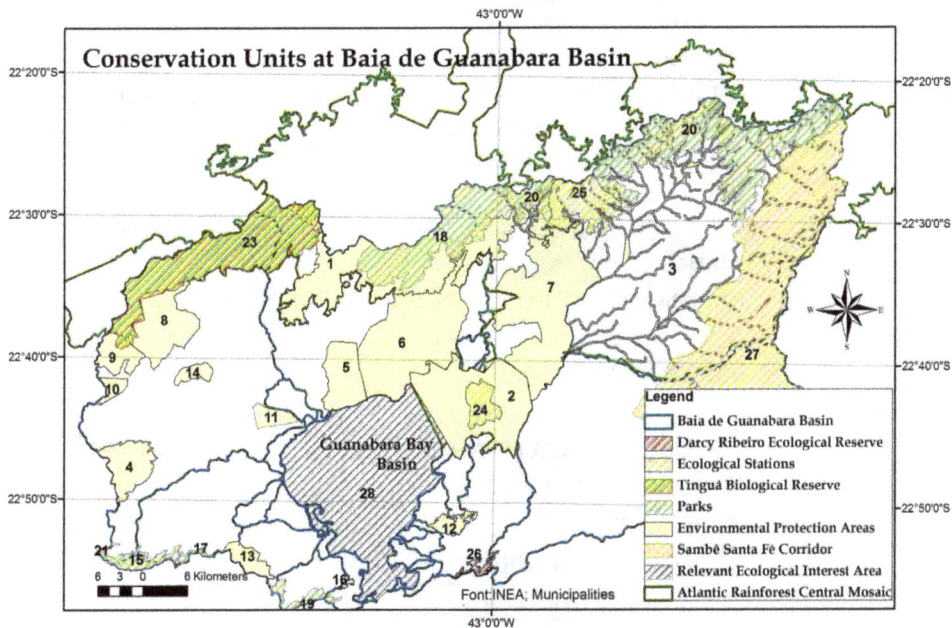

Fig. 5. Main Conservation Units in the Guanabara Bay Basin

The National System of Conservation Units (SNUC) was created in Brazil by the Federal
Law No. 9985/2000, which includes the main categories of Protected Areas as follows:

- Area of Environmental Protection (APA): it is a rather large area characterized by a
  considerable population density and with abiotic, biotic, aesthetic, or cultural features
  of great importance, above all for the quality of life and human wellness. Protecting
  biological diversity, regulating the settlement processes, and ensuring the sustainable
  use of natural resources are among its main aims.
- Biological Reserve: it aims at strictly safeguarding the natural aspects within its borders,
  avoiding direct human interference or environmental changes, through measures to
  recover altered ecosystems and management actions necessary to recover or maintain
  the natural balance, biological diversity, and natural ecological processes.
- Ecological Station: it aims at safeguarding nature and carrying out scientific research
  activities.
- National Park: it aims at preserving natural ecosystems of great beauty and ecological
  importance, giving the opportunity to carry out scientific research activities or
  developing environmental education and interpretation activities, as well as promoting
  recreational activities at direct contact with nature and ecological tourism.

- Area of Considerable Ecological Interest (ARIE): not very large area, with a scarce population density and extraordinary natural features of great importance at a regional and local level.

Management Responsability	ENVIRONMENTAL PROTECTED AREAS (APAS)
Federal	1. Petrópolis; 2.Guapemirim
State	3. Guapi-Macacú River Basin; 4. Gericinó- Mendanha
Municipal	5. Estrela; 6. Suruí; 7. Guapi-Guapiaçu; 8. Tinguá; 9. Rio D'Ouro; 10. Tinguazinho; 11. São Bento; 12. Engenho Pequeno; 13. Pretos Forros; 14. Retiro; 15. Pedra Branca; 16.São José; 17-Morro do Valqueire
	PARKS
Federal	18. National Park of Serra dos Órgãos; 19. National Park of Tijuca Forest
State	20. State Park of Três Picos; 21. State Park of Pedra Branca
Municipal	22. Municipal Park of Barbosão
	BIOLOGICAL RESERVE
Federal	23. Tinguá
	ECOLOGICAL STATIONS
Federal	24. Guanabara Ecological Station
State	25. Paraíso Ecological Station
	ECOLOGICAL RESERVE
Municipal	26. Darcy Ribeiro Reserve
	CORRIDOR
State	27. Sambê Santa Fé
	ARIE
Municipal	28. Guanabara Bay

Table 2. Main Conservation Units in the Basin

The Brazilian Forest Code (Law No. 4771/1965) defines the limits were set on the use of property, where existing vegetation must be respected and considered of common interest to all, except for the removal of vegetation for public service interests provided there are environmental licenses and compliance with established environmental compensation. According the Brazilian Forest Code and the Resolutions of CONAMA (National Environmental Council) numbers 302 and 303, Permanent Protection Areas (APPs): are protected areas, covered or not by native vegetation, for the purpose of preserving water resources, landscape, geological stability, biodiversity, the gene flow of wild fauna and flora, protecting the soil and ensuring the well being of the human population. The APPs include mangrove swamps, riparian vegetation, sand dune scrubs "restingas", regions above 1,800 meters of altitude and hillsides with slopes above 45 degrees. According to the Brazilian Code Legal Reserves (LRs) are areas located within a farm, with exception to permanent preservation areas, necessary for sustainable uses of natural resources, conservation and rehabilitation of ecological processes, conservation of biodiversity, and the shelter and protection for native fauna and flora. Current Brazilian law provides that the Legal Preservation should be 80% in the Amazon, 35% in savanna regions in Amazonian states, and 20% in other regions in the country. The reforestation should be done with species native to the area.

The new Forest Code in Brazil indicates some changes in regards to the Permanent Protection Areas (APP) and Legal Reserves (WWF, 2011). According to current legislation, at least 30 meters from banks and rivers, steep slopes, hilltops and wetlands should be protected. Thus, those who deforest need to restore vegetation. Under the new code, the minimum protection may be reduced to at least 15 meters, and meadows cease to be considered APP. In relation to Legal Reserves, properties of up to 4 taxed modules (varies among different municipalities) do not need to have a reservation, which will be mandatory only for properties that exceed four modules. The amendments to the Brazilian Forest Code may have an important negative effect on Brazil's capacity to reduce emissions from deforestation and forest degradation. The proposed changes will effectively allow more land to be converted for agricultural purposes in Areas of Permanent Preservation, such as hillsides (inclusively forest land 45% in slope or over) and riversides. In addition, existing cultivation of some products including grapes, apples and coffee will continue to be allowed in areas designated as Permanent Protection Areas (APP). The bill provides an amnesty for some small landowners, and may encourage illicit practices. This new proposal will lead to serious consequences in decreasing of urban and peri-urban water supplies in the face of accelerating population and economic growth. In addition, deforestation and land clearing pose serious problems to the carbon cycling to McPherson (1998), urban forests can reduce atmospheric CO, in two ways. Trees directly sequester CO, as woody and foliar biomass while they grow. Also, trees around buildings can reduce the demand for heating and air conditioning, thereby reducing emissions associated with electric power production (Mc Pherson, op cit).

5.3 Remnant vegetation in the conservation units of strict preservation uses and sustainable uses

Figures 6 to 8 and Table 3 showed the remnant vegetation in Conservation Units of integral protection and of sustainable use in the Guanabara Bay Basin. The greatest loss of vegetated areas was observed in conservation units of sustainable use. In environmental protected areas "APAS", there was a loss of 20.23 percent in vegetation class between 1985 and 2010, the equivalent of 100.371 square kilometers. In the strict protection units as parks it was observed the vegetation loss of 11.27 percent which represents a decrease of 39.43 square kilometers in vegetated areas. The Tinguá Biological Reserve has been decreasing its vegetated in 5.23 percent along the study periods.

As a Conservation Unit of Sustainable Use, the Environmental Protected Area of Guapi-Macacu River Basin, established in 2002, contributes to the supply of drinking water to nearly 2.5 million inhabitants living in six municipalities in the State of Rio de Janeiro (Da costa 2007). This basin has suffered several interventions as the built of the Channel of Imunana with the purpose of draining the frequently flooded adjacent areas (Da Costa op cit).

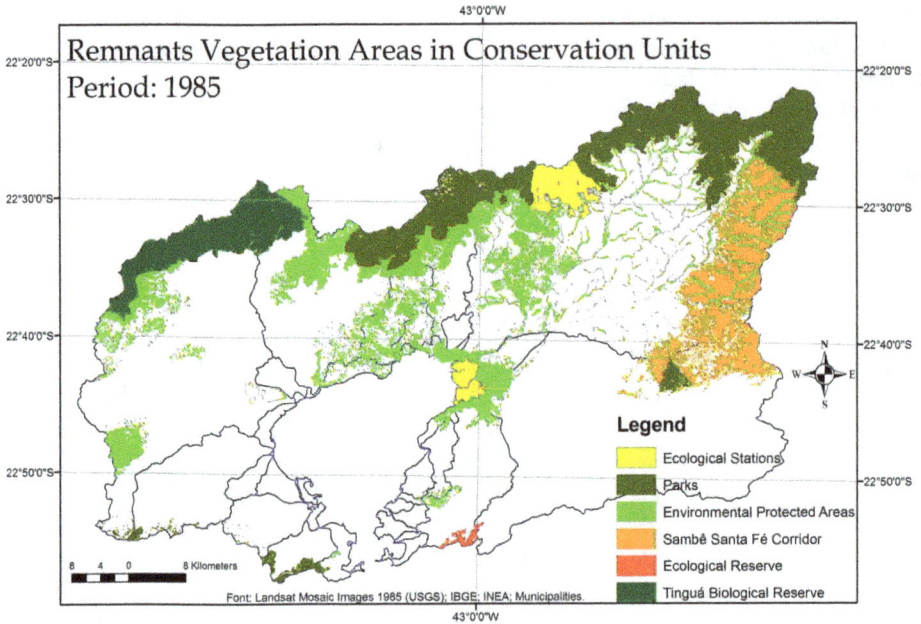

Fig. 6. Remnants vegetation areas in conservation units of Guanabara Bay (1985).

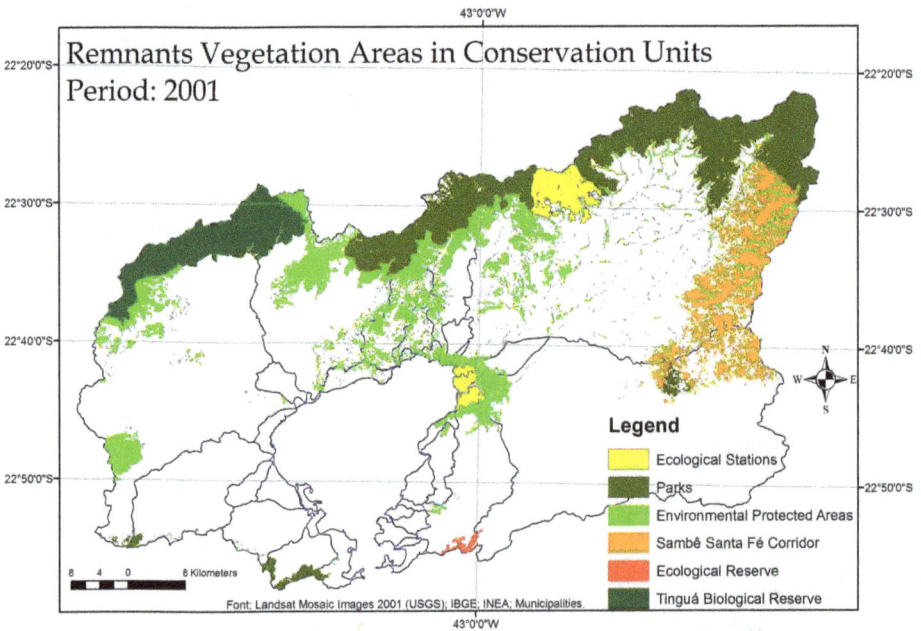

Fig. 7. Remnants vegetation areas in conservation units of Guanabara Bay (2001).

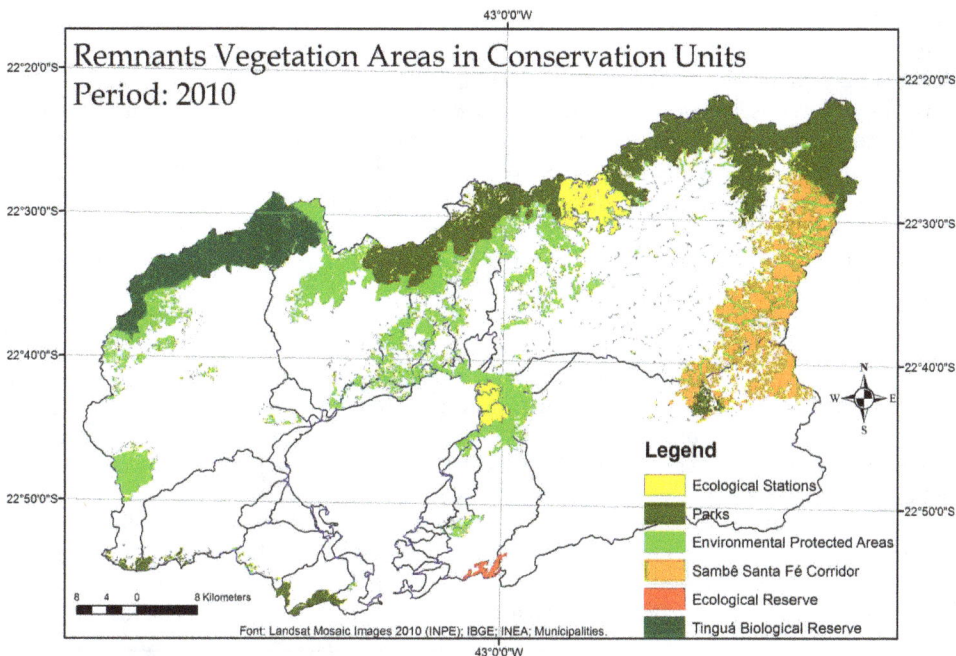

Fig. 8. Remnants vegetation areas in conservation units of Guanabara Bay (2010).

It is evident in the Figures 7 and 8 that the higher forest cover rate is associated with the
riparian vegetation along the rivers. Results show the disappearance of large part of riparian
vegetation along the Guapi-Macacu river banks. According to CONAMA (National
Environmental Council) Resolution No. 9 from 1996, riparian forests are considered
corridors linking forest remnants, thus increasing landscape connectivity. In addition, forest
fragmentation pattern of Guapi-Macacu river basin appeared to be associated with
topography and slope. With the implementation of the Metropolitan Arch and the
Petrochemical Complex it is expected a rapid anthropogenic increase moved by the process
of building infrastructure networks in urban areas.

The magnitude and extent of human impacts have altered biodiversity, hydrologic
connectivity (Pringle,2001), conservation of aquatic ecosystems and also water supplies in the
medium and long term. According to Pedreira et al., as cited in Hernández (2010), the main
environmental pressures in water quality in the Guanabara Basin are: inappropriate land-use
activities, specifically, removal of majority of the original vegetation cover, removal of riparian
forest, unplanned urban sprawl, lack of sewage treatment and improper supervision of
industrial activities; causing steep erosion and river siltation. Despite the promulgation of
wide-reaching legislation, including Law 9,433/1997, which institute the National Water
Resource Policy and defined the legal and administrative framework for the National Water
Resource System and CONAMA Resolution 357/2005, which established the classification of
water bodies and the conditions for effluents discharging, the water pollution is steadily
increasing. In critical areas surrounding the Guanabara Bay (particularly along its NW coast),
less than 60% of the population has access to adequate sewage treatment, only about 10% of
the total sewage is treated before being released into the Bay, the rest being released untreated

into the Bay's tributaries (Bidone & Lacerda, op cit). According to Hespanhol (2008) in terms of water resource management, it is therefore fundamental, especially in urban areas, that we abandon the outmoded orthodox principles and implement a new paradigm based on the key-words of water conservation and reuse, as only thus will it be possible to minimize the costs and environmental impacts associated with the new channeling projects.

In the mountain regions, the most affected areas in terms of vegetated loss in the Guapi-Macacu Basin were observed in National Park of Serra dos Órgãos and also State Park of Três Picos. About eleven percent of the park's vegetation has been lost between 1985 and 2010, which represents an area of thirty three square kilometers. According to Goncalves et al. (2009) to reconcile conservation and land-use one of the alternatives is to establish buffer zones around protected areas, within which human activities are subjected to specific rules and restrictions. Brazil's Conservation Units National System (SNUC) determines that protected areas should be surrounded by buffer zones where human activity is restrict, but the established size of the buffer seems arbitrary (Alexandre et al, 2010). In 1990, the National Environment Council (CONAMA) Resolution number 13 had already defined a buffer zone of 10 kilometers around protected areas, where any activity that may affect the biota should be licensed (CONAMA, 1990). In 2010 the resolution No. 13 was revoked by Resolution No. 428/2010, which reduced the buffer zone to 3 kilometers for licensing the enterprises of a significant environmental impact , located from the edge of Conservation Units, where the buffer zone is not established with the exception of private reserves (RPPN), the Environmental Protection Areas (APAs) and consolidated urban areas.

| Conservation Units | Vegetation Remnant Areas (Square Kilometers) | | | Variation (%) |
	1985	2001	2010	1985-2010
Environmental Protection Areas	496.193	427. 573	395.822	-20.23
Parks	349.687	338.979	310.257	-11.27
Ecological Reserve	4.141	3.685	4.120	-0.48
Biological Reserve	151.083	149.767	143.172	-5.23
Ecological Station	63.550	61.297	59.185	-6.87
Sambê Santa Fé Corridor	255.330	206.108	198.720	-22.17

Table 3. Vegetation Remnant Areas (km^2) in the Conservation Units at Baia de Guanabara Basin

5.4 Remnant vegetation according to altitude

Guanabara Bay Basin is characterized by a large number of small ponds (over 40), usually <100 km2. The river profiles are characterized by a strong slope change in a few tens of miles to a relief of hills before reaching the coastal plain (Bidone & Lacerda, 2003). The upper part of the basin occurs in the oceanic ridge of the Serra do Mar, a mountain system, with a maximum of 2,000 to 2,200 m and consists of a block of cracks inclined to the north-northeast (Cabral et al, 2007). The southwestern plains, where vegetation has been extensively cleaned, are characterized by plan terrain to gently sloping terrain.

Figures 9 to 11 and Table 4, show the areas of remnant vegetation in relation to altimetry. Between 1985 and 2010 Guanabara Bay Basin lost 382.174 square kilometers of vegetated area. The vegetation is less impacted at higher altitudes ("serras") probably due to difficulty for agricultural purposes.

According to Freitas et al (2010) the highest deforestation and fragmentation occurred in less declivous areas, where there are more roads, and more intensive land use. The results corroborate those previously reported by Freitas op cit, with a significant decrease in 62.01 percent of vegetated areas in the lowlands, the equivalent of an area of 110.689 square kilometers.

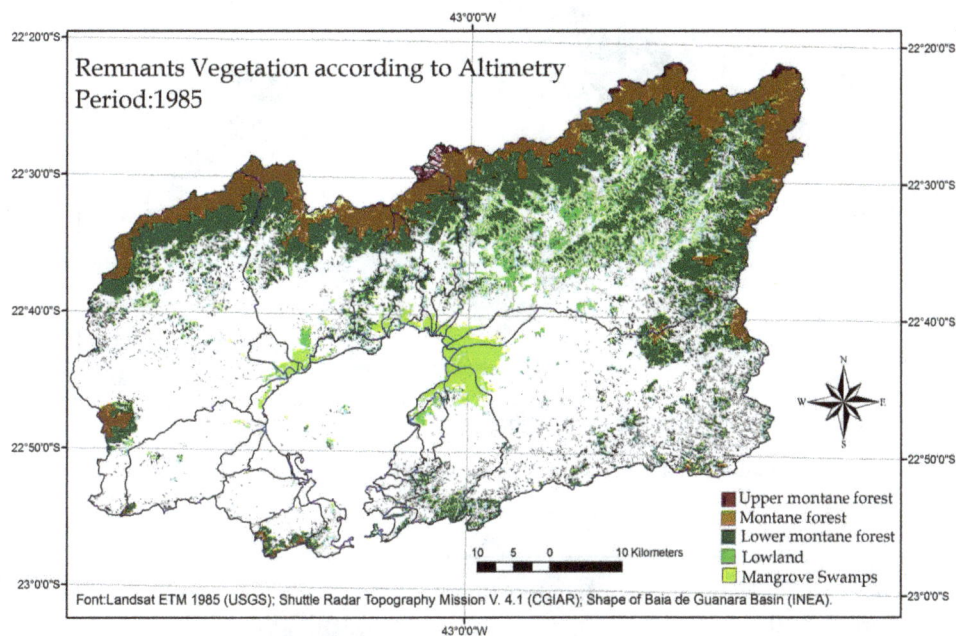

Fig. 9. Remnant vegetation according to altimetry (1985)

The mangrove area has decreased from 86.308 km2 in 1985 to 67.397 km2 in 2010. The main causes of mangrove degradation in the Guanabara Bay Basin include population pressure, agriculture, as well as pollution. On a positive note, between 2001 and 2010 there was a small increase in mangrove area, probably due planting and replanting initiatives. However, the Landsat images do not allow more detailed assessments in relation to degradation stage of mangroves that continually receive different kinds of waste coming from urban, commercial and industrial activities.

The major extension of vegetation remnants was observed in higher altitudes ("serras"). Areas with steep slopes are less used and is much more likely to remain forested (Silva, 2007). Although montane forests and upper montane forests have been losing areas along the study periods, mainly due to intentional fires or those for clearing land for pastures. In September 2010, intentional fires in the State Park of Três Picos destroyed 80 hectares of pasture lands, natural forests and fields of altitude called "campos rupestres". In August 2011 a fire broke out 30 hectares of forest in the National Park of Serra dos Órgãos. The major cause of the vegetation loss in the Guanabara Basin was due to urban development without planning.

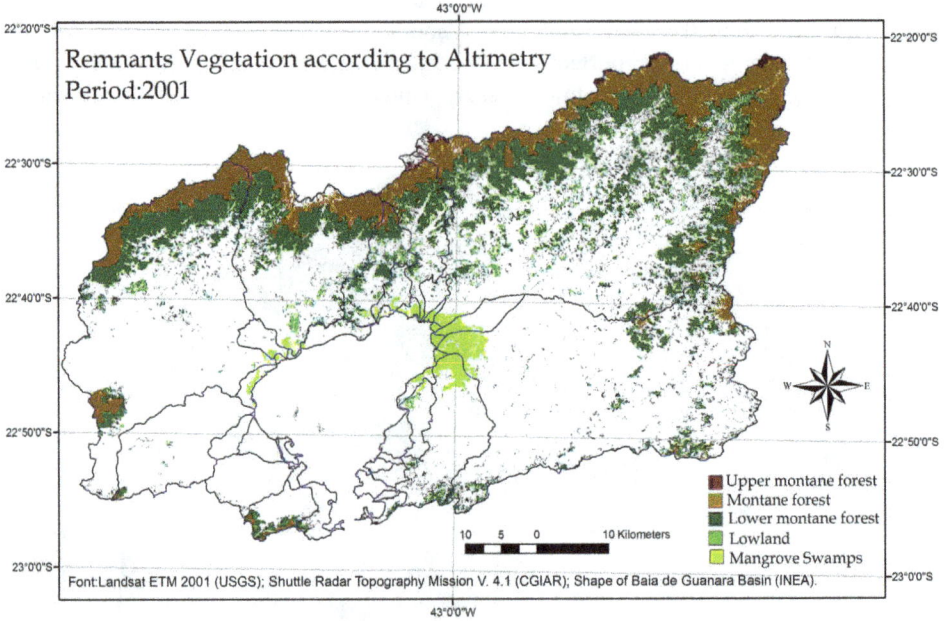

Fig. 10. Remnant vegetation according to altimetry (2001)

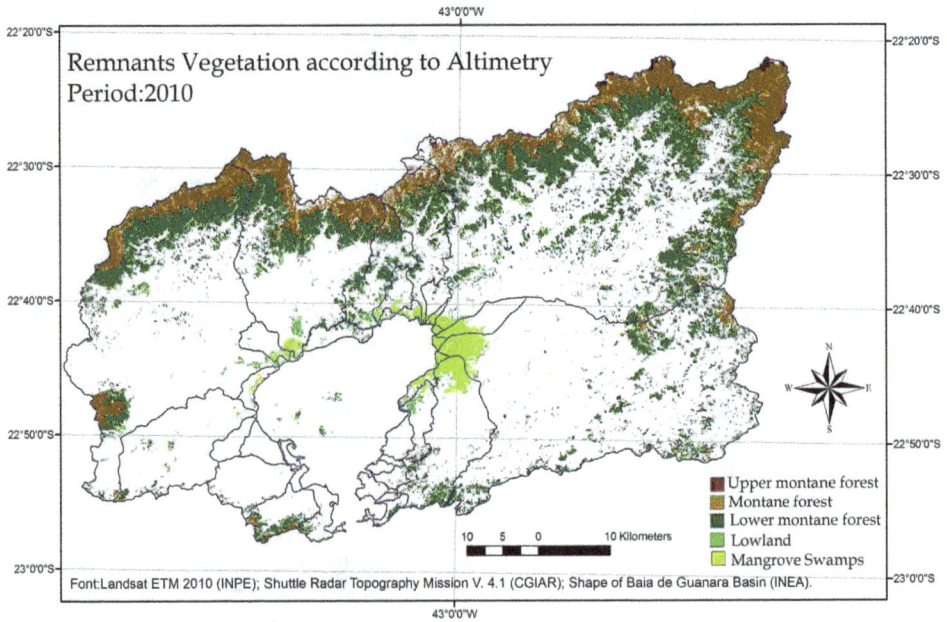

Fig. 11. Remnant vegetation according to altimetry (2010)

Area (Km²)	1985	2001	2001	Variation (%) 1985-2010
Mangrove swamps and wetlands (0-7m)	86.308	67.397	67.792	-21.31
Lowland (0-40m)	178.497	90.401	67.808	-62.01
Lower montane forest (40-500m)	884.356	691.698	672.229	-23.98
Montane forest (500-1500m)	405.536	396.678	364.694	-10.07
Upper montane forest (1500-2200m)	17.90	15.05	10.1	-43.52

Table 4. Remnant Vegetation in relation to altimetry

5.5 Vegetation fragmentation in the Atlantic Rainforest Central Mosaic

Landscape mosaics are described by the landscape components of patches, corridors, and the surrounding matrix (Forman, 1995). Factors such as patch size and shape, corridor characteristics, and connectivity work together to determine the pattern and process of the landscape (Forman, op cit). Franklin et al. (2002) has proposed four requisites for building situational definitions of habitat fragmentation: (1) what is being fragmented, (2) what is the scale(s) of fragmentation, (3) what is the extent and pattern of fragmentation, and (4) what is the mechanism(s) causing fragmentation. According to Franklin op cit., fragmentation at the range-wide scale can affect dispersal between populations, fragmentation at the population scale can alter local population dynamics, and fragmentation at the home range scale can affect individual performance measures, such as survival and reproduction. The topography can also influence patterns of forest fragmentation and forest cover, as previously demonstrated in several regions, including the Brazilian Atlantic Forest region (Silva et al., 2007; Freitas et al., 2010). Fahrig (2003) defined four effects which influence the fragmentation process on habitat pattern:(a) reduction in habitat amount; (b) increase in number of patches; (c) reduction in patch size; and (d) increase of isolation between patches. However, fragmentation measures vary widely; some include only effect (e.g., reduced habitat amount or reduced patch sizes) whereas others include two or three effects but not all four (Fahrig, op.cit). The large connected area (corridor) allows the exchange of genetic material with large populations.

The study area is included in the Atlantic Rainforest Central Mosaic with a large and contiguous corridor observed in mountain region. The figures 12 to 14 show that the six classes of vegetation fragments, varied from very small fragments (<0,03ha) and small fragments (<80ha) to a unique and very large vegetation block of 87,241 hectares observed in mountain region in 1985. This contiguous forest block is composed by different Conservations Units of integral protection and of sustainable use as National Park of Serra dos Órgãos, State Park of Três Picos, Paraiso Ecological Station, Tinguá Biological Reserve and part of Sambê Santa Fé Corridor and Petrópolis Environmental Protected Area. Between 1985 to 2010 the continuous vegetation block has been reduced in 26.44 percent, the equivalent of 23,066 hectares of vegetated area.

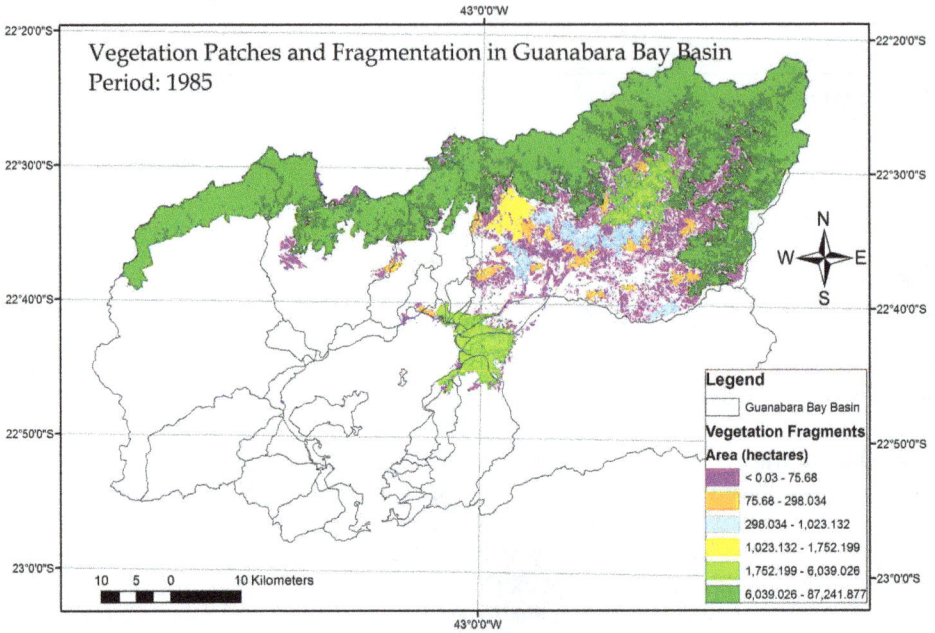

Fig. 12. Vegetation Fragmentation in Guanabara Bay Basin (1985)

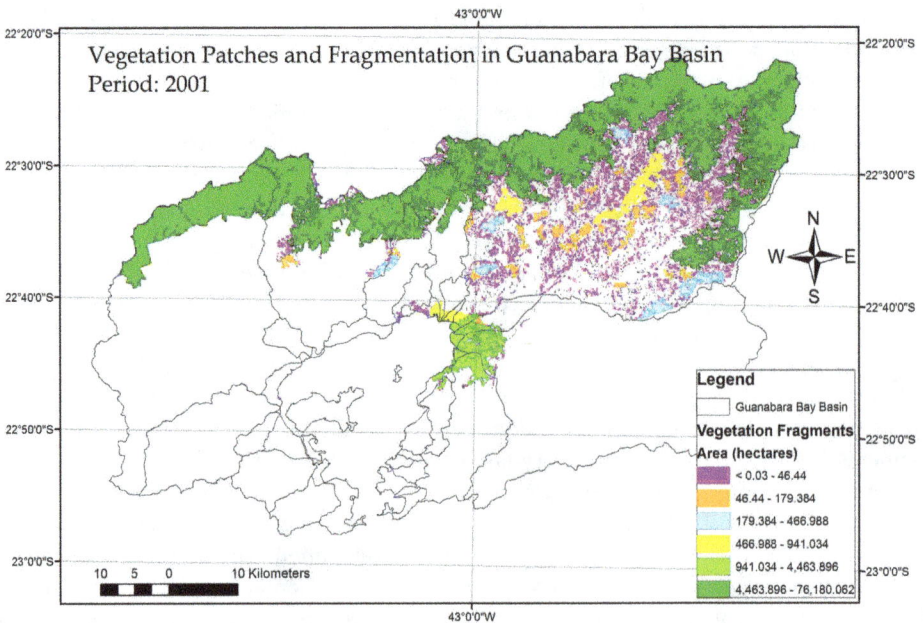

Fig. 13. Vegetation Fragmentation in Guanabara Bay Basin (2001)

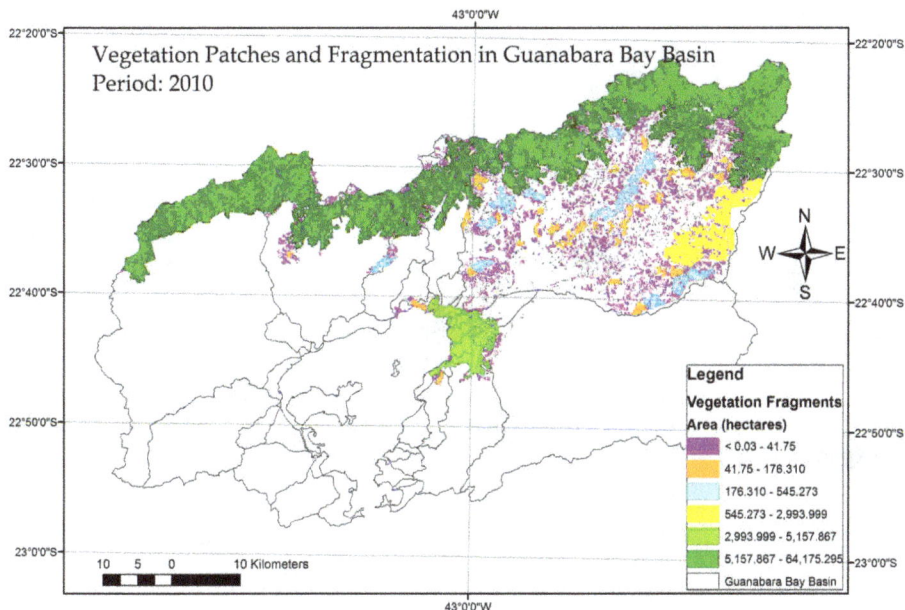

Fig. 14. Vegetation Fragmentation in Guanabara Bay Basin (2010)

In the same period, there was also a lack of continuity between forest fragments that were previously connected. Fragment size and connectivity are among the key landscape factors that affect species survival in fragmented landscapes (Carvalho et al., 2009). Metzger et al. (2009) suggested that fragment size is usually related to the amount and diversity of resources, which directly influence the size and number of resident populations.

The study shows that the largest reduction in size of forest patches was observed on the plains. Although the vegetation patch in lowlands is extremely important to allow connection between highlands, it was observed that vegetation fragments become smaller and more widely spaced. Figure 12 shows a vegetation patch of 4,648 hectares in the central part of the Guapi-Macacu River Basin , which was reduced to some small patches disconnected in 2010. Also, the number of the bigger fragments declined while the smaller ones increased, which means that during that study period a strong fragmentation took place. According to Freitas et al. (2010) higher density of roads is a primary predictor of forest fragmentation and deforestation. As shown in Figure 14, there was a fragmentation in the south and southwest of the Sambê Santa Fé Corridor in some small patches of vegetation. According to the analysis in 1985 the Corridor of Sambe Santa Fé had a large-continuous patch of 188.200 square kilometers. In 2001 this large patch was reduced to 142.117 square kilometers and several fragments of different sizes, which will bring serious problems for the local biodiversity. The National Park of Serra dos Órgãos has lost approximately 20 percent of the vegetated area.

5.6 Anthropogenic occupation in Guanabara Bay Basin in relation to slope classes

The topography determines the expansion of roads and the land use activities, which will impact the forest cover (Freitas et al, 2010). The thematic maps overlay of anthropogenic class and slope layers (Figure 15) showed that the human occupation occurs mainly in the slope classes between 0 and 3 percent, less than 10 meters of altimetry. The anthropogenic class occupied an area over 600 square kilometers in plan terrain, which is subject to risk of flooding, especially in rainy periods. In April 2010, heavy rain caused destruction and death in the State of Rio de Janeiro. The anthropogenic growth is also evident in slopes between 3 to 8 percent, with 268 square kilometers occupancy in gentle sloping terrain. In steep slopes and near streams, where it is difficult to grow crops and accessibility is limited, forest is commonly found (Teixeira et al., 2009). In addition, in most cases the Atlantic Forest region is located in sites where access is difficult (Cabral et al., 2007; Silva et al.,2007). According to Freitas op cit, forests far from land use (buildings and agriculture) and major cities are more likely to be preserved and regenerated. The results corroborate those from authors cited above with low occurence of anthropogenic occupation on slopes between 20 to 45 percent and 45 to 75 percent with higher amount of remnant vegetation. This fact is explained by the difficulty of occupying the higher slopes and the lack of infraestructure which restrict the urban expansion.

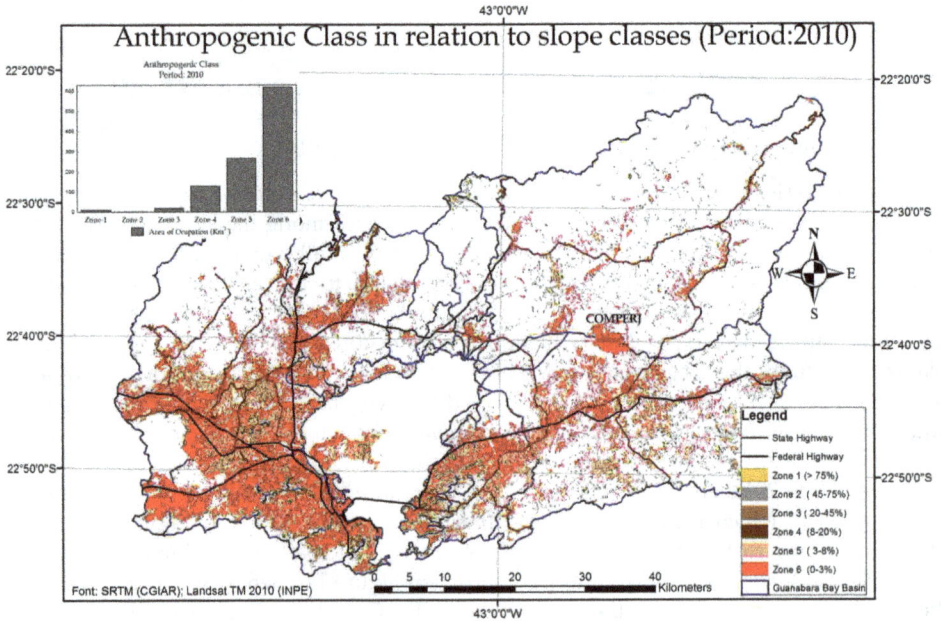

Fig. 15. Anthropogenic Occupation in relation to slope classes.

Figure 16 shows that the anthropogenic occupation in plan terrain also occurs in buffer zones that must be preserved as the Permanent Protection Areas (APP), along the rivers, in the buffer zones of riparian vegetation. According to Naiman & Décamps (1997) riparian zones play essential roles in water and landscape planning, in restoration of aquatic systems, and in catalyzing institutional and societal cooperation for these efforts. Rivers and their adjoining riparian zones are considered to be the most important corridors for movements of animals in natural landscapes (Forman & Godron 1986). Furthermore, human alteration of riverine ecosystems involves not only changes to flow regimes but also simultaneous changes in hydrologic connectivity (Nilsson et al. 2005).

The human occupation is also observed in mountain regions, inside buffer zones of strict protection conservation units as Parks and Reserves. The Figure 16 also shows a drastic reduction of seven kilometers in buffer zones, which may allow the expansion and implementation of large-scale infrastructure projects and also a rapid urban occupation in the eastern part of Guanabara Bay Basin. There is a concern with the rapid urban growth in Itaboraí, Cachoeiras de Macacu and Guapemirim which could bring serious damage to surrounding pristine vegetation, riparian vegetation and also wetlands.

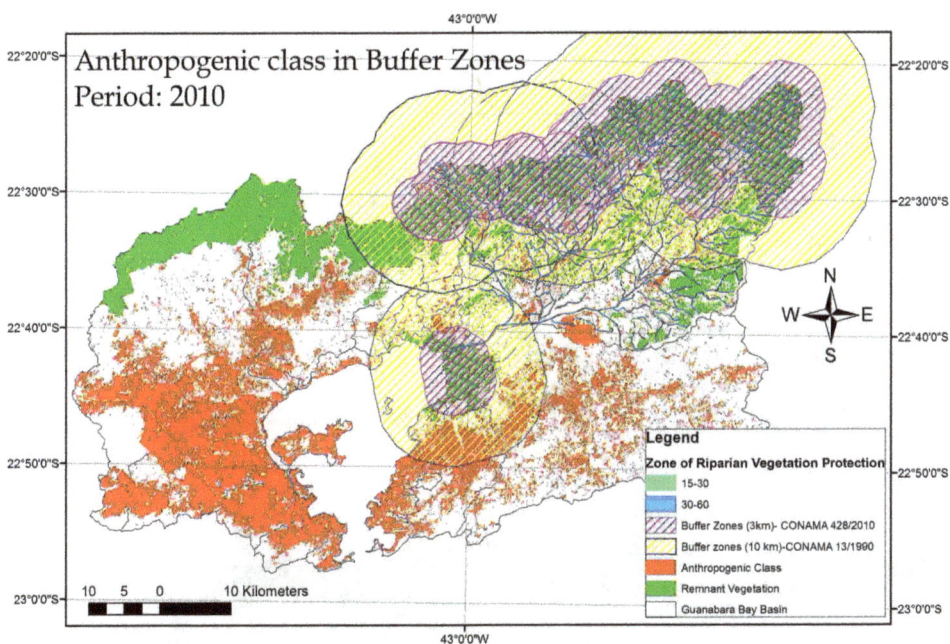

Fig. 16. Anthropogenic occupation in buffer zones

5.7 Critical areas for protection

The Conservation Units in the Atlantic Rainforest Central Mosaic are becoming fragmented and have lost a great amount of vegetated areas along the study period. It was identified two critical areas that Conservation or restoration actions become more urgent: Sambê Santa Fé Corridor and Guapi-Macacu River Basin. The major rupture in the continuity of Sambê Santa Fé Corridor is observed in southwestern part of the corridor (Figure 17a,b,c) with highly fragmented landscapes into small and isolated fragments. In 1985 it was observed the close proximity of the patches and also the amount of vegetation areas was higher than in 2010. Figure 17c shows that the patches were reduced in habitat amount, increase in isolation among patches and reduction in patches size.

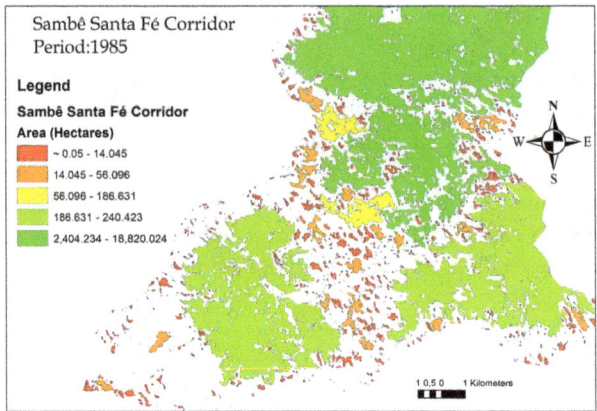

(a)

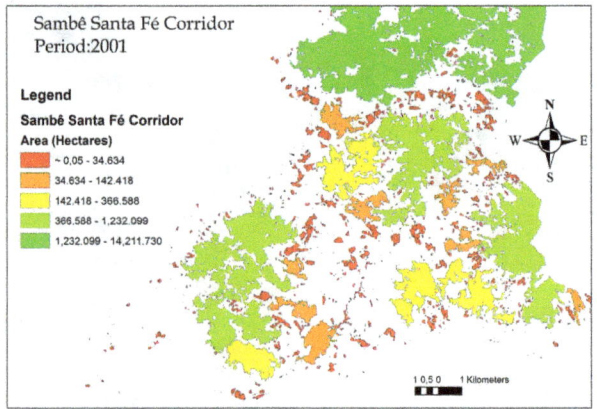

(b)

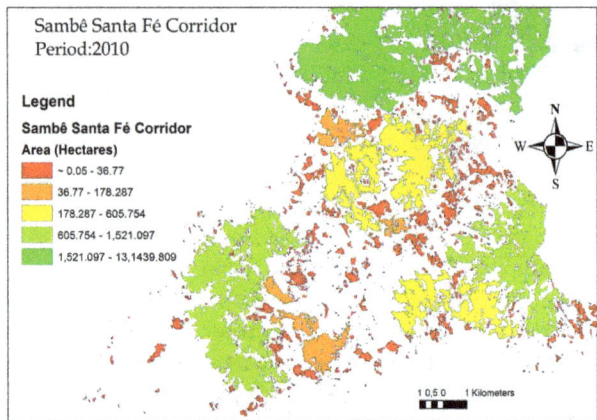

(c)

Fig. 17. Sambê Santa Fé Corridor

(a)

(b)

(c)

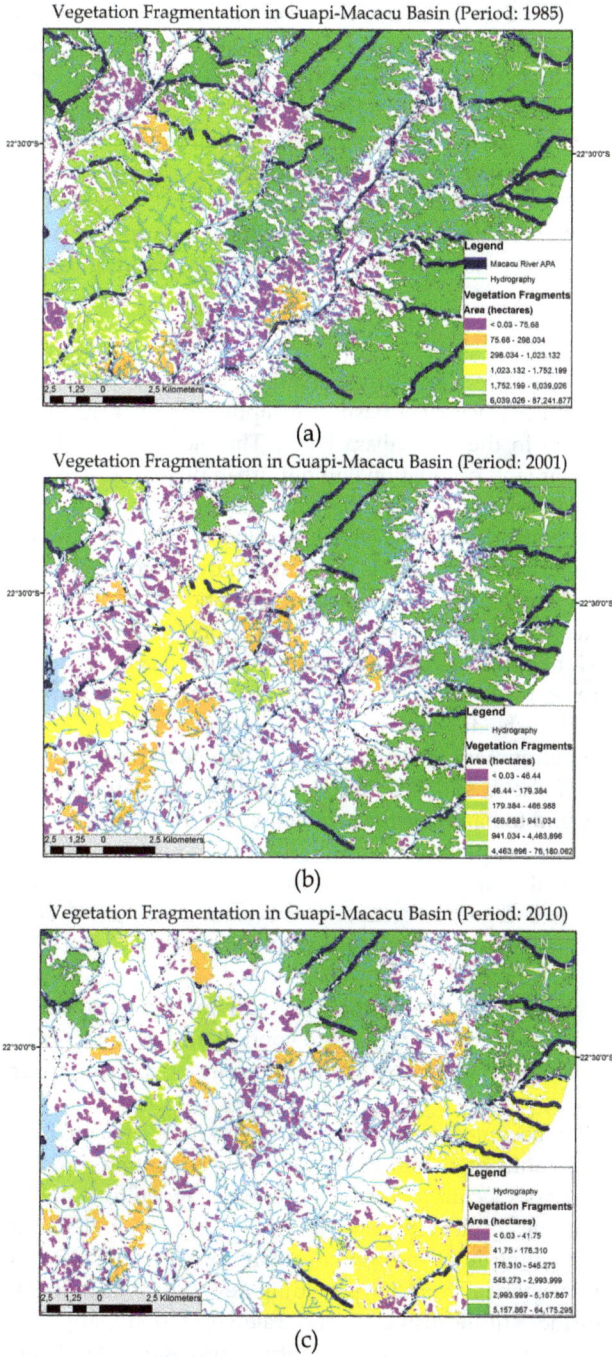

Fig. 18. Guapi-Macacu River Basin (1985, 2001 and 2010), respectively.

The vegetation patches in Guapi-Macacu River Basin are being fragmented along the years. Figures 18a,b,c shows a decrease in amount of vegetated area among the study periods. In 2010 it was observed that the vegetation fragments became smaller and isolated and the riparian vegetation is much less evident. Fig 18c shows that riparian vegetation had a severe decreasing in the central part of Guapi-Macacu. Human-centered attitudes toward water have deteriorated many riverine ecosystems, implying that the derived benefits have brought considerable environmental and social costs (Nilsson, 2007). Therefore, enhancements of river connectivity require thorough analysis and, ideally, should be carried out in concert with rehabilitation of flow dynamics (Nilsson op cit.).

6. Conclusion

Forests are cleared, degraded and fragmented in the Guanabara Basin. The riparian vegetation in the Guapi-Macacu river basin is disappearing over the years and it may soon affect the water supply in the Guanabara Basin. The most serious threat comes from the disorderly and irregular land occupation without urban planning.

People yet ignore the importance of the riparian vegetation and this negligence will cause in a near future serious problems in the available ground water and consequently in the water supply. Degraded riparian vegetation leaves surrounding ecosystems vulnerable to some disturbances as flood and drought. The existing vegetation in the riparian zone needs to be kept intact or protected by law. There must be strongly enforced laws to limit urban occupation in the river bank and avoid activities which deplete the riparian vegetation.

Despite its water reserves, Brazil now runs the risk of losing their most precious natural resources: water and forests, due to disorderly process of land occupation and irresponsible degradation of the environment. Everything that happens in one point of a river basin will influence the total of the basin. The irresponsible use of natural resources is bringing drought, which is bound to handicap the production of vital resources for sustaining human population, as is already happening in some parts of the world (Moraes, 2009).

There is a need for the stakeholders to establish conservation initiatives and share experiences in order to safeguard the riparian vegetation, mangrove swamps and Atlantic Forest remnants of the Guanabara Bay Basin. The forests are essentials for maintaining a drinking-water provision.

It is necessary urgent efforts to restore the remaining forests, with reforestation initiatives, especially in the headwaters of rivers. And to connect and expand the remnants of forests that are already fragmented which reduces the capacity of species to disperse through the landscape.

The Landsat satellite becomes very useful for working with medium scale maps for distinguishing higher ranks of classification. Moderate resolution remote sensing is widely used in a variety of sectors including land use planning, agriculture, and forestry. However, this approach cannot be applied directly in large scale maps. Thus, it is necessary an integrated study combining high resolution satellite images with data from other sources (thematic maps and numerical data) for providing new possibilities for land-cover analysis of the Basin. The further studies using a large-scale vegetation mapping for verifying with detailed the different succession stages of Atlantic Forest and its biodiversity. According to National Environmental Council (CONAMA) number 6, 1994 and Law n. 11.428 (December,

2006) different types of land use and natural vegetation cover are categorized based on ecological succession.

Water production for life also depends on protected areas. A connected administration (Federal, State and Municipal) of different protected areas is essential. Parks, reserves and environmental protected areas have to plan together and also work together against fires, environmental disasters, irregular occupations, trafficking and poaching of animals. Impacts in higher places affect lower places. So those the opposite. It is very important the interaction between public and private agencies and also non Governmental Organizations NGOs, universities, and research organizations aiming at define land use objectives and restrictions.

The restoration of Guapi-Macacu River Basin is not easy and involves efforts by different sectors of society to the implementation of integrated alternatives such as: technical, ecological, and also environmental education programs and policies. It is also urgent a linking of remnant vegetation of lowlands with that of highlands aiming at guarantee the water supply in the basin and also the integrity of the ecosystem.

Sambê Santa Fé Corridor restoration should be considered a high priority conservation action, since the corridor contributes to maintain the water quality and quantity of Guapi-Macacu watershed.

In Brazil were created some Conservation Units and ecological corridors to preserve biodiversity and restore landscape connectivity. Many units have been established for over 20 years and still do not have a management plan which difficult the environment zoning. Some NGO´s and public and private organs are currently involved with Atlantic Forest conservation efforts. Although, these efforts is not sufficient to preserve/conserve the Protection Areas and its remnant vegetation. This fact is mainly because the political and economic interests have a higher priority than environmental ones.

Urban planning is essential, especially in the areas of influence of Rio de Janeiro Petrochemical Complex (COMPERJ) that are expanding rapidly. The urban expansion in the eastern part of the basin may cause serious damages to remnant vegetation , especially with the reduction of buffer zones, including strict protection in areas as parks and ecological stations. The application of public and private investments in water reuse is also urgent to meet additional population demand.

With the increasing of world population it is necessary a change in way of life in order to seek what is really important for our survival by using different sources of sustainable energy as well as making use of non conventional materials. Humans can no longer continue exploiting and destroying the forests because in the end, we will be no water, no food and no prospect of survival in our planet.

7. Acknowledgment

Fundação Carlos Chagas Filho de Amparo à Pesquisa do Estado do Rio de Janeiro (FAPERJ)- Auxílio APQ1, Brazillian Institute of Spatial Research (INPE); U.S. Geological Survey (USGS); Consortium for Spatial Information- CGIAR; Instituto Brasileiro do Meio Ambiente e dos Recursos Naturais Renováveis – IBAMA; Instituto Estadual do Ambiente (INEA); Reserva da Biosfera da Mata Atlântica (RBMA); Programa de Despoluição da Baía de Guanabara (PDBG); Instituto Brasileiro de Geografia e Estatística (IBGE).

8. References

Alexandre B., Crouzeilles R. & Grelle, C. E. V. (2010). How Can We Estimate Buffer Zones of Protected Areas? A Proposal Using Biological Data. *Natureza & Conservação* 8,2, pp165-170

Amador, E. S. (1997). *Baía de Guanabara e Ecossistemas Periféricos: Homem e Natureza*. Edição do Autor. Reprodução e Encadernação: Reproarte Gráfica e Editora Ltda. Rio de Janeiro. Brasil. 539 pp.

Atkinson, P.M. (2004). Spatially weighted supervised classification for remote sensing. *International Journal of Applied Earth Observation and Geoinformation 5*, (July 2004), 277-291

Bidone, E.D. & Lacerda L.D. (2004). The use of DPSIR framework to evaluate sustainability in coastal areas. Case study: Guanabara Bay Basin, Rio de Janeiro, Brazil. *Reg Environ Change 4*, (March 2003),5–16

Brasil. (1965). *Lei Federal 4.771, 15 de setembro de 1965*. July 12, 2009. Available from: <http://www.planalto.gov.br/ ccivil_03/leis/L4771.htm>

Brasil. (1997). *Lei Federal 9.433, 08 de janeiro de 1997*. July 31, 2009. Available from: < http://www.jusbrasil.com.br/ legislacao/1027020/lei-9433-97>

Brasil. (2000). *Lei n° 9.985 de 18 de julho de 2000*. March 28, 2009. Available from: http://www.planalto.gov.br/ ccivil_03/leis/L9985.htm

Brasil (2006). *Lei n⁰· 11.428 de 22 de dezembro de 2006*. August 12, 2011. Available from: http://www.planalto.gov.br/ ccivil_03/ _ato2004-2006/2006/lei/l11428.htm

Cabral, D. de C; Freitas, S.R. Fiszon, J.T. (2007). Combining sensors in landscape ecology: imagery-based landfarm-level analysis in the study of human-driven forest fragmentation. *Sociedade & Natureza*, 19, 2, pp69-87

Camara, G., Souza R.C.M., Freitas U.M., Garrido J. (1996). SPRING: Integrating remote sensing and GIS by object-oriented data modelling. *Computers & Graphics*, 20, 3, (May-Jun, 1996), pp395-403

Carreiras, J.M.B., Pereira, J.M.C., Shimabukuro, Y.E. (2006). Land-Cover Mapping in the Brazilian Amazon using SPOT-4 vegetation data and machine learning classification methods. *Photogrammetric Engineering and Remote Sensing, 72,8*, pp897-910

Carvalho, F. M.V, Marco Júnior P. de,Ferreira, L. G. (2009). The Cerrado into-pieces: Habitat fragmentation as a function of landscape use in the savannas of central Brazil. Biological Conservation 142 (2009) 1392–1403

CIBG–Centro de Informação da Baía de Guanabara. (2010). *Panorama da Baía. Origem e Evolução*. August 12, 2010. Available from: <http://www.cibg.rj.gov.br/>

CITES *(Convention on International Trade in Endangered Species of Wild Fauna and Flora)*. (1975). November 09, 2010. Available from: http://www.cites.org/index.html

CONAMA (Conselho Nacional de Meio Ambiente). (1990). *Resolução n°13 de 6 de dezembro de 1990*. January 5, 2011. Available from <http://www.mma.gov.br/port/conama/res/res90/res1390.html>

CONAMA (Conselho Nacional de Meio Ambiente). (1994). *Resolução n° 6. 4 de maio de 1994*. March 5, 2011. Available from: http://www.ciflorestas.com.br/arquivos/lei_resolucao_061994_9446.pdf

CONAMA (Conselho Nacional de Meio Ambiente). (2002). *Resolução n° 302. 13 de maio de 2002*. February 3, 2011. Available from: <http://www.mma.gov.br/port/conama/res/res02/res30202.html>

CONAMA (Conselho Nacional de Meio Ambiente). (2002). *Resolução n⁰ 303 de 20 de março de 2002*. February 3, 2011. Available from:

<http://www.aesa.pb.gov.br/legislacao/resolucoes/conama
/303_02_preservacao_permanente. pdf>
CONAMA (Conselho Nacional de Meio Ambiente). (2005). *Resolução n° 357 de 17 de março de
2005*. January 10, 2010. Available from:
http://www.mma.gov.br/port/conama/res/res05/res35705.pdf
CONAMA(Conselho Nacional de Meio Ambiente). (2010). *Resolução n° 428 de 17 de dezembro
de 2010*. August 06, 2011. Available from:
http://www.mma.gov.br/port/conama/legiabre.cfm?codlegi=641
Da Costa, J. R. (2008). Impactos Ambientais na bacia hidrográfica de Guapi-Macacu e suas
conseguências para o abastecimento de água nos municípios do leste da Baía
Guanabara. CETEM (Centro de Tecnología Mineral. Retrieved from:
http://www.cetem.gov.br/publicacao/series_sgpa/sgpa-10_final.pdf
Foody, G.M. (2002). Status of land cover classification accuracy assessment. *Remote Sensing of
Environment* 80, (July 2001), 185-201. ESRI, 2008. *ArcView 9.2*. California, U.S.A.:
Redlands.
Fahrig, L. (2003). Effects of habitat fragmentation on biodiversity. *Annu. Rev. Ecol. Evol. Syst.*
(2003). 34, pp487–515
Forman, R. T. T. (1995). Some general principles of landscape and regional ecology.
Landscape Ecology 10, 3, pp133-142.
Franklin, A.B.; Noon, B.R. & Luke, G. T. (2002). What is habitat fragmentation? *Studies in
Avian Biology*, 25, pp:20-29.
Freitas, S.R. Hawbaker, T. J. Metzger J. P. (2010). Combining sensors in landscape ecology:
imagery-based and farm-level analysis in the study of human-driven forest
fragmentation Roads and topography can determine patterns of land use and
distribution of forest cover, particularly in tropical regions. *Forest Ecology and
Management* 259, (2010), pp410–417
FUNDAÇÃO SOS MATA ATLÂNTICA/INPE. (2009). *Atlas dos remanescentes florestais da
Mata Atlântica e ecossistemas associados no período de 2005-2008*. Fundação SOS Mata
Atlântica/INPE. January 10, 2011. Available from:
< http:// mapas. sosma.org.br/site_media /download/atlas%20mata%20atlantica-
relatorio2005-2008.pdf
Grimm, N. B., Foster, D., Groffman, P., Morgan Grove, J., Hopkinson C. S., Nadelhoffer K. J.
, Pataki D. E. & Peters D. P.C. (2008). The Changing Landscape:ecosystem
responses to urbanization and pollution across climatic and societal gradients.
Front Ecol Environ. 6, 5, pp264–272.
Hespanhol, I. (2008). *A new paradigm for urban water management and how industry is coping
with it*. In: Jimenez, B. & Asano, T. (Eds). *Water reuse: an international survey of
current practice, issues and needs*. IWA Publishing, London. Scientifi c and Technical
Report 20, pp. 467-482, Retrieved from:
<http://www.scielo.br/pdf/ea/v22n63/en_v22n63a09.pdf>
Hernández, J.H.A. (2010). Water Quality Monitoring System approach to support Guapi-
Macacu River Basin Management, Rio de Janeiro, Brazil. Masters Thesis.
(Environmental Science). Universidad Autonoma de San Luis Potosí. México.2010,
95f. Retrieved from:< http://www.alice.cnptia.embrapa.br /bitstream/doc
/875522/1/Thesis19092010coorientacao Rachel.pdf
IBAMA – Instituto Brasileiro do Meio Ambiente e dos Recursos Naturais Renováveis. (2002).
Modelo de Valoração Econômica dos Impactos Ambientais em Unidades de Conservação.
IBAMA, 66p. Retrieved from: http://www.ibama.gov.br/phocadownload/cnia/5-
valeconomicauc.pdf

INPE/DPI (2011) *SPRING user's manual* . April 25, 2011. Available from:
  <http://www.dpi.inpe.br/spring/ portuguese/manuals.html>

IBG – Instituto Baía de Guanabara. *Unidades de conservação*. January 2010. Available from:<
  http://www.portalbaiade guanabara.org.br/portal/baiadados.asp>

Jarvis, A.; Reuter, H.I; Nelson, A.; Guevara, E. (2008). *Hole-filled SRTM for the globe Version 4.*
  March 09, 2011. Available from: <http://srtm.csi.cgiar.org>

Mc Pherson, G. (1998). Atmospheric Carbon Dioxide Reduction by Sacramento's Urban
  Forest. *Journal of Arboriculture* 24, 4, pp215-223

Medeiros, B. F. (2009). The favela and its touristic transits. *Geoforum* 40, (2009), pp580–588

Metzger, J. P., Martensen, A. C., Dixo, M. Bernacci, L.C., Ribeiro, M. C., Teixeira, A.M.G.,
  Pardini, R.(2009). Time-lag in biological responses to landscape changes in a highly
  dynamic Atlantic Forest region. *Biological Conservation*, 142, (March 2009), pp1166-1177.

Moraes, L.A. F de. (2009). A visão integrada da Ecohidrologia para o manejo sustentável dos
  ecossistemas aquáticos. *Oecologia Brasiliensis* .13, 4, pp676-687

Naiman, R. J., and Décamps H. (1997). The ecology of interfaces: riparian zones. *Annual
  Review of Ecology and Systematics*. 28, pp621-658.

Nilsson, C., C. A. Reidy, M. Dynesius, and C. Revenga.(2005). Fragmentation and flow
  regulation of the world's large river systems. *Science*. 308, pp405-408.

Nilsson, C., R. Jansson, B. Malmqvist, and R. J. Naiman (2007). Restoring riverine
  landscapes: the challenge of identifying priorities, reference states, and techniques.
  Ecology and Society 12, 1. 16. [online] URL:
  http://www.ecologyandsociety.org/vol12/iss1/art16/

Niterói (2011). *Lei Municipal 2.856, 27 de julho de 2011*. September 10, 2011. Available from:
  <sraengenharia.com/site/legislacao/lei-2856-de-250711/>

Pringle, C.M. (2001). Hydrologic Connectivity and the Management of Reserves: a Global
  Perspective. *Ecological Applications*, 11,4, (March 2000), pp. 981–998

Richards, J.A. (1993). *Remote Sensing Digital Image Analysis: An Introduction* (2rd revised and
  enlarged edition), Springer-Verlag, ISBN 3-540-58219-3, New York

Silva, W.G., Metzger, J.P., Simões, S.B Simonetti C. A. (2010). Relief influence on the spatial
  distribution of the Atlantic Forest cover on the Ibiúna Plateau, SP. *Braz. J. Biol.*, 67,3,
  pp 403-411

Teixeira, A.M.G., Soares-Filho, B.S., Freitas, S.R., Metzger, J.P. (2009). Modeling landscape
  dynamics in an Atlantic Rainforest region: implications for conservation. *Forest
  Ecology and Management* 257, (October, 2008), pp1219–1230

USGS - *Global Visualization Viewer*. 04 June, 2010. Available from:
  <http://landsat.usgs.gov/Landsat_Search _and_ download.php>

Tuominem, J. , Lipping, T., Kuosmanen, V., Haapanen R. (2009). *Remote Sensing of Forest
  Health*. InTech, ISBN 978-953-307-003-2, Viena, Austria.

Veloso, H. P., Rangel Filho A. L. R., Lima J. C. A. (1991). Classificação da vegetação brasileira
  adaptada a um sistema universal. Fundação Instituto Brasileiro de Geografia e
  Estatística (IBGE), ISBN 85-240-0384-7, Rio de Janeiro.

Waleed, R.; Grealish. (2004). Mapping arable soils using GIS-based soil information database
  in Kwait. *Management of Environment Quality* 15,3, pp. 229-237.

Wang , X.; Pullar D.(2005). Describing dynamic modeling for landscapes with vector map
  algebra in GIS . *Computers & Geosciences*, 31, 8, (February, 2005) pp956-967.

WWF. (2011). *Código Florestal:Entenda o que está em jogo com a reformada nossa legislação
  ambiental*. August 30, 2011. Available from:
  <http://assets.wwfbr.panda.org/downloads/cartilha_codigoflorestal_20012011.pdf>

10

Geospatial Analysis of Deforestation and Land Use Dynamics in a Region of Southwestern Nigeria

Nathaniel O. Adeoye[1,*], Albert A. Abegunde[2] and Samson Adeyinka[2]
[1]*Department of Geography, Obafemi Awolowo University, Ile-Ife,*
[2]*Department of Urban & Regional Planning, Obafemi Awolowo University, Ile-Ife,*
Nigeria

1. Introduction

Deforestation is a complex phenomenon as there is little agreement about the components and the processes involved in it. FAO, (2001a) defined deforestation as 'the conversion of forest to another land use or the long-term reduction of the tree canopy cover below the minimum 10 percent threshold'. The world's original forest area, estimated at about 6 billion hectares, has declined steadily and about one-third of the forests have been lost during the past few hundred years (Sharma et al., 1992). The forests and woodlands of North Africa and the Middle East, for example, declined by 60 percent; those of South Asia by 43 percent; of tropical Africa by 20 percent; and of Latin America by 19 percent (Houghton et al. in WRI, 1987).

Although the world's forest area has been declining for centuries, it is in the last half of 20th century that the process was accelerated to an alarming rate (Osemeobo, 1991; Federal Environmental Protection Agency, 1992; Jaiyeoba, 2002). Since the 1960s, there has been a major change in the rate at which the forests are cleared. FAO, (1997) reported annual rates of deforestation in developing countries at 15.5 million hectares for the period 1980-1990 and 13.7 million hectares for the period 1990 - 1995. The total forest area lost during the 15-year period was more than 220 million hectares, much larger than the total land area of Mexico. Between 1950 and 1983, the area of Africa's woodlands and forests declines by 23% from 901 to 690 million hectares. Between 1981 and 1985 tropical African countries such as Nigeria, Cote d'lvoire, Sudan and Zaire were losing their forest at annual rates of 4.0%, 2.5%, 5.0% and 3.7%, respectively (IBRD/World Bank, 1992). In absolute terms, tropical forests in Africa are being lost at the rate of 3.7 million hectare a year with over half of the deforestation in West Africa alone (UNEP, 2002). The rate of deforestation in Nigeria in the 1980s was of the order of 400,000 ha yearly, while afforestation was a mere 3,200 ha. At such rates, Nigeria's remaining forest area would disappear by 2020 (Jaiyeoba, 2002).

Deforestation and forest degradations are now widely recognized as one of the most critical environmental problems facing the human society today with serious long term economic, social and ecological consequences. This issue has received much attention from policy makers to general public in recent years with vivid images of cleared forests and burning trees around

---

*Corresponding Author

the world. One of the consequences of deforestation is that the carbon originally held in forests is released to the atmosphere, either immediately if the trees are burned, or more slowly as unburned organic matter decays (Moutinho and Schwartzman, 2005). As reported by Diaz, et al., (2002), tropical deforestation in the Amazon alone is responsible for 2/3 of the Brazilian greenhouse gas emissions and it is estimated that 200 million tons of carbon, not including emissions from forest fires, are released annually into the atmosphere.

The effect of deforestation on biodiversity and climate change has been the subject of scientific studies and many documentaries of media. Achard et al., 2002; Houghton, 2003; Fearnside and Laurance, 2004, revealed in their studies that Global deforestation and forest degradation rates have a significant impact on the accumulation of greenhouse gases (GHGs) in the atmosphere. The Food and Agriculture Organization (FAO, 2001) estimated that during the 1990s 16.1 million hectares per year were affected by deforestation, most of them in the tropics. The Intergovernmental Panel on Climate Change (IPCC) calculated that, for the same period, the contribution of land-use changes to GHG accumulation into the atmosphere was 1.6 ± 0.8 Giga ($1\text{ G} = 109$) tonnes of carbon per year (Prentice et al., 2001), a quantity that corresponds to 25% of the total annual global emissions of GHGs. The United Nations Framework Convention on Climate Change (UNFCCC), in recognizing climate change as a serious threat, urged counties to take up measures to enhance and conserve ecosystems such as forests that act as reservoirs and sinks of GHGs. The Kyoto Protocol (KP), adopted in 1997, complements the UNFCCC by providing an enforceable agreement with quantitative targets for reducing GHG emissions. Besides, Aina and Salau, (1992) and Adesina, (1997) reported that forest loss leads to loss of wildlife habitats and extinction of plant and animal species that play important roles in maintaining a balance in the environment.

From the foregoing, it becomes obvious that the world tropical forest including Nigeria is depleting fast because of human influence. This is a problem at the macro and micro-level; such depletion is the result of government activities such as road development, arable farming, and land clearing for pasture (Osemeobo, 1991; Taylor et al., 1994; and Olofin, 2000). Statistical estimates have also shown that there is a negative correlation between exploitation of the forest and conservation in Nigeria that is, the annual rate of forest los is greater than the rate of conservation (Osemeobo, 1990).

The state of forests in general and tropical forests in particular, has been drawing the increasing disturbing attention of the world community. For instance, in the tropics, where about 2.5 billion people depend on natural forest resources for many economic and environmental goods and services, the depletion of forests has been posing threat to their means of livelihood.

Recently, the United Nations initiated a global awareness through its Global Environmental Outlook 2000 (GEO, 2000). In the developed countries of Europe and America, this awareness is high and it is the cause of several policies and strategies aimed at environmental preservation and conservation. In developing countries the awareness is just emerging. Presently 115 nations have Environmental Protection Agencies (EPA) and more than 215 international environmental treaties have been signed on issues bothering on global warming, biodiversity, ocean pollution, ozone layer depletion, and export of hazardous wastes (Ibah, 2001).

In Nigeria, government is also taking steps to correct the nation's degenerated environmental condition. One of such efforts is the establishment of Federal Environmental Protection Agencies (FEPA), (now a department under the Ministry of Environment), with its branches in all the states to monitor environmental quality including the forest resources. Very recently, and considering the need for environmental preservation, the government has established the Federal Ministry of Environment to oversee the country's environmental problems. Many Non – Governmental Organizations (NGOs) have also sprung up to discuss environmental degradation and proffer solutions (Ogunsanya, 2000).

The waves of concern on the state of our forest resources and the condition of the environment, have translated into a number of researches. Among the academics, studies have clearly revealed that forests worldwide have been and are being threatened by uncontrolled degradation and conversion to other types of land uses, influenced by increasing human pressure due to uncontrolled increase in human population resulting in agricultural expansion; and environmentally harmful mismanagement, including, for example, lack of adequate forest-fire control and anti-poaching measures, unsustainable commercial logging, overgrazing and unregulated browsing, harmful effects of airborne pollutants, economic incentives and other measures taken by other sectors of the economy (Houghton et al. in WRI, 1987; Sharma et al., 1992; Olofin, 2000; IPCC, 2000; FAO, 2001a; WRI, 2001; Jaiyeoba, 2002). Scholarly writings have also tried to explain the dimension and severity on global environmental changes (Arokoyu, 1999; Ogunsanya, 2000 and Jaiyeoba, 2002).

Studies have also explored the causal factors of deforestation. For example, there is a general agreement that deforestation is due to drought, forest fire, use of fuel wood, spread of extensive agriculture, and rapid urbanization, among others, (Areola, 1994; Olofin, 2000; Meyer and Turner, 1992; Taylor et al., 1994 and Smith, 1993). There are also substantial and growing works on resource management and conservation for the purpose of improving the environment (Areola, 1994; Smith, 1993; Mitchell, 1989; Munro et al., 1986; Olokesusi, 1992; Reed, 1996 and Rees, 1990).

In 2001, FAO published the Global Forest Resources Assessment 2000 (FRA 2000), which was largely based on information provided by the countries themselves and a remote sensing survey of tropical countries. It was supplemented by special studies undertaken by FAO. Among the outputs were two new global forest cover maps, estimates of forest cover, deforestation rates and forest biomass for each country, and several specialized studies on such topics as forest management and forest fires (FAO, 2001).

But Boroffice, (2006) argued that the often-quoted rates of deforestation for Nigeria were based on mere estimates or surrogate data rather than empirical studies. Most of the vegetation maps produced by international organizations, such as FAO, for the country are nothing more than broad generalizations which are not usually in tandem with local realities and are therefore, of little use to local authorities for planning purposes. Thus, the rate of forest loss at both local and national levels is not known with any accuracy.

To further the frontiers of knowledge on the state of the forest resources, which is still an inconclusive issue and to establish the emergence land use pattern from the depleted forest area, this study therefore, examines and analyses the extent of forest loss and land use dynamics; measures the rates of change over the periods of 25 years (1978-2003) in a part of southwestern Nigeria using a set of remotely sensed images and geographic information system (GIS) technology. There is a consensus among scientists that satellite images and GIS provide a reliable means for adequate and regular monitoring of forest estate. According to Ayeni (2001), application of GIS in environmental monitoring in developing world is still at its infancy yet; it has been extensively used in Europe and North America. Recognizing the importance of reliable tool for forest monitoring both for environmental sustainability as well as economic well being, there is, therefore, need to explore the tool in the developing world.

The specific objectives of study are to:

i.   identify the categories of land use and land cover from remotely sensed images;

ii.   measure the areal extent of each of the land uses/covers

iii. assess the rate(s) of change in land use and forest area and compare the rates of change over time; and

iv.   analyze the temporal patterns of land use and changes in forest coverage area over the period 1978-2003

2. The study area

Studies on deforestation can be carried out in any part of the world where growth in human population has taken place, which still has potential for further growth and development. The choice of the study area is however, guided by the primary objective of the study to examine the occurrence of deforestation, which is better perceived in area where there has been rapid urbanization as the case of the study area. The study area is a region located in Southwestern Nigeria, well known for its dense forest resources and fertile soils. However, as a result of rapid population growth, which led to the development of urban centres and increased farming activities, much of the forest areas have been converted to farmlands, perhaps to meet the needs of the teaming population. The emerging concern about the disappearance of forests in the study area therefore presents the reason for a study of this nature. Coincidentally there is a preponderance of various types of data from which the type of study envisaged here can be expected with minimum difficulty. Besides, the concern for sustainability provides the impetus for the choice of the area.

The study area spans part of Osun and Ekiti States (Figure1) and lies within latitude 7° 30′ and 7° 45′ North of equator and longitude 4° 30′ and 5° 00′ East of the Greenwich (Figure 2). It is about 866.25 sq kilometres rectangle in size. On the Nigeria topographical map, it is found in Ilesa sheet 243. The relief is rugged with undulating areas and granitic out-crops in several places. The notable ones among the hills are the Efon-Alaaye hill to the east of the study area, and domed hills in Ilesa area. The climate is of the lowland Tropical rain forest type with distinct wet and dry seasons. Many factors are responsible for the removal of forest resources in the area; among them are intensive agricultural practices, the establishment of more local government areas in recent years, the development of tertiary institutions and location of industrial plants as the case of Ilesa, which led to the influx of people to the area and subsequent development of housing estate and infrastructures.

According to 1991census figure, the study area is a populated area. Because of the dense population, the area has witnessed great structural development and growth, which in effect brought negative impact to the natural resources. The area of major population concentration is Ilesa with population of 130,321 based on 1991 census figure. Other populated areas include Ijebu-Ijesa, Efon-Alaaye, to mention but a few.

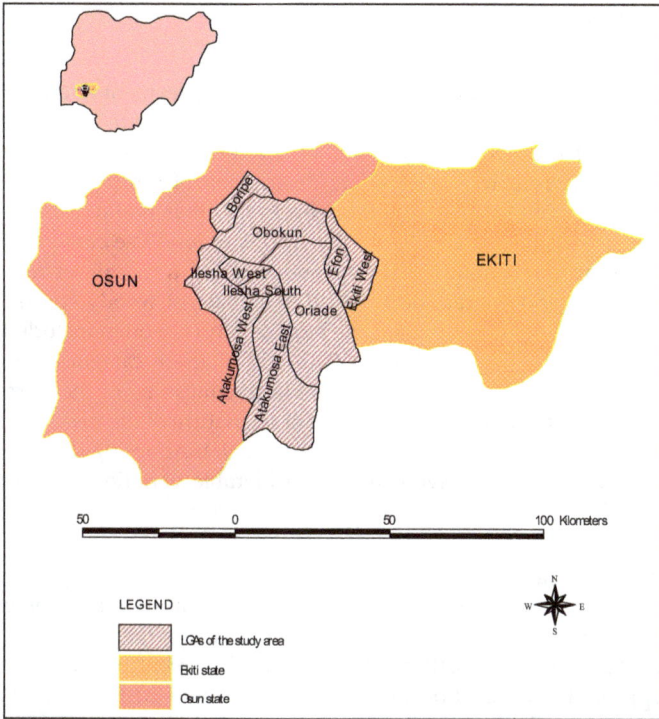

Fig. 1. The LGAs of the study area

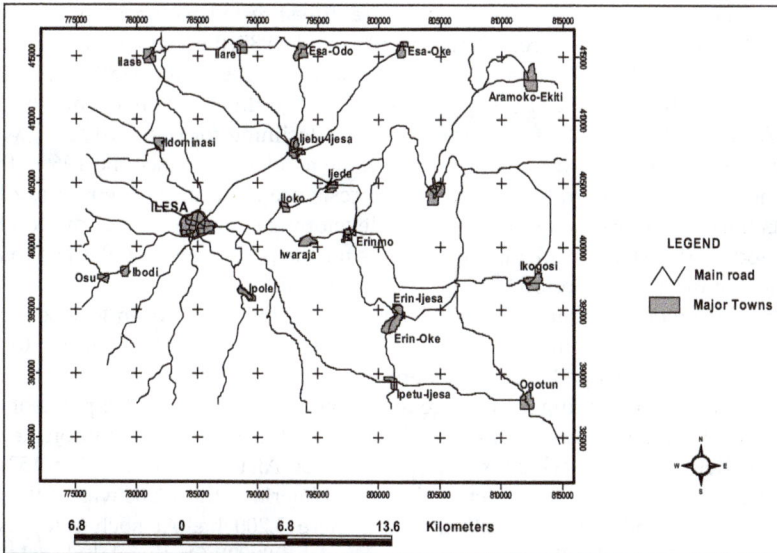

Fig. 2. Major communities of the study area

3. Conceptual clarification

3.1 Deforestation

Deforestation is a complex phenomenon as there is little agreement about the components and its process. FAO (2001a) defined it as 'the conversion of forest to another land use or the long-term reduction of the tree canopy cover below the minimum 10 percent threshold'. Deforestation is also referred to as "complete destruction of forest canopy cover through clearing for agriculture, cattle ranching, plantations, or other non-forest purposes" (Poor, 1976; and Mayaux and Malingreau, 1996). Other forms of land-use changes, such as, forest fragmentation (altering the spatial continuity and creating a mosaic of forest blocks and other land cover types), and degradation (selective logging of woody species for economic purposes that affects the forest canopy and the biodiversity) are often included in estimating deforestation. The characterization of forest into one of these categories depends on the temporal and spatial scale of observation. The subjective meaning of the term deforestation is thus not only linked to a value system but also to the nature of the measurement designed to assess it (Poor, 1976; and Mayaux and Malingreau, 1996). Adopting different perspectives of deforestation in data analysis have caused considerable variations in estimation of the area of forest cleared.

3.2 Facts about deforestation

The world's original forest area, estimated at about 6 billion hectares, has been declining steadily. About one-third of the forests have been lost during the past few hundred years (Sharma, et al., 1992). Between 1850 and 1980 about 15 percent of the earth's forests and woodlands disappeared as a result of human activities. While the forests and woodlands of North Africa and the Middle East, declined by 60 percent; those of South Asia by 43 percent that of tropical Africa declined by 20 percent; and of Latin America by 19 percent.

Depletion of forests is of particular significance because in the tropics, about 2.5 billion of her people depend directly or indirectly on natural forest resources for many economic and environmental goods and services. Between 1980 and 1985 the estimated annual rate of tropical deforestation was 0.6 percent or 11.4 million hectares (FAO, 1988). Recent studies estimates deforestation in the tropics at a rate of 17 to 20 million hectares annually (Rowe et al, 1992). Although the world's forest area has been declining for centuries, it is within the last half of the 20th century that the process became accelerated. Since the 1960s, there has been a major change in the rate at which the forests are cleared. A recent study by FAO (1997) puts the annual rates of deforestation in developing countries at 15.5 million hectares for the period 1980-1990 and 13.7 million hectares for the period 1990 - 1995. The total forest area lost during the 15-year period was more than 29.2 million hectares.

In 1999, the FAO reported that 10.5 per cent of Africa's forest had been lost between 1980 and 1995, the highest rate in the developing world and in sharp contrast to the net afforestation seen in developed countries. Forest loss between 1990 and 2000 was more than 50 million hectares, representing an average deforestation rate of nearly 0.8 per cent per year over this period (FAO, 2001a). Between 1990 and 2002, a total of 12 million hectares of forests were deforested, of which sub-regional West Africa accounted for 15% of the countries (UNEP, 2002). The rate of deforestation in Nigeria in the 1980s was of the order of 400,000 ha yearly, while re-afforestation was a mere 3,200 ha. At such rates, Nigeria's remaining forest area would disappear by 2020 (Jaiyeoba, 2002). On the global scale, the rate of tropical deforestation is not known with any accuracy (World Resources Institute, 2003).

This informs gap in our knowledge. However, certain factors have been advanced to be responsible for deforestation and forest degeneration.

3.3 Causes of deforestation

Deforestation occurs for many reasons but it is important to distinguish between the causes that are directly related to deforestation and those that are underlying. The direct causes are those activities (by individuals, corporations, government agencies, or development projects), which clear forests. Underlying causes are those behind the direct causes, which motivate the direct causes. (http://www.wrm.org.uy/deforestation/UN report.html). Direct causes include commercial timber production, clearing of land for agriculture and urban expansion, and harvesting of wood for fuel and charcoal. These activities also open up forests by the construction of access roads to logging sites, fragmenting the forests and facilitating further clearance, resource extraction, and grazing by locals and commercial organizations (Rowe et al., 1992; UNDP, UNEP, World Bank and WRI, 2000; State of the World's Forests, 2001). According to the United Nations Framework Convention on Climate Change, the overwhelming direct cause of deforestation is agriculture. Subsistence farming is responsible for 48% of deforestation; commercial agriculture is responsible for 32% of deforestation; logging is responsible for 14% of deforestation and fuel wood removals make up 5% of deforestation (UNFCCC, 2007).

Indirect (underlying) causes of deforestation include population growth, policies, agreements, legislation, lack of stakeholder participation and market factors that encourage the use of forest products, leading to loss, fragmentation or degradation (Rodgers, Salehe and Olson, 2000). Other causes of forest loss are conflict, civil wars and lack of good governance (Verolme and Moussa, 1999).

In Africa as in all parts of the world, deforestation were caused by a combination of natural and human factors, the chief of which has been the conversion of forest lands to agricultural land (Adesina, 1997; Rowe et al., 1992). As Williams (1990) reported, the introduction of new crops and new methods of exploitations around 1600 radically altered tropical forests. It was reiterated that forests were cleared to make way for cash crops such as, rubber in Malaysia and Indonesia, coffee in Brazil, tea in India and China, sugar cane in the Caribbean, tobacco and oil palm in Asia.

Ola-Adams (1981) attributed the removal of forested areas to intensive agricultural practices. It was reported that approximately 2,000 hectares of western edge of Ogbesere forest reserve in Nigeria had been cut over and replaced by permanent agriculture. The study further revealed that in some other parts of the high forest, several areas of forest estate were being de-reserved for the establishment of agricultural crops. Besides, in the current efforts to diversify the country's economy, large areas of high forest zones were being cleared and planted with food and tree crops; 45,845 ha. for food crops; 10,000 ha. for oil palm; 73,000 ha. for cocoa and about 140,000 ha. for rubber plantations.

There are many causes of contemporary deforestation, among them are corruption of government institutions (Burgonio, 2008; WRM, 2003), the inequitable distribution of wealth and power (Global Deforestation, 2006), population growth (Marcoux, 2000), and overpopulation (Butler, 2009; Stock and Rochen, 2009), and urbanization (Karen, 2003). Globalization is often viewed as another root cause of deforestation (YaleGlobal, 2007; Butler, 2009), though there are cases in which the impacts of globalization (new flows of labour, capital, commodities, and ideas) have promoted localized forest recovery (Hecht, et al, 2006).

Experts do not agree on whether industrial logging is an important contributor to global deforestation (Angelsen and Kaimowitz, 1999; Laurance, 1999). Some argue that poor people are more likely to clear forest because they have no alternatives, others that the poor lack the ability to pay for the materials and labour needed to clear forest (Angelsen and Kaimowitz, 1999). One study found that population increases due to high fertility rates were a primary driver of tropical deforestation in only 8% of cases (Geist and Lambin, 2002). However, a shift in the drivers of deforestation over the past 30 years has been noted. Whereas deforestation was primarily driven by subsistence activities and government-sponsored development projects like transmigration in countries like Indonesia and colonization in Latin America, India, Java, and so on, during late 19th century and the earlier half of the 20th century. By the 1990s the majority of deforestation was caused by industrial factors, including extractive industries, large-scale cattle ranching and extensive agriculture (Rudel, 2005; Butler and Laurance, 2008).

3.4 Effects of deforestation

Deforestation causes extinction, changes to climatic conditions, desertification, and displacement of populations (Sahney, et al, 2010). It is a contributor to global warming (Fearnsidel and Laurance, 2004) and is often cited as one of the major causes of the enhanced greenhouse effect. According to the Intergovernmental Panel on Climate Change, deforestation mainly in tropical areas, could account for up to one-third of total anthropogenic carbon dioxide emissions (IPCC Fourth Assessment Report). But recent calculations suggest that carbon dioxide emissions from deforestation and forest degradation (excluding peatland emissions) contribute about 12% of total anthropogenic carbon dioxide emissions with a range from 6 to 17% (Van der Werf, et al, 2009). Scientists also state that, Tropical deforestation releases 1.5 billion tones of carbon each year into the atmosphere (Defries, 2007).

Reducing emissions from the tropical deforestation and forest degradation (REDD) in developing countries has emerged as new potential to complement ongoing climate policies. The idea consists in providing financial compensations for the reduction of greenhouse gas (GHG) emissions from deforestation and forest degradation" (Wertz-Kanounnikoff and Alvarado, 2007).

Deforestation results in declines in biodiversity. Estimate shows that rainforest is losing 137 plant, animal and insect species every single day due to deforestation, which equates to 50,000 species a year (Rainforest Facts, 2010). Leakey and Roger, (1996) state that tropical rainforest deforestation contributed to the ongoing Holocene mass extinction. But scientific understanding of the process of extinction is insufficient to accurately make predictions about the impact of deforestation on biodiversity (Pimm, et al, 1995). Most predictions of forestry related biodiversity loss are based on species-area models, with an underlying assumption that as the forest declines species diversity will decline similarly. However, many such models have been proven to be wrong and loss of habitat does not necessarily lead to large scale loss of species. Species-area models are known to over-predict the number of species known to be threatened in areas where actual deforestation is ongoing, and greatly over-predict the number of threatened species that are widespread (Pimm, et al, 1995).

3.5 Control

Major international organizations, including the United Nations and the World Bank, have begun to develop programs aimed at curbing deforestation. The blanket term Reducing

Emissions from Deforestation and Forest Degradation (REDD) describes these sorts of programs, which use direct monetary or other incentives to encourage developing countries to limit and/or roll back deforestation. Funding has been an issue, but at the UN Framework Convention on Climate Change (UNFCCC) Conference of the Parties-15 (COP-15) in Copenhagen in December 2009, an accord was reached with a collective commitment by developed countries for new and additional resources, including forestry and investments through international institutions, that will approach USD 30 billion for the period 2010 - 2012 (UNFCC, 2009). Significant work is underway on tools for use in monitoring developing country adherence to their agreed REDD targets. These tools, which rely on remote forest monitoring using satellite imagery and other data sources, include the Center for Global Development's FORMA (Forest Monitoring for Action) initiative (FORMA, 2009) and the Group on Earth Observations' Forest Carbon Tracking Portal, (GEO FCT, 2010). Methodological guidance for forest monitoring was also emphasized at COP-15 (UNFCC, 2009).

4. Methodology

4.1 Data acquisition and image processing

Remotely sensed data of different sources were used for this study (Figures 1, 2, 3, & 4). This is because of the constraint of the availability of field data in this part of the world. These are Landsat MSS, acquired on March 15, 1978 of 80m spatial resolution; SPOT XS, obtained on May 19, 1986 and SPOT XS acquired on November 28, 1994, of 20m spatial resolution respectively; NigeriaSat_1 acquired on September 23, 2003, of 32m spatial resolution (Table 1).

S/no	Data type	Date/Year obtained	Spatial Resolution	Swath	Spectral bands	Agency
1.	Landsat MSS	March 15, 1978	80m	185km	4 Bands [Blue, Green, Red & Near infrared]	FORMECU, Abuja, Nigeria
2.	SPOT XS	May 19, 1986	20m	60km	3 Bands [Green, Red & Infrared]	RECTAS, Ile-Ife, Nigeria
3.	SPOT XS	November 28, 1994	20m	60km	3 Bands [Green, Red & Infrared]	FORMECU, Abuja, Nigeria
4.	NigeriaSat-1	September 23, 2003	32m	600km	3 Bands [Near-infrared, Red & Green]	NASRDA, Abuja, Nigeria

Source: Author's field survey

Table 1. Attributes of the images used for the Study

Ideally, studies such as this would be better conceived if images were acquired twice a year to allow for seasonal variation in foliage coverage. In southwestern Nigeria for instance, the rainy season spans the period of eight months that is, between March and October while the dry season starts from November and lasts until February. The images used are both acquired in rainy season (Landsat MSS, 1978, SPOT XS, 1986 and NigeriaSat-1, 2003) and dry season (SPOT XS, 1994). These differences in the date of acquisition may cause disparity in the results. However, they were used because of the spectral information of the study area they contain.

The data were extracted as a sub-scene from the original dataset. For the purpose of temporal land use/cover change detection, a common window covering the same geographical coordinates of the study area was extracted from the scene of the images obtained. The sub-map operation of ILWIS 3.2 Academic allows the user to specify a rectangular part of a raster map to be used. To extract the study area from the whole scene of the images obtained, the numbers of rows and columns of the area were specified. While Landsat MSS 1978 contained 600 lines and 733 columns pixels, SPOT XS 1986 and 1994 consist of 1373 lines and 2005 columns pixels, respectively, when NigeriaSat_1 2003 has 1083 lines and 1150 columns of pixels.

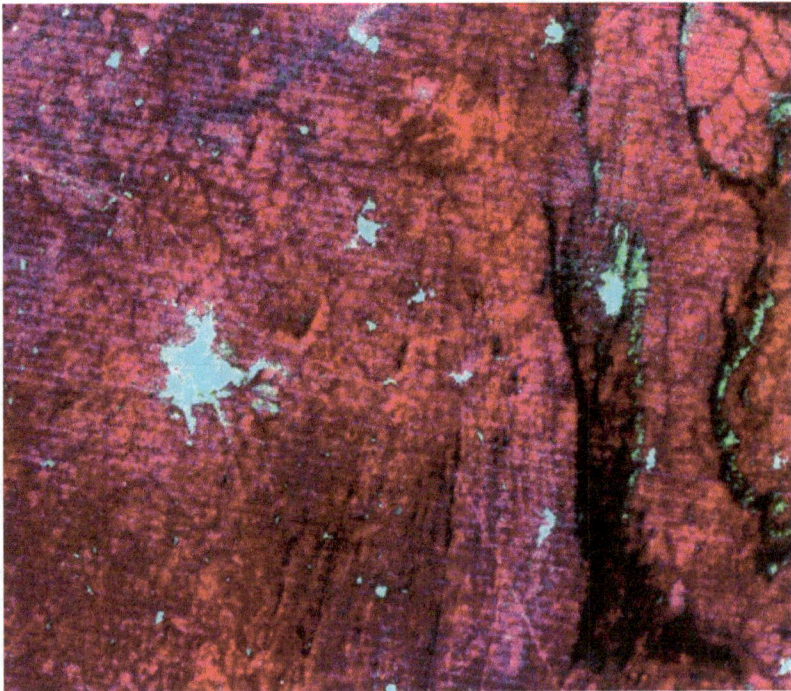

Fig. 3. Landsat Multispectral (MSS) 1978

The false colour composite was used for all the image data to relate colours and patterns in the image data to the real world features. For Landsat MSS 1978, channel 1 was assigned to red plane, channel 2 to green plane, and channel 3 to blue plane. This makes the band Red, Blue, Green (RBG-123) false colour composite. For SPOT XS 1986 and 1994, channel 3 was assigned red plane, channel 2 to green and channel 1 to blue plane. The band combination then consisted of Blue, Green and Red (BGR-321) to produce false colour composite. For NigeriaSat_1 2003 false colour composite, channel 1 was assigned to red plane, channel 2 to green plane and channel 3 to blue plane. This puts the band combination as Red, Green and Blue (RGB-123).

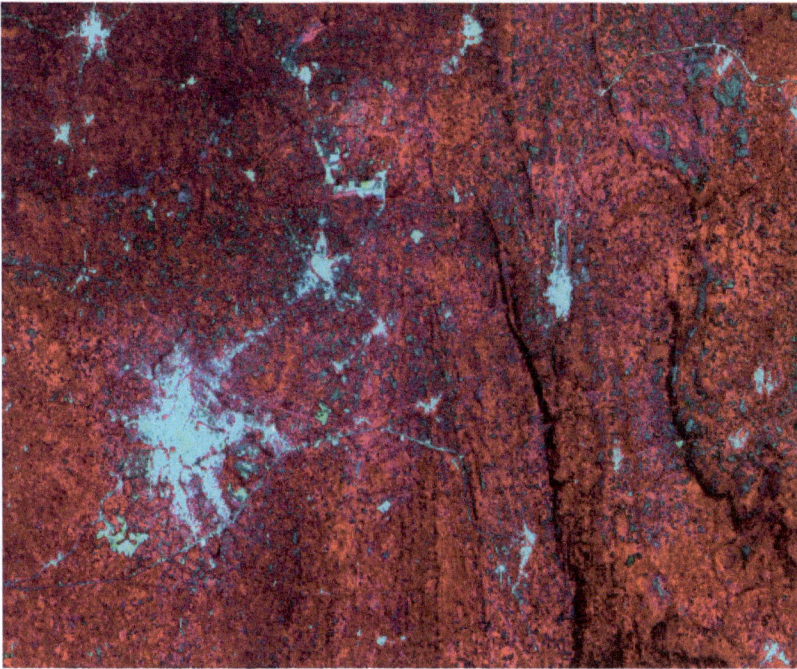

Fig. 4. SPOT Satellite XS 1986

With constraints such as spatial, spectral, temporal and radiometric resolution, relatively simple remote sensing devices cannot record adequately the complexity of the Earth's land and water surfaces. Consequently, error creeps into the data acquisition process and can degrade the quality of the remotely sensed data collected. Therefore, it is necessary to preprocess the remotely sensed data before the actual analysis. Radiometric and geometric errors are the most common types of errors encountered in remotely sensed imagery. The commercial data provider has removed the radiometric and systematic errors of the data used. However, while the unsystematic geometric distortion remains in the image. The geometric errors were corrected using ground control points (GCP).

The process of georeferencing in this study started with the identification of features on the image data, which can be clearly recognized on the topographical map of the study area and whose geographical locations were clearly defined. Stream intersections and the intersection of the highways were used as ground control points (GCP). The latitude and longitude of the GCPs of clearly seen features obtained in the base map were used to register the coordinates of the image data used for the study. All the images were georeferenced to Universal Transverse Mercator projection of WGS84 coordinate system, zone 31N with Clarke 1880 Spheroid. Nearest-neighbor re-sampling method was used to correct the data geometrically. A correlation threshold was used to accept or discard points. The correlation range was within limits that is, 1 pixel size. The x and y corrections were below 0.5 pixel.

Fig. 5. SPOT Satellite (XS) 1994

Fig. 6. NigeriaSat-1 2003

4.2 Classification

In this study, the satellite images were classified using supervised classification method. The combine process of visual image interpretation of tones/colours, patterns, shape, size, and texture of the imageries and digital image processing were used to identify homogeneous groups of pixels, which represent various land use classes of interest. This process is commonly referred to as training sites because the spectral characteristics of those known areas are used to train the classification algorithm for eventual land use/cover mapping of the remainder of the images.

To validate the tonal values recorded on the satellite images with the features obtained on the ground and also to know what type of land use/cover is actually present, the study engaged in ground truthing to five communities of the study area. These are Ilesa, Ijebu-Ijesa, Efon-Alaaye, Iloko-Ijesa and Erin-Oke (Figure 2). Before the ground truthing, map of the study area was printed and was used as guide to locate and identify features both on ground and on the image data. The geographical locations of the identified features on the ground were clearly defined. These were used as training samples for supervised classification of the remotely sensed images. Eight categories of land uses and land covers were clearly identified during ground truthing. These are forest/secondary re-growth, agro-forestry, arable farmlands, fallow/shrub, bare soils, water body, bare rocks and built-up areas/settlements.

4.3 Accuracy assessment

Determination of meaningful change categories was conducted by evaluating the classification accuracy. Every classified pixel has accuracy for a particular land use/cover type. The most common and typical method to assess classification accuracy Error Matrix (sometimes called a confusion matrix or contingency table) was used to assess the accuracy assessment for this study. Error matrix compares, on a category-by-category basis, the relationship between known referenced data and the corresponding results of an automated classification. Such matrices are square, with the number of rows and columns equal to the number of categories whose classification accuracy is being assessed (Jensen, 1996).

5. Results and discussion

5.1 Accuracy assessment of satellite images

The accuracy assessment of four temporal data shows that most land use types were classified with acceptable level of accuracies. The low classification accuracies found in Arable farmlands, Agro-forestry and Fallow/Shrub classes was due to the similar spectral characteristics in them and the prevailing season, which posed constraint to the classification process. However, the overall accuracy of the land use categories makes the study reliable for planning. The average reliability of Landsat MSS 1978 was 57.24% while the overall accuracy was 76.20%; SPOT XS 1986 average reliability was 66.46% and the overall accuracy of 67.43%; SPOT XS 1994 average reliability was 65.04 % while the overall accuracy was 60.88%. NigeriaSat_1 2003 average reliability was 63.25 % and the overall accuracy was 72.05% (Tables 2, 3, 4 and 5).

	Agrofo-restry	Arable farmlands	Bare soils	Exposed rock	Forest	Settlement	Shrub/Fallow	Water body	ACCURACY
Agroforestry	216	4	0	12	82	0	190	0	0.43
Arable farmlands	13	1801	302	74	1309	0	2782	10	0.29

	Agrofo-restry	Arable farmlands	Bare soils	Exposed rock	Forest	Settlement	Shrub/Fallow	Water body	ACCURACY
Bare soils	3	1460	2417	44	271	74	359	6	0.52
Exposed rock	0	82	77	4580	443	0	1	73	0.87
Forest	70	549	50	2274	45232	0	3011	48	0.88
Settlement	0	0	52	0	0	2695	0	0	0.98
Shrub/Fallow	653	882	54	6	3809	0	5809	0	0.52
Water body	0	55	26	129	274	0	1	33	0.06
RELIABILITY	0.23	0.37	0.81	0.64	0.88	0.97	0.48	0.19	

Average Accuracy = 56.92 %
Average Reliability = 57.24 %
Overall Accuracy = 76.20 %

Table 2. Matrix of land use/land cover for 1978

	Agroforestry	Arable farmland	Bare soils	Exposed rock	Forest	Settlement	Shrub/fallow	Water body	ACCURACY
Agroforestry	2027	0	1	0	1	0	29	0	0.82
Arable farmland	0	1254	241	10	0	0	38	544	0.57
Bare soils	113	5000	3227	238	78	68	1000	3002	0.21
Exposed rock	0	173	367	4052	21	0	29	2303	0.50
Forest	214	15	146	146	23942	0	4431	4	0.72
Settlement	0	3	41	0	0	22488	0	0	0.85
Shrub/fallow	226	559	959	32	511	0	11444	21	0.80
Water body	0	195	85	107	0	0	3	1381	0.78
RELIABILITY	0.79	0.17	0.64	0.88	0.98	1.00	0.67	0.19	

Average Accuracy = 65.68 %
Average Reliability = 66.46 %
Overall Accuracy = 67.43 %

Table 3. Matrix of land use/land cover for 1986

	Agro-forestry	Arable farmlands	Bare soils	Exposed rock	Forest	Settlement	Shrub/fallow	Water body	ACCURACY
Agro-forestry	3763	0	0	0	124	0	109	0	0.78
Arable farmlands	0	33	2	1	0	0	3	0	0.82
Bare soils	5	7825	11370	1222	238	113	2717	274	0.41
Exposed rock	0	139	493	15659	1890	0	940	19489	0.35
Forest	711	4	0	678	22169	4	1071	141	0.87
Settlement	0	0	54	0	0	25244	0	0	0.86
Shrub/fallow	137	1031	466	120	861	3	8163	7	0.70
Water body	0	5	34	64	8	0	33	1926	0.89
RELIABILITY	0.82	0.00	0.92	0.88	0.88	1.00	0.63	0.09	

Average Accuracy = 71.28 %
Average Reliability = 65.04 %
Overall Accuracy = 60.88 %

Table 4. Matrix of land use/land cover for 1994

	Agro-forestry	Arable farmland	Bare rock	Forest	Bare soils	Settlement	Shrub/fallow	Water body	ACCURACY
Agroforestry	1333	0	0	54	0	0	44	0	0.93
Arable farmland	0	5485	363	0	464	0	1662	44	0.68
Bare rock	0	928	13507	0	34	30	19	1530	0.84
Forest	2244	0	10	11412	0	0	4558	26	0.63
Bare soils	1	5174	41	4	10517	720	6118	124	0.46
Settlement	0	174	22	0	1438	13267	7	632	0.85
Shrub/fallow	1475	504	64	657	1351	0	24127	123	0.85
Water body	0	86	207	0	4	58	0	263	0.43
RELIABILITY	0.26	0.44	0.95	0.94	0.76	0.94	0.66	0.10	

Average Accuracy = 70.97 %
Average Reliability = 63.25 %
Overall Accuracy = 72.05 %

Table 5. Matrix of land use/land cover for 2003

5.2 Land use change between 1978 and 2003

The overall results of the study indicate that the area of forestland has been continuously declined, while the area of shrub/fallow, built-up area (settlements) and waterbody was proportionally increased (Table 6).

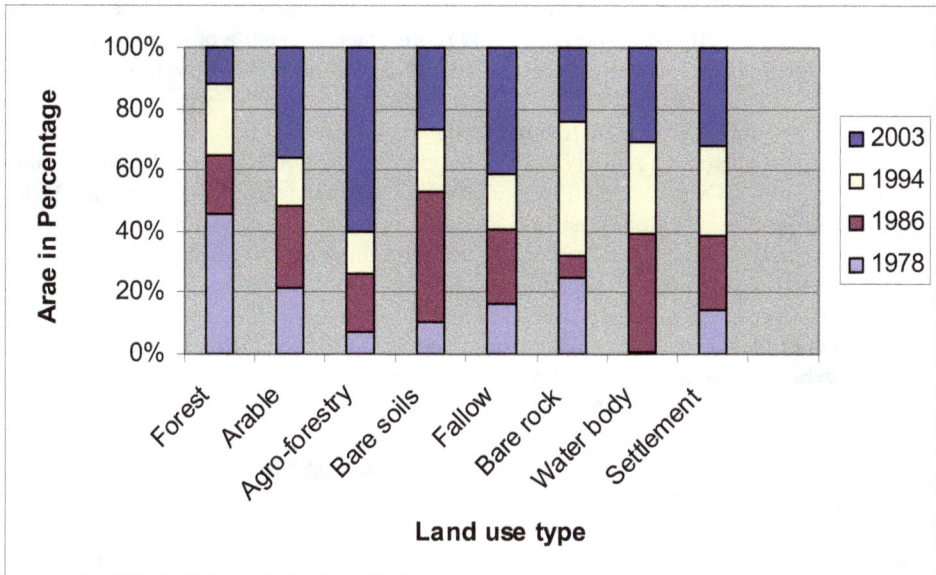

Source: Author's Data Analysis

Fig. 7. Land use/land cover change 1978-2003

Categories of Land Use (LU)	1978 LU (ha.)	% of LU 1978	1986 LU (ha.)	% of LU 1986	% Change in LU 1978-1986	1994 LU (ha.)	% of LU 1994	% Change in LU 1986-1994	2003 LU (ha.)	% of LU 2003	% Change in LU 1994-2003	Overall change (hectares)
Forest	72310.93	60.40	30587.80	28.11	-57.70	36635.75	36.52	-19.77	18841.70	14.44	-48.57	-53,469 (Decrease)
Arable farmlands	9441.88	7.89	11830.08	10.87	25.29	6830.37	6.81	-42.27	15923.20	12.21	133.13	6,482.32 (Increase)
Agro forestry	558.76	0.47	1498.87	1.38	168.23	1080.13	1.08	-27.95	4774.30	3.66	342.03	4,215.54 (Increase)
Bare rock	7340.62	6.13	2246.29	2.06	-69.40	12979.33	12.94	477.81	7282.28	5.58	-43.89	-58.34 (Decrease)
Bare soils	2213.28	1.85	9347.52	8.59	322.30	4348.26	4.33	-53.48	5814.88	4.46	33.73	3,601.60 (Increase)
Water body	139.36 1.39 km²	0.12	5782.70 57.83 km²	5.31	4060.43	4617.84 46.18 km²	4.60	-20.15	4608.72 46.09 km²	3.53	-20.00	4,469.36 (Increase)
Fallow	23397.91	19.54	35330.25	32.47	51.00	26804.25	26.72	-24.13	59436.85	45.57	121.74	36,038.94 (Increase)
Settlement	943.39	0.79	1567.60	1.44	66.17	1947.94	1.94	24.26	2089.57	1.60	7.29	1,146.18 (Increase)

Table 6. Land use change between 1978 and 2003

5.3 Forest status between 1978 and 2003

The area of forest in the study area was becoming smaller with time (Table 6). In 1978, forest area was 72,310.93 hectares, which constituted 60.40% of the entire land use. By 1986, the areal extent had decreased from 72,310.93 to 30,587.80 hectares, a decline of 57.7% within the period of 9 years. In 1994, there seems to be an increase in the areal extent of the resources. This is because in 1986, it was 30,587.80 hectares but in 1994 it increased to 36,635.75 hectares. This increase was not an expansion in the coverage but re-growth of the forest area. In 2003, the forest resources in the study area had almost disappeared when the areal extent was reduced to 18,841.70 hectares. This represented a decline of 48.57% within the period of 10 years. From the analysis, a total area of 53,469 hectares of the forest resources was removed.

The decline in the forest area as discovered in the study confirms the report of FAO (2001) that tropical forest including Nigeria was on the decline. According to the United Nations Food and Agriculture Organization (FAO), 93,900 km² of forest were cleared per year during the 1990s, with annual rates of forest loss (positive for all continents with tropical forests): Africa 0.8%, Asia 0.1%, Oceania 0.2%, North and Central America 0.1%, and South America 0.4% (FAO, 2001). The study also upholds the recognition of Cassel-Gintz, and Petschel-Hels (2001) that discovered deforestation in their recent study and concluded that it is one of the core problems of global environmental change. UNDP, UNEP, World Bank and WRI, (2000) associated extensive deforestation with loss of biodiversity, climate change, watershed degradation and these pose threat to cultural survival of indigenous population, economic loss among others. According to the estimates by the Food and Agricultural Organization (FAO 1983), Nigeria through careless exploitation and husbandry, destroys about 600,000 ha of her forest every year, as against the reforestation efforts of about 25,000 ha a year, which replenishes only about 4 percent of the loss. At such rates, Nigeria's remaining forest area would disappear by 2020 (Jaiyeoba 2002).The general effect of timber exploitation mostly from the high forests which cover only about 12.41 million ha of the country's 91.1 million ha of land space, (i.e. about 13.5%), is that the forest areas are fast reducing to savanna vegetation. Consequently, the forest stability is disrupted and its ecosystem has been seriously disturbed.

The disturbances are not only in terms of its ability to regenerate through the natural process, but some species of trees and fauna are being endangered (Okafor 1988).

In recent times, precisely June 1992, the Earth Summit in Rio de Janeiro, Brazil more correctly known as the United Nations Conference on Environment and Development (UNCED) was convened. There was an agreement on a set of "Principles for a global consensus on the management, conservation and sustainable development of all types of forests" and devoted a full chapter of its Agenda 21 to "Combating deforestation". Since this conference, there has been debate on the phenomenon and conscious efforts are being made to carefully use and protect natural resources mostly in the developed countries. But there seems to be limited enthusiasm from developing countries, including Nigeria, where ineptitude, particularly as regards the implementation of environmental laws, has been the greatest bane to a sustainable environmental practice (Lines et al.1997; Ibah. 2001). For instance, between 1995 and 1999 the military junta in Nigeria showed nonchalant attitude to the sustainability of the forest resources. This gave room to illegal lumbermen to penetrate the forest in the country, both reserved (protected) and the opened forest (unprotected), as the case of the study area, thus declined the area of forest nationwide. In poor and developing countries where often the administrative structures are either too weak and/or corrupt to enforce rules and regulations effectively and where tax and incentive systems do not work, forest clearing and environmental degradation has proceeded unabated.

5.4 Arable farmland status from 1978 to 2003

Arable farming is one of the prominent economic activities among the people of the study area as it is reflected in Table 6. The farmlands occupied 9,441.88 hectares, which represented 7.89% of the entire land use in 1978. In 1986, the land area of arable farmlands expanded from 9,441.88 to 11,830.08 hectares, making an increase of 2,388.20 hectares. But there seems to be a decrease in the cultivated area in 1994. This is due to the period at which the image was obtained that is, November, the beginning of dry season in southwestern Nigeria. At this time, most crops had been harvested except few such as, cassava, maize, among others which are left for proper maturation. Thus, majority of the farmlands are abandoned till the next planting season and they are sometimes overgrown with shrubs during this period. Between 1994 and 2003, arable farmlands increased from 6,830.37 to 15,923.20 hectares, representing an increase of 133.13%.

A close assessment of arable farmland areas between 1978 and 2003 shows that, 6,482.32 hectares of the land area was gained. This is negligible if we are going to solve the problem of food insecurity. However, the expansion of farmlands and declining in the forest area affirms the claim of Ola-Adams, (1981) and Williams, (1990). In their studies it was discovered that forests were removed to pave way for food and tree crops. Besides, the United Nations Framework Convention on Climate Change (UNFCCC, 2007) asserted that the overwhelming direct cause of deforestation is agriculture. In their discovery, subsistence farming was responsible for 48% of deforestation; commercial agriculture 32%; logging 14% and fuel wood removals make up 5% of deforestation. This was the observation of the study during the ground truthing. Besides, arable farmlands were scattered all over the study area especially around the settlements and some newly deforested areas and there were no large scale mechanized farming in the area. Again, arable farmlands were common features around the mountainous terrain and crops such as, maize, yam, cassava, rice, among others were mainly grown in the study area.

5.5 Status of agro-forestry between 1978 and 2003

Agro-forestry, otherwise known as tree crop plantation in the study only constituted 0.47% that is, 558.76 hectares of the entire land use in 1978 (Table 6). The improved spatial resolution of the image data (SPOT 1986 and 1994) however, brought a significant difference to what was recorded in Landsat MSS 1978. Also, variation in the period at which the image data were taken influenced the differences in the pixel statistics of the two images. It should be recalled that while SPOT XS 1986 was obtained in May, the beginning of rainy season in southwestern Nigeria, SPOT XS 1994 was acquired in November, the beginning of dry season in the area. In May, tree canopies are just springing up due to the long periods of dryness. On the other hand, the tree canopies are still fresh in November despite the inception of the dry period. This difference in the time period is capable of influencing the canopy characteristics of the farmlands, consequently affecting the result. The estimated area of agro-forestry in 2003 was 4,774.30 hectares, which represented an increase of 342.03%. Although agro-forestry is not commonly found in the study area due to poor topography in some areas, the tree crop plantations are well grown in some areas around Ilesa, Erin-Oke, Aramoko-Ekiti (Figure 2). Because of the economic value of the crops, many deforested areas have been replaced with the cocoa plantations in particular, which confirms the findings of Ola-Adams, (1981) and Williams, (1990).

Although research in agro-forestry especially in Africa, has paid a significant quantum of emphasis on this concept for about three decades, it again is adaptation of earlier system of slash and burn and perennial forest gardens, although on a more scientific basis. The benefits of incorporating trees, especially fast growing legume species, lies in increasing soil fertility, obtaining complimentarity in resource use, reducing environmental stress and protecting crops.

5.6 Bare rock status from 1978 to 2003

The pixel statistics in Table 6 shows that the underlying rocks were being exposed in the study area. For instance, the area occupied by the exposed rock was 7,340.62 hectares in 1978, which was 6.13% of the total land use. The proportional decreased in 1986 was due to the prevailing season that is, wet season. During this period, the whole landscape was covered with vegetation making it difficult for rock to be seen as rock. This constraint in effect influenced the recorded areal extent of 2,246.29 hectares, which represented a decrease of 69.40%. In 1994, the area occupied by rock increased to 12,979.33 hectares, because of the prevailing season that is, dry season. At this time, the vegetation had dried up thus, making exposed rocks visible for classification. The year 2003 showed a decrease of 43.89% in the area of bare rocks. This could be because of the prevailing season at the time the image was acquired that is, September, the peak period of raining season in southwestern Nigeria. It is instructive to note that the rocks of the study area show great variation in grain size and in mineral composition, which enhanced the growth of dense vegetation whenever there is enough moisture. During the field survey, some quarry sites were sighted, which could in effect lead to the reduction in the size of the bare rocks in the study area.

5.7 Status of bare soils from 1978 to 2003

There are continuous exposure of soils to intense solar radiation and direct precipitation in the study area. This is shown by the expansion of bare soils discovered in the study. For example, in 1978, the area occupied by bare soils was 2,213.28 hectares representing 1.85% of the entire land use area. This increased to 9,347.52 hectares in 1986, representing 322.30% increase over the period of 9 years (1978 - 1986). The bare soils decreased by 53.48% in 1994 and expanded by 33.73% in the year 2003. Between 1978 and 2003, the estimated area of bare soils expanded from 2213.3 hectares to 5814.9 hectares, which put an increase at 3601.6 hectares (Table 6).

Although the prevailing season partly influenced the trend, it was obvious some land areas were void of vegetation in the study area. Continuous exposure of soils could pose serious problem to agricultural sector and consequently continuous food insecurity. This is because clearing of forest lands contribute to severe soil erosion, soil infertility and flooding. The multiplier effect on the economy could be severe if quick action is not taken. In the first place, soil erosion would reduce agricultural land area, which in effect could induce food shortages and out-migration of the able-body from the rural community to the urban environment. This claim is already manifesting in most rural communities in Nigeria including the study area. Arokoyu (1999) reported that "environmental devastation has led to the loss of the means of livelihood of people, fall in agricultural outputs, out-migration of able-bodied youths, and engendered social rifts and intensified confusion. Presently, a community in the study area (Efon-Alaaye) has already been devastated by soil erosion and government had sunk millions of Naira to redeem the community from being swept away by erosion. This situation is capable of discouraging the able-bodied youths from the community and induces migration to urban centres, thereby increasing urbanization process.

5.8 Water body status from 1978 to 2003

The drainage basin area in 1978 was 1.39 km². This increased to 57.83 km² in 1986, which was equivalent to 4060.43% increase. This increment was dictated by the improved spatial resolution of SPOT XS 1986, which is 20m as against 80m spatial resolution of Landsat MSS 1978. In SPOT XS 1986, streams, rivers, ponds and dams were more clearly visible, which enhanced the area calculation. In 1994, the drainage basin area calculated reduced by 20.15%, which consequently put the total area at 46.18 km². The reason for this can be traced to the prevailing season. At the time the image data was obtained, the rain had just ended in southwestern Nigeria and the tree canopy was still fresh and dense. Most streams and rivers were covered with riparian forest thus, making it difficult for rivers to be assigned to class ID for eventual classification. In 2003, the area of water body was 46.09 km², which was still a reduction in the coverage of water body. The reason for this is connected to the prevailing season that is, wet season and some noises in the image data, which in effect reduced its optimum visual systems. The pixel statistical result of water body in the study does not reflect the situation on ground and does not support the present claim of climate change. High spatial resolution images is therefore, recommended for study on hydrology.

5.9 Status of fallow/shrub from 1978 to 2003

Bush fallow is a periodic relocation of farmland for the purpose of allowing soil to regain its fertility. Table 6 shows that the area of fallow has been increasing over the years, the highest being recorded between 1994 and 2003. For example, between 1978 and 1986, the fallow area increased from 23,397.9 hectares to 35,330.3 hectares, which puts the percentage change within the period of 8 years at 51%. This therefore implies that per year, the land areas that were left to fallow in the study area were 1,492 hectares. The spatial coverage, however, decreased from 35,330.3 hectares in 1986 to 26,804.2 hectares in 1994, which put decline at

24.13%. This indicated that 8,526.1 hectares of fallow lands were used up within the period of 9 years. But between 1994 and 2003, the trend changed as fallow area increased from 26,804.2 hectares in 1994 to 59,436.8 hectares in 2003. This shifted the percentage change over the period of 9 years from 24.13 to 121.74.

The increase in the areal extent of shrub/fallow and a decrease in the forest area suggested that part of the deforested areas were abandoned to secondary re-growth. According to Aweto (1990), in Nigeria, the area previously characterized by continuous forest cover has been converted into secondary re-growth vegetation, mainly as a result of shifting cultivation and lumbering. In study area, some deforested areas were left uncultivated thereby converted to secondary re-growth while some farmlands were left to fallow thus, created a large expanse of fallow lands in the area.

5.10 Built-up area status from 1978 to 2003

The development of settlements in the study area has been rather gradual. This is because the study area comprises of many rural settlements and few urban centres and why major development takes place in the urban settlements, little or no development takes place in the rural settlements. This accounted for the proportion of 943.39 hectares in 1978. Although there was an increase in the areal extent in 1986, it only amounted to an increase of 66.17% over the period of 8 years. Settlements also expanded in their areal extent between 1986 and 1994, (i.e. 1567.6 to 1,947.9 hectares). This shows that within those periods, all the settlements expanded by 380.3 hectares, which constituted 24.26% increase. Between 1994 and 2003, settlements expanded by 142 hectares, which makes the total area covered by settlements to increase from 1,947.9 hectares, to 2,089.6 hectares. This shows that annually, between 1994 and 2003, 15.78 hectares of lands were gained for settlements, which represented 7.29% growth between 1994 and 2003.

The expansion of settlements, due to increased in human population and decrease in forest area shows that forests were been destroyed in the study area to pave way for human habitation. This confirms the findings of the Mather, (1990) and Harcourt, (1992) that reported an inverse relationship between population and forest cover. Geist and Lambin, (2002) also revealed that population increases due to high fertility rates were a primary driver of tropical deforestation in only 8% of cases.

6. Conclusion

The study, through the capability of GIS technology and remote sensing data revealed a steady decline in forest area and land use intensification with the expansion in farmlands, fallow ground and built up/residential areas. This indicates that forests were being converted to agricultural use and housing estate. However, the disappearance of forest resources could pose serious threat to biodiversity. For instance, there has been an underlying assumption that as the forest declines, species diversity will decline similarly. Many such models have been proven to be wrong and loss of habitat does not necessarily lead to large scale loss of species. Study that will give more insight on the process of biodiversity extinction resulted from deforestation will be needed since the scientific understanding of the process of extinction is insufficient to accurately make predictions about the impact of deforestation on biodiversity. Moreover, it should be borne in mind that forest resources preserve ecosystem, offers economic and social opportunity for people. Besides, forests foster medicinal conservation. There is therefore need to protect our forest since unprotected forest will disappear faster than the protected one. Demarcating some forest zones as forest reserve areas could do this. Also, the indigenous participation and the involvement of the community people together with the forest guards in forest monitoring will go a long way to salvaging our forest from total depletion. The approach of GIS in this study was found useful as adequate tool for regular and up-to-date monitoring of forest and earth resources.

7. Acknowledgements

The impetus received from Prof. Bola Ayeni (Department of Geography, University of Ibadan, Nigeria) and Prof. F. A. Adesina (Department of Geography, Obafemi Awolowo University, Ile-Ife, Nigeria) in contributing a chapter to this book was appreciated. The authors valued the cooperation of the staff of RECTAS, Obafemi Awolowo University, Ile-Ife; NASRDA, Abuja and GIS unit of the Department of Geography where the satellite images used for this study were collected. Finally, the privilege offered the authors by the CEO of the publisher to contribute a chapter to this book was treasured.

8. Definition of terms

Deforestation:

Deforestation is the removal of a forest or stand of trees where the land is thereafter converted to a non-forest use. Examples of deforestation include conversion of forestland to farms, ranches, or urban use. Deforestation occurs for many reasons: trees or derived charcoal are used as, or sold, for fuel or as timber, while cleared land is used as pasture for livestock, plantations of commodities, and settlements.

Land use:

Land use is characterized by the arrangement, activities and inputs people undertake in a certain land cover type to produce change or maintain it (FAO, 2005). Land use is the specific activity a piece of land is put into. Various land use patterns emerge after the land has been subjected to use over time. In the rural area for instance, the type of land use include farming, plantation, grazing, etc.

Land cover:

The definition of land cover is fundamental, because in many existing classifications and legends, it is confused with land use. Land cover is the observed (bio) physical cover on the earth's surface (FAO, 2005).

Biodiversity:

Biological diversity or biodiversity means the variety of plant and animal life at the ecosystem, community or species level and even at the generic level, Biodiversity is most commonly measured and reported at species level with characteristics such as species riches (number of species), species diversity (types of species) and endemism (uniqueness of species to a certain area) being the most useful elements for comparison (UNEP, 2002).

Environmental degradation:

Environmental degradation can be defined as any modification of the environment that implies a reduction or loss of its physical and biological quantities caused by natural phenomena or human activities ultimately representing a decrease in the availability of ecosystem, good and services to human populations (Landa et al., 1997).

Spatial & spectral resolution:

The spatial resolution refers to the size of the area on the ground that is summarized by one data value in the imagery. This is the Instantaneous Field of View (IFOV). Spectral resolution refers to the number and width of the spectral bands that the satellite sensor detects (Eastman, 2001).

9. References

Achard, F., Eva, H. D., Stibig, H. J., Mayaux, P., Gallego, J., Richards, T. and Malingreau, J. P. (2002). Determination of deforestation rates of the world's humid tropical forests, *Science* 297: 999-1002.

Adesina, F.A. (1997). Vegetation Degradation and Environmental Changes in the Tropics, *Ife Research Publications in Geography* 6: (1 & 2), 68-78.

Aina, T.D. and Salau, A. T. (1992). *The Challenge of Sustainable Development in Nigeria*, Ibadan, Nigerian Environmental Study/Action Team, (NEST).

Angelsen, A., and Kaimowitz, D. (1999). "Rethinking the Causes of Deforestation: Lessons from Economic Models", *The World Bank Research Observer, 14:1*, Oxford University Press, pp. 73-98. http://ideas.repec.org/a/oup/wbrobs/v14y1999i1p73-98.html.

Areola, O. (1994): Geography Sense and National Development, *The Nigerian Geographical Journal*, New series 1: 20-35.

Arokoyu, S.B. (1999). "Environmental Degradation, Resource Alienation and Peasant Activities: A case study of Eleme Local Government Area of Rivers State, Nigeria", Abstracts of Papers Presented at the 42nd Annual Conference of the Nigerian Geographical Association, Held at Ogun State University, Ago- Iwoye, Nigeria May 16-20th, In: Onakomaiya S.O and K.T. Gbadamosi (eds.) *Geographical Perspectives on Nigerian Development in the Next Millennium*, 88-89.

Aweto, A.O. (1990). "Plantation forestry and forest conservation in Nigeria." *The Environmentalist*. 10:127-34, Summer.

Ayeni, Bola (2001). Application of GIS; A paper presented at the 44th Annual Conference of Nigerian Geographical Association, at University of Ibadan, July.

Boroffice, R.A. (2006). Press Conference Address at International Stakeholders' Workshop on Geo-Information System-Based forest monitoring in Nigeria (*GEOFORMIN*). March 27-30, 2006 at NUC Auditorium, Maitama, Abuja.

Burgonio, T. J. (2008). "Corruption Blamed for Deforestation", Philippine Daily Inquirer, January 3, http://newsinfo.inquirer.net/breakingnews/nation/view_article.php?article_id=1 10193.

Butler, R. A. (2009). "Impact of Population and Poverty on Rainforests", *Mongabay.com / A Place Out of Time: Tropical Rainforests and the Perils They Face*. http://rainforests.mongabay.com/0816.htm. Retrieved May 13, 2009.

Butler, R. A. (2009)."Human Threats to Rainforests—Economic Restructuring". *Mongabay.com / A Place Out of Time: Tropical Rainforests and the Perils They Face*. http://rainforests.mongabay.com/0805.htm. Retrieved May 13, 2009.

Butler, R. A. and Laurance, W. F. (2008). "New Strategies for Conserving Tropical Forests",*Trends in Ecology & Evolution, Vol. 23, No. 9*. pp. 469–472.

Cassel-Gintz, M., and Petschel-Hels, G. (2001). GIS-Based Assessment of the Threat to World Forests by Patterns of no Sustainable Civilization Nature Interaction, *Journal of Environmental Management* 59:279–298.

Defries, R. (2007). "Earth Observations for Estimating Greenhouse Gas Emissions from Deforestation in Developing Countries", *Environmental Science and Policy*, 06/02/07.

Diaz, M.C., M. del C. V., D. Nepstad, M. J. C. Mendonça, R. M. Seroa, A. A. Alencar, J. C. Gomes, R. A. Ortiz (2002). Prejuízo oculto do fogo: custos econômicos das queimadas e incêndios florestais da Amazônia. Instituto de Pesquisa Ambiental da Amazônia e Instituto de Pesquisa Econômica Aplicada, Belém, Brazil. Disponível em: http://www.ipam.org.br. Cited in Moutinho P. and Schwartzman S. 2005. *Tropical Deforestation and Climate Change*, Amazon Institute for Environmental Research, Brazil, p. 111.

Eastman J. R. (2001). *IDRISI Guide to GIS and Image Processing* Vol. 1, Worcester M. A., USA, Clark Labs, Clark University. P.17-34.

Fearnside, P. M. and Laurance, W. F. (2004). Tropical Deforestation and Greenhouse-Gas Emissions, *Ecological Applications*, Volume 14, Issue 4 (August 2004), pp. 982–986.

Federal Environmental Protection Agency, Nigeria (1992). "Achieving Sustainable Development in Nigeria" National Report for the United Nations Conference on Environment and Development. Rio de Janeiro, Brazil. pp. 1-12

Food and Agricultural Organization (FAO) (1983). Forest Product Prices, No. 46. 1963-1982, *FAO* of the United Nations, Rome.

Food and Agricultural Organization (FAO) (1988). An Interim Report on the State of Forest Resources in the Developing Countries, Forest Resources Division, Forestry Department, FAO of the United Nations, Rome.

FAO - Food and Agriculture Organization. Global Forest Resources Assessment. 2000. Summary Report. Food and Agriculture Organization website http://www.fao.org/forestry/fo/fra/index.jsp.

FAO. (2001). Global Forest Resources Assessment 2000 main report. FAO Forestry Paper No. 140. Rome (*www.fao.org/forestry/fo/fra/main/index.jsp*).

Food and Agricultural Organization (FAO) (1997). *Africover Landcover Classification*. Rome.

Food and Agricultural Organization (FAO), (2001a). "*Forest Resources Assessment 2000*" FAO, Rome.

Food and Agricultural Organization (FAO), (2005). *Land Cover Classification System, Classification Concepts and User Manual*, Environment and Natural Resources Series 8, FAO, Rome, p. 3

Forest Monitoring for Action (FORMA, 2009): Center for Global Development: Initiatives: Active, Cgdev.org (2009-11-23). Retrieved on 2010-08-29.

Geist, H. J. and Lambin, E. F. (2002). "Proximate Causes and Underlying Driving Forces of Tropical Deforestation", *BioScience, Vol. 52, No. 2*. pp. 143–150. http://www.freenetwork.org/resources/documents/2-5Deforestationtropical.pdf.

GEO FCT Portal, (2010). http://www.portal.geo-fct.org. Retrieved on 2010-08-29.

Global Deforestation, (2006). *Global Change Curriculum*, University of Michigan Global Change Program, January 4, 2006.

Harcourt, C. (1992). Tropical Moist Forests. In: Groombridge, B. (ed.), *World Conservation Monitoring Centre, London*: Chapman and Hall.

Hecht, S. B., Kandel, S., Gomes, I., Cuellar, N. and Rosa, H. (2006). "Globalization, Forest Resurgence, and Environmental Politics in El Salvador", *World Development Vol. 34, No. 2*. pp. 308–323. http://www.spa.ucla.edu/cgpr/docs/sdarticle1.pdf.

Houghton R A. (2003). Revised Estimates of the Annual net Flux of Carbon to the Atmosphere from Changes in Land use and Land Management 1850-2000. *Tellus*, 55 : 378-390.

Ibah, L. (2001). "The Nigerian Environment Score Sheet" *The Punch*, Thursday, January 11, p.29

IPCC Fourth Assessment Report, Working Group I Report, "The Physical Science Basis", Section 7.3.3.1.5 (p. 527).
http://www.ipcc.ch/pdf/assessment-report/ar4/wg1/ar4-wg1-chapter7.pdf

IPCC, (2000). Land Use, Land-use Change, and Forestry. *Intergovernmental Panel on Climate Change*, Cambridge: Cambridge University Press.

Jaiyeoba, I. A. (2002). "Environment", Africa Atlases: Atlas of Nigeria, Paris: Les Edition, JA, pp. 122-123.

Jensen, J. R. (1996). *Introductory Digital Image Processing: A Remote Sensing Perspective*. United State of America: Prentice- Hall, Inc., P.197.

Karen, (2003). "Demographics, Democracy, Development, Disparity and Deforestation: A Crossnational Assessment of the Social Causes of Deforestation". *Paper presented at the annual meeting of the American Sociological Association, Atlanta Hilton Hotel, Atlanta, GA, Aug 16, 2003.*
http://www.allacademic.com/meta/p_mla_apa_research_citation/1/0/7/4/8/p1 07488_index.html. Retrieved May 13, 2009.

Landa, R., Meave, J. and Carabias, J. (1997). Environmental Deterioration in Rural Mexico: An Examination of the Concept", *Ecological Applications* 7(1): 316 – 329.

Laurance, W. F. (1999). "Reflections on the Tropical Deforestation Crisis", *Biological Conservation, Volume 91, Issues 2-3*. pp. 109–117.
http://studentresearch.wcp.muohio.edu/BiogeogDiversityDisturbance/Reflection sDeforestCrisis.pdf.

Le Quéré, C. and Scholes, R. J. (2001). The Carbon Cycle and Atmospheric Carbon Dioxide, Pages 183-237, *in:* Houghton, J. T. Ding, Y. Griggs, D. J. Noguer, M. van der Linden, P. J., Dai, X. Maskell, K.and Johnson, C. A. (eds.), Climate Change 2001:Tthe Scientific Basis, Contribution of Working Group I to the Third Assessment Report of the Intergovernmental Panel on Climate Change, Cambridge University Press, Cambridge, United Kingdom.

Leakey, R. and Roger L. (1996). *The Sixth Extinction: Patterns of Life and the Future of Humankind*, Anchor, ISBN 0-385-46809-1.

Lines, C., Bolwell, L and Norman, M. (1997). *Geography – Study Guide GCSE*, London: Letts Educational, pp. 152 –153.

Marcoux, A. (2000). "Population and Deforestation", *SD Dimensions*, Sustainable Development Department, Food and Agriculture Organization of the United Nations (FAO), August 2000. http://www.fao.org/sd/WPdirect/WPan0050.htm.

Mather, A.S. (1990). *Global Forest Resources*. London: Elhaven Press,

Mayaux, P. and Malingreau, J. P. (1996). Central Africa Forest Cover Revisited: An Iterative Approach Based on a Multi-satellite Analysis, Submitted to *Ambio*, July.

Meyers, W.B. and Turner, B.L. II, (1992). "Human Population Growth and Global Land use/Land cover Change", in: *Annual Review of Ecology and Systematics*, No. 23, pp. 39-61.

Mitchell, B. (1989). *Geography and Resource Analysis*. U.K: Longman Group Ltd., Longman Scientific and Technical.

Moutinho P. and Schwartzman S. (2005). Tropical Deforestation and Climate Change, Amazon Institute for Environmental Research, Brazil.

Munro, D.A. et al. (1986). Learning from Experience: A state of the Art Review and Evaluation of Environmental Impact Assessment. Canada: *Audits Supply and Services*, Ottawa;

Ogunsanya, A. A. (2000). *Contemporary issues in Environmental Studies*, Ilorin: Haytee Press & Publishing Co. Ltd. III.

Okafor, F. C. (1988). Rural Development and the Environment Degradation versus Protection. In P. O. Sada & F. O. Odemerho (Eds.), Environmental issues in Nigerian Development. Ibadan, Nigeria: Evans Brothers.

Ola-Adams, B.A. (1981). Strategies for Conversation and Utilization of Forest Genetic Resources in Nigeria, *The Nigeria Journal of Forestry* Vol.11 (2): pp. 32-39.

Olofin E. A. (2000). Geography and Environmental Monitoring for Effective Resource Management, *The Nigerian Geographical Journal*, New series Vol. 3 & 4: 5 – 8.

Olokesusi, F. (1992). Environment Impact Assessment in Nigeria: Current Situation and Future Directions for Future, *Journal of Environmental Management*. 35: 163-171.

Osemeobo, G. J. (1990). "Land use Policies and Biotic Conservation: Problems and Prospects for Forestry Development in Nigeria." *Land Use Policy.* 7:314-22. Oct

Osemeobo, G. J. (1991). "A Financial Analysis of Forest Land use in Bendel, Nigeria" *Forest Ecology and Management,* vol. 40, May 31.

Pimm, S. L, Russell, G. J, Gittleman, J. L, Brooks, T. M. (1995). "The future of biodiversity", *Science* 269 (5222): 347–341. doi:10.1126/science.269.5222.347. PMID 17841251.

Poor, D. (1976). *Ecological Guidelines for Development in Tropical Rain Forests, IUCN*, Morges

Prentice, I. C., Farquhar, G., Fashm, M., Goulden, M., Heimann, M., Jaramillo, V., Kheshgi, H., Le Quéré, C. and Scholes, R. J. (2001). The Carbon Cycle and Atmospheric Carbon Dioxide, Pages 183-237, *in:* Houghton, J. T. Ding, Y. Griggs, D. J. Noguer, M. van der Linden, P. J., Dai, X. Maskell, K.and Johnson, C. A. (eds.), Climate Change 2001:Tthe Scientific Basis, Contribution of Working Group I to the Third Assessment Report of the Intergovernmental Panel on Climate Change, Cambridge University Press, Cambridge, United Kingdom.

Rainforest Facts, (2010). http://www.rain-tree.com/facts.htm (2010-03-20). Retrieved on 2010-08-29.

Reed, D. (1996). Sustainable Development, In: Reed, D. (ed.) *Structural Adjustment, The Environment, and Sustainable Development.* London, Earthscan Publication Ltd, pp. 25-44.

Rees, W.E. (1990). The Ecology of Sustainable Development, *The Ecology*, vol. 20: 18-23.

Rodgers, A., Salehe, J., and Olson, J. (2000). Woodland and Tree Resources on Public Land, In: Temu, A.B., Lund, G., Malimbwi, R. E., Kowero, G.S., Klein, K., Malende, Y. and Kone, I. (eds.), *Off-forest Tree Resources of Africa, A Proceedings of a workshop* held at Arusha, Tanzania, 1999, The African Academy of Sciences (AAS).

Rowe, R.N., Sharma, N.P. and Browder, J. (1992). Deforestation: Problems, Causes and Concerns. In: Sharma, N.P. (ed.), Managing the World's Forests: Looking for Balance Between Conservation and Development. IOWA, Kendall/Hunt Publishing Co. 33-45.

Rudel, T. K. (2005). "Tropical Forests: Regional Paths of Destruction and Regeneration in the Late 20th Century", Columbia University Press, ISBN 023113195X.

Sahney, S., Benton, M. J. and Falcon-Lang, H. J. (2010). "Rainforest Collapse Triggered Pennsylvanian Tetrapod Diversification in Euramerica", *Geology* 38 (12): 1079–1082.

doi:10.1130/G31182.1.

http://geology.geoscienceworld.org/cgi/content/abstract/38/12/1079.

Sharma, N.P., Rowe, R., Openshaw, K., and Jacobson, M. (1992). "World Forests in Perspective". In: Sharma, N.P. (ed.), Managing the World's Forests: Looking for Balance Between Conservation and Development. IOWA, Kendall/Hunt Publishing Co. pp. 17-26.

Smith, L. G. (1993). *Impact Assessment and Sustainable Resource Management,* United Kingdom: Longman Group Ltd.

Stock, J. and Rochen, A. (2009). "The Choice: Doomsday or Arbor Day". http://www.umich.edu/~gs265/society/deforestation.htm. Retrieved May 13, 2009.

Taylor, D. M., Hortin D., Parnwell M. J. G. and Marsden T. K. (1994). The Degradation of Rainforests in Sarawak, East Malaysia, and its Implications for Future Management Policies, *Geoforum.* 25 (3): 351 –369.

http://news.mongabay.com/Butler_and_Laurance-TREE.pdf.

UNEP, (2002). *Africa Environment Outlook Past, Present and Future Perspectives,* Earthprint Ltd., England. 130.

UNEP, UNEP, World Bank, & WRI. (2000). World resources 2000–2001: People and Ecosystems: The Fraying Web of Life. Washington D.C: World Resources Institute.

UNFCC, (2009). "Copenhagen Accord of 18 December 2009".

http://unfccc.int/files/meetings/cop_15/application/pdf/cop15_cph_auv.pdf. Retrieved 2009-12-28

UNFCC , (2009). "Methodological Guidance".

http://unfccc.int/files/na/application/pdf/cop15_ddc_auv.pdf. Retrieved 2009-12-28.

UNFCCC (2007). "Investment and Financial Flows to Address Climate Change", UNFCCC *Int.* UNFCCC. p. 81.

ttp://unfccc.int/files/essential_background/background_publications_htmlpdf/a pplication/pdf/pub_07_financial_flows.pdf.

Van der Werf, G. R., Morton, D. C., DeFries, R. S., Olivier, J. G. J., Kasibhatla, P. S., Jackson, R. B., Collatz, G. J. and Randerson, J. T. (2009). "CO_2 Emissions from Forest Loss", *Nature Geoscience* 2 (11): 737–738. doi:10.1038/ngeo671.

Verolme, H.J.H. and Moussa, J. (1999). Addressing the Underlying Causes of Deforestation and Forest Degradation- Case Studies, Analysis and Policy Recommendations, *Biodiversity Action Network,* Washington, D.C.

Wertz-Kanounnikoff, S. and Alvarado, L. X. R. (2007). Bringing 'REDD' into a New Deal for the Global Climate, *Analyses, no. 2,* 2007, Institute for Sustainable Development and International Relations.

Williams, M. (1990). Forests, In: Turner B.L., Clark, W.C., Kates, R.W., Richard, J.F., Matthews, J.T. and Meyers, W.B. (eds.), *The Earth as Transformed by Human Action, Global and Regional Changes in the Biosphere over the Past 300 years.* Cambridge University Press, Cambridge, pp.179-201.

World Bank, (1992). *Annual Report 1992,* Washington, D. C., World Bank.

World Rainforest Movement (WRM, 2003). WRM Bulletin Number 74, September 2003, http://www.wrm.org.uy/bulletin/74/Uganda.html.

World Resources Institute, (1987). *World Resources 1987,* New York, Basic Books.

YaleGlobal, (2007). "The Double Edge of Globalization", *YaleGlobal Online,* Yale University Press June 2007. http://yaleglobal.yale.edu/display.article?id=9366.

11

Unsupervised Classification
of Aerial Images Based on the Otsu's Method

Antonia Macedo-Cruz[1], I. Villegas-Romero[3],
M. Santos-Peñas[2] and G. Pajares-Martinsanz[2]

[1]Hidrociencias, Campus Montecillo,
Colegio de Postgraduados, Montecillo, Estado de México,
[2]Facultad de Informática, Universidad Complutense de Madrid,
[3]Universidad Autónoma Chapingo,
[1,3]México
[2]España

1. Introduction

Remote-sensing research focusing on image classification has long attracted the attention of the remote-sensing community because classification results are the basis for many environmental and socioeconomic applications. However, classifying remotely sensed data into a thematic map remains a challenge because many factors, such as the complexity of the landscape in a study area, selected remotely sensed data, and image-processing and classification approaches, may affect the success of a classification [1].

In forest management, a number of activities are oriented towards wood production or forest inventories with the aims of controlling parameters of interest such as diameter of trees, height, crown height, bark thickness, canopy, humidity, illumination, $CO2$ transformation among others, always with the goal of environmental sustainability with high social impact. The unsupervised classification of aerial image offer solutions for monitoring production in forest trees while the same time costs are minimized. Also with Unmanned Aerial Vehicles (UAV) equipped with an appropriate image classification system, have become powerful tools for early fire forest detection and posterior monitoring. This technology has also been applied for crop monitoring under wireless sensor network architecture.

Clustering is the task of categorizing objects having several attributes into different classes so that the objects belonging to the same class are similar, and those that are broken down into different classes are not. Clustering is the subject of active research in several fields such as statistics, pattern recognition, machine learning, data mining, information science, agriculture technology and spatial databases. A wide variety of clustering algorithms have been proposed for different applications [1], [2].

Classification and segmentation in agriculture and forest management is an interesting topic but not new. There are many classification approaches that are oriented toward the identification of textures in agricultural and forest images. Most of them can be grouped as follows.

- Currently, many of the agriculture, livestock and forestry are planned using spatial analysis tools, seeking different specific objectives [3]. In this sense, the images acquired by remote sensors provide the necessary spatial resolution to obtain information about objects, areas, or phenomena on the earth's surface, at different scales. These sensors measure the intensity of the energy emitted or reflected by the objects by means of the electromagnetic spectrum [1].

- There are many segmentation techniques reported in the literature [4, 5]. Most color image segmentation techniques are usually derived from methods designed for graylevel images. Processing each channel individually by directly applying graylevel where the channels are assumed independent and only their intra-spatial interactions are considered [6]. Another option is decomposing the image into luminance and chromatic channel: after transforming the image data into the desired (usually application dependent) color space, texture features are extracted from the luminance channel while chromatic features are extracted from the chromatic channels, each in a specific manner [7]. Reference [8] and [9] show combining spatial interaction within each channel and interaction between spectral channels and gray level texture analysis techniques are applied in each channel, while pixel interactions between different channels are also taken into account.

Based on the presented considerations and in order to tackle the classification problem addressed in this paper, we have designed a new automatic strategy based on the thresholding Otsu´s method is proposed. The first step consists in the thresholding of each R, G and B channels into two parts based on within-class and between-class variances suggested by Otsu [10]. This allows to classify each pixel to one part of each channel, so that conveniently combined the a pixel should be classified as belonging to one of the eight possible classes. Although in this paper we only use eight classes, the method can be easily extended to more classes, as described in section 2.2.2, even we can achieve until twenty seven. This makes the main contribution of this paper.

Additionally, one major advantage of this algorithm is that it does not need to know how many classes are required to be clustered in advance, as it is required for most supervised clustering processes. The termination criterion is established based on the within-class variance, according to the Otsu's method.

The proposed method is compared against the well-known Fuzzy c-means clustering [11], [12]. The prediction of the correct number of clusters is a fundamental problem in unsupervised classification problems. Many clustering algorithms require the definition of the number of clusters beforehand. To overcome this problem, various cluster validity indices have been proposed to assess the quality of a clustering partition [13]-[16]. Five cluster validation indices have been used in our tests, they are: Dunn's [15]-[19], Davies-Bouldin [15], [19]-[21], Calinski-Harabasz [15], [21]-[24], Krzanowski and Lai [22], [25]; and Hartigan [21], [22], [26]. We have used five because there is not relevant conclusions in the literature about their performance, depending on the application their behavior could vary considerably. Based on the above indices we have verified the best performance of our approach against the MS method, particularly in the images where water bodies are present.

The remainder of the paper is organized as follows. In Section 2, materials and methods; two issues will be addressed, unsupervised classification of color images and five of cluster validation indexes. Section 3, result and discussions; Conclusions are presented in Section 4.

2. Materials and methods

2.1 Study area

In the present investigation 16 color aerial photographs in digital format were used, owned by the Institute of Geography of the Autonomous University of Mexico, taken in October 1997. The photographs correspond to the catchments of the river La Sabana, Guerrero, with spatial resolution of 1:19500, and three-band spectral resolution visible and radiometric resolution RGB from 0 to 255 levels. As an example, Fig. 1 displays a representative image of the set of images analyzed in this work. As we can see it contains several textures which must be classified as belonging to a cluster.

Fig. 1. Land cover images in RGB color model

2.2 Classifier based on the theory of the Otsu's method
2.2.1 Brief description of the Otsu's method

Otsu's method [10], one of the most widely used thresholding techniques in image analysis, has showed great success in image enhancement and segmentation. As mentioned before, it is an automatic thresholding strategy; we exploit the automatic capability for designing the unsupervised classification strategy justifying its choice.

This research sought the optimal threshold (single or multiple) for each of the spectral bands of the color image by applying the algorithm modified [27, 28]:

The number of pixels with gray level i is denoted f_i giving a probability of gray level i in an image of

$$p_{i=}\frac{f_i}{N} \tag{1}$$

Then, the probability distributions for each class is

$$w_k = \sum_{i \in C_k} pi \qquad (2)$$

The w_k is regarded as the zeroth-order cumulative moment of the kth class C_k. and the means for classes is

$$u_k = \sum_{i \in C_k} \frac{i.pi}{w_k} \qquad (3)$$

In the case of bi-level thresholding, Otsu defined the between-class variance of the thresholded image as:
Where

$$\sigma_B^{2*} = w_1(u_1 - u_T)^2 + w_2(u_2 - u_T)^2 \qquad (4)$$

$$\mu_T = \sum_{k=1}^{M} \omega_k \mu_k \qquad (5)$$

And

$$\sum_{k=1}^{M} w_k = 1 \qquad (6)$$

Assuming that there are M-1 thresholds, {t1, t2, ..., tM-1}, which divide the original image into M classes: C1 for [1,..., t1], C2 for [t1+1, ..., t2], ... , Ci for [t i-1+1, ..., t i], ..., and CM for [tM-1+1, ..., L], the optimal thresholds {t1*, t2*, ..., tM-1*} are chosen by maximizing σ_B^2 as follows

$$\{t_1{}^*, t_2{}^*, ..., t_{M-1}{}^*\} = \arg\ Max\{\sigma_B^2(t_1, t_2, ..., t_{M-1})\} \qquad (7)$$

$$1 \le t_1 < ... < t_{M-1} < L$$

then

$$(\sigma_B^2) = \sum_{k=1}^{M} w_k u_k^2 - u_T^2 = \sum_{k=1}^{M} w_k u_k{}^2 \qquad (8)$$

A threshold value t_{Otsu} developed by Otsu is the one that maximizes $var_{between-class}$, or equivalently minimizes $var_{within-class}$.

$$t_{Otsu} = \arg\left\{Min_{1 \le t \le L}(Var_{within-class}^t)\right\}$$
$$t_{Otsu} = \arg\left\{Max_{1 \le t \le L}(Varet_{between-class}^t)\right\} \qquad (9)$$

where L is the number of gray levels in each band, in our images L is 256 because each pixel is represented with eight bits.

2.2.2 Unsupervised classification strategy by within-class and between-classes spectral variances

There are three steps in the proposed classification strategy. First, the assignment process, that consists in assigning one of the possible classes to each pixel. Second, the codification of each cluster, which is identified by a label. Finally, a regrouping process so that very similar classes are merged into one.

2.2.2.1 Assignment process

Given a pixel i located at (x, y) in the original RGB image. Its three spectral components in this space are obtained, namely $R(x, y) = i_r$, $G(x, y) = i_g$ and $B(x, y) = i_b$.

As already mentioned, the thresholding methods split the histogram into two regions. As there are three spectral components, six sub-regions are obtained. If necessary, successive thresholding can be applied to each spectral channel. The second thresholding produces three partitions per channel. If a third thresholding is applied, four regions per component are obtained and so on. Therefore, assuming that eventually the number of thresholds per channel is M, there will be t_{R1}, t_{R2}, ... t_{RM} , thresholds for channel R, and in the same way, t_{G1}, t_{G2}, ..., t_{GM} for component G, and t_{B1}, t_{B2}, ..., t_{BM}, for component B. Based on this, each pixel i can be coded as $\tilde{i}_s$ according to its spectral components by Equation (10):

$$\tilde{i}_s = \begin{cases} 0 & if \ i_s \leq t_{s1} \\ 1 & if \ t_{s1} < i_s \leq t_{s2} \\ 2 & if \ t_{s2} < i_s \leq t_{s3} \\ \quad \vdots \\ M & if \ i_s > t_{sM} \end{cases} \tag{10}$$

where s denotes the spectral component, i.e., $s = R$, G or B, and t_{si} are the consecutive thresholds.

For example, it is known that in the RGB colour space values are in the range $[0, 255]$. So, considering the spectral component R with two thresholds, $t_{R1} = 120$ and $t_{R2} = 199$, a pixel will be coded as 0, 1, or 2, if its spectral value R is smaller than 120, between 120 and 199, or greater than 199, respectively.

2.2.2.2 Cluster labelling

Once the whole image has been coded, the next step is the labelling of the existing classes. If M thresholds haven been obtained, there are $n = M + 1$ histogram partitions per channel, and therefore the number of possible combinations is n^d, where d is the number of spectral components, i.e., $d = 3$ in the RGB colour space. This number of combinations represents the number of classes. Each cluster is identified by its label. Every pixel is assigned its corresponding label according to Equation (11). So, given the pixel $i \equiv (x, y)$ with codes $\tilde{i}_R$, $\tilde{i}_G$, and $\tilde{i}_B$, its label will be given by $\tilde{P}_i$ as follows:

$$\tilde{p}_i = n^2 \tilde{i}_R + n \tilde{i}_G + \tilde{i}_B \tag{11}$$

2.2.2.3 Merging process

Let C_k be the number of clusters obtained by the classification procedure, where k identifies a class between 1 and n^d, each class containing N_k pixels of the original image. It could be said that each class is defined by a tri-dimensional vector (d = 3). The elements of that vector are the spectral components of the pixels according to the RGB colour model, i.e., $i_R \equiv (i_R^k, i_G^k, i_B^k)$ for the pixel $i \equiv (x, y)$, if the pixel and its spectral components belong to class C_k. For each class, the average value of the membership degrees to that class is calculated by Equation (12):

$$\mu_k \equiv \left(\mu_R^k, \mu_G^k, \mu_B^k \right) = \frac{1}{N_k} \sum_{i_k \in C_k} i_k \qquad (12)$$

Based on the potential of Otsu's method, it is possible to estimate the within-class and the between-classes spectral variances, denoted by σ_k and σ_{kh} respectively, according to Equations (1) and (16). Obviously, σ_k is only related to class C_k and, as expected, σ_{kh} involves the two classes C_k and C_h, $k \neq h$:

$$\sigma_k = \frac{1}{d \cdot N_k} \sum_{i_k \in C_k} \left[\left(i_R^k - \mu_R^k \right)^2 + \left(i_G^k - \mu_G^k \right)^2 + \left(i_B^k - \mu_B^k \right)^2 \right]^{1/2} \qquad (13)$$

$$\sigma_{kh} = \frac{1}{d} \left[\left(\mu_R^k - \mu_R^h \right)^2 + \left(\mu_G^k - \mu_G^h \right)^2 + \left(\mu_B^k - \mu_B^h \right)^2 \right] \qquad (14)$$

Based on those variances, some classes can be fused due to their spectral similarities. The similarity is a concept defined as follows. Given the clusters C_k and C_h, for $k \neq h$, they are merged if $\sigma_k \geq \sigma_{kh}$ $\sigma_h \geq \sigma_{kh}$. This is based on the hypothesis that if a good partition is already achieved, the classes obtained are properly separated, without overlapping, and then no further fusion is required. On the contrary, if classes overlap, the between-class variance σ_{kh} is greater than the individual within-class variances, σ_k and σ_j. This re-clustering process is repeated until all the between-class variances are greater than their corresponding within-class variances. Without lost of generality, if two classes are merged, the resulting fused class will be re-labelled with the name of the class with the smaller variance value. This does not affect the classification process because only labels are modified.

After the fusion process, it must be checked if more clusters are necessary. This is carried out on the basis that if after the combination process no class has been fused, it means that more clusters are needed. A new clustering process starts again with the number of thresholds increased by one. This is repeated until a fusion occurs.

2.3 Fuzzy C-Means clustering

Fuzzy c-means clustering (FCM) is a data clustering technique wherein each data point belongs to a cluster to some degree that is specified by a membership grade. This technique was originally introduced by Jim Bezdek [11], as an improvement on earlier clustering methods.

The FCM algorithm attempts to partition a finite collection of elements $X = \{x_1, x_2, ..., x_n\}$ into a collection of c fuzzy clusters with respect to some given criterion.

The FCM algorithm, processes n vectors in p-space as data input, and uses them, in conjunction with first order necessary conditions for minimizing the FCM objective functional, to obtain estimates for two sets of unknowns.

The unknowns in FCM clustering are:

1. A fuzzy c-partition of the data, which is a $c \times n$ membership matrix $U = \{\mu_{ik}\} \in V_{cn}$ with c rows and n columns. The values in row i give the membership of all n input data in cluster i for $k=1$ to n ; the k-th column of U gives the membership p of vector k (which represents some object k) in all c clusters for $i=1$ to c. Each of the entries in U lies in $[0,1]$; each row sum is greater than zero; and each column sum equals 1.

2. The other set of unknowns in the FCM model is a set of c cluster centers or prototypes, arrayed as the c columns of a $p \times c$ matrix V. These prototypes are vectors (points) in the input space of p-tuples. Pairs (U,V) of coupled estimates are found by alternating optimization through the first order necessary conditions for U and V. The objective function minimized in the original version measured distances between data points and prototypes in any inner product norm, and memberships were weighted with an exponent $m>1$

That is:

As $X = \{x_1, x_2, ..., x_n\}$ and the set all V_{cn} real matrices of dimension $c \times n$, with $2 \leq c < n$. Can be obtained a matrix representing the partition follow $U = \{\mu_{ik}\} \in V_{cn}$. The basic definition FCM for m > 1 is to minimize the following objective function:

$$\min z_m(U;v) = \sum_{k=1}^{n}\sum_{i=1}^{c} \mu_{ik}^m \left\| x_k - v_i \right\|_G^2 \tag{15}$$

G is a matrix of dimension pxp symmetric positive definite

$$\left\| x_k - v_i \right\|_G^2 = (x_k - v_i)^t G (x_k - v_i) \tag{16}$$

Where

$$v_i = \frac{1}{\sum_{k=1}^{n}(\mu_{ik})^m} \sum_{k=1}^{n}(\mu_{ik})^m x_k \qquad i = 1,...,c \tag{17}$$

$$\mu_{ik} = \frac{\left(\dfrac{1}{\left\| x_k - v_i \right\|_G^2}\right)^{2/m-1}}{\sum_{j=1}^{c}\left(\dfrac{1}{\left\| x_k - v_j \right\|_G^2}\right)^{2/m-1}} \qquad i = 1,...,c;\ k = 1,...,n \tag{18}$$

The exponent m is known as exponential weight and reduces the influence of noise when getting the centers of the clusters. The higher the m > 1, the greater this influence. More details on fuzzy c-means clustering [11, 12].

2.4 Methods for cluster validation

Evaluation of clustering results (or cluster validation) is an important and necessary step in cluster analysis, but it is often time-consuming and complicated work [16].

The procedure of evaluating the results of a clustering algorithm is known under the term cluster validity. In reference [15] two kinds of validity indices are showed: external indices and internal indices. A third is added in reference [29], based on relative criteria. The first is based on external criteria. This implies that we evaluate the results of a clustering algorithm based on a pre-specified structure, which is imposed on a data set and reflects our intuition about the clustering structure of the data set. The second approach is based on internal criteria. We may evaluate the results of a clustering algorithm in terms of quantities that involve the vectors of the data set themselves. The third approach of clustering validity is based on relative criteria. Here the basic idea is the evaluation of a clustering structure by comparing it to other clustering schemes, resulting by the same algorithm but with different parameter values.

To evaluate the proposed classification method, five cluster validation techniques are applied, based on internal criteria. These indices are used to measure the "goodness" of the result of the grouping; comparing the proposed classification method against the old pattern recognition procedure called Fuzzy c-means clustering.

2.4.1 Dunn's index

This index identifies sets of clusters that are compact and well separated. For any partition $U \leftrightarrow X : X_1 \cup ... X_c$ where X_i represents the ith cluster of such partition, the Dunn's validation index, D, is defined as:

$$D(U) = \min_{1 \le i \le c} \left\{ \min_{\substack{1 \le j \le c \\ j \ne i}} \left\{ \frac{\delta(X_i, X_j)}{\max_{1 \le k \le c} \{\Delta(X_k)\}} \right\} \right\} \tag{19}$$

where $\delta(X_i, X_j)$ defines the distance between clusters X_i and X_j (intercluster distance); $\Delta(X_k)$ represents the intracluster distance of cluster X_k, and c is the number of clusters of partition U. The main goal of this measure is to maximize intercluster distances whilst minimizing intracluster distances. Thus large values of D correspond to good clusters. Therefore, the number of clusters that maximizes D is taken as the optimal number of clusters, c.

2.4.2 Davies-Bouldin index

As the Dunn's index, the Davies-Bouldin index aims at identifying sets of clusters that are compact and well separated. The Davies-Bouldin validation index, DB, is defined as:

$$DB(U) = \frac{1}{c} \sum_{i=1}^{c} \max \left\{ \frac{\Delta(X_i) + \Delta(X_j)}{\delta(X_i, X_j)} \right\} \tag{20}$$

where U, $\delta(X_i, X_j)$, $\Delta(X_i), \Delta(X_j)$ and c are defined as in equation (20). Small values of DB correspond to clusters that are compact, and whose centers are far away from each other. Therefore, the cluster configuration that minimizes DB is taken as the optimal number of clusters, c.

2.4.3 Calinski and Harabasz index
The index of Calinski and Harabasz is defined by:

$$CH(k) = \frac{B_k / (k-1)}{W_k / (n-k)} \tag{21}$$

where k denotes the number of clusters, and B_K and W_k denote the between and within cluster sums of squares of the partition, respectively. Therefore an optimal number of clusters is then defined as a value of k that maximizes CH(k).

2.4.4 Krzanowski and Lai index
The index of Krzanowski and Lai is defined by:
where;

$$KL(k) = \left| \frac{diff_k}{diff_{k+1}} \right| \tag{22}$$

$$diff_k = (K-1)^{2/p} W_{k-1} - k^{2/p} W_k \tag{23}$$

and p denotes the number of features in the data set. Therefore a value of k is optimal if it maximizes KL(k).

2.4.5 Hartigan index
The index of Hartigan is defined by:

$$Han(k) = \log\left(\frac{B_k}{W_k} \right) \tag{24}$$

where, B_K and W_k denote the between and within cluster sums of squares of the partition, respectively. Therefore a value of k is optimal if it minimizes Han(k).

3. Results and discussion

In accordance with the objectives and methodology used in this research encouraging results were obtained regarding the proposal to adopt the criterion used in Otsu's method to the process of clustering and unsupervised classification of colour images, and the application of cluster validity methods by five cluster validation indices, compare the results with those generated by the old pattern recognition procedure, Fuzzy c-means clustering.
In the present paper three issues are addressed. First, our proposed RGB unsupervised classification method. Secondly, to compare results we apply an old pattern recognition procedure, the Fuzzy c-means clustering. Third, five of cluster validation indices will be proposed to evaluate the quality of clusterings. To demonstrate the effectiveness of our proposed RGB unsupervised classification method, using 16 digital images from colour aerial photographs in which they can observe different land cover objects such as buildings,

streets, roads, tree crop plots, temporary plots of crops, pastures, water bodies etc.. Due to limited space, only the results of one experiment are included.

3.1 Unsupervised classification method by theory of the Otsu's method (results)

The proposed methodology is based on the method of Otsu. First a single thresholding of each of the bands of the RGB image, creating two classes per band as RGB in forming our startup account with eight classes. However, not all are representative, so that once segmented in this way the image is inserted through a sorting process exhaustive analysis of the variation between classes and within classes.

According to the characteristics of land cover images, which must be considered objects with different heterogeneous properties in size, shape and spectral behavior, we make the classification of the image labeled and grouped by simple thresholding, using the comprehensive analysis of variances between classes and within classes, to group and classify objects in the image.

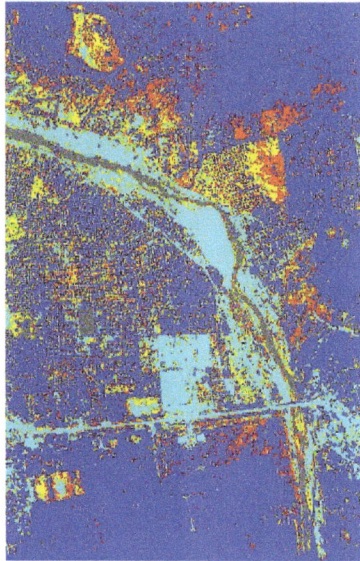

Fig. 2. Image classified by the proposed classification Otsu method, where the optimal number of class is five.

As a result of this process, the classifier automatically grouped the different objects of the landscape into five classes, generating the classified image showed in Fig. 2, and the number of pixels per class, you can see in Table 1. With size of 1 542 288 pixels per band, where 53.1% of the surface of the image contains vegetation cover, the image represented by the blue color, and 14.8% contains bodies of water identified by the gray color. The 8.3% contain natural green grass, identified by the red color, 11.7% contains dry natural grass in the image identified by the yellow color, 12% contain areas without vegetation identified by the cyan color. Among the latter the accumulated deposits on the river bank, the effect of Hurricane Pauline, as well as streets and buildings are considered.

Classes	Pixels	Color
Plant Coverage	818,893	Blue
Water Bodies	228,830	Gray
Natural green grass	128,199	Red
Dry natural grass	180,696	Yellow
Area without vegetation coverage	185,670	Cyan
TOTAL	**1,542,288**	

Table 1. distribution of the classification of Fig. 2

3.2 The Fuzzy c-means clustering (results)

As a result of this process, the classifier automatically grouped the different objects of the landscape into three classes, generating the classified image showed in Fig. 3, and the number of pixels per class you can see in Table 2. With size of 1 542 288 pixels per band, where 23% of the surface of the image contains plant coverage, identified by the green colour; 36% of the surface of the image contains trees, identified by the blue colour; 28% of the surface of the image contains natural green grass, identified by the red colour; 9% of the surface of the image contains dry natural grass, identified by the yellow colour; and 4% of the surface of the image contains the accumulated deposits on the river bank, the effect of hurricane Pauline, the image represented by the cyan colour.

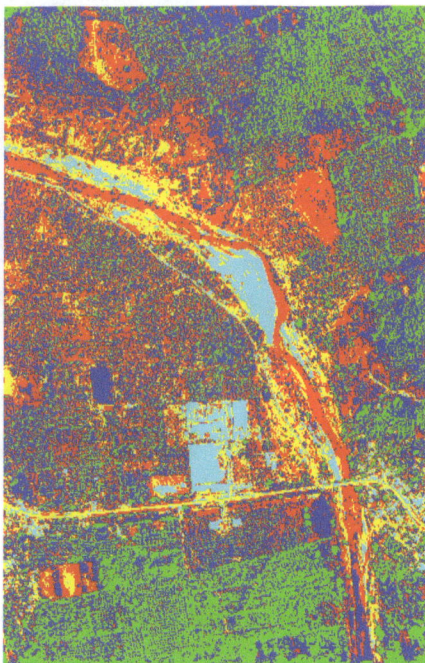

Fig. 3. Image classified by Fuzzy c-means clustering; where the optimal number of class by DB is five.

As showed in visual analysis (Fig. 2 and 3), our proposal clearly identifies the water bodies, whereas by Fuzzy c-means clustering classifier water bodies with grass coverage are confused.

Classes	Pixels	Color
Plant Coverage	358,176	green
Trees	549,978	Blue
Natural green grass	424,131	Red
Dry natural grass	143,474	Yellow
Area without vegetation coverage	66,529	Cyan
TOTAL	1,542,288	

Table 2. Distribution of the classification of Fig. 3

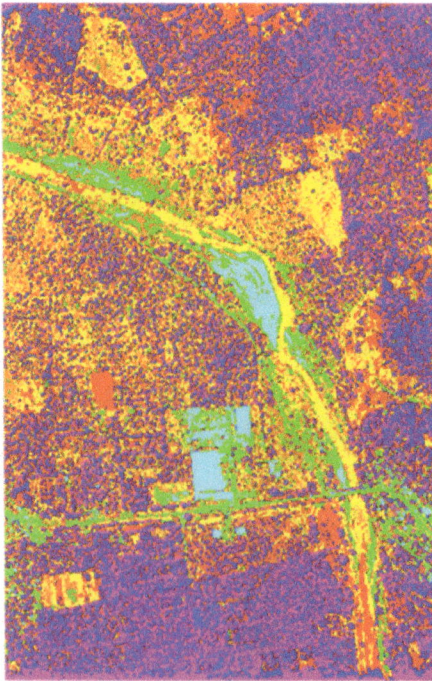 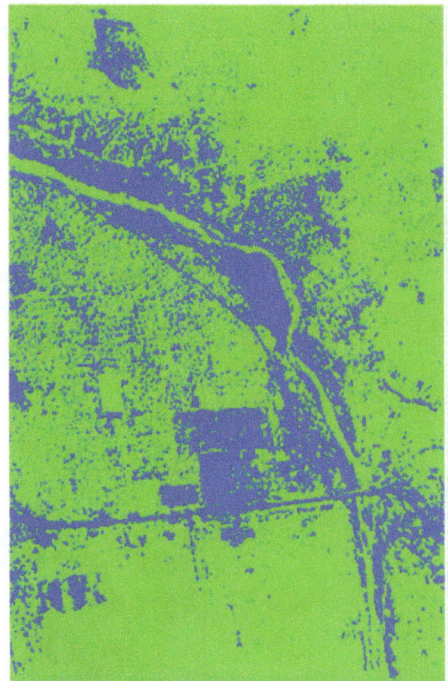

(a)   (b)

Fig. 4. Image classified by Fuzzy c-means clustering; where the optimal number of class is seven, by CH and Han, in (a); and two by Dunn and KL, in (b).

3.3 Quantitative results to validate the optimal number of clusters

Since clustering is an unsupervised method and there is no a priori indication for the actual number of clusters presented in a data set, there is a need of some kind of clustering result validation.

The results of the classification of colour images from aerial photographs by the proposed Otsu's method, and the method known as Fuzzy c-means clustering are evaluated using five levels of validation. These indices are detailed in Section 2.3. Since the results of the Fuzzy c-means clustering algorithm, requires as input, the number of clusters, so it runs for 2.3, ..., 6; Fig. 3 corresponds to the optimal number of groups according to DB, who is the same as the proposed method.

No. Class	2	3	4	5	6	7	Optimo
DB	0.660	0.597	0.576	**0.535***	0.605	0.584	*minimum*
CH	2.222	3.018	3.317	3.600	3.810	**3.891***	*maximum*
Dunn	**2.374***	1.417	1.199	1.154	1.020	0.926	*maximum*
KL	**57.660***	37.527	29.976	25.080	21.922	20.104	*maximum*
Han	36.324	18.041	11.896	8.577	6.639	**5.494***	*minimum*

* optimal

Table 3. Results obtained from 5 internal indices to validate the optimal number of clusters generated by Fuzzy c-means clustering.

Davies-Bouldin Index says: the cluster configuration that minimizes DB is taken as the optimal number of clusters, therefore for this methodology, the optimal number of clusters is 5.
Calinski and Harabasz Index says: an optimal number of clusters is then defined as a value of k that maximizes CH(k), therefore for this methodology, the optimal number of clusters is 7.
The Dunn's Index says: the value that maximizes D is taken as the optimal number of clusters, therefore for this methodology, the optimal number of clusters is 2.
Krzanowski and Lai index says: a value of k is optimal if it maximizes KL(k), therefore for this methodology, the optimal number of clusters is 2
Hartigan index says: a value of k is optimal if it minimizes Han(k). therefore for this methodology, the optimal number of clusters is 7.
With the executed validation indices for Fuzzy c-means clustering, we want to find a match on the number of clusters of the new classification proposed. This was achieved with the Davies Bouldin index. Therefore, it indicates that the number of classes generated by the new classifier is optimal.

4. Conclusions

Otsu's method improved and changes implemented in this research can get the optimal threshold value as a basis for segmenting and classifying images in RGB color model, using a method of unsupervised classification.
Under the principle of the concept of within-class and between-class variances as suggested by Otsu; the algorithm automatically regrouped and merged different values of the groups obtained from the image in RGB colour domain and once the within-class variance is less than the between-class variance for each clustered class, the algorithm is finished.

Since this algorithm does not need to know how many classes are required to be clustered, five cluster validity indices have been proposed to validate if the number of clusters classified by Otsu is suitable. Davies Bouldin index indicates that the number of classes generated by the new classifier is optimal.

Also the classification made by the proposed method is better than that Fuzzy c-means clustering, as showed in Fig. 2, our method properly classified water bodies, whereas Fuzzy c-means clustering confuses the water bodies with vegetation.

This method unsupervised of image classification, can be widely used as support in decision-making in aspects of the environment by diagnosing areas of interest such as the loss of tree cover due to deforestation or fires, crop areas, Water Bodies, etc.. by virtue the proposed methodology to classify areas of different coverage density including areas where the soil surface has been exposed to erosion.

5. Acknowledgment

The authors would like to thank the World Bank from the Robert S. McNamara Fellowships Program (RSM) to support this research on: "Técnicas Inteligentes de Reconocimiento Aplicadas a la Clasificación del Uso del Suelo".
The Carolina Foundation and SRE_Mexico by co-financing for doctoral studies.
The "Colegio de Postgraduados", Mexico, for permission to join the Teachers' Training Sub-Investigators.
The UNAM by providing the aerial images, which is essential material for this investigation.

6. References

[1] D. Lu, and Q. Weng. 2007. A survey of image classification methods and techniques for improving classification performance, Int. J. Remote Sensing 28(5):823-870.

[2] A. K. Jain, M. N. Murty and P. J. Flynn. 1999. Data clustering: a review, ACM Computing Surveys, Vol. 31 (3):264-323.

[3] E.R. Hunt; Cavigelli, M.; Daughtry, C.T.; McMurtrey, J.; Walthall, S.L. Evaluation of digital photography from model aircraft for remote sensing of crop biomass. Precision Agriculture 2005, 6 (4), 359-378.

[4] X. Xie, and Majid Mirmehdi. 2007. Colour Image Segmentation using Texems, Annals of the British Machine Vision Association, (6):1-10.

[5] X. Xie, 2008. A Review of Recent Advances in Surface Defect Detection using Texture Analysis Techniques, Electronic Letters on Computer Vision and Image Analysis, Special Issue on Quality Control by Artificial Vision, vol. 7(3): 1 -25.

[6] M. Haindl, and V. Havlicek. 2002. A simple multispectral multiresolution Markov texture model. In International Workshop on Texture Analysis and Synthesis, pp: 63-66.

[7] M. Dubuisson-Jolly, and A. Gupta. 2000. Color and texture fusion: Application to aerial image segmentation and GIS updating. Image and Vision Computing, 18:823-832.

[8] C. Palm, 2004. Color texture classification by integrative co-occurrence matrices. Pattern Recognition, 37(5):965-976

[9] M. Mirmehdi, and M. Petrou. 2000. Segmentation of color textures. IEEE Transactions on Pattern Analysis and Machine Intelligence, 22(2):142-159.

[10] N. Otsu, 1979. A threshold selection method from gray-level histogram. IEEE Transactions on System Man Cybernetics 9(1):62-66.

[11] J. C. Bezdek, 1981. Pattern Recognition with Fuzzy Objective Function Algorithms. Plenum Press, New York, NY, USA .

[12] J. C. Bezdek , J. Keller, R. Krisnapuram, N. Pal. 2005. Fuzzy Models and Algorithms for Pattern Recognition and Image Processing (The Handbooks of Fuzzy Sets). Springer, 776 pp.

[13] M. Halkidi, Y. Batistakis and M. Vazirgiannis. 2002a. Cluster validity methods: part I, SIGMOD Rec., Vol. 31(2):40-45.

[14] M. Halkidi, Y. Batistakis and M. Vazirgiannis. 2002b. Cluster validity methods: part II, SIGMOD Rec., Vol. 31, (3). 19-27.

[15] K. Wang, B. Wang, and L. Peng. 2009. CVAP: Validation for Cluster Analyses. Data Science. Journal 8(20), 88-93

[16] M. Halkidi, Y. Batistakis, and M. Vazirgiannis. 2001. On Clustering Validation Techniques. Intelligent Information Systems Journal, 17 (2-3): 107-145.

[17] J. Dunn. 1974. Well separated clusters and optimal fuzzy partitions. J.Cybernetics, Vol. 4, pp. 95-104

[18] J.C. Bezdek, and N.R. Pal. 1998 .Some new indexes of cluster validity, IEEE Transactions on Systems, Man and Cybernetics, Vol. 28 (B):301-315

[19] N. Bolshakova, and F. Azuaje. 2003 Cluster validation techniques for genome expression data. Signal Processing, 83(4): 825-833.

[20] D.L. Davies, and D.W. Bouldin. 1979. A cluster separation measure. IEEE Transactions on Pattern Recognition and Machine Intelligence, Vol. 1(2):224-227

[21] E. Dimitriadou, S. Dolnicar, and A. Weingessel. 2002 An examination of indexes for determining the Number of Cluster in binary data sets. Psychometrika, 67(1): 137-160.

[22] S. Dudoit, and J. Fridlyand. 2002. A prediction-based resampling method for estimating the number of clusters in a dataset. Genome Biology, 3(7): 0036.1-21.

[23] G. Shu, B. Zeng, Y. P Chen, and O. H. Smith. (2003) Performance assessment of kernel density clustering for gene expression profile data. Comparative and Functional Genomics, 4(3): 287-299.

[24] R. Calinski, and J. Harabasz. 1974. A dendrite method for cluster analysis. Commun Statistics 3:1-27.

[25] W. Krzanowski, and Y. Lai. 1985. A criterion for determining the number of groups in a dataset using sum of squares clustering. Biometrics. 44:23-34.

[26] J.A. Hartigan. 1985. Statistical theory in clustering. J Classification., 2:63-76.

[27] Macedo, A.; Pajares, G.; Santos, M. Unsupervised classification with ground color cover images. Agrociencia 2010, 44, 711-722.

[28] Macedo-Cruz, A.; Pajares, G.; Santos, M.; Villegas-Romero, I. Digital image sensor-based assessment of the status of Oat (Avena sativa L.) Crops after frost damage. Sensors 2011, 11, 6015-6036.

[29] F. Kovács, C. Legány, and A. Babos. 2005. Cluster validity measurement techniques. Proceedings of the 6th International Symposium of Hungarian Researchers on Computational Intelligence, Budapest, Nov. 2005, 18-19.

12

Bunjil Forest Watch
a Community-Based Forest Monitoring Service

Chris Goodman
Object Consulting Pty Ltd,
Australia

1. Introduction

Imagine the power of an Internet-enabled social network that tracked disturbances to the world's most precious forests. Independent observers could expose failings in forest management and help improve governance.

This scenario is not far-fetched, although satellite-monitoring technology has to be made more accessible to non-technical, grass-roots organisations that are independent of official agencies. The good news is that new and organic forms of social organisation and activism are possible by merging the blogosphere with new public tools such as CrowdMap.com. (Ushahidi, 2011). One example is an interactive map developed to show land grabs linked to political elites in Sarawak (Malaysia Today, 2011).

This essay proposes a free public online service that provides non-expert conservation groups in remote locations with alerts about recent forest disturbances in their area. It also explains how such a service might work. Many of the required technical components already exist in various forms.

Local conservation groups living in remote forest areas should not need to understand all the technologies behind the service; nor have advanced computing resources or broadband at their disposal. They should be able to just sign-up to receive free, timely reports about recent disturbances in their area, in their own language, on their phone.

Under the proposal, local groups would control which areas are monitored by subscribing to the service. On receiving reports about a disturbance, the groups would perform enforcement activities according to their judgment and circumstances. They would also provide feedback by responding to the reports.

It is critical that the complex collection and processing of remotely sensed data be completely automated.

The proposal puts the public at the centre by actively encouraging the participation of volunteer observers to perform the routine task of regularly checking new images obtained via satellites. The volunteer observers need not be experts nor have any other connection with the local group other than a common desire to preserve the forest. The service itself would provide all the training for volunteers to become competent observers.

Removing barriers to participation allows the service to be widely deployed. It is envisaged that environmental NGOs would promote the free service to communities in forest regions, while soliciting volunteers from among their international support base.

A system that can provide this service would need to combine:

- Automated detection, collection, processing and delivery of new satellite images;
- A public online volunteer observer registration system.
- Automated distribution of observation tasks to volunteers;
- An open registration system to add new protected areas;
- Automated delivery of reports to remote conservation groups;
- And a process for local groups to respond to the reports with on-the-ground information.

These features must all be integrated by a task-based workflow system. The workflow issues messages to volunteers when new images are ready and to local groups when new reports are made. It promotes regular interaction by actively prompting users to complete tasks and by providing encouragement for completing tasks. It also prompts the local groups to provide feedback on the reports they receive.

The proposed monitoring service would build on the accomplishments of existing deforestation monitoring systems, but differ in a number of ways:

- It is geared towards early detection and intervention in user-selected areas, rather than a complete regional analysis;
- It is completely online and cloud hosted, so there are no infrastructure requirements for users;
- It is non-institutional, relying instead on online relationships and reputations;
- It formally separates observation and response into separate roles;
- It is completely workflow based. New data triggers tasks to create reports, which create new tasks;
- It is tightly integrated with online training, wikis, blogs and discussion forums;
- It relies on continuous user feedback for quality control and ground data collection.

Volunteers would be able to register with minimal barriers to entry and then be encouraged to develop their skills and knowledge with online training and networking with other volunteers.

This design is not an argument against automated detection. On the contrary, algorithms that can highlight deforestation and degradation assist volunteers to identify disturbances and help them know when to raise an alert.

Nor is this an argument against developing capacity among local groups to perform their own monitoring. Using volunteers as a resource has several benefits. First, they already have familiarity with and access to modern computers, monitors, broadband bandwidth and social networks. Second, they belong to different networks than the local groups. These may be crucial for exposing corruption and lobbying internationally. Third, there are likely to be many more volunteers in urban areas willing to spend time monitoring, and this allows the local group to spend scarce resources on activities such as verification, enforcement and reporting. Finally, the service may foster a greater awareness of and connection to deforestation issues among the volunteers, as well as develop invaluable cross-cultural relationships between and among the local groups and the volunteers.

This alert service would not work without complementary conservation strategies. Strategies include protected areas, UN-REDD, sustainable development, land reform and anti-corruption programs. National monitoring programs would also be required for systematic coverage and to measure net verifiable national reductions as required for REDD (Nepstad, et al., 2009).

REDD stands for Reduced Emissions from Deforestation and Degradation. This UNFCC program is based on the idea that developed countries wishing to reduce climate change can pay developing countries to reduce CO_2 emissions from deforestation or forest degradation through the implementation of policies such as strengthened law enforcement, fire management or sustainable forest management. The framework requires measuring the existing carbon stored in the forest and estimating what would be emitted under a business as usual scenario. A project to avoid those emissions is proposed and at the end of a set time period, the actual emissions are measured and compared to what would have happened. The reduced emissions have a financial value that can be traded in carbon markets. Some of the value is hopefully transferred to the locals as income for preserving the forest. Redd-Monitor.org has a good introduction to REDD and its many controversies. REDD is important but not essential to this service.

Just as REDD threatens to recentralise forest governance (Phelps et al., 2010), this service may help democratise forest monitoring away from national forest departments where the capability and governance is not yet in place, and towards grass-roots organisations.

There are several challenges to achieve this. One critical ingredient is regular, low-cost access to recent satellite images - and automated processing of those images into a format volunteers could reliably decode.

To eliminate costs to end users, the system should be based on open-source software, cloud-hosted, and have free regular access to timely satellite data. The solution needs to focus on simplicity of use and hide as much complexity as possible behind a well-designed web-application.

2. Purpose

"Never depend upon institutions of government to solve any problem. All social movements are founded by, guided by, motivated and seen through by the passion of individuals."

Margaret Mead

This tool could provide a complementary self-selecting targeted approach to monitoring areas of high conservation value wherever a local group wishes to protect their forests from external threats.

The main purpose of the service is to provide local conservation groups with timely information about forest disturbances in their area and to provide them with increased opportunities to respond quickly to recent deforestation, particularly illegal logging and land clearing. A recent study of Sumatra and Kalimantan found that at least 6.5% of all forest cover loss had occurred in land where clearing was banned, and a further 13.6 % where it was legally restricted (Broich et al., 2011).

A secondary purpose is to develop networks between people working to conserve remote areas and 'environmentalists' in populated, digitally-connected areas.

3. Who are the users?

"Enforcement against illegal deforestation is clearly a state function, but civil society can provide a formidable assist with timely, high-quality, user-friendly information."

Three Essential Strategies for Reducing Deforestation
(Aliança da Terra et al., 2007)

This proposal separates forest monitoring into two main roles, the *volunteer observers*, who regularly review the latest satellite images and the *local groups*, who rely on alerts when disturbances are detected. Other roles include the sponsors & NGOs, satellite data providers, and developers.

3.1 Local groups

"Community Forest Management (CFM) establishes formal systems between communities and Forest Departments in which communities have the right to controlled amounts of forest products from a given parcel of forest and in return agree to protect the forest and manage it collectively. Mostly these parcels are relatively small, from 25 to 500 hectares, being managed by groups of 10 to 50 households. A number of countries have used CFM very effectively to reverse deforestation and degradation processes"

GOFC-GOLD Sourcebook (GOFC-GOLD, 2010)

The service needs to be promoted to community-based conservationists who may live in remote forest communities and who may be difficult to reach via conventional marketing channels. The service must be distributed to the networks used by local groups and use a language they share.

Local groups could be environment advocacy groups, rangers protecting a park, indigenous people protecting their land or community based forest managers. The local groups might be participating in a REDD project, or other programs.

Groups may be isolated both physically and politically. Over 1150 rural activists have been killed in conflicts related to land in Brazil alone, according to Catholic Land Pastoral (Dangle, 2011). Murder convictions are rare and even rarer for those that hire the gunmen. Fighting deforestation is also dangerous in Indonesia and Malaysia. In Papua New Guinea, where more than 800 languages are spoken in one of the most biodiverse regions on earth, deforestation is running at over 1.5%pa, most of it illegal.

"States with rain forests are often unable to collect optimal revenue from the massive profit earned by timber companies that harvest state forests because this profit already has a hidden destination. Heads of state and their political supporters are siphoning off these moneys to become phenomenally wealthy."

David Brown, PhD Dissertation (Brown 2001)

As much as possible, the solution must remove the barriers for local groups to have access to timely reports. The groups cannot be assumed to have expertise in remote sensing, but may be able to interpret maps, directions or coordinates. Computer literacy cannot be assumed, but access to a mobile phone is almost universal. Access to smart-phones, GPS and phone cameras is becoming increasingly common but is not yet universal. Some literacy in the predominant national language is required by at least one member of the group or a trusted partner. The capacity to visit, investigate and record deforestation events in their locality is important. The ability to prevent or discourage deforestation in some way is also important.

Engaging local groups would be the first bottleneck to expanding the reach of the service. Enhancing the monitoring capability could expose enforcement bottlenecks in that region. Other capacity constraints such as computation, memory and bandwidth are easier to overcome. It is unlikely to be difficult to recruit sufficient volunteers as each volunteer could potentially review an image 180km on a side (Goodman, 2010).

The 'user-experience' for local groups should be designed to be sensitive to local and regional cultural norms and languages.

3.2 Volunteers

The volunteer observers sign up to monitor satellite images on behalf of a local group. The volunteers may be distributed around the world with no direct connection to the local group other than through the monitoring service. They must have adequate time and Internet access.

The Internet, as a low cost medium with global reach, can facilitate the formation of global virtual communities - compensating for a lack of critical mass of activists in a given country (Ackland et al., 2006), (Chadwick & Howard, 2009). Creating a critical mass is an even greater challenge for remote communities in developing tropical countries.

Community Based Forest Monitoring could rapidly engage environmental activists who are already active users of the Internet. It may be quicker to develop the observational capability among digitally connected volunteers, and develop collaborative networks, than building the capacity in remote communities.

To design the volunteers' "user-experience" it is necessary to understand their reasons for participating. Volunteers may be motivated and inspired to be active by ecological experiences and connections with nature; a sense of personal responsibility; a desire to change the world and feeling that they could make a difference; by fear and anxiety about ecological crisis and commitments to justice; or by influential people and social networks (La Rocca, 2004). It also has to be cool. Barriers to becoming active include time available: lack of skills or confidence; alienation or lack of opportunity. Challenges for keeping volunteers active include making the work enjoyable and meaningful; making a difference and responsibility to the local groups.

Each volunteer's participation is sustained by the regular tasks assigned to them and by feedback from the local group.

Volunteers are not necessarily living in the same country as the local group, although this might become common in tropical countries with advanced urban populations such as Brazil, Malaysia and Indonesia. They may even come from among the local group. Volunteers should ideally share a common language with the local group they serve. The volunteer user-experience needs to at least cater for English, French, Spanish, Portuguese and Bahasa speakers to cover the main tropical forest regions.

Volunteers need to recognise the limitations on the local groups' ability to combat illegal loggers, especially the great danger, difficulty reaching sites, and limited law enforcement.

3.3 Environmental NGOs

Non-Government Organisations could promote the volunteering opportunity to their members. They also provide a narrative structure to the regular tasks and feedback.

It is possible (but not necessary) that local groups and volunteers enter into agreements to preserve the forest. These could be through a NGO. The service forms a backbone of information exchange and monitoring that may support the terms of the agreement by building trust among the parties.

NGOs may wish to rebrand the service as their own. Associating the service with their trusted reputation gives credibility to the service while at the same time the service extends their offering and builds their networks.

Environmental NGOs with strong regional networks among local groups are important in promoting the service through the local groups' networks. These NGOs may even partner with local groups who need assistance with subscriptions, communications or translations in some regions.

International NGOs with large member and supporter networks are important for promoting the service to potential volunteers. NGOs constantly struggle to find new and meaningful ways to engage with their supporters, beyond asking for donations. By allowing volunteers to 'adopt' a threatened area, the NGO provides an opportunity for supporters to feel they are contributing in a direct and meaningful way.

3.4 Sponsors

Sponsors may provide a financial incentive to the local group to preserve the forest. It is not essential for the financial agreement to be integrated into the monitoring service. Sponsors can add incentives for subscribing to the system. Project sponsors may be affiliated with volunteers, NGOs or local groups or in a combination.

Sponsors or NGOs may target then reach out to local groups in areas identified as high risk (Sales et al., 2011).

Sponsors are also critical to financing the development, operation and maintenance of the system.

4. Existing deforestation monitoring systems

Before describing the proposed service in detail a review of existing deforestation monitoring programs is presented.

Detecting deforestation and forest degradation from space by observing changes in light reflected from the canopy is not a straightforward task. Nevertheless, detecting deforestation from space has developed over several decades and is now considered routine (Asner, 2009). Detection of forest degradation is harder but also possible.

Brazil has the largest and most systematic use of remote sensing for environmental protection of any country. Some notable operational systems include DETER and PRODES by the Brazil Space Agency, INPE; SAD by Brazilian not-for-profit IMAZON; and CLASlite by the Carnegie Institution for Science which is also focused primarily on the Amazon.

Between 2005 and 2008, PRODES indicated deforestation in the Amazon had slowed compared to what it would have been without the detection and enforcement. However, more recent data from INPE indicates deforestation rates have accelerated by 27% from August 2010 to April 2011 (BBC, 9 May 2011 & The Guardian, 12 June 2011).

4.1 INPE: DETER & PRODES

"If you are going to do prevention and enforcement, you need to be there as rapidly as possible."
                    Gilberto Câmara, Director of INPE quoted by Alexei May in NYT (May 2008)

The Brazil Space Agency INPE runs DETER and PRODES. The newer DETER system can detect large scale illegal logging in near real time while PRODES has higher resolution but results are only updated annually.

DETER provides an update every 15 days and sends alerts to Brazil's Ministry of Environment enforcement agency IBAMA, and police. Loggers are fined and sometime have their property confiscated. DETER relies on a range of satellites including Advanced Spaceborne Thermal Emission and Reflection Radiometer (ASTER) and the China-Brazil Earth Resources Satellite CBERS-2.

The Brazil Space Program for monitoring the Amazon commenced during the Cold War as a military operation, but was later re-purposed to facilitate economic development and expansion. Now the technology has further developed and applied to the detection of deforestation. IT systems tend to reflect the organisation that created it (Rajão & Hayes, 2009). Originally designed by specialists for use by officials and agencies, it is now evolving into an open system.

The technology and capability developed by INPE should have global applicability, although it is currently only deployed over the Amazon region. Expanding deforestation detection systems beyond Brazil's national borders is challenging. Especially to protect areas covered by reluctant, indifferent or corrupt government agencies; where the government has limited jurisdiction; and where the country has no capacity to access and interpret the results or to enforce protection.

INPE committed in 2008 to making the data and technology publicly available, through the Data Democracy Initiative of the Committee on Earth Observation Satellites (CEOS). The commitment extends to governments of developing nations. The *CBERS for Africa* project will provide CBERS images to African countries as part of the Group on Earth Observation.

INPE software has also been released as open source as the SPRING library and INPE has released code for applications TerraLib, TerraView and Marlin built on SPRING. However, a non-specialist would be unlikely to figure out how to extract meaningful data from the current systems.

4.2 IMAZON's deforestation alert system

Instituto do Homem e Meio Ambiente da Amazônia (IMAZON) developed *Systema De Alerta de Desmatamento* (SAD) to monitor deforestation in the Amzon. SAD reports monthly. Like DETER it uses the low spatial resolution (250m) images from the Moderate Resolution Imaging Spectrometer (MODIS) aboard ASTER and publishes data on Amazon deforestation rates each month. Unlike DETER, SAD uses Normalised Difference Fraction Index (NDFI) to detect not only deforestation but also forest degradation. This picks up a lot more land that is degraded. Both SAD and DETER results have been challenged by powerful opponents and withstood rigorous analysis.

ImazonGEO is an open-source open-data Spatial Data Infrastructure (SDI) from Imazon that integrates remote sensing with law enforcement (Souza et al., 2009).

Neither SAD nor DETER are good at detecting deforestation less than 25ha (Escada et al., 2011). PRODES is better at detecting small-scale disturbances, but has low temporal resolution. One technique to improve detection capacity is to combine data from higher spatial resolution sensors with high temporal resolution sensors. This involves using the older but higher resolution images to extract better information from the newer but lower resolution images. Cloud cover can affect temporal resolution by preventing the satellite from capturing a clear image. Cloud cover particularly affects the humid tropics. Access to a range of sensors on different satellites can improve the frequency of capturing cloud free images.

4.3 FORMA

Forest Monitoring for Action (FORMA) is a prototype system by the Centre for Global Development that achieves good resolution using Time Series or Trajectory Based Methods based on the MODIS Vegetation Continuous Field product (VCF) to look at long term trends in change in NDVI. FORMA can detect deforestation the size of a football field. It also detects fires in near real time. The 2009 prototype covers Sumatra and is updated monthly (Hammer et al., 2009).

4.4 CLASlite

Carnegie Landsat Analysis System Lite (CLASlite) [claslite.ciw.edu] is an automated satellite mapping approach that performs statistical analysis on raw satellite images to detect sub-pixel changes in forest cover. While broad-scale clear felling is easy to detect, CLASlite can also detect selective logging down to one or two trees. It is able to distinguish undisturbed forest from recent degradation and regrowth.

After calibration, pre-processing, atmospheric correction, and cloud masking steps, CLASlite will analyse the spectrum reflected in each pixel. Vegetation that photosynthesizes has a different spectral signal from dead trees, rocks or soil. A 'Monte Carlo' analysis then produces a range of possible combinations that converge on the most likely explanation for the data (Asner, 2009). By determining the fractional cover from canopy, dead wood and bare surfaces, CLASlite can provide maps of the forest's composition, including where it has been disturbed. If a tiny red reflection is picked up indicating bare earth, and that signal forms a line over several pixels, the most likely explanation would be a road.

Detecting new logging tracks early increases the opportunity to combat deforestation and degradation, as these are often the first indication of more extensive logging to come.

For input, CLASlite can use a wide variety of satellite imagery including: Landsat 4 and 5 Thematic Mapper (TM), Landsat 7 Enhanced Thematic Mapper Plus (ETM+), ASTER, Earth Observing-1 Advanced Land Imager (ALI), *Satellite pour l'Observation de la Terre 4 and 5* (SPOT), and MODIS.

4.4.1 Applicability

Originally designed for lowland tropical forest, the CLASlite detection method has been tested on imagery from Borneo, Madagascar, the Hawaiian Islands and Mozambique (Asner, 2009).

To generically detect deforestation and disturbance, the method needs to identify changes in forest canopy cover without being overly sensitive to variation in forest type (Asner, 2009). Results show that very different forests can be directly assessed and compared anywhere in the world by the system (Asner, 2009).

4.4.2 Licencing

CLASlite was created by Greg Asner and the Department of Global Ecology, Carnegie Institution for Science. CLASlite is supported by the Gordon and Betty Moore Foundation, the John D. and Catherine T. MacArthur Foundation, and the endowment of the Carnegie Institution for Science.

According to the end user license agreement, the Carnegie Institution intends to work with third parties in the dissemination and use of CLASlite for the purpose of conducting environmental studies and monitoring. The user agreement protects Carnegie's proprietary information, restricts copying and requires attribution to Carnegie in all reports. Results obtained from the use of CLASlite, are to be used for non-profit purposes only. The technology is provided for free to governments and others in Latin America (Tollefson, 2009).

The software could have additional utility if published as a modular library with an Application Programing Interface (API). This would allow it to be mashed-up into new applications. Porting the software to Google Earth Engine should demonstrate this.

4.4.3 Deployment and outreach

A 2008 grant from the Moore Foundation has allowed the CLASlite team to provide training and technology transfer in most tropical forest nations in the Andes-Amazon region, stretching from Venezuela and Guyana across to Peru, Ecuador, and Bolivia (Carnegie, 2008).

The capacity building program provides one-day workshops and technical support to government, academic and NGOs in the region. The aim is for each country to build up its own remote-sensing team (Regalado, 2010).

The CLASlite user website is intended as a space for collective knowledge building to improve forest monitoring and management in the Andes-Amazon region.

4.4.4 Constraints

"While the principal advantage of CLASlite is that it opens options to users who are not necessarily specialists, it is necessary to have people who know exactly what can and cannot be done with CLASlite. I don't know if I would call this a difficulty, but it is a characteristic shared with other approaches to monitoring deforestation."

Manuel Peralvo, Ecuador (CLASlite, 2009)

Although CLASlite is presently only available to non-commercial institutions in the Andes-Amazon Region, it demonstrates that forest monitoring can become an everyday activity that no longer requires huge investments in computers or expertise (Knapp, 2008).

CLASlite presently requires some user technical training to install and maintain. To what extent could the preparation of images for CLASlite and the creation of maps be automated? The steps for preparing CLASlite input are not trivial. They depend on the satellite sensor and rely on third party software such as ENVI or ERDAS to:

- Geo-reference the image to a UTM projection (WGS-84 ellipsoid);
- For LANDSAT, resample the thermal band to match the spectral image resolution;
- Reorder bands, if necessary;
- Save the image to GeoTIFF or ENVI format;

These steps are largely repetitive for regularly monitoring new images of the same location, so an automated image workflow could be contemplated. How well would CLASlite or its successor perform unsupervised is unclear.

A new version of CLASlite is being integrated with Google Earth Engine.

4.5 Other monitoring technologies

The state of the art in forest monitoring is advancing on many fronts. New satellites and sensors are increasing resolution and hyper-spectral bands. Forest monitoring systems must remain adaptable to operationalize new sensors and algorithms as they become available.

The PALSAR (Phased Array L-Band Synthetic Aperture Radar) sensor abort the Japanese ALOS satellite also showed it is possible to detect forest disturbances even through cloud cover (Kellndorfer, 2008). Unfortunately the satellite failed in April 2011. This capability is useful against illegal loggers who use persistent cloud cover to hide their operations.

Airborne LiDAR (Light Detection And Ranging) is opening up a new dimension particularly for the estimation of carbon in a forest.

Much effort is now focussed on political, financial and technical systems for valuing and measuring, reporting and verifying carbon stored in forests for the UN-REDD program. Using LIDAR this is technically possible but still difficult as it introduces carbon inventories, which are sensitive to forest type. The required political and international financial institutions add further complexity. REDD projects work on longer-term financial cycles and may operate better at regional and national levels. The technology for REDD is significantly more complex than the detection of recent deforestation (easy) or forest degradation (harder but established). REDD should promote considerable innovation in forest carbon monitoring.

There are now many tools and standards from which to build a forest monitoring system. TerraLib, TerraView and Marlin are based on INPE's SPRING open-source library. OpenLayers and Geographic Resources Analysis Support System (GRASS GIS) from the Open Source Geospatial Foundation (OSGeo) are also useful open-source toolkits.

The Kings College London, KCL Geodata portal contains a collection of useful tools for environmental monitoring. [sites.google.com/site/consmapping]

4.6 Google earth engine

Google demonstrated a prototype of Google Earth Engine at the IPCC COP15 in Copenhagen in December 2009. Earth Engine is a project of Google.org, Google's philanthropic arm. It is supported by technology partners Greg Asner, the developer of CLASlite, Carlos Souza Jr. developer of IMAZON's SAD and others. Financial sponsors include Google itself as well as the Gordon & Betty Moore foundation, which also sponsored CLASlite.

As well as simplifying access to images, the Earth Engine will include algorithms that can transform the raw images into deforestation maps.

This engine will allow Google's vast storage, computational and bandwidth resources to be harnessed to provide post-processed images in a user friendly format. The problem of creating and maintaining IT infrastructure for distributing, storing and viewing large data sets is solved by moving the application to Google's cloud. Users need only have a web browser. The engine will "Facilitate transparency and security to their data and results. Because the data, analysis and results reside online, they can also be easily shared and independently verified." (Google.org, 2010)

Google Earth Engine is expected to include (GOFC-GOLD, 2010):

- Integrated access to many satellite data products;
- A means to request additional data from public databases;
- Tools for creating spatial and temporal mosaics of the data products;
- Built-in mapping and monitoring algorithms;
- Atmospheric correction, if desired (See Asner, 2009);
- CLASlite forest-view, forest-cover or forest-change maps;
- Imazon SAD functionality;
- An API to introduce new algorithms;
- A geoviewer such as Google Earth browser plug-in;
- Google Fusion Tables;
- Just-in-time computation.

Both CLASlite [code.google.com/p/claslite] and Imazon's open-source Spatial Data Infrastructure (SDI) [code.google.com/p/imazon-sad] are being ported to Google Earth Engine.

The prototype catalogue currently includes access to incomplete archives of many Earth monitoring satellites including LANDSAT, Terra & Aqua and various products from MODIS. These include Surface Reflectance images at multiple frequency bands, and mosaics or composite products such as MODIS Enhanced Vegetation Index (EVI) and Burn Area Index (BAI).

The beta site demonstrates the potential with a high-resolution forest map of Mexico created in record time. Despite the potential, a casual visitor to the prototype may be disappointed as it does yet not contain recent images. To detect recent deforestation, the catalogue must be updated continuously. The solution must include a means to add to the catalogue on demand, and to load other data products.

There is a lack of documentation regarding the prototype. An API has been foreshadowed but no product roadmap has been announced.

All data will be ortho-rectified on import. This makes it possible to mix images from other sensors. A multi-sensor approach can adapt to missing or poor quality data leading to improved continuity. Ortho-rectified maps can be published to Google Maps or other viewers. This also makes it easier to combine the output with other maps and data.

It is unclear what form of open access will be granted to Google Earth Engine. The company has announced it is giving away 10 million hours of CPU per year, but not said how this will be metered. Developer access to code and user access to applications has yet to be defined. Implementing applications SAD and CLASlite into an API promises to create a next generation Spatial Data Infrastructure with unprecedented storage and computation power, increased usability, and universal accessibility for civil society.

5. Bunjil forest watch

Bunjil Forest Watch is a web service proposed by the author for the rapid detection and reporting of deforestation. Existing technology would provide the building blocks from which such a service might be built.

This application does not attempt to measure, report and verify, just to detect, and let people on the ground verify and report back. This is simpler than the technology required for REDD. It is designed as an alarm bell to disrupt deforestation as it occurs and to assist advocacy. It emphasises speed of detection and intervention over systematic regional cover.

The system proposed here is likely to work best on smaller scales. Many conservation issues apply at landscape scales where changes of geology or hydrology lead to unique ecosystems and conservation hot-spots. Sites that may be too big to easily monitor on the ground, may not be so big that local groups are unable to do field inspections or enforcement.

The architecture comprises modules to collect satellite data; to process the images into maps; to enlist volunteers; to create and distribute tasks; to facilitate observation tasks; to generate and distribute disturbance reports and to ensure the integrity of the processes.

Before describing each architectural component of Bunjil Forest Watch, a story about how it could work will help tie the pieces together. Imagine the end user's perspective:

"As a conservationist, wishing to protect my land from illegal clearing, I want to know about any changes to my land as soon as possible so that I can respond to them. I don't know much about satellites, and I can't afford to pay someone to continually check for new data. If a service could just email me to tell me when something changes and where, it would help me protect my land."

The local group learns of the existence of the forest watch service through their networks. A member may either visit the web site and self-subscribe or have a partner in a regional environment organisation set up the subscription on their behalf.

The main web site has a link in multiple languages to take the user through the online subscription process to add a site to the monitoring list. The subscription records the name and contact details for the group, information about the area they wish to monitor and why. The borders of the area of interest can be defined by outlining them on a map. It is essential that this process not intimidate users by asking questions they may be unable to answer. Also, it should provide a help button to ensure that potential subscribers get assistance to complete the process.

Anyone can register to be a volunteer observer. Observers may be recruited via environmental NGOs. During registration the observer is asked to commit to promptly completing any observation tasks sent to them, and to complete the online training. A volunteer may be assigned to one or more parts of a conservation area.

New subscriptions may need to be processed by an expert group who can choose data products and processing appropriate for the biogeography of the Area Of Interest (AOI). Automated mapping algorithms may require tuning with locally-relevant training data and forest definitions in order to produce maps that reflect different definitions of forests, deforestation and degradation.

These experts may be drawn from among the volunteers, developers, environmental NGOs academics or remote sensing professionals. Selecting the appropriate satellite sensors, bands, algorithms and layers is complex. However, this need only be done once to set up the initial settings for a periodic monitoring service. When new images become available, the system must be able to process them automatically, adjusting for clouds, rainfall, haze, and seasons, while still presenting useful images to the observer. The goal should be to reduce the amount of expert input as much as possible through automation. The expert group may also need to vet subscriptions.

Once a subscription is established the system must discover any new imagery that covers the conservation area of interest as soon as it is published. This calls for an automated 'satellite spider' to troll the databases of earth monitoring providers round the clock and to check the metadata describing new images to determine if they are relevant to an area of interest. Satellite images are large, and require processing before they are useful for detecting changes to forest canopy. The spider does not download the images, just the metadata necessary to check if they are relevant. When a new relevant image is found, the spider signals the processing engine.

The processing engine fetches the new image, and performs the necessary adjustments, such as atmospheric correction; cloud detection; ortho-rectification and forest-cover spectral analysis. The actual steps depend on the satellite, sensor, and band and whether these steps have already been completed by the data provider.

Once the image has been processed into a map showing forest cover, it is necessary to access older images or maps of the same place so that a comparison can be made. This is used to detect recent changes, either manually by an observer or using a forest-change algorithm. Finally the processing engine signals to the task manager.

The Task Manager creates a new observation task and emails or tweets the volunteer using the following template:

Dear <Name>, A new image is available for <Conservation Area>. The image was created on <Date>. You task is to look for recent changes to forest cover and file a report by <Date Due>. Click on this link to start your task before <Date> or click here if you are unable to start this task.

The link opens a web application showing both the new and old images covering the area of interest. The area to be inspected is clearly outlined. Any areas where the forest-change algorithm has detected recent change are also highlighted. The observer is repeatedly tasked with reviewing the same area on a regular basis, and only needs to identify whether the change indicates a disturbance in that area. Basic training in image classification can be provided as an online course with completion being part of the registration process for observers.

The volunteer can mark or outline any disturbance she sees using a small set of mark-up tools. She then adds a description to the place-mark, such as 'new road' 'fire', 'crops' or 'clearing'. These place-marked descriptions are automatically collated into a disturbance report. When finished, she reviews the report and clicks send. If no disturbances are observed she files an empty report to complete the task. If unsure about a change, she may mark the location with a question for a more experienced observer to review and complete the task.

On completing her task she immediately receives encouraging feedback and a summary of her cumulative activity and credits, as well as a list of other tasks that may be attempted.

Other tasks can include online training, or assisting other volunteers. These interactions reinforce the volunteers' motivation to remain active.

If any disturbance is reported, the system sends an email or text message alerting the local group. The direct contact details of local groups and volunteers are shielded in the system. The report includes the description and coordinates of each disturbance. If the group has Internet access, they may also review the raw images and maps in a web application, or a low-resolution version if bandwidth is limited.

The local group acts on the report according to their judgement and resources. The local group may also forward the report to an enforcement agency, or send a ranger to investigate on the ground. They should not be pressured into actions by volunteers from the relative safety of a foreign country.

The service asks the local group to respond to each report. The response can be via return email or text, or online. Each response is stored with the report to ensure transparency and to assist with accuracy and other issues. While a no comment response is allowed in some circumstances, the groups are encouraged to include in the responses the accuracy and utility of the report, what investigations were performed and describe any steps taken to deter future disturbances. This feedback helps the service improve. The local group may also update the integrated wiki of their area at any time.

6. Architecture

This section describes the main components of the proposed system.

Because the system aims to provide a free service to local groups, while keeping volunteers engaged, usability is a primary concern. Difficult to follow instructions and slow or erratic responses must be avoided. All interactions must be self-explanatory and support multiple languages.

6.1 Subscription manager

This module handles requests from a local group to protect a conservation area and manages the steps in the subscription process. It also manages the local group's secure online account. Each subscription must include at least the following data before reports can be generated:

- Coordinates of the area of interest;
- Contact details to send reports;
- Report media supported (Fax, SMS, MMS, Email, Hi Def Web, Smart-phone);
- A name for the Local Group;
- Preferred language and other languages spoken;
- A unique name of the area of interest, such as a park or local name;
- Aims of the monitoring project;
- A commitment to use the reports to prevent deforestation.

The subscription manager may also capture:

- Sponsoring NGO & contact details, if applicable;
- The local group's access to technology such as GPS, broadband or smart-phones;

A means to hide the base location and identity of the local group behind an intermediary, such as an NGO should be available when requested.

The Subscription Manager also generates a geo-wiki for each subscription and encourages the local group to add further information to complete a profile of their land. To make subscriptions easier, this need not be completed immediately. This wiki template has sections for:

- Describing the area, e.g. geography, ecology, history, threats;
- Identifying any priority conservation areas within the area of interest, such as critical habitats;
- Uploading photos, stories, expert reports, and biographies of locals (if safe to do so);
- Visitor stories and photos.

The site geo-wiki is moderated by the local group (or their NGO partner) so anyone can contribute. Each subscription also has a message forum dedicated to it, for sharing intelligence about the local area. This would be in the local group's language. The moderator also has the option to request the boundary of the area of interest be nominated for inclusion in the World Database on Protected Areas [protectedplanet.net]. Conversely, if the protected area is already recorded in the database, then the subscription manager should be able to import this definition.

6.2 Satellite data service

There are many high resolution satellites from multiple providers. Satellite sensors such as Landsat TM and ETM+ (USA), Terra ASTER (USA-Japan), CBERS-2 (China-Brazil), SPOT MSS (France), and IRS-2 (India) provide data required for high resolution mapping of deforestation, logging, and other tropical forest disturbances (Defries et al., 2009). Each sensor has its own characteristics, making it more difficult to compare scenes.

The coordinates of subscribed areas may be submitted to the satellite operator's mission planners to increase the probability or frequency that the site will be scanned.

The application does not access the satellites directly, but relies on public data providers such as USGS, ESA, JAXA and INPE.

For rapid detection, the properties to look for in a satellite data service are:

- Covers the area of interest;
- Sensor has a suitable resolution and frequency bands;
- New images are continuously acquired at reasonable frequency;
- Data is (or can be) geometrically and radiometrically corrected;
- New data is processed and published as soon as it is captured;
- A notification service is available, containing meta-data describing the images, so that a spider can be configured to discover images;
- Meta-data is published using open standards;
- Free access to the public or at least copy-controlled access for not-for-profits at no cost.

If free public access is not available, it may still be possible to negotiate restricted access that allows the application to create and display deforestation maps without sharing the raw data files.

6.3 Satellite spider

The satellite spider is a component of the application that continually looks for new images covering each subscribed area of interest. The spider maintains the coordinates of all active subscription areas. It trawls the image meta-data on online databases. It requires a separate 'plug-in' for each satellite data service supported. For example, USGS supports an RSS service that allows the spider to read an XML data feed describing the latest LANDSAT data [landsat.usgs.gov/Landsat7.rss]. The spider must convert from latitude and longitude of the area of interest to the path and row of the satellite [glovis.gov]. Other providers may offer a Web Map Service (WMS). The spider runs 24/7 but does not actually download or process any images. Each time a new image is found that is relevant to monitoring an area of interest, the spider sends a request to the processing engine to download it and create a map.

6.4 Image processing engine

The image processing engine must download images found by the spider, process them automatically into forest-change maps and store them in a map server.

The processing algorithms used are no different from those developed for existing detection systems, i.e. ortho-rectification, geo-registration, atmospheric correction, cloud removal, NDVI spectral analysis or 'unmixing'.

The processing engine has significant bandwidth and computation requirements and must be cloud hosted to avoid infrastructure maintenance.

The Google Earth Engine API promises to provide much of this functionality. Alternative toolkits are available to do processing. Amazon.com could be used for cloud storage and computation.

While a variety of sensors, data products and algorithms exist, the processing engine should aim to present a consistent display, independent of these factors. For example, a standard colour scheme and legend could be used to classify disturbance types. For a non-expert, the most easily understood imagery is high-resolution natural colour such as seen on Google Earth. However they are not updated regularly on Google Earth. Also, this is not the best format for detecting forest disturbance. Normalised Difference Vegetation Index (NDVI) images can complement visual images. Fire alerts should be integrated into the community based monitoring service. Existing fire detection systems based on MODIS, such as Indofire (Landgate, 2009) could be interrogated daily and send alerts to the volunteer and local group if a fire is detected near the area of interest. Fires are highly correlated with deforestation in many countries, depending on agricultural practices.

Once the processing engine has created the map it signals the workflow system to create an observation task.

6.5 Map server

The Processing Engine stores new maps in a map server. The map server must keep a time-series archive of forest-change maps and images for each area of interest, as well as disturbance reports, place-marks and corresponding responses from local groups. The map server does not need to store raw satellite data that is available elsewhere.

When a volunteer is conducting an observation task, the Observation Portal requests maps from the map server to be displayed using an established API. Responsiveness to these data requests is important to the volunteer's experience of the Observation Portal.

6.6 Volunteer manager

This module handles requests from volunteers to register, and it manages their account and profile. Their real identity should be secured by the system. The manager records which languages are spoken and the preferred countries or regions in which the user may wish to volunteer.

The Volunteer Manager keeps track of tasks assigned and completed and any training the volunteer completes. It suggests available e-courses in which the volunteer can enroll.

Volunteers may self-register and deregister, and edit most of their profile, but not their qualifications, ratings or feedback.

Volunteers must pass some basic training and assessment, and agree to a code of conduct before qualifying. Observation tasks are only assigned to qualified volunteers.

6.7 Observation portal

The Observation Portal is a web application for comparing images and annotating disturbances. It could be built on existing geographic display tools such as OpenLayers or Google Maps.

The Observation Portal must be configured to assist the volunteer complete the observation task. The link embedded in the volunteer's email (or tweet) should open the portal at the correct coordinates and zoom, and display the correct layers to show the latest image and differences to previous images. Also displayed are any enhancements, such as fire detection, or outputs of automatic deforestation detection systems if available. The date each image was captured, the task id and due date are shown by default. The boundaries of the area of interest and relevant park boundaries should be displayed.

It is crucial this is displayed automatically as soon as the user clicks the task-link in their email. Expecting volunteers to navigate to an online database, find, order and download images, then load them into the viewer would lead to a high failure rate. These defaults must either be setup automatically, or be preset by an expert when a subscription is created.

Correct initial settings allow the novice volunteer to concentrate on observation rather than learning and adjusting the tools. Zooming, panning and selecting from a small set of layers are essential for the task.

The user has options to display meta-data if desired, such as the sensor, band and processing steps. They may also call up earlier reports, images or maps that outline earlier disturbances, or predefined elements such as roads or management boundaries. Protected area boundaries may be accessed from online resources such as the World Database on Protected Areas [protectedplanet.net] via the Subscription Manager.

The user experience is simplified down to just the graphic elements necessary to help the novice user to complete the current task, rather than present a rich and complex GIS interface.

Why use human observers to compare the before and after images when automatic detection is possible?

Firstly, because automatic detection can only create a map. It must still be interpreted by humans. Although the algorithm could send alerts automatically, the local group may need to analyse to check for false alarms. Secondly, volunteers may also become advocates. By exposing illegal deforestation to a globally connected audience, they increase awareness and engagement beyond national and bureaucratic hierarchies. New international networks among local groups and volunteers may help embolden local authorities and communities to better protect forests.

One way to visually compare images or maps is to view them side-by-side. One service that demonstrates this technique is AnotherEarth.org (Firth, 2011). It displays two Google Earth javascript windows side-by-side in a browser. Both new and old imagery is displayed from an identical viewpoint. Panning or zooming one window will pan or zoom the other to the same viewpoint. Because the images are geo-referenced, aligning old and new images can be automated and differences easily observed. Another technique is the before-after rollover developed by Andrew Kesper at the Australian Broadcasting Corporation to show the changes following natural disasters (Kesper, 2011). Other techniques include flicking or fading between images. Research on user preference and performance is required to determine the best methods.

The Observation Portal also presents a reporting panel to the user. This contains a simple set of screen icons for drawing point, line or polygon place-marks. Each new place-mark or polygon includes a standard form for classifying disturbance types as well as free text annotations. The date, observer and task id is recorded automatically in each place-mark. The observer may create multiple place-marks for a single disturbance report, and may edit them until the task is sent. Once sent, the report can only be edited under version control.

Completing the task is simply a matter of pressing 'Send', as the Workflow System automatically distributes the report, and ensures it is recorded and a response received.

The reporting panel is implemented using a standard GIS toolkit.

6.8 Report manager

"A final important element is the portal for integration of ground-sampled data into this platform; including data from smartphones used in trials in community-based forest monitoring"

REDD Sourcebook (GOFC-GOLD, 2010)

This module manages the collection, storage and transmission of disturbance reports created by volunteers in the Observation Portal. It delivers the report to the local group, using the agreed contact details and method (email, text). The report manager also requests and manages feedback from the local group to rate the accuracy of each report and record any activities the local group made in response to the disturbance.

The report manager archives all the reports relating to each protected area. Both the reports and associated feedback are stored in the geo-wiki for the area of interest, together with any maps created by the processing engine. They may be used for future research or tuning the system. They can be searched geographically, visually, temporally, by disturbance type or by volunteer.

The Report Manager can translate standard fields in the report if there is a mismatch between the language of the report and the recipients. Automatic translation within major languages could be integrated. Other translations would need to be referred to a regional NGO.

A disturbance report will contain:

- a reference to the originating task - so all task parameters are archived;
- the handle of the volunteer who completed the task, which links to their public profile;
- the time and date the report was completed;
- geo-referenced place-marks created by the volunteer;
- completed forms and annotations created by the volunteer for each place-mark;
- severity of the disturbance – from minor to serious;
- observer's confidence of the accuracy of the disturbance.

Once a local group responds, the report may also contain:

- veracity response: indicates whether the observation indicated a real change or not;

- accuracy response: how accurate were the coordinates in the report. This may help detect alignment issues;
- conservation value of the information - whether it helped reduce impacts;
- comments or photos uploaded by the local group or NGO;
- actions taken, such as referrals to law enforcement.

A response may be entered directly by the local group if they have web or smart-phone access, or by the NGO. The response may also add comments to the volunteer's report.

Even when the local group has no GPS or other capability to record coordinates in their response, their text or photos can still be referenced back to the original disturbance report, and therefore indirectly geo-referenced.

Local groups can also forward reports to enforcement agencies. Carlos Souza of Imazon describes a reporting system that will "allow users to … be able to identify cases of illegal deforestation that are being judged, send requests, as protests, to prioritize cases with the environmental agencies and courts, monitor the length of the proceedings, and receive alerts about the status of the process. We hope that this type of geo-wiki tool can engage civil society in order to accelerate the cases and bring positive pressure on the enforcement system to properly punish violators. This is important, because the application of enforcement law represents the major bottleneck to stopping illegal deforestation in the Brazilian Amazon" (Souza et al., 2009).

6.9 Workflow system

The workflow is the central control process that keeps the system alive by responding to external events and ensuring tasks are created, assigned and completed. The workflow ensures new data is processed, tasks are assigned and reports reach the people on the ground quickly and reliably. A good workflow system allows business rules and process logic to be encoded in a flexible but rigorous language, rather than buried in code.

Yet Another Workflow Language (YAWL) is one open-source workflow language and implementation (Hofstede et al., 2008).

Workflow systems are now commonplace in corporate enterprises to encode and automate business processes. Examples include defect management systems. Workflow systems can also facilitate virtual organisations, such as open-source software collaboration teams. A workflow typically describes an artefact as it is transferred from one stage to another.

For this forest monitoring application, the flows would describe a raw or processed satellite image, an observation task, or a report. The workflow system assigns tasks to people or groups; manages the scheduling and email notification system; and supports resolution, escalation and exception processes.

A new task is triggered when a new forest-change map is ready. The system assigns an observation task to the selected volunteer. The observation task may be reassigned to another volunteer if it remains in a queue for too long, or if the first volunteer requests a review. When a disturbance report is created, a new task ensures that it is sent to and acknowledged by the correct local group, and also ensures the local group sends a response within a reasonable time.

Within this basic flow are many alternate possibilities, exceptions and error cases. A flexible workflow engine allows core behaviours to be reconfigured to suit ad-hoc organisations as they evolve. There is no reason to limit Bunjil Forest Watch to a single instance. Different groups may find reasons to build and deploy variations to meet unforeseen needs.

6.10 The social network

A community-based forest monitoring system needs a wiki to encourage and strengthen community ties and share information. Social network software must be integrated into the solution to allow volunteers to interact with other volunteers, local groups, experts and the public, and to develop monitoring skills.

The social network tools consist of a geo-wiki for each local group and subscribed area, a public profile page and optional blog for each volunteer, a general wiki for the application user guide, FAQ and support in multiple languages, and a discussion forum for posting questions and defects.

Many of the interactions are automatically generated by workflow tasks. A user's reputation is updated each time one of their reports is verified. Reports and photos uploaded by local groups may be automatically added to the site wiki and geo-located with the original disturbance report. Other interactions are initiated by users or prompted by the workflow, to keep the geo-wiki up-to-date.

The social network could be implemented by customising an existing open-source Content Management System (CMS), such as Radiant or Wordpress. There are many mass-market social network sites; however a high level of integration and customisation is required to integrate with the workflow and keep the focus on the tasks. The social network must support volunteers and local groups to combat deforestation.

The site geo-wiki could well be integrated into the UNEP's *ProtectedPlanet* [protectedplanet.net] or *Atlas of Our Changing Environment* [na.unep.net/atlas/google.php]. These global Wikis already have much of the functionality required to define a community-based protected area, add to a blog, and upload geo-located photos. Using existing infrastructure is easier than creating a new site, provided the workflow integration can be achieved. The social network could also link to content in relevant environment sites such as Mongobay.com or GloboAmazonia.com.

Each volunteer has a profile page and optional blog for sharing stories about the observing experience. A volunteer's real identity can be hidden, but their online reputation as a volunteer is tracked, including the punctuality, reliability and accuracy of their reports.

6.11 Online learning

Bunjil Forest Watch will rely on a Learning Management System (LMS) to manage online and collaborative training for volunteers. This includes a wiki, forum, course material, videos and exams. The main purpose of the online learning module is to improve volunteers' observation skills and knowledge. The need for training users has been identified by both the CLASlite and IMAZON teams.

The training describes a volunteer's responsibilities and shows examples to help illustrate the kind of satellite images or maps the student will be likely to encounter and the sorts of changes to look out for. The training also shows how to create a report.

To qualify as an observer, the volunteer must complete an online test where they review pre-analysed images and correctly identify threats. Further training is available to retain and increase competency and to broaden knowledge in subjects relevant to forest monitoring, for example forest ecology; remote sensing; sustainable development or cross cultural communication. *Instituto de Pesquisa Ambiental da Amazônia* – IPAM, have an online course on *The Amazon Rainforest and Climate Change.* [ipam.org.br/curso/login]

Volunteers must complete some online training to qualify as an observer. Only qualified observers receive real assignments and can send reports. This is to discourage uncommitted and unreliable users.

Volunteers receive credits as they complete training modules. They also receive credits as they successfully perform observation tasks. Credits increase the volunteer's grading. This allows volunteers to perform more critical tasks and assist or review others.

The Learning Management System manages the syllabus of courses; enrolment in e-courses; and serves the training and examination material. The LMS can also be based on open-source software for example Chamilo [chamilo.org], Wikiversity [wikiversity.org] or Khan Academy [khanacademy.org].

Volunteers can also create course material for other users.

Local groups can also access the training. Unlike volunteers, they are not required to complete training to receive reports.

7. Discussion

7.1 Mobiles and Smart-phones

Local groups are not required to have knowledge of GIS, or remote sensing, but often will have access to a mobile phone and be within mobile range. Africa has already achieved 50% mobile phone penetration, rising at 20% pa. "Smart" phones with GPS and camera are now a mass-market technology in developed countries, but may remain too expensive for many local groups in developing countries for some years. However, a phone "app", would greatly improve the utility of the service. It would incorporate the deforestation report, the original forest change map, and a GPS to direct the group directly to the disturbance. Using camera, messaging and GPS it would be simpler to file a response that is automatically and reliably cross-referenced and geo-located in time and space.

Open Data Kit [opendatakit.org] by the University of Washington and Google.org is an open-source multilingual suite of mobile data collection tools for the Android platform. Woods Hole Research Centre has trialled the technology for collecting data from REDD forest plots.

Smart- phones will greatly assist collecting reliable evidence for both scientific and enforcement purposes. The smart-phone app could be included as an enhancement, but making access to a smart-phone a requirement for participation risks limiting access to many groups.

7.2 Accuracy

"There has to be trust in the forest-monitoring data, and these nations have to see them as their own ...There's this face-to-face collaboration that is really critical."

Dan Nepstad, quoted in *Nature* (Tollefson, 2009)

Volunteers, even with auto detection algorithms, still face a big challenge to correctly interpret the imagery, and distinguish a significant disturbance from artefacts such as data errors, seasonal variations or variations in viewing conditions. The generation of images is unsupervised and the viewer self-trained. This increases the likelihood of errors.

Quality control is important to reduce incorrect reports. False positives create unnecessary work and travel for the local group while false negatives mean disturbances are not picked up. Opponents of the local group can use errors or inconsistencies to undermine the reports. Maps prepared automatically for alerting may not reach the standards required for long-term REDD Monitoring, Reporting & Verification (MRV). But they may not be required to since a more rigorously controlled analysis can still be created later to prosecute a case.

The usual way to measure the accuracy of a new system is to compare the results with known data. A pilot of the system can be chosen to overlap with an established regional monitoring program and the results can be compared.

Additionally, as this system is repeatedly monitoring for change over fixed but relatively small areas, there is an opportunity to introduce self-correcting feedback. Ensuring the local users of the service report back to the observer on the quality of the reports gives a very good guide to the accuracy and performance of the system.

This is why local groups must rate the quality of each report they receive. The responses are automatically collated and generate statistics on each volunteer as they become more experienced. The reputation of the observers and local groups accumulate with each transaction. This feedback helps tune the service to identify poorly performing volunteers, unresponsive local groups, common false alarms and system biases or failures.

7.3 Public access

While the LANDSAT archive and other data are publicly available online, a conservation monitoring service may also need access to higher resolution imagery. High-resolution visual images may be easier for non-experts to understand. However there are trade-offs in costs, temporal resolution (frequency) and bandwidth. Many deforestation detection systems combine multiple sensors to make a statistical estimate of where deforestation is occurring. However, the critical factor for alerts is timeliness. The main business model for most high-resolution satellite providers is providing data to governments. Providing data for deforestation in remote areas, with suitable checks and balances, would not undermine this model and may enhance the reputation of the provider. For example, the GeoEye Foundation provides access to IKONOS (0.5m) resolution imagery to NGO's for humanitarian purposes.

After access to data, the next greatest issue is access to the technology to process the data. Fortunately, the owners and creators of the worlds most advanced forest monitoring systems share this aim. Increasingly, research is being published in open journals accessible to non-institutional scientists.

Another hurdle is access to computing capacity to produce the images. It is significant, but not unachievable. The cloud computing paradigm for IT infrastructure is applicable to this application. There may be infrastructure providers prepared to offer free access during low-demand for not-for-profit applications. Google's aforementioned commitment to donate ten million hours of CPU time per year is promising.

The interactive sessions must run as on-demand services, but these are data rather than computationally intensive.

7.4 Costs

The cost of computation continues to fall while speed continues to increase but removing all costs to local groups and volunteers is essential.

The registration and subscription processes ensure that no image is downloaded or processed unless there is a local group prepared to protect the area and there is a volunteer prepared to observe the images on a regular basis. This avoids wasted computation and data transfer as areas outside the area of interest are masked out of the calculation.

Obtaining regularly updated imagery at low cost is one of the key challenges. In 2008 the USGS made a decision that was a watershed for open access by providing free online LANDSAT data. Unfortunately, barriers to accessing the data remain, especially in Africa where international bandwidth is limited (Roy et al., 2010). Barriers include limited tertiary education, especially in remote-sensing fields; conflicting national interests and priorities; inadequate awareness of potential uses; insufficient infrastructure and high data costs.

The cost of implementing and maintaining the service must also be addressed. It is likely to be developed in a series of prototypes of increasing functionality.

7.5 Displacement

Although protected reserves reduce deforestation rates, they may not eliminate deforestation in the reserve completely (Clark et al., 2008). Displacement of deforestation occurs when an area is protected but its surroundings are not. If the area is reserved as a carbon bank then displacement is undesirable as emissions are merely moved rather than reduced. However displacement may be desirable if the purpose of the park is to protect a unique biodiverse area. For example, encouraging palm plantation corporations to shift expansions to areas already degraded by earlier logging, without destroying more primary forest, benefits both biodiversity and climate goals.

8. Conclusion

We can encode the motto *think global act local* into the DNA of our next generation of Earth observation infrastructure. This promises to open a new global front to combat illegal deforestation and degradation.

The solution described here could be built from existing open-source components, hosted on cloud infrastructure. More and more satellite imagery is freely available on public databases, and methods to process the images are advancing. This paper has described one way that these available resources could be put to better use.

9. References

Ackland, R.; O'Neil, M.; Bimber, B.; Gibson, R. & Ward, S. (2006). New Methods for Studying Online Environmental-Activist Networks. Virtual Observatory for the Study of Online Networks (VOSON). *26th International Sunbelt Social Network Conference*, Vancouver. April 2006.
    http://voson.anu.edu.au/papers/environmental_activists_methods_presentation.pdf

Aliança da Terra; Amigos da Terra; Instituto Centro de Vida; IMAZON; IPAM; Instituto SocioAmbiental; Nucleo de Estudos e Prática Jurídica Ambiental; Faculdade de Direito -- Universidade Federal de Mato Grosso; Woods Hole Research Center. (2007). Three Essential Strategies For Reducing Deforestation.
    http://www.whrc.org/policy/pdf/cop13/3-Strategies-Dec-07.pdf

Asner, G.P.; Knapp, D.E.; Balaji, A. & Paez-Acosta, G. (2009). Automated mapping of tropical deforestation and forest degradation: CLASlite. *Journal of Applied Remote Sensing* 3:033543.

BBC (9 May 2011). Brazil: Amazon rainforest deforestation rises sharply.
    http://www.bbc.co.uk/news/world-latin-america-13449792.

Broich, M.; Hansen, M.; Stolle, F.; Potapov, P.; Margono, B. & Adusei, B. (2011). Remotely sensed forest cover loss shows high spatial and temporal variation across Sumatra and Kalimantan, Indonesia 2000–2008. *Environ. Research Letters* V6 (January-March 2011) doi:10.1088/1748-9326/6/1/014010

Brown, David W. (2011). Why Governments Fail to Capture Economic Rent: The Unofficial Appropriation of Rain Forest Rent by Rulers in Insular Southeast Asia Between 1970 and 1999. PhD. Dissertation.

Carnegie. (2008). Carnegie Science, press release, Dec 4 2008, carnegiescience.edu/news/tropical_forest_carbon_monitoring_gets_big_boost

Chadwick, A. & Howard, P. (2009). Handbook of Internet Politics, *Routledge*

Clark, S.; Bolt, K. & Campbell, A. (2008). Protected areas: an effective tool to reduce emissions from deforestation and forest degradation in developing countries? Working Paper, *UNEP World Conservation Monitoring Centre*, Cambridge, U.K. www.unep-wcmc.org/medialibrary/2010/10/05/2a7f53e5/Clark_et_al_2008.pdf

CLASlite (2009) Peralvo, M., quoted in CLASlite Capacity Building website, Accessed Oct 2011. claslite.ciw.edu/en/action/training.html

Dangl, B. (2011). The bloody cost of Amazon deforestation. *The Guardian*. (June 12 2011) www.guardian.co.uk/commentisfree/cifamerica/2011/jun/12/brazil-amazon-rainforest

DeFries, R.; Asner, G.; Achard, F.; Justice, C.; Laporte, N.; Price, K.; Small, C. & Townshend, J. (2009). Monitoring Tropical Deforestation For Emerging Carbon Markets. In: *Reduction of Tropical Deforestation and Climate Change Mitigation*, Mountinho P. & Schwartzman S. (Eds).

Escada, M.; Maurano, L.; Rennó, C.; Amaral, S. & Valeriano, D. (2011). Evaluation of data from Early Warning Systems Amazon: DETER and SAD, '*Avaliação de dados dos Sistemas de Alerta da Amazônia: DETER e SAD*', Anais XV Simpósio Brasileiro de Sensoriamento Remoto - SBSR, Curitiba, PR, Brasil, 30 de abril a 05 de maio de 2011

Firth, J. (2011). Another Earth. anotherearth.org

GOFC-GOLD. (2010). A sourcebook of methods and procedures for monitoring and reporting anthropogenic greenhouse gas emissions and removals caused by deforestation, gains and losses of carbon stocks in forests remaining forests, and forestation. *Global Observation of Forest and Land Cover Dynamics*. www.gofc-gold.uni-jena.de/redd/sourcebook/Sourcebook_Version_Nov_2010_cop16-1.pdf

Goodman, C. (2010). Bunjil – A social network for proactive monitoring of tropical rainforests. *Telecommunications Journal of Australia*. 60 (1): pp. 4.1 to 4.16. DOI: 10.2104/tja10004. Available online.

Google.org. (2010). Presentation to WorldBank, undated. Available Online web.worldbank.org/WBSITE/EXTERNAL/TOPICS/EXTARD/0,,contentMDK:22555701~pagePK:210058~piPK:210062~theSitePK:336682,00.html

Hammer, D.; Kraft, R. & Wheeler, D. (2009). Forest Monitoring for Action - Rapid Identification of Pan-tropical Deforestation Using Moderate-Resolution Remotely Sensed Data. *Centre for Global Development* www.cgdev.org.

Hofstede, A.; Aalst, W.; Adams, M. & Russell, N. (2008) Modern Business Process Automation: YAWL and its Support Environment. Updated 2011. Available Online at: www.yawlfoundation.org

IUCN. World Database on Protected Areas. A joint initiative between IUCN and UNEP WCMC. www.protectedplanet.net,

Kellndorfer, J. (2008). New Eyes in the Sky: Cloud-Free Tropical Forest Monitoring for REDD with the Japanese Advanced Land Observing Satellite (ALOS). Woods Hole Research Centre, *UNFCC-COP15*. Bali, Dec 2007. www.whrc.org/policy/pdf/cop13/Bali_ALOS.pdf

Kesper, A. (2011). *Australian Broadcasting Corporation*. www.abc.net.au/news/specials/japan-quake-2011/

Knapp, D. (Dec 4, 2008). quoted in Mongobay, 'Rainforest canopy-penetrating technology gets boost for forest carbon monitoring', *Mongobay* Dec 4, 2008.
    news.mongabay.com/2008/1204-asner.html

La Rocca, S. (2004). Making a Difference, Factors that influence participation in grassroots environmental activism in Australia. PhD Dissertation. Available online at *The Change Agency*, www.thechangeagency.org/01_cms/details.asp?ID=13

Landgate. (2009). Indofire, *West Australian Land Authority*.
    indofire.landgate.wa.gov.au/indofire.asp

Malaysia Today. (March, 2011). Malaysia Today: WE RELEASE THE LAND GRAB DATA!
    malaysia-today.net/archives/archives-2011/38921-we-release-the-land-grab-data

May, A. (May 25, 2008). Brazil Rainforest Analysis Sets Off Political Debate. *New York Times*.
    www.nytimes.com/2008/05/25/world/americas/25amazon.html?pagewanted=all

Nepstad, et al. (2009). The End of Deforestation in the Brazilian Amazon, *Science* 4 December 2009: 326 (5958), 1350-1351. [DOI:10.1126/science.1182108]

Phelps, J.; Webb, E. & Agrawal A. (2010). Does REDD+ Threaten to Recentralize Forest Governance? *Science* (16 April 2010): Vol. 328 no. 5976 pp. 312-313 DOI: 10.1126/science.1187774, www.sciencemag.org/content/328/5976/312.summary

Rajão, R. & Hayes, N. (2009). Conceptions of control and IT artifacts: an institutional account of the Amazon rainforest monitoring system. *Journal of Information Technology*, 24 (4) pp. 320-331],
    ufmg.academia.edu/raonirajao/Papers/127750/Conceptions_of_control_and_IT_a
    rtefacts_an_institutional_account_of_the_Amazon_rainforest_monitoring_system

Regalado, A. (2010). New Google Earth Engine. *ScienceInsider*, 3 December 2010.
    news.sciencemag.org/scienceinsider/2010/12/new-google-earth-engine.html

Roy, D.; Ju, J.; Mbow, C.; Frost, P. & Loveland, T. (2010). Accessing free Landsat data via the Internet : Africa's challenge, *Remote Sensing Letters*, (Volume 1, Issue 2, 2010), DOI:10.1080/01431160903486693, pages 111-117 Available online

Sales, M.; Souza, C. & Hayashi, S. (August 25, 2011). Risk of Deforestation. Edition 2, August 2011-July 2012. IMAZON, '*Boletim Risco de Desmatamento Agosto de 2011 a Julho de 2012*' Reported in 'Model predicts deforestation hot spots in Brazilian Amazon' Moukaddem, K. mongabay.com news.
    mongabay.com/2011/0825-moukaddem_imazon_forecast.html

Souza, C.; Pereira, K.; Lins, V.; Haiashy, S. & Souza, D. (2009). Web-oriented GIS system for monitoring, conservation and law enforcement of the Brazilian Amazon. *Earth Sci Inform*. 2:205–215. DOI 10.1007/s12145-009-0035-6
    www.springerlink.com/content/e1g673k21k168591/fulltext.pdf

Tollefson, J. (2009). Climate: Counting carbon in the Amazon. *Nature* 461, 1048-1052 (2009) | doi:10.1038/4611048a , Published online 21 October 2009.
    nature.com/news/2009/091021/full/4611048a.html

Ushahidi (2011). CrowdMap. Ushadihi.org.

13

Deforestation and Waodani Lands in Ecuador: Mapping and Demarcation Amidst Shaky Politics

Anthony Stocks[1], Andrew Noss[2],
Malgorzata Bryja[2] and Santiago Arce[2]
[1]Idaho State University,
[2]Wildlife Conservation Society – WCS
USA

1. Introduction

One of the major forces of deforestation around the tropics is the chipping away of forested areas for pastures by agricultural peasants who are difficult to control by remote central governments (Colchester 1998) and by loggers who enjoy the same advantages of working in isolated areas as colonists and who tend to bring roads into the forest. The major danger to many forests is fire, made more likely by the agricultural colonization that follows road construction (Nepstad et al. 2001). In the Ecuadorean Amazon fire is not a major threat, probably because of the year-round rainfall regime that maintains high levels of humidity, and road construction is driven first by oil exploration and exploitation activities that in turn facilitate access and settlement by colonists and loggers (Bromley 1972; Viña et al. 2004). Ecuador's 1964 Law of Agrarian Reform and Colonization classified large portions of Amazon as unoccupied, allowing colonists to claim 50 ha plots along roads, directly promoting deforestation by requiring proof of improvements to establish legal land titles (Bilsborrow et al. 2004; Bremner & Lu 2006; Fuentes 1997; Kimerling 1991).

In many parts of the world, however, the tropical forests have potential allies in the form of indigenous people who have inhabited the forest for millennia and are anguished about seeing it degraded and cut down. In fact it is estimated at present that 85% of the world's areas designated for biodiversity conservation are inhabited by indigenous peoples, whereas outside of the parks and nature preserves, the world's remaining pristine forested habitats are nearly all occupied by indigenous peoples (Alcorn 2000; Colchester 2001; Schmidt & Peterson 2009; Weber et al. 2000). This is true of the Ecuadorean Amazon in particular (GeoPlaDes 2010). In fact, conservation-minded outsiders have only a few choices it they want protection for these habitats. They can try to protect the forests while excluding the indigenous people – treating them essentially as fauna and making enemies of them – or they can assist them as allies (Colchester 2000, 2004; Schwartzman et al. 2000). The latter choice carries its own problems. Not all indigenous people want to save the forest, given their current assessment of costs and benefits of doing so; a certain amount of discrimination is necessary. Those who do usually want to either own the land (Colchester 2000), or, in the case of protected areas, to have signed legal agreements with governments giving them use and management (or co-management) rights which carries its own problems with coordination, bureaucracy and political will on the part of the state.

Although recognition of ownership and/or control of large tracts of land by private individuals or groups is easy for conservationists in the developed world, it becomes much more problematic in remote forest frontiers where indigenous people may be less visible, forested areas are often not densely populated and conservationists may have closer relationships with governments than with indigenous peoples. Hence, efforts in counter-mapping have become common in the past 20 years in order to identify the areas occupied by and claimed by indigenous peoples (Chapin &Threlkeld 2001; Peluso 1995; Poole 1998; Stocks 2003; Stocks & Taber 2011; Wainwright & Bryan 2009).

Are indigenous people, once they have control and appropriate resources, able to protect forested habitats? Many would say that it is a mistake to equate indigenous occupation with conservation (e.g., Redford & Stearman 1993; Redford & Sanderson 2000). While this may be true as a general statement, the specific local outcome depends on a number of factors that include indigenous levels of organization, the kinds of resources available, their sense of place connected with a history of occupation, their own economy and the political power of competitors– industrial or individual – that also seek to occupy the forest. Evidence from a few sources in the neo-tropics indicates that when indigenous people are protecting a historic homeland with some outside help, they tend to be more successful in maintaining forests than colonist populations faced with essentially the same economic realities (Colchester 2000; Lu et al. 2010; Nepstad et al. 2006; Ricketts et al. 2010; Soares et al. 2010; Stocks et al. 2007). Indeed, in the Brazilian Amazon, the inhibitory effect on deforestation of various kinds of protected areas (strict protection, sustainable use, indigenous lands and military areas) is greatest in the category of indigenous lands (Soares et al. 2010). In the Ecuadorean Amazon this feature of indigenous lands is particularly notable in the A'í (Kofán) case (Borman et al. 2007) and the protection by the A'í of their forests on titled (*resguardo*) lands is also true in Colombia. One recent study with a broad world-wide sample argues that forests owned and/or controlled by local communities, tend to have less deforestation. Livelihood benefits and carbon storage both increase when the historic dwellers own the land or can otherwise control land use (Chhartre & Agrawal 2009).

This chapter is premised on the idea that the Waodani people of Ecuador have the essential prerequisites of ownership, sense of historic occupation, threats from colonists and outside assistance to halt the deforestation of their titled territory; and that physical demarcation of their land is a necessary step in the protection of the forest. However, the particular kind of political organization they currently exhibit, the nature of the colonists that are deployed along their frontiers, industries within their territory, and overlaps with national protected areas make demarcation a somewhat different exercise than one would think. The rest of this paper explores the history of demarcation and the consolidation of land under their control and process of providing technical assistance to them in the face of what we have called "shaky politics." The dilemma of working with the Waodani is not uncommon in conservation. Conservationists of tropical forests typically encounter groups with essentially egalitarian political philosophies who find the establishment of hierarchical bureaucracies difficult to maintain and impossible to make function in the ways that western bureaucracies operate with top-down control. What makes the Waodani case an outlier relates to the recentness of their permanent contact with western civilization and the powerful forces arrayed against their own control and management of their titled territories. There is something to learn about working with indigenous people in conservation from this case.

2. The Waodani people and their territory

The Waodani (also known as Huaorani or Auca) were the last indigenous people in Ecuador to be "integrated" into the western world, with two clans—the Tagaeri and Taromenane— remaining in "voluntary isolation" to this day. Their language has no known relatives and at the time of contact only included two words borrowed from other languages (Trujillo León 1996); their DNA confirms their extended isolation from neighboring peoples (Cardoso et al. 2008; González-Andrade et al. 2009). A tiny population of no more than 500 people defended with their spears through random and dispersed raids a vast territory of over 20,000 km^2 between the Napo and Curaray rivers from all outsiders—Kichwas, colonists, oil companies, the Ecuadorean military, and missionaries (Cabodevilla 1999; Holt et al. 2004; Kane 1993). They did not participate in historical trade networks and emphasized extreme closure from all outsiders whom they call "cowode" or cannibals (Rival 1999). Their political structure was egalitarian—centered on the "nanicabo" or family longhouse—with no classes, no chiefs or leaders. Gender roles were also flexible (Holt et al. 2004). A man could become a leader for a specific event, and when the event passed so did his leadership. The intensely independent and individualistic social system led to frequent divisions of nanicabos and/or moves to distance themselves from conflicts and to escape retaliation. The primary mechanism for social control was peer pressure, with the ultimate threat of death by spearing (Rival 1999; Yost 1991).

Under increasing pressure from oil companies exploring their territory for oil reserves and from colonists in the early 1950s, the Waodani responded with spearing raids, at the same time intensifying an internal cycle of violence that produced the highest rate of death by homicide for any indigenous group anywhere (Beckerman et al. 2009), and induced several young women to escape to neighboring Kichwa farms where they became slaves (Yost 1981). An attempt by Summer Institute of Linguistics (SIL) missionaries to make peaceful contact resulted in five missionaries being speared in 1956—an event that brought worldwide attention to the plight of the Waodani (Eliot 1958). Two SIL women, one sister and one wife of the murdered missionaries, subsequently worked with one of the escaped Waodani women to learn the language, and together the three made peaceful contact with a first Waodani clan in 1958 (Yost 1981). SIL intensified their efforts and brought several clans together in an area known as the "*Protectorado*", in the west of the historically occupied Waodani territory, and made the first formal territorial claim to the Ecuadorean government in 1964 (Yost 1979). CONFENIAE (Confederación de las Nacionalidades Indígenas de la Amazonía Ecuatoriana) provided further backing for this territorial demand, resulting in the first Waodani land title in 1983 for 66,000 ha of the Protectorado which effectively cleared the way for oil activities and the opening of the vía Auca further east (Kane 1993, Kimerling 1991). The second and most important land title was awarded to the Waodani in 1990 and also represented a favor to oil companies, converting a large portion of the Yasuní National Park (declared in 1979) to Waodani territory, but under the condition that the Waodani did not oppose oil activities on their new lands (Kimerling 1991; Rival 1992), and recognizing the Auca road and collateral colonization as a massive cut into Waodani territory.

The legally recognized lands of what number today approximately 3500 Waodani indigenous people in Ecuador therefore amount to around 700,000 hectares in three separate land titles given at three points in history with formal ownership vested in somewhat different ways (Figure 1 and Table 1). Additionally, a number of Waodani communities are located in the Yasuni National Park, while two Waodani clans remain formally "uncontacted" and range widely within the titled territory and the park. All in all, the Waodani inhabit over 1.5 million hectares of land in the rainy upper Amazon and Andean foothills, arguably the world's most biodiverse forest (Bass et al. 2010).

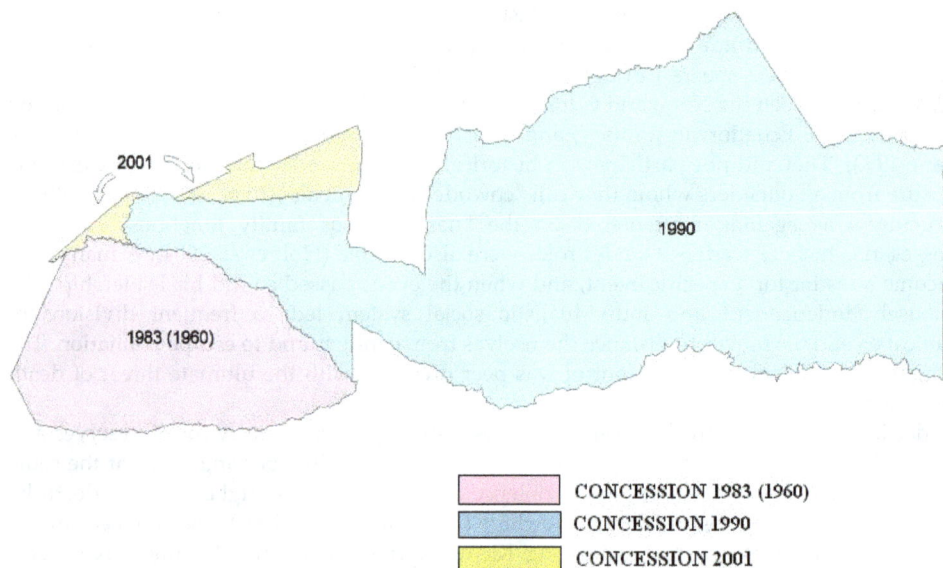

Fig. 1. Waodani territory titles (Lara et al. 2002b)[1]

Year	By whom and to whom	Extension (ha)
1960 (1983)	IERAC (Instituto Ecuatoriano de Reforma Agraria y Colonización) to Waodani Ethnic Group, specifying the "community organizations" Tiweno, Tzapino, Wamono, Kiwaro, Dayuno, Toñampade	66,570
1990 / 1998	IERAC to the "Waorani Ethnicity", specifying the community organizations identified as: Kewediono, Damointado, Nuevo Tiweno, Kenaweno, Nuevo Golondrina, Cononaco, Owanamo, Tagaeri, Tiwino and Yasuní	612,560 / 613,750
2001	INDA (Instituto Nacional de Desarrollo Agrario) to the Organización de Nacionalidades Waodani de la Amazonía Ecuatoriana (ONHAE)	29,019

Table 1. Waodani territory titles (Lara et al. 2002b)

---

[1]The size of the first title is inaccurately represented in this map. The first territory has never been accurately georeferenced.

The process of land titling has accompanied an equally ambiguous process of organizational development for the Waodani. The Protectorado experience, while it brought an end to the more overt internal violence, was traumatic as it combined the initial contact with the west (sedentary communities, change in diets and technologies, dependence on outsiders for subsistence and new material goods, change in health conditions) with intense social change because of the concentration of the population in a small area, a new generalized leadership by a woman who acted as cultural broker, controlling the flow goods and services from missionaries and other outsiders (replacing the traditional situational leadership by men), and new relationships with Kichwas through marriage that facilitated access by Waodani to external resources while at the same time permitting access by Kichwas to Waodani natural resources and land (Holt et al. 2004; Yost 1981; 1991). As early as 1972, however, the Waodani in the Protectorado began returning to areas they had previously occupied to the west, in some cases settling near oil operations in order to benefit from work and gifts provided by the companies.

3. The development of Waodani political representation

The first formal Waodani political organization, ONHAE (Organización de la Nacionalidad Huaorani del Ecuador), was not constituted until 1990, by the first young Waodani men who had received formal education in Spanish, and supported by the oil company Maxus which required a formal counterpart to secure its operations in Waodani territory. In 1993 Maxus signed a ground-breaking 20-year agreement with ONHAE (although numerous oil companies work in Waodani territory this remains the only agreement signed with the Waodani organization and benefitting the Waodani people as a whole; subsequent agreements have only benefitted the communities within the oil company's area of operations), providing resources for the organization itself and its leaders, as well as health, education and community development resources (Rival 2000). This agreement has been maintained by the Spanish company, Repsol, which took over Maxus operations in 1996. At the same time, however, in line with indigenous rights agendas and indigenous political organizations including CONFENIAE and CONAIE (Confederación de Nacionalidades Indígenas del Ecuador), ONHAE founders also expressed the new organization's mission as preventing oil exploitation and road construction in Waodani territory, emitting a series of declarations to that effect since 1991 (Lara et al. 2002a; Rival 1992). It appears, therefore, that the Waorani might have chosen the card that had been dealt. Even Ziegler-Otero (2004: 6) agrees that 'to preserve any semblance of cultural self-determination, indigenous people must be capable of negotiating.' Negotiating and making a deal with Maxus offered the Waorani at least some tangible – though temporary - benefits in the context where rejecting those benefits would have left the Waorani empty-handed. Indeed, there are facts indicating that the Waorani were aware of the dilemma and they chose the pact with Maxus as 'a lesser evil'. As Aviles (2008:42-43) states, ONHAE representatives first attracted the attention of the public by marching in Quito and denouncing both Maxus and Petroecuador (Ecuadorian oil company). The fact that the same people ended up signing the deal with Maxus shortly afterwards demonstrates the ultimate powerlessness of the Ecuadorian indigenous people in their dealings with the state and the external sources of revenues. This turn of events also shows the reason why dealings with the Waorani have been so frustrating for many international environmentalists. They resent ONHAE's perceived lack of principles and feel that the leadership has 'signed agreements with the Ecuadorian state and the oil companies, in apparent contradiction of their organizational positions and public statements' (Ziegler-Otero:1).

ONHAE and its successor organization, NAWE (Nacionalidad Waodani del Ecuador), since 2007 therefore represent a radically new form of political organization for the Waodani, structured as required by external "western" agencies including the government and the oil companies as well as NGOs. A president and other leaders ("dirigentes" for education, health, territory, tourism, etc.) are elected by representatives of the various communities at assemblies held irregularly every 3-18 months. The communities themselves, now numbering over 40, are new social and political structures, though in practice they have not greatly altered the traditional nanicabo social system, despite boasting presidents and other leaders. Thus new communities continue to form, often with only one or two families who have moved away from another community because of disagreements or to gain access to new resources including external assistance. Often the community presidents are not the de facto authority in the community, and a number of "big men" act as local leaders negotiating with ONHAE/NAWE, with other influential Waodani individuals and communities, and with external actors. To the degree that external actors (government representatives, oil companies, NGOs) make agreements with these "big men", obviously the strength of ONHAE/NAWE is undermined, and the generation of a Waodani conscience and unity falters.

External actors assume that ONHAE/NAWE intermediate on behalf of the Waodani people, as a form of representative democracy whereby the leaders make decisions that express the will of the people, and sign agreements including land titles with ONHAE/NAWE on behalf of the Waodani people as a whole (Ziegler-Otero 2004). In contrast, the Waodani people themselves consider ONHAE/NAWE's primary role to be the negotiation with oil companies (primarily Maxus/Repsol because individuals and communities negotiate directly with the other oil companies whose concessions overlap with the particular community's land) and the administration of the oil company-financed projects (distribution of benefits including health services, school lunches and other education services, and other goods and services). A traditional mark of Waorani leadership is the ability to get goods from outsiders and the Maxus/Repsol negotiation reaffirms the traditional orientation of their leadership (Ziegler-Otero 2004:129). The people in the communities do not expect ONHAE/NAWE to make other kinds of decisions on their behalf (High 2006). Thus, one primary challenge for today's NAWE is to provide – in addition to the social services – technical services that add value to the organization and involve them in overall conservation and development planning for Waodani holdings.

Whereas fire is not the danger in eastern Ecuador that it poses in Brazil, nonetheless colonization and road-building fuel deforestation. For the Waodani, the colonization has historically been Kichwa migrants settling along the Napo River on the north who gradually have deforested land towards the interior, and Kichwa and other migrants settling along the Curaray River on the south with the same effect (Trujillo León 1996). Oil production has driven colonization (Kichwa, Shuar, and mestizo colonists) by penetrating the very heart of the Waodani titled territory with two roads from north to south: one, the Via Auca since 1980 which has sparked uncontrolled settlement and the other, the Via Maxus since 1993 with relatively tight control over settlement by the Maxus and Repsol oil companies in turn (Finer et al. 2009; Kane 1993; Rival 1992; Villaverde et al. 2005). On the west, the major impacts have come from the expansion of the national agricultural frontier toward the Amazon with numerous roads connecting the rivers that flow from the Andes to join eventually with the Amazon River. Urban settlement near Waodani borders is common. On the east the titled territory is bordered by the Yasuní National Park which, at present, remains relatively un-colonized and is, as indicated above, the location of numerous Waodani communities. The two clans out of contact range both in the park and in the eastern part of the titled territory (Figures 2 & 3).

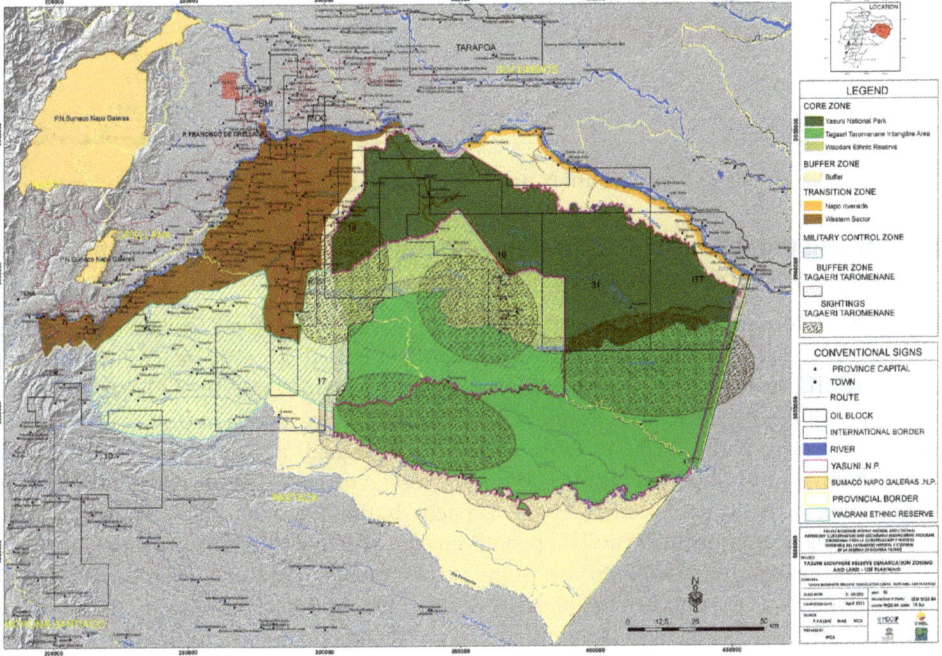

Fig. 2. The Waodani territory within the Yasuní Biosphere Reserve (MAE, MGDF, UNESCO, WCS 2011).

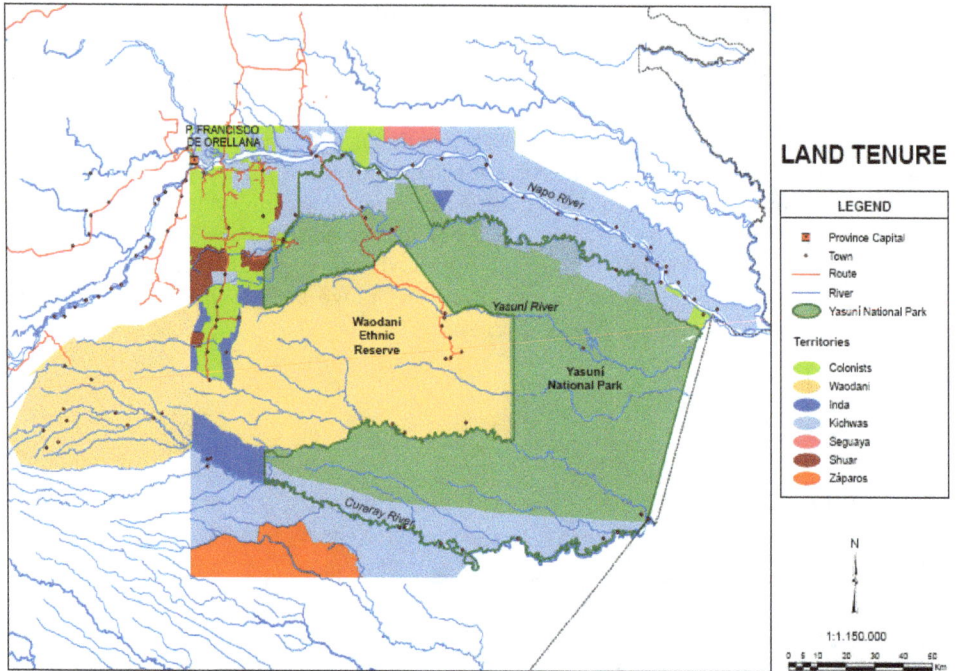

Fig. 3. The Waodani territory, Yasuní National Park, and neighbors (WCS elaboration).

4. WCS and the IMIL project: Strategies for shoring up the political structure and supporting conservation

The Wildlife Conservation Society (WCS) has been working with the Waodani organization NAWE since 2007, under the USAID-financed "Integrated Management of Indigenous Lands" (IMIL) Project. The project provides continuation to a previous phase of USAID funding through the Chemonics-administered CAIMAN (Conservación de Áreas Indígenas Manejadas) project (2002-2007). WCS and NAWE signed a formal memorandum of understanding, and subsequently a sub-grant agreement, in order to promote territorial consolidation, institution-strengthening, capacity-building, and alliances with other organizations to support the Waodani people, culture and territory. The principal territorial concern expressed by NAWE to WCS and in a strategic planning exercise (Vallejo & Burbano 2008) was to complete the physical demarcation of the territory begun years before with CONFENIAE assistance (Kimerling 1991) and also advanced under the CAIMAN project (Chemonics International 2007). The principal institutional priority expressed by NAWE was the incorporation of Waodani technical staff to the organization, and the training of this staff. WCS therefore discussed with NAWE ways to address both issues through the consolidation of a technical team of Waodani mapping technicians who are capable of collecting field information with a GPS (Global Positioning System) unit and generating basic maps in ArcView. Additionally the IMIL project has approached the organizational strengthening of NAWE through providing USAID funds that NAWE can use to carry out its own contract negotiation with outside technical actors. The outcome so far indicates a marked improvement in NAWE planning and responsibility for projects, a small but significant step in consolidating the territory and gaining community confidence. For example, NAWE technicians trained under the IMIL project have assumed responsibility for developing and implementing agreements with the Ministry of Environment's Socio Bosque program (described further below) as well as community management plans; one has been elected community president, and others were hired by the Ministry of Justice in its program to protect the Tagaeri Taromenane Intangible Zone (described further below).

The demarcation process successfully completed the remaining 89 km of territorial boundary between Waodani and neighboring Kichwa communities. These boundaries followed more or less those established in the legal titles, with local adjustments made by consensus with the Kichwa neighbors according to history of use and prior verbal agreements. The Waodani did not consider it necessary to demarcate the boundary with the Yasuní National Park as their elders, at least, consider the park to be part of their territory.

In addition, during the boundary demarcation process WCS discussed with NAWE and the technicians the role of community mapping as a tool for territorial management, finding that ONHAE/NAWE had already recognized its importance in their territorial management plan (Lara et al. 2002a) and in their strategic plan (Marchán 2006) developed previously but not implemented. An enormous challenge facing the Waodani is the imposition of external management systems and boundaries on their territory—with at least eight active oil exploration and exploitation concessions overlapping Waodani territory. One assessment describes the titled Waodani territory as under the administration of oil companies with the approval of the government (Lara et al. 2002a).

Also overlapping Waodani territory are a series of protected areas which each restrict Waodani actions: the Yasuní National Park was created in 1979, with significant boundary adjustments in 1989 and again in 1992; the Yasuní Biosphere Reserve was declared in 1989 (including the entire Waodani territory); and the Tagaeri-Taromenane intangible zone was declared in 1999 and formally demarcated in 2007 (Finer et al. 2009; Lara et al. 2002a; Villaverde et al. 2005). The Waodani communities currently located within the Yasuní National Park are not permitted under Ecuadorean law to obtain land titles (Ecolex 2003).

While the Waodani at times express their ignorance of these boundaries including the Yasuní National Park, the Tagaeri-Taromenane Intangible Zone, and even the border with Peru (Randi Randi 2003), the borders in practice mean that other actors are managing significant portions of Waodani territory. In addition, the 2008 Ecuador Constitution ratifies the government's rights to sub-surface resources including oil and minerals, but also forest resources and environmental services (Bremner & Lu 2006; Reed 2011). Thus the Waodani are legal owners but not actual administrators of their territory (CARE 2002).

Community mapping therefore represents a tool whereby the Waodani can visualize and negotiate with others their own land and resource use plans, zoning, and vision for their territory (Alcorn 2000; Eghenter 2000; Peluso 1995).

The first community that requested assistance from NAWE and WCS technicians in undertaking a community mapping effort was Kewediono (= Keweriono / Que'hueriono) motivated by the association which unites five neighboring communities and which has developed a world-class and internationally acclaimed eco-tourism project—the Huaorani Ecolodge - supported by Tropic Tours in Nature. These communities had previously defined a strict conservation area surrounding the lodge itself, and wanted maps to illustrate and reinforce these management decisions. The subsistence hunting area is the bulk of the territory, and is not subject to logging or deforestation for cultivation, while less than 1% of the territory has been deforested for farms and settlements (Custodio et al. 2008). The second set of communities requesting support is led by Gadeno (Gareno) which also boasts a tourism project, though one operated privately under concession to the Gadeno big man. The distribution of zones is similar to that for Kewediono (Table 2), with the detail that Gadeno also identified a small sacred area where the first Waodani had settled in this region (Custodio et al. 2009) and the mapping effort revealed an unsuspected zone where no hunting was permitted according to the tradition of the community of Meñempade, located on the same map as Gadeno.

Communities	Total area (ha)	Conservation %	Hunting %	Deforested %
Kewediono, Kakatado, Wentado, Apaika, Nenkipade	59,900	7	92	<1
Gadeno, Konipade, Meñepade, Dayuno	26,702	8+6	85	<1
Kiwado	26,703	74	24	<2
Toñampade	39,346	57	41	<2
Teweno	17,514	62	37	1
Guiyedo, Ganketa, Timpoka	59,494	67	32	1
Daimontado	20,957	75	19	6

Sources: Custodio et al. 2008; 2009; Arce et al. 2009; 2011; Landivar et al. 2010; Espín et al. 2010; Stocks & Espín 2010.

Table 2. Community mapping exercises with Waodani communities.

Subsequent mapping efforts took place after the launching of the government of Ecuador's innovative Socio Bosque program, whereby the government signs 20-year agreements with private (individual as well as collective) land-owners, paying an annual incentive per hectare of native forest protected (de Koning et al. 2011, Reed 2011). Although the program's only requirement is that forest not be cleared, allowing subsistence hunting or community tourism uses (commercial hunting is illegal nationwide), the communities' interest in using the maps to join the Socio Bosque program may have increased the proportion of conservation area formally identified in the mapping process. The communities may also combine the strict conservation area with the subsistence area in their proposals to Socio Bosque. Kiwado became the first Waodani community to join Socio Bosque in May 2011, assigning 24,000 ha of its territory that includes most of the conservation and subsistence hunting areas identified in the earlier mapping exercise (Arce et al. 2009).

In addition to the mapping itself, which emphasizes Waodani language place names and sites of cultural importance, each community established its own regulations for the use areas it had defined — for example strict conservation / tourism, conservation / subsistence hunting, sacred, settlement and cultivation. The Guiyedo case is the first mapping effort with communities located inside the Yasuní National Park rather than the Waodani titled territory (Stocks & Espín 2010), and is intended by the communities to serve as a tool for negotiating a co-management agreement with the National Park for the areas utilized by the communities, as well as for joining the Socio Bosque program. Lacking a title, the co-management agreement is a pre-requisite for Socio Bosque. The northern boundary recognized by the communities is the Tiputini river, a boundary agreed with neighboring Kichwa communities, though both Waodani and Kichwa venture across the boundary. The recognition of the Yasuní park boundaries is also conditional on whether or not individuals perceive that the park and Ministry of Environment are providing benefits. Guiyedo is also the first community mapping exercise to include locations where signs of the Tagaeri-Taromenane uncontacted groups have been found by the Waodani.

Withal, the mapping work with communities has identified the hunting turfs probably associated with the first Nanicabo settlements in each of the areas. These turfs (loosely called 'community territories') are well-known by community elders and each contains areas identified with at least some hunting restrictions. In a situation of "shaky politics" at the level of the ethnic group, the turfs or territories reassure communities that their own management will be respected. Additionally demarcation at the level of the titled territory assures local communities that their land claims are taken seriously by the central authorities and that they share a common border with other Nanicabo communities. The ambiguity with regard to their neighbors of other ethnicities is somewhat resolved by the series of inter-community border agreements. In terms of strategy, the training of a technical team fielded by the central organization, NAWE, has proved to be successful, both in terms of getting the work done competently in association with professionals contracted by NAWE and in terms of community perceptions of added value to NAWE itself. The community conservation areas, now firmly geo-referenced, are the basis for agreements with the Pre-REDD+ program, Sociobosque, and will undoubtedly play a significant role in later REDD+ programs. Certainly it cannot be argued that these lands are not under pressure from deforestation by Kichwas and colonists entering Waodani territory. The efforts of NAWE and the communities to develop and implement local boundary and territory monitoring programs will continue to be critical in protecting them.

5. Conclusions

The central issue in this chapter has been an attempt to avoid deforestation in one of the world's most productive and critically threatened habitats through consolidating the hold of indigenous people on the land and by working with an indigenous organization with a tenuous hold on legitimacy and a limited mandate from communities in order to improve resource management and conservation. This is a challenge that most serious conservationists encounter at some time in their career if they work around forest frontiers. There is no easy recipe for success or guarantee that the forest will not disappear in a determined number of years, but our experience is that the probability of encountering a relatively intact forest in, say, 50 years is greatest if indigenous people are in control of the outcome in areas they have historically inhabited, and greater still if they still maintain language and customs connected to their historic past. The work IMIL has done with the Waodani so far has, arguably, increased both the definition of the land controlled by them in general and specifically by community – land that includes large areas of no-commercial-hunting zones – and their ability to relate to a central ethnic organization that can be trusted to provide valuable assistance in the defense of land.

Because of the economy of the Waodani, the land they control is 99% or more in forest, so from a carbon perspective any investment in stability has benefits beyond the forest patches actually counted as conservation land. Such situations should be obvious targets for financing through global carbon markets as REDD+ proposes to do.

From the point of view of a conservationist, the key to working with the Waodani is the recognition of their sovereignty over land and resources. This recognition should not carry with it the assumption that the Waodani people are "cultural" ecologists. Actually, they are and their ecological knowledge runs deep, but the social expression of their own ecological adaptation involved levels of violence no longer tolerated in the Ecuadorian state. The modern adaptation to multiple cash sources and the heritage of decentralized political control has left them more vulnerable than most groups, so the importance of a central organization with some input into resource management is emphasized by the situation. The rub will come if the state tries to exercise its own constitutionally-granted rights to the forests in imposed ways that do not recognize Waodani sovereignty. One hopes that wise heads prevail in the government ministries that deal with the Waodani. If the Waodani are assisted to manage the forests in ways that strike them as culturally appropriate, everyone wins.

6. References

Alcorn, J.B. 2000. Borders, rules and governance: mapping to catalyse changes in policy and management. Gatekeeper Series no. 91. London: IIED. 22 pp.

Arce, S., P. Landivar & A. Stocks. 2009. Mapeo participativo y zonificación de la comunidad de Kiwado. Technical Report #16. Puyo: NAWE, WCS.

Arce, S., G. Bryja & D. Espín. 2011. Mapeo participativo y zonificación de la comunidad de Damointado. Technical Report #21. Puyo: NAWE, WCS.

Aviles, M.D. 2008. Narratives of resistance: an ethnographic view of the emergence of the Huaorani women's association in the Ecuadorian Amazon. Master's thesis. Gainesville, Florida: University of Florida.

Bass, M.S., M. Finer, C.N. Jenkins, H. Kreft, D.F. Cisneros-Heredia, S.F. McCracken, N.C.A. Pitman, P.H. English, K. Swing, G. Villa, A. di Fiore, C.C. Voigt & T.H. Kunz. 2010. Global conservation significance of Ecuador's Yasuní National Park. PLoS ONE 5(1):1-22.

Beckerman, S., P.I. Erickson, J. Yost, J. Regalado, L. Jaramillo, C. Sparks, M. Iromenga & K. Long. 2009. Life histories, blood revenge, and reproductive success among the Waorani of Ecuador. PNAS 106(20):8134-8139.

Bilsborrow, R.E., A.F. Barbieri & W. Pan. 2004. Changes in population and land use over time in the Ecuadorian Amazon. Acta Amazonica 34(4):635-647.

Borman, R., C. Vriesendorp, W.S. Alverson, D.K. Moskovits, D.F. Stotz & Á. del Campo. 2007. Ecuador: Territorio Cofán Dureno. Rapid Biological Inventories: 19. Chicago: The Field Museum.

Bremner, J. & F. Lu. 2006. Common property among indigenous peoples of the Ecuadorian Amazon. Conservation and Society 4(4):499-521.

Bromley, R.J. 1972. Agricultural colonization in the upper Amazon basin: the impact of oil discoveries. Tijdschrift voor Econ. en Soc. Geografie 63:278-294.

Cabodevilla, M.A. 1999. Los Huaorani en la historia de los pueblos del Oriente. Coca: CICAME. 488 pp.

Cardoso, S.M., A. Alfonso-Sánchez, F. González-Andrade, L. Valverde, A. Odriozola, A.M. Pérez-Miranda, J.A. Peña, B. Martínez-Jarreta & M.M. de Pancorbo. 2008. Mitochondrial DNA in Huaorani (Ecuadorian amerindians): a new variant in haplogroup A2. Forensic Science International: Genetics Supplement Series 1:269-270.

CARE. 2002. Monito ome Ecuador Quihuemeca (Nuestra tierra en el Ecuador): propuesta para una circunscripción territorial Huaorani. Quito: Proyecto SUBIR. 28 pp.

Chapin, M. & B. Threlkeld. 2001. Indigenous landscapes: A Study in Ethnocartography. Arlington, VA: Center for the Support of Native Lands, 2001.

Chemonics International. 2007. Helping indigenous nationalities in Ecuador conserve their territory and culture. Quito: United States Agency for International Development.

Chhartre A. & A. Agrawal. 2009. Trade-offs and synergies between carbon storage and livelihood benefits from forest commons. Proc of the Natl Acad Sci USA 106:17667-17670.

Colchester, M. 1998. Who will garrison the fortresses? A reply to Spinage. Oryx 32(3):11-13.

Colchester, M. 2000. Self-determination or environmental determinism for indigenous peoples in tropical forest conservation. Conservation Biology 14:1365-1367.

Colchester, M. 2001. Global policies and projects in Asia: indigenous peoples and biodiversity conservation. Publication 130. World Wildlife Fund, Washington, D.C.

Colchester, M. 2004. Conservation policy and indigenous peoples. Cultural Survival Quarterly 28(1):17-23.

Custodio, S., A. Stocks, G. Bryja & G. Remache. 2008. Mapeo participativo y zonificación del territorio de la Asociación Keweriono. Technical Report No. 7. Puyo: NAWE y WCS.

Custodio, S., A. Stocks & G. Bryja. 2009. Mapeo participativo y zonificación de las comunidades del sector de Gareno (Gareno, Konipare, Meñepare y Dayuno). Technical Report No. 8. Puyo: NAWE y WCS.

de Koning, F., M. Aguiñaga, M. Bravo, M. Chiu, M. Lascano, T. Lozada & L. Suárez. 2011. Bridging the gap between forest conservation and poverty alleviation: the Ecuadorian Socio Bosque program. Environmental Science & Policy 14:531-542.

Ecolex. 2003. Informe de identificación y caracterización de conflictos en los territorios Awá, Cofán y Huaorani. Quito: Ecolex. 42 pp.

Eghenter, C. 2000. Mapping peoples' forests: the role of mapping in planning community-based management of conservation areas in Indonesia. Peoples, Forests and Reefs (PeFoR) Program Discussion Paper Series. Washington, D.C.: Biodiversity Support Program. 40 pp.

Eliot, E. 1958. Through gates of splendor. New York: Harper & Row.

Espín, D., P. Landivar & G. Bryja. 2010. Mapeo participativo y zonificación de la comunidad de Teweno. Technical Report #18. Puyo: NAWE, WCS.

Finer, M., V. Vijay, F. Ponce, C.N. Jenkins & T.R. Kahn. 2009. Ecuador's Yasuní Biosphere Reserve: a brief modern history and conservation challenges. Environmental Research Letters 4:1-15.

Fuentes C., B. 1997. Huaomoni, Huarani, Cowudi: una aproximación a los Huaorani en la práctica política multi-étnica ecuatoriana. Quito: Abya Yala.

GeoPlaDes. 2010. Estudio Multitemporal de Cobertura Vegetal y Uso del Suelo entre los Años 1990, 2010 y Proyección al 2030 a Escala 1:50.000 para el Centro y Sur Oriente de la Amazonía Ecuatoriana y Sistematización de la información sobre los aspectos socio-económicos y culturales para facilitar la toma de decisiones en proyectos de desarrollo y conservación. Quito: TNC, WCS, USAID.

González-Andrade, F., L. Roewer, S. Willuweit, D. Sánchez & B. Martínez-Jarreta. 2009. Y-STR variation among ethnic groups from Ecuador: mestizos, Kichwas, Afro-Ecuadorians and Waoranis. Forensic Science International: Genetics 3:e83-e91.

High, C. 2006. Oil development, indigenous organisations, and the politics of egalitarianism. Cambridge Anthropology 26(2):34-46.

Holt, F.L., R.E. Bilsborrow & A.I. Oña. 2004. Demography, household economics, and land and resource use of five indigenous populations in the northern Ecuadorian Amazon: a summary of ethnographic research. Occasional Paper. Chapel Hill, North Carolina: Carolina Population Center.

Kane, J. 1993. With spears from all sides. The New Yorker, September 27, 1993:54-79.

Kimerling, J. 1991. Disregarding environmental law: petroleum development in protected natural areas and indigenous homelands in the Ecuadorian Amazon. Hastings International & Comparative Law Review 14:849-903.

Landivar, P., A. Stocks, D. Espín & G. Bryja. 2010. Mapeo participativo y zonificación de la comunidad de Toñampade. Technical Report #17. Puyo: NAWE, WCS.

Lara, R., E. Pichilingue, R. Narváez & G. Sánchez. 2002a. Monito Ome--Los Huaorani protegemos nuestro territorio. Quito: Ecociencia, Proyecto SUBIR.

Lara, R., E. Pichilingue, R. Narváez, M. Moreno, G. Sánchez & P. Hernández. 2002b. Plan de manejo del territorio Huaorani. Quito: Proyecto CARE/SUBIR, EcoCiencia y ONHAE.

Lu F, C. Gray, R.E. Bilsborrow, C.F. Mena, C.M. Erlien, J. Bremner, A. Barbieri & S.J. Walsh. 2010. Contrasting colonist and indigenous impacts on amazonian forests. Conservation Biology 24:881-5.

Marchán, J. 2006. Plan Estratégico de Vida de la Nacionalidad Waorani. Puyo: ONHAE, AMWAE.

Nepstad, D., G. Carvaljo, A.C. Barros, A. Alencar, J.P. Capobianco, J. Bishop, P. Moutinho, P. Lefebre, U.L. Silva Jr. & E. Prins. 2001. Road Paving, fire regime feedbacks, and the future of Amazon forests. Forest Ecology and Management 154:395-407.

Nepstad D, S. Schwartzman, B. Bamberger, M. Santilli, D. Ray, P. Schlesinger, P. Lefebvre, A. Alencar, E. Prinz, G. Fiske & A. Rolla. 2006. Inhibition of Amazon deforestation and fire by parks and indigenous lands. Conservation Biology 20:65-73.

Peluso, N.L. 1995. Whose woods are these? Counter-mapping forest territories in Kalimantan, Indonesia. Antipode 27(4):383-406.

Poole, P. 1998. Indigenous lands and power mapping in the Americas: merging technologies. Native Americas XV(4):34–43.

Randi Randi. 2003. Diagnostico sociocultural Awá, Cofán y Waorani. Quito: Randi Randi.

Redford, K.E. & S.E. Sanderson. 2000. Extracting humans from nature. Conservation Biology 14(5):1362-1364.

Redford, K. & A.M. Stearman. 1993. Forest-Dwelling Native Amazonians and the Conservation of Biodiversity: Interests in Common or in Collision? Conservation Biology 7(2):248-255.

Reed, P. 2011. REDD+ and the indigenous question: a case study from Ecuador. Forests 2:525-549.

Ricketts, T.H., B. Soares-Filho, G.A.B. da Fonseca, D. Nepstad, A. Pfaff, A. Petsonk, A. Anderson, D. Boucher, A. Cattaneo, M. Conte, K. Creighton, L. Linden, C. Maretti, P. Moutinho, R. Ullman & R. Victurine. 2010. Indigenous lands, protected areas, and slowing climate change. PLoS Biology 8(3):e10000331.

Rival, L. 1992. Huaorani y petroleo. Pp. 125-179 in G. Tassi (ed.). Naufragos del mar verde: la resistencia de los Huaorani a una integración impuesta. Quito: Abya Yala & CONFENIAE.

Rival, L. 1999. Prey at the center: resistance and marginality in Amazonia. Pp. 61-79 in: S. Day, E. Papataxiarchis & M. Stewart (eds.). Lilies of the field: marginal people who live for the moment. Boulder: Westview Press.

Rival, L. 2000. Marginality with a difference, or how the Huaorani preserve their sharing relations and naturalize outside powers. Pp. 244-260 in: P.P. Schweitzer, M. Biesele & R. K. Hitchcock (eds.). Hunters and gatherers in the modern world: conflict, resistance, and self-determination. New York: Berghahn Books.

Schmidt, P. & M.J. Peterson. 2009. Biodiversity conservation and indigenous management in the era of self-determination. Conservation Biology 23:1458-1466.

Schwartzman, S., A. Moreira & D. Nepstad. 2000. Rethinking Tropical Forest Conservation: Perils in Parks. Conservation Biology 14(5):1351-1357.

Soares-Filhoa, B., P. Moutinho, D. Nepstad, A. Anderson, H. Rodriguesa, R. Garcia, L. Dietzsch, F. Merry, M. Bowman, L. Hissa, R. Silvestrini, and C. Maretti. 2010. Role of Brazilian Amazon protected areas in climate change mitigation. Proceedings of the American Academy of Sciences. www.pnas.org/cgi/doi/10.1073/pnas.0913048107.

Stocks, A. 2003. Mapping dreams in Nicaragua's Bosawas biosphere reserve. Human Organization 62:65-78.

Stocks, A., B. McMahan & P. Taber. 2007. Indigenous, colonist and government impacts on Nicaragua's Bosawas Reserve. Conservation Biology 21:1495–1505.

Stocks, A. & D. Espín. 2010. Mapeo participativo y zonificación de las comunidades de Guiyedo, Timpoka y Ganketa en el Parque Nacional Yasuní. Technical Report #19. Puyo: NAWE, WCS.

Stocks, A. & P. Taber. 2011. Ironies of conservation mapping. In K. Offen and J. Dym, eds. Mapping Latin America (in press). Chicago: University of Chicago Press.

Trujillo León, J. 1996. The Quichua and Huaorani peoples and Yasuní National Park, Ecuador. Pp. 75-92 in: K.H. Redford & J.A. Mansour (eds.). Traditional peoples and biodiverity conservation in large tropical landscapes. Washington, DC: TNC.

Vallejo, I. & A. Burbano. 2008. Planificación estratégica: NAWE. Technical Report #6. Puyo: NAWE, WCS.

Villaverde, X., F. Ormaza, V. Marcial & J. P. Jorgensen. 2005. Parque Nacional y Reserva de Biósfera Yasuní: historia, problemas y perspectivas. Quito: Grupo FEPP, WCS.

Viña, A., F.R. Echavarría & D.C. Rundquist. 2004. Satellite change detection analysis of deforestation rates and patterns along the Colombia-Ecuador border. Ambio 33(3):118-125.

Wainwright, J. & J. Bryan. 2009. Cartography, territory, property: postcolonial reflections on indigenous counter-mapping in Nicaragua and Belize. Cultural Geographies 16:153-178.

Weber, R., J. Butler & P. Larson. 2000. Indigenous peoples and conservation organisations: experiences in collaboration. World Wildlife Fund (USA), Washington, D.C.

Yost, J.A. 1979. El desarrollo comunitario y la supervivencia étnica: el caso de los Huaorani, Amazonía Ecuatoriana. Quito: Instituto Lingüístico de Verano. Cuadernos Etnolingüísticos No. 6. 29 pp.

Yost, J.A. 1981. Twenty years of contact: the mechanisms of change in Wao ("Auca") culture. Pp. 677-704 in: N.E. Whitten Jr. (ed.). Cultural transformations and ethnicity in modern Ecuador. Urbana: University of Illinois Press.

Yost, J.A. 1991. People of the forest: the Waorani. Pp. 95-115 in: M. Acosta-Solis (ed.). Ecuador in the shadow of volcanoes. Quito: Ediciones Libri Mundi.

Ziegler-Otero, L. 2004. Resistance in an Amazonian community: Huaorani organizing against the global economy. New York: Berghahn Books.

Part 3

Preventing Deforestation

14

Preserving Biodiversity and Ecosystems: Catalyzing Conservation Contagion

Robert H. Horwich[1], Jonathan Lyon[1,2], Arnab Bose[1,3] and Clara B. Jones[1]

[1]Community Conservation,
[2]Merrimack College,
[3]Natures Foster
[1,2]USA
[3]India

1. Introduction

The natural world is in a chronic state of crisis and under constant threat of degradation, primarily by anthropogenic factors. In general, current conservation strategies have failed to effect long-range solutions to the rapid loss of biodiversity (Persha et al., 2011). Deforestation continues despite efforts by mainstream (top-down) conservation programs (Persha et al., 2011; Schmitt et al., 2009), and the effectiveness of large-scale protected areas has, at best, a mixed record of success (Brockington et al., 2008; Persha et al., 2011). Scientific disciplines, in particular, ecology and conservation biology, continue to emphasize threats to biodiversity (Schipper et al., 2008), to debate conservation priorities (Brooks et al., 2006), to advance unproven strategies (SSC, 2008), and to offer no more than hypothetical solutions to pressing problems (Milner-Gulland et al., 2010; Turner et al., 2007). The bulk of the scientific community remains tangential to the conservation needs of communities in habitat countries, with a critical lack of input and connectivity between the extensive scientific literature and ground-level practices (Milner-Gulland et al., 2010).

Resurgence of the "fortress conservation", "protectionist" narratives (commitment to conservation programs at the expense of indigenous and other local people) promoting a 19th century wilderness ideal free of humans remains a cornerstone of much conservation thought, policy, and planning. As pointed out by Brockington et al. (2008), commitment to community-based conservation "has been downplayed from being an *approach* to conservation to becoming a *component* to justify and legitimate interventions to create new protected areas or interventions to conserve specific species". This "back to the barriers" movement (Hutton et al., 2005), supported by many conservation biologists (Kramer et al., 1997; Oates, 1999; Terborgh, 1999), has been accompanied by an increase in conservation funding, with large conservation organizations reverting back to protectionist landscape conservation and away from community-based (ground-level or bottom-up) resource management (Hutton et al., 2005).

In his discussion of the ongoing conflicts between indigenous peoples' movements and conservation organizations, Dowie (2009) noted: "When, after setting aside a 'protected' land mass the size of Africa, global biodiversity continues to decline and the rate of species extinction approaches one-thousand times background levels, the message seems clear that there might be something terribly wrong with this plan... A better strategy might be simply to turn more human beings into true conservationists...." Community conservation projects, at the core, are based on the strategy of turning more human beings into conservationists (see Persha et al., 2011). The approach pursues this goal by working with people living in species-rich landscapes, assisting them to form networks with one another, with community-based organizations, with non-government organizations, and with government agencies for the protection of biodiversity and ecosystems (Brockington et al., 2008). When implemented according to field-tested procedures, community conservation provides an effective ground-level solution to environmental degradation and the loss of biodiversity (Horwich & Lyon, 2007; Horwich et al., 2011). Indeed, 60-85% of conserved areas are inhabited by people who are potential conservationists and who are necessary components of success (Brockington et al., 2008; Horwich & Lyon, 2007; Horwich et al., 2011; Persha et al., 2011).

One flaw inherent to debates over the community conservation approach entails the type of questions being asked: Is community conservation successful? Who should be responsible for protecting natural resources? These are not, however, the truly relevant questions. Community conservation is *one solution* to environmental degradation, deforestation and the loss of global biodiversity. The truly relevant question is: Why aren't *all* conservationists, scientists, in particular, conservation biologists, and non-government organizations actively incorporating successful community conservation models into their mission statements, policies, and programs (see Persha et al., 2011)? Community conservation projects are growing in number and in success (Borrini-Feyerbend et al., 2004; Dowie, 2009; Horwich & Lyon, 2007; Horwich et al., 2011; IUCN, 2003). Indigenous and other local groups are gaining political power and expertise, becoming conservation activists, and, in some instances, regaining management of homelands (Dowie, 2009). Indeed, many communities have initiated sustainable conservation projects (Pathak et al., 2004). There is also recent evidence that community-managed tropical forests show lower and less variable annual deforestation rates than do the traditional protected areas (Porter-Bolland et al., 2011 In Press) with potential for reducing carbon loss effecting global climate change (Soares-Filho, 2010). However, the effects of community and indigenous managed projects in terms of their geographic scale, recognition by professional conservationists, including many non-governmental organizations, as well as their economic and political influence, are not yet sufficient to mitigate the deleterious effects of increasing environmental degradation and escalating loss of biodiversity. This condition persists, in part, because regional and governmental entities, as well as non-governmental organizations, have failed to include indigenous and other community stakeholders as partners (Persha et al., 2011). There is a need for national and regional governments and non-governmental organizations to network with community-based organizations having the mission, goals, and objectives to initiate, facilitate, train, and empower communities in habitat countries to preserve and manage local resources (Horwich & Lyon, 2007; Horwich et al., 2010; Persha et al., 2011).

Unsuccessful project outcomes have been minimized by the *Community Conservation* (www.communityconservation.org) model developed over time by trial-and-error but by now tested in the field and proven to be a valid and reliable procedure with utility for other community-based programs, as demonstrated by the cases discussed below. Table 1 gives a list of 23 projects that *Community Conservation, Inc.* has either initiated or contributed significantly to in its earlier stages. Three points are important to note from this table. Most important is that the highest level of community participation has occurred by encouraging the creation of community-based organizations to manage or contribute to the project. This also implies a high level of community empowerment. In Assam, India, the Manas Biosphere Reserve is now being protected by a network of 14 community groups (Horwich et al, 2010, Horwich et al., 2011). In this regard it is notable that the communities played a significant role in having UNESCO recently remove the "World Heritage Site in danger" listing. In the cloud forests of Peru, the Yellow Tailed Woolly Monkey Project has been stimulating community groups to create community reserves under Peruvian law. In Papua New Guinea, the Tree Kangaroo Conservation Program has created a community group that is a federation of over 26 clans. The second point is that community groups have stimulated or contributed to the creation of new protected areas or act as complementary protectors of public and private lands. Thirdly, communities can play a major role in regional or landscape protection as is occurring in the Golden Langur Conservation Project in the Manas Biosphere Reserve in Assam, India (Horwich et al. 2010), the Tree Kangaroo Conservation Program in Papua New Guinea (Ancrenaz et al, 2007) and what is evolving in the Yellow Tailed Woolly Monkey Project in the cloud forest of Peru (Shanee et al., 2007).

Project Name *indicates government-run project	Country	Start Date	Community Groups Formed	Protected Area Existed	Protected Area Created	Area (hectares)
Ornate Box Turtle Conservation*	USA	1991				-
Ferry Bluff eagles	USA	1988	1		1 private	?
Kickapoo Valley Reserve	USA	1991	1		+	3,440
Valley Stewardship Network	USA	1994	1			-
Blue Mounds Area Project	USA	1995	1			6,000
Badger Army Plant Lands	USA	1997	1		+	3,040
Community Baboon Sanctuary	Belize	1985	1		1 private	5,120
5 Blues Lake National Park	Belize	1991	1		1	1,680
Gales Point Manatee Project	Belize	1991	1		1	28,000
Temash River National Park	Belize	1994	1		1	14,921
CA River Turtle Conservation	Belize	1994	1			-
Punta Laguna	Mexico	1989		+		4,400
Chacocente	Nicaragua	1997		+		4,224
Spider Monkey Conservation	El Salvador	2001		+		606
Tree Kangaroo Conservation	Papua New Guinea	2001	1		1	82,000
Manas Biosphere	India	1998	14	+		285,000
Golden Langur Kakoijana	India	1998	2	+		1,700
Ngabe Tribal Lands	Costa Rica	2007	1	+		12,000
Tulear South	Madagascar	2007	1		2	2,600
Peru Cloud Forest	Peru	2009	8		8	35,000
Homeland of the Crasne*	Russia	1994	1	+		35,200
Nariva Swamp	Trinidad	1997	1	+		8,192
Cape 3 Points	Ghana	2011	1	+		5,100

Table 1. Updated list of *Community Conservation* projects with selected information.

Our community-based model, comprised of nine social stages, progresses as follows: (1) initial contacts with community leaders and elders to *catalyze* informal communication with village inhabitants, providing opportunities to openly and candidly discuss the significance of their resources and benefits to be gained from cooperative and participatory initiatives for conservation of their natural resources fostering (2) informal relationship-building in indigenous and other local communities leading to (3) participatory education providing (4) a window of opportunity for local conservation leaders to emerge (5) who invoke support from others, in our cases, the majority of villagers fostering (6) development of a formal infrastructure and plans within the possibilities of each community context. Eventually, (7) as the cases presented in this article show, the lessons and activities implemented and learned in the initial target village diffuse through local networks of communication (e.g., hearsay; printed information from enlightened schoolchildren to their kin) broadening matrices of conservation activists through other modes of social transmission and problem-solving (e.g., observational learning; imitation; education; brainstorming sessions and focus groups; contacts by the target community to members of other villages inviting them to inspect their conservation efforts and to attend planning sessions, lectures, seminars, and public events; informal and formal visits from target community members to other villages). The first seven social stages of our community-based, bottom-up model have the potential to foster (8) diffusion from the target village to other communities and, beyond, to regional entities through a process that we term *conservation contagion* consolidating the *horizontal network* (e.g., community-based organizations and communities: Berkes, 2004). Finally, (9) educational initiatives, lobbying, and relationship-building with entities in *vertical networks* (e.g., regional, governmental, non-governmental, and international entities: Berkes, 2004) have the potential for linkage, creating multidimensional, multi-scale partnerships benefiting all stakeholders. Our formulations, catalysis and community contagion, are detailed below.

The objective of this Chapter is to illustrate how community conservation, when carried out in the field, using tried and proven methods, is an effective solution to reducing deforestation and the loss of biodiversity and consequent climate change and reduced carbon emmissions. Section 2 defines the philosophies, concepts and practices in small projects that lead to successful results in contrast to the preponderance of large Integrated Conservation and Development Projects (ICDPs) whose contrasting philosophies, concepts and practices have resulted in mainly failures at high costs. Section 3 introduces the concept of conservation contagion and how, when it is stimulated, can lead to regional change. Section 4 complements the other sections with examples of successful projects from Belize, India and Namibia, illustrating successes that led to regional change with increased community protection resulting in increased reforestation and increases in focal species. Section 5 discusses lessons learned from the process of community conservation and its examples. Finally section 6 gives policy implications for future successful possibilities.

2. What makes a successful community conservation project (CCP)

Community conservation or community-based conservation projects under a number of names have been developed over the past two decades as important alternatives to the traditional protected areas that exclude humans. *Community Conservation, Inc. (CC)*, and other non-governmental organizations are project identifiers used in the present chapter to designate ground-level initiatives developed over the past two decades as important alternatives to traditional conservation organizations, historically prioritizing protected areas independent of human interests and often excluding indigenous and other local groups from targeted areas ("fortress conservation"). These community conservation projects have been based on ethical, theoretical, and practical arguments of conservation practitioners and social scientists (Borrini-Feyerabend, 1996; Broad, 1994; Brosius et al., 1998; Davey, 1998; Gadgil & Guha, 1993; Ham et al., 1993; Johnson, 1992; Oates, 1999; Stolton & Dudley, 1999). However, in recent years there has been growing criticism of community-based conservation programs and a call for renewal of protected areas that exclude local communities from the programs and their management (Brandon et al., 1998; Inamdar et al., 1999; Robinson, 1993; Terborgh, 1999).

While many critics of community conservation projects are biologists (Oates, 1999; Terborgh, 1999), social scientists have also criticized these projects (Belsky, 1999; Brechin et al., 2002). Most of the criticism has been directed toward large integrated conservation and development projects (ICDPs) while the successes of small community-based projects have been overlooked (Horwich & Lyon, 2007). Even practitioners of ICDPs, however, have been disappointed by their limited achievements (Robinson & Redford, 2005) and are attempting to learn from their failures to maximize future success (McShane & Wells, 2004; Rhoades & Stalling, 2001). What seems clear at present is that ICDPs have been adhering to a faulty paradigm and have much to learn from community conservation and community-based forestry paradigms (Shepard, 2004).

The philosophies of community conservation, community-based conservation and integrated conservation and development projects "originated from a shift in protected area management away from keeping people out by strict protection and toward more sympathetic treatment of local communities, including efforts to share benefits from the conservation of biodiversity" (Wells et al., 2004). However, the philosophy and methods by which activities are carried out are extremely different between large ICDPs and smaller community conservation projects. These different postures have resulted in the lumping of unjustified criticism of our community conservation projects (Belsky, 1999, 2000) resulting in curtailed progress. Thus, it is important to differentiate the two organizational models.

Community conservation projects are centered on conservation of natural resources and the role they play in the lives of indigenous and other local, rural peoples. Dealing with people on a one-to-one basis at a community scale has been a primary focus of *Community Conservation's* projects. Optimally, the limited resources of community conservation projects, especially finances, are best distributed over a long time-period in order to use them efficiently and prudently. While natural resource conservation is central to *Community Conservation's* priorities, it is followed closely by the needs of rural indigenous and other local populations, not only economic, but also social ones, emphasizing quality of life.

There are some fundamental differences between how successful community conservation projects and most ICDPs approach a project (Table 2). Community conservation projects are holistic. Flexibility and the expectation of change are important to their success; thus, practitioners must be adaptable, learning from problems as well as successes. Regional planning can be accomplished by building on, and expanding from, small community projects, from the specific projects to the general goals, objectives and mission that fits the composite project and region. Community conservation projects focus more on doing than on planning. The primary role of community conservationists is as catalysts to educate, motivate, and reinforce residents and communities in habitat countries to protect and conserve their natural resources both for themselves and for the rest of the world. While projects should be long-term and on-going, the catalytic role generally ends once community members are prepared to assume responsibility for programs in their communities and regions. ICDPs, instead, generally have been pervasive in time and space, with no plans to exit their initiatives or to transfer leadership to indigenous and other local residents (Sayer & Wells, 2004).

Community Conservation	**ICDPs** (from McShane and Wells 2004)
• People as solution	• People as problem
• Community scale	• Regional scale
• Small annual budget ($20,000)	• Large annual budget ($800,000)
• Long-term / on-going	• Definite end
• Community as managers	• Top-down
• Low planning / high implementation	• High planning / low implementation
• Flexible and adaptable	• Rigid
• Socially sustainable	• Not socially sustainable
• Predominantly successful	• Predominantly failures

Table 2. Major differences between community conservation projects (CCPs) and integrated conservation and development projects (ICDPs).

2.1 Rural people are the solution not the problem

Another difference between community conservation projects and ICDPs concerns significant philosophical differences resulting in very different approaches and practices. ICDPs conceptualize local communities as generators of habitat degradation (McShane & Newby, 2004; McShane & Wells, 2004) based on the premise that humans utilize natural resources and may abuse them when community and governmental regulating systems break down, characteristic of the "tragedy of the commons" (Feeney et al., 1990; Hardin, 1968). However, if rural indigenous and other local people are seen as threats, they will have a greater probability of living up to that expectation. In contrast, when people depend on resources, they may be educated to understand that they must not over-exploit resources without losing them. Because they use them, they have knowledge of and appreciate them. Living on site, they can better protect the resources. But there are also many outside forces competing with rural residents for resources. Thus, by giving rural indigenous and other local people entitlement and responsibility over their resources, many will see the importance of biodiversity conservation. Indeed, in almost all cases, when we have asked local rural people for their help in protecting their natural resources, they have responded very favorably and effectively.

2.2 Community scale at a personal level

Another philosophical difference between community conservation projects and ICDPs is that ICDPs approach conservation on a large scale, possibly because donor agencies think financially in those terms. However, major outlays of money and other resources for short time-periods have proven to be ineffective, inducing greed, waste, mismanagement and corruption. It may also highlight differences between the community base and corrupt western affluence (Gezon, 1997). More importantly, in order to work successfully with small rural communities, initiatives must cooperate at the lowest levels of community and regional organization, engaging personal, face-to face communication involving the thoughts, emotions, beliefs, attitudes and values of indigenous and other local stakeholders who are the voices of most decision-making, problem-solving and negotiations. The rural poor embody the same desires to assume responsibility for and to manage their own affairs as others, regardless of economic position. As a result, community involvement is an important ingredient for all conservation projects. Even when protecting large landscapes, community conservation projects can be effective at the regional scale by dividing initiatives into ground-level components for the retention of person-to-person interactions fostering trust and friendship. This process is exemplified by the case study of a community-based project in Assam, India (Horwich et al., 2010).

2.3 Community conservation projects – below the "conservation radar"

While ICDP enthusiasts have undergone some soul searching because of the backlash and criticism from biologists and sociologists, they have rightly begun to learn from ICDP failures (McShane & Wells, 2004). Unfortunately, small community conservation projects have been neglected and are "below the radar" of the mainstream conservation community (see Horwich & Lyon, 2007) not withstanding published accounts of successes as well as problems of such projects (Horwich, 1990a, 1998, 2005; Horwich & Lyon, 1988, 1995, 1998, 1999; Horwich et al., 1993; Lyon & Horwich, 1996). Despite documented successes of community-based initiatives, funding for these projects has also been under the conservation radar, remaining at very low levels.

Similarly, published articles have also ignored benefits to local communities from conservation of forests and their resources upon which indigenous and other local groups may depend (Shepard, 2004). Shepard (2004) notes " conservation organizations themselves form their own international environment in which they talk to and argue with one another. Because they spend so much of their time in this company, there is too little exchange of ideas with those engaged in the forestry and poverty worlds." She goes on to state that "ICDPs and the conservation bodies that manage them have been out of the mainstream of changing thought about forest management and rural livelihoods and now risk getting stuck with an old-fashioned paradigm. Though it will always be difficult to persuade people to abandon some of their old assumptions, it is now urgent that they consider doing so." (Shepard, 2004).

2.4 Community participation

Level of participation - Communities are complex, heterogeneous groups of people with conflicting goals, aims, and desires. Complexities based on gender, politics, class, patronage, ethnicity, age, social standing and religion often have complex social histories that include exploitation, marginalization, and conflict (DuPuis & Vandergeest, 1996). For the last decade, the terms of community conservation, community participation, community-based conservation became buzz words as community conservation projects and ICDPs were in vogue. However, what has never often been clearly differentiated is that community involvement can occur at many levels in a continuum from top-down management through informal and formal consultation to formal advisory committees and ultimately to community co-management and indigenous management (see Horwich & Lyon, 2007; Horwich et al., 2004) as discussed by Arnstein (1969), Berkes (1994), Barrow (1996) and Stevens (1997). Thus in any project, the level of community participation must be clearly identified. The main thrust in the projects that we sponsor is to facilitate, particularly by persuasion and education, creation of an empowered community-based group capable of continuing a project once we have left it (Table 1). The lowest levels of government involvement and the highest levels of community participation, as represented by indigenous and other local community co-management, allow for strong partners in decision-making and project control. Co-management allows for government checks and balances and support; as well, indigenous and other local management is currently working well at one of our projects, the Community Baboon Sanctuary in Belize, discussed below.

Incentives for community participation - Top-down government management or private ownership are not the only ways to conserve and protect natural resources. Historically, many successful indigenous communal systems were working before Europeans came to dominate most natural landscapes. Singleton (1998) mentions institutions used by the Pacific Northwest Native Americans before the Europeans reached those shores. There have since been many struggles of poor people to regain those rights (Guha, 1989). Thus community conservation efforts have the potential to redefine and restructure managerial systems. This provides an enormous incentive for community participation and as has been shown by many forest projects (Poffenberger & Gean, 1996), and other Self Help development programs (Wilson, 2002), villagers have responded by being empowered to assume responsibility for their resources. On the other hand, efforts of ICDPs, offering a wealth of developmental possibilities to these same rural residents, often failed generally because of the limitations discussed below.

Uphoff and Langholz (1998) present a model of three basic categories of incentives for people to conserve or exploit protected resources: 1) legal, 2) financial or 3) social/cultural. According to these authors, if initiatives incorporate all of these elements, they have a high probability of being adopted. Projects lacking the features have a low likelihood of being adopted. The authors noted that rewards from ICDPs' conservation tactics "amounted to tacit bribes" for getting villagers to adopt new practices depending upon an infusion of outside resources. These "bribes" seemed to undermine community practices for the conservation of resources. An integrated balance of these three factors is needed to induce conservation,. Uphoff and Langholz (1998) make a strong case for the importance of social values leading to stewardship of natural resources, echoing what we have found in our experiences with community-based conservation. Although money was one motivator in our project at the Community Baboon Sanctuary in Belize, community members consistently demonstrated pride in their conservation efforts, especially their flagship "baboons" (the local term for black howler monkeys, *Alouatta pigra*). In Assam, the pride of achievement was integral to project success exemplified by local leaders reporting their accomplishments in developing Self Help Groups and forestry committees. These entities continue to provide important social functions for meeting and talking, for practicing thrift, and for receiving information on a short term basis (Wilson, 2002). More importantly, they have "a more powerful purpose: to gain and share information, to take social action, and to link to government resources."(Wilson, 2002). It is that power that has stimulated some of the local forestry groups in Assam to actively protect their forests for themselves and for the pride to convey their actions to project partners. There is also no surprise that rural villagers also appreciate natural areas for their beauty and greenness and enjoyed walking and sitting in the shade and seeing wildlife (Allendorf, 2007).

Despite what we have found, the conservation community has focused on poverty and economic incentives as a prime motivator of poor rural communities. Indeed, a great deal of money has been spent on ICDPs and "despite this high level of investment and effort, we can only point to some individual, localized successes. Taken as a whole, we have had little impact on stemming or even slowing the rising tide of biodiversity loss." (Kiss, 2004). In addition, little money from the large grants ever reached the beneficiaries on the ground level (Sayer & Wells, 2004). Indeed, when we look at the levels of funding in comparison with small successful community conservation project budgets, the failures in spending are extensive, almost obscenely wasteful.

From 1990 to 2004, Kiss (2004) notes that the World Bank has supported 226 conservation-related projects internationally with a total budget of $2.65 billion (from a variety of funding sources). In terms of prudent financing, $2.65 billion divided by 226 projects and by 14 years gives an average annual project budget of some $837,547. In comparison, two small successful community conservation projects annually averaged $12,035 (Community Baboon Sanctuary) for six years and $22,367 (Golden Langur Conservation Project in Assam, India) for seven years. ICDPs spent over 40 times that of the community conservation projects. While it is difficult to assess the total impact of all the World Bank projects relative to the two community conservation projects noted, it is clear that the scale of funding is dramatically different.

Despite this financial information, many of the tenets that hold for successful community conservation are paradoxical. However, research has shown that rural people do not always act rationally and in their own interests (Ariely, 2010). Indian rural villagers were asked to do various tasks for three levels of pay. Those who could earn the equivalent of one day's pay or two weeks pay did not differ. However, those who could earn the equivalent of 5 months pay did the task significantly worse. Ariely (2010) noted that using money to motivate people could be counter-intuitive. For tasks requiring cognitive ability, low to moderate performance-based incentives can help. But if financial incentives are too high, the attention to the reward becomes distracting and creates stress that reduces the level of performance.

2.5 Project implementation

Planning versus implementation – ICDPs have invested a great deal of time and energy in project preparation, often executed by outside experts. These practices increase budgets and restrict program flexibility (Sayer & Wells, 2004). *Community Conservation* projects, in contrast, initiate ground-level efforts immediately, with some research, preparation and planning integrated into the early stages of a project. From the beginning, formal, expert knowledge and information have been united with local expertise throughout every stage of our programs as documented in the cases detailed in the present report.

Flexibility and adaptability – While the ICDP paradigm attempts to reduce uncertainty through over-planning and preparation (Sayer & Wells, 2004), the community conservation approach is resilient and capable of adapting to change. Although general planning is necessary, too much emphasis on planning and accompanying financial investments have plagued ICDPs and often left them with a legacy of inflexibility (Sayer & Wells, 2004). Further, it is likely that one disadvantage of the mainstream ICDP model is that it establishes unrealistic expectations and levels of resource-access for community members. With generalized goals from the initiation of our projects that are adaptable to the features of different local projects, we maximize flexibility and influence by relying primarily upon local resources, providing models that can be sustained by indigenous and other local groups over the long-term. In addition, since resource efficiency was necessary, particularly due to our limited funding, we developed contingency plans in the event that opportunistic responses were required.

Funding and project length – Many mainstream ICDPs arguably wasted large amounts of money, often because of utilization of templates applied to all situations regardless of differences from locale to locale and project to project and also because, seemingly paradoxically, promiscuous infusion of resources unsuited to differing project scales has the effect of compromising the planning, efficiency, "goodness-of-fit", and effectiveness of programs. Furthermore, resources often fail to impact the intended beneficiaries and their resources, directed, instead, into staff and consulting fees, centralized planning, and problem-solving and decision-making divorced from local community realities, requirements, criteria, and contingencies (Gezon, 1997; Sayer & Wells, 2004). In some cases, large amounts of resources were invested in small regions for short periods of time, draining critical resources for wider and more judicious efforts (Sayer & Wells, 2004). ICDPs generally forecast 3-5 years per project, a standardized time-frame often too brief, on the one hand, or unnecessarily extended, on the other, for successful implementation of their goals and objectives. Furthermore, ICDPs often fail to project and plan exit strategies, sometimes leading to abrupt termination of or unproductively extending involvement with community projects (Sayer & Wells, 2004). Our community-based conservation projects, instead, utilize modest funding consistent with local economies, maximizing project realism, "goodness-of-fit", and successful long-term persistence. It is our position that both community conservation projects and ICDPs entail costs as well as benefits and that cooperation among these networks has the potential to minimize the disadvantages and maximize the advantages of all tactics and strategies.

Project sustainability – While ICDPs were initiated with the hope of financial sustainability (Wells et al., 2004), our community conservation projects often have, as a goal, partial sustainability. In developing countries, entrance and user fees earned, for example, by ecotourism, classes on ethnobotany, or small businesses (e.g., restaurants serving and selling traditional foods) might promote village sustainability, pride in efforts supporting local conservation programs, expansion of the local economy, and imitation by other groups. Success of programs is not necessarily correlated with predictions from conservation biology or other scientific approaches, e.g, mathematical modeling, because factors critical to short- and long-term successes often arise as spontaneous, condition-dependent events. For example, the residents of the Community Baboon Sanctuary suggested a plan for sustainability based on mandating residents to take turns as local guides, denoting a more mature stage of project implementation as well as a potential template for other community conservation projects. With an increase to 6000 tourists, visiting the Community Baboon

Sanctuary, base-level financial sustainability became a reality, with additional programs requiring and stimulated by short-term grants (Horwich & Lyon, 1998). In reality, community or government protected areas in developing countries still need richer nations to infuse conservation efforts with financial and other resources (Wells et al., 2004; Balmford & Whitten, 2003); alternatively, other creative ideas such as trust funds or direct payments need to be considered (Kiss, 2004). Social sustainability of most *Community Conservation, Inc.* projects occurred because of sufficient, motivated, and prepared social capital combined with social and other (e.g., modest and targeted economic) incentives (Uphoff & Langholz, 1998; Wilson, 2002). Although the Community Baboon Sanctuary has experienced significant challenges over time, it has persisted for over twenty-six years because community-based conservation became a guiding ethic in the minds and lives of local people. When the Community Baboon Sanctuary reached a turning-point in 1998, a local group of women formed the Woman's Conservation Group to ensure its continuity and long-range stability (Horwich et al., 2011).

Integrating conservation and development - While there is no question that, consistent with the mission of ICDPs, conservation and development should be integrated, community-based organizations have not developed explicit proposals for how the integration should be structured and implemented (Sayer & Wells, 2004). ICDPs maintained the false assumption that helping communities develop economically would lead to the conservation of natural resources (McShane & Newby, 2004). However, the opposite often happened as discussed above. A holistic approach of integrating conservation and development is, *ceteris patibus*, effective because economics is only one of the incentives communities respond to. A holistic plan may be a highly productive approach to integrating social incentives, development and conservation through facilitating the development of networks consisting of ground-level associations (individual and group) of community-based organizations and non-governmental organizations based on trust.

Financial donors - While donors provide the essential financing for successful projects and want their contributions recognized, it is to the benefit of community-based organizations if they provide a mentoring rather than a "hands-on" or otherwise controlling posture, particularly because community-based operations often require opportunistic responses to local conditions and processes that cannot easily be detailed in an application for funding and for short-range assessment by funding agencies. Donors to ICDPs often had controlling interests in these projects (Sayer & Wells, 2004). *Community Conservation, Inc.'s* project finances have been modest, falling under the "conservation community radar" as noted earlier. As a result, donor agencies generally did not attempt to maintain control of their investments; however, they rarely publicized successes of the community conservation projects, emphasizing, instead, large, "flagship" funding recipients likely to reinforce powerful and influential national and international networks.

Non-governmental organization role – Non-governmental organizations can facilitate the empowerment of community-based organizations by providing seed money or by acting as a financial and tactical mentor. They can also provide management training for the community-based organizations (Horwich et al., 2004). Training would optimally be a combination of long-term mentoring combined with, for example, short-term seminars (Bernstein, 2005). Although others are skeptical that local communities are the best managers of their natural resources (McShane & Newby, 2004), *Community Conservation* takes the position that indigenous and other local populations are, with training and other support (e.g., interactive networking), capable, even ideal, stewards of biodiversity in and surrounding their traditional lands because of the nature and degree of their historical connections with and investments in these habitat domains.

A powerful and effective function of non-governmental organizations is to catalyze and facilitate communities to protect their natural environment (Horwich, 1990a; Horwich et al., 2004) and to help them create new community-based organizations whose mission is environmental conservation (Agrawal & Gibson, 1999). Non-governmental organizations should retreat after the community-based organization develops capacities to function independently. However, non-governmental organizations should continuously monitor the progress of projects, providing additional advice and support when required and solicited. In 1989, subsequent to catalyzing a community conservation project around the Temash River in Belize, a biologically important mangrove habitat, the government was not supportive of our organization's contingency plans. By 1997, three years after the zone was designated a protected area, our re-catalyzing the project through a meeting involving all stakeholders (Producciones de la Hamaca & Community Conservation Consultants, 1998) led to a newly formed indigenous Belizean non-governmental organization managed by Belize nationals in cooperation with local communities, creating a comprehensive project (Horwich, 2005; Caddy et al., 2000).

2.6 Stimulating regional change from the bottom up

In an effort to revise the role of ICDPs, it has been suggested that they are well situated to work at a landscape scale, integrating with networks at this level (Robinson & Redford, 2004). However, small community conservation projects should not be excluded from these plans since they can be employed at a regional scale by initiating a number of community level projects within a region and having them collaborate, eventually unifying them into a larger project or federation that would include all of the participating community-based organizations, non-governmental organizations and government agencies (Horwich et al., 2010). An ultimate goal should be integration of networks and clear delineation of functions (e.g., managerial, social, political, economic, etc.) at all levels of organization.

Even though ICDPs often have functioned on the false assumption that national governments would embrace the idea of community involvement and pass laws and regulations to facilitate the community involvement in natural resource management (McShane & Newby, 2004), working with governments and their agencies to revise old and develop new policies remains an important objective. In Belize, successes of and publicity about the Community Baboon Sanctuary led other motivated communities to adopt that model, sometimes with modifications, creating their own conservation project, adapted to their own circumstances. Additionally, communities lobbied the Belize government to create new sanctuaries, using existing laws to create community-based sanctuaries and protected areas and the Belize government eventually adopted a community co-management policy (Meerman, 2005). Because the government of Belize recognized significant progress at community levels, forestry and fisheries agencies contracted with community-based organizations and non-governmental organizations to sign memoranda of understanding. Currently, approximately twenty village and town groups have signed or are in the process of negotiating such agreements (Young & Horwich, 2007) (Figures 1 & 2).

In Assam, India, the Golden Langur Conservation Project, initiated by *Community Conservation*, Natures Foster and Green Forest Conservation in 1998, targeted the full Indian range of the golden langur in western Assam. Through village meetings, seminars and general meetings the project brought civil authorities, non-governmental organizations and communities together. As communities participated in the project they formed community-based groups leading to conservation contagion. Early on, the project focused on the Manas Biosphere, the largest main habitat of the golden langur in India. The project brought attention to the Manas Biosphere through a series of four celebrations throughout the Biosphere finally resulting in participation of 20,000 and then 35,000 attendees. A number of community groups formed and began carrying out forest protection patrols that were later funded by the Bodoland Territorial Council. Eventually, the project stimulated the organization of a network to meet and discuss protection of the entire Biosphere. Collectively, this led to regional change resulting in an increase of forest renewal and an increase of the targeted species throughout its Indian range from 1500 to now over 5600 golden langurs (Horwich et al., 2010, 2011). There are also indications of an increase of both the elephant population (Ghosh, 2008b) and the tiger population (Anonymous, 2011). This community protection network played a major role in the UNESCO delisting Manas as a World Heritage Site in danger.

3. Catalyzing conservation contagion

Community conservation practitioners must view themselves as catalysts to stimulate and guide indigenous and local people, building their capacity and encouraging them to assume responsibility for creating solutions to the natural resource challenges in their communities. In the parlance of chemistry, catalysts (e.g., *Community Conservation* personnel) reduce the so-called energy of activation (i.e., ignorance of, passivity to, disinterest in, or resistance to conservation initiatives) to stimulate a reaction (e.g., progressive and sustainable conservation programs). In our experience, programmatic initiatives have the potential to spread to other communities and, often, throughout regions by a process of diffusion that we term *conservation contagion*. Sometimes conservation contagion diffuses more broadly, influencing protectionist policies at the country-wide or even international levels. Initially, these objectives may require non-governmental organizations to take a prominent, leading role. However, as a community becomes increasingly empowered and independent, the practitioner-organizers become less and less visible, and members of the community emerge as primary leaders. Although our 9-stage social model incorporates a number of catalytic factors facilitating the occurrence of conservation contagion processes (e.g., building trust with village members, infusing information *via* local networks), in reality, on the ground, they are intimately integrated with the social and cultural process, what Berkes (2004) terms "social-ecological systems". Thus, successful development, management, and stabilization of horizontal networks are not automatic, linear processes but complex and dynamic ones over time and space. Our case studies from Belize, Namibia, and India exemplify individuals, community-based organizations, and non-governmental organizations *catalyzing* indigenous and other local people to participate in conservation initiatives followed by conservation contagion resulting from application of our 9-stage model. Catalyzing conservation contagion alone does not guarantee program success without application of all stages in our model that, although a dynamic process, significantly tempers the programmatic, unstructured, and unpredictable state that Westley et al. (2006) call, "getting to maybe".

Conservation contagion may sound like a fuzzy or murky concept, but, when one sees the results, it is an important phenomenon that deserves study and the awareness of community conservation practitioners. The word contagion has both a negative meaning, as in the spreading of disease, and a more neutral meaning, as the rapid communication of an influence. In a sense, the phenomenon of trends that are communicated rapidly may be thought to spread like epidemics with three characteristics: 1) contagiousness, 2) small causes can have large effects and 3) change happens not gradually but at one dramatic moment (Gladwell, 2002). Given these characteristics, research on such phenomena would be difficult. However, studies on human networks (Christakis & Fowler, 2009) may give us some insight into conservation contagion.

Since we have not carried out any research on the communications, social networks, or person-to-person interactions important in creating instances of conservation contagion we have observed, we can give only a start to understanding conservation contagion by listing and discussing some anecdotal observations that both seem to play a role in the phenomenon and show that contagion is occurring. We have identified eight facets: 1) copying of methods to initiate other projects, 2) person-to-person contacts, 3) presentations and responses to them, 4) formation of new community conservation organizations, 5) requests by communities or community groups to join an existing project, 6) knowledge of a project and requests for help by other communities, 7) large crowds at events and 8) creating project publicity within the country. Examples of some of these will be described and discussed in the following sections on Belize and Assam, India.

4. Examples of successful community conservation projects

4.1 Belize

In the early 1980's, cooperative organization and planning among Horwich, Lyon, and members of a community in Belize District, northwest of Belize City, led to the establishment of the Community Baboon Sanctuary dedicated to preservation of the endangered black howler monkey (*Alouatta pigra*) and its moist tropical forest habitat (Horwich & Lyon, 1988). This project led to development of a local ecotourism industry (Horwich et al., 1993). By 1990, the Community Baboon Sanctuary spawned creation of the country's first rural museum, managed semi-independently under the umbrella of the Belize Audubon Society (BAS). At this time, Horwich and Lyon formed *Community Conservation* (Horwich 1990a, 2005; Horwich & Lyon, 1995, 1998, 1999; Young & Horwich, 2007), initiating what would become an international network of ground-level, bottom-up programs.

In the Community Baboon Sanctuary project, *Community Conservation* and other non-governmental organizations functioned as catalysts, first by initiating locally-based programs and, subsequently, by educating village leaders about the short-, mid-and long-range value of biodiversity preservation. At a later stage, community leaders employed their legitimate authority to influence village residents and to disseminate information, often through delegates. Once this stage successfully affected participation in conservation efforts and established community-based infrastructures devoted to conservation, committed villagers were recruited to serve as workers dedicated to managing and sustaining conservation programs (e.g., as office workers or guards to eject poachers from protected lands). Outside organizers (e.g., *Community Conservation* personnel) were subsequently freed to devote attention to additional community-based conservation enterprises while continuing to maintain contact with and provide support to villages when required and solicited. In some cases, a small cadre of organizers maintained a physical presence in or near communities with viable conservation infrastructures, continuing to serve as facilitators and advisors, at least over the short term. In other instances in Belize, former organizers remained in villages as researchers or employees.

In the 1990s, through a process of conservation contagion, formal publicity and informal dissemination of information about the Community Baboon Sanctuary stimulated and influenced many rural Belizean communities to cooperate, exerting pressure on central government for participation in their conservation initiatives. These events ultimately influenced the Belizean government to create a series of protected areas (Young & Horwich, 2007) as well as a country-wide network of Special Development Areas including indigenous and local groups inhabiting the newly protected landscapes (McGill, 1994). In 1994, the Belize government coordinated an ecotourism seminar (Vincent, 1994), published a community tourism booklet, and created a video on community-based tourism. In addition, the Belize Departments of Forestry and Fisheries began to negotiate informal and, later, formal co-management agreements with communities (Young & Horwich, 2007). Cooperative development of the Community Baboon Sanctuary catalyzed a series of reactions (conservation contagion from local level to central government: a "vertical network") arising from our community-based conservation model. By 1991, communities led by St. Margaret's Village, adjacent to Five Blues Lake, lobbied the government of Belize to create Five Blues Lake National Park and other protected areas in response to the country's rapidly developing ecotourism industry (Young & Horwich, 2007). In 1992, the Minister of Tourism, Glenn Godfrey, embraced the Gales Point Project, subsequently leading government to include it in a new Special Development Area, a step preliminary to creation of a protected area in the region (McGill, 1994).

In 1997, the Inuit Council of Canada visited Belize's southern Toledo District, an area of Mayan concentration, to coordinate a seminar on co-management with the Kekchi Council of Belize (an indigenous Mayan organization), citing the Community Baboon Sanctuary as an example of community co-management. Eight years earlier in 1989, however, *Community Conservation* gathered signatures of local governments in three Toledo villages to initiate a cooperative plan for a Toledo Biosphere Reserve, composed of the Temash River, an important mangrove habitat, the Columbia Forest Reserve, and the Sapodilla Cayes (Horwich, 1990b). At that time, there was community support for the plan, but regional politicians showed no interest.

Following the central government's creation of the Sarstoon-Temash National Park in 1994, independent of input by communities impacted by the plan, Horwich traveled to Toledo in 1997 to re-catalyze and revive a component of the Toledo Biosphere initiative. A strategy was devised whereby Horwich would work with Judy Lumb, a resident of Belize with ties to the Garifuna community in southern Belize, to organize a conference on community co-management for the Sarstoon-Temash National Park stakeholders (see Producciones de la Hamaca & Community Conservation Consultants, 1998). From that event, the Sarstoon-Temash Institute of Indigenous Management (SATIIM), an indigenous-based non-governmental organization created by Mayans and Garifuna, was developed to coordinate impacted communities and to manage the National Park using our model.

Based on the success and aftermath of the conference, Horwich and Lyon obtained a United Nations Development Project (UNDP) grant for the Belize-based Protected Areas Conservation Trust (PACT, 1998) to create a community co-management park system. In contrast to the Namibian case described below, PACT's steering committee significantly modified project goals, leading to failure and abandonment of plans for the initiative (Catzim, 2002). However, a concrete result of our community-based conservation efforts in Belize was the central government's eventual adoption of community co-management as national policy (Meerman, 2005) and signing of agreements with communities to co-manage at least a dozen protected areas distributed throughout Belize - both terrestrial and marine (Figure 1). Thus, implementation of the *Community Conservation* model generated the process of a conservation contagion that proved successful beyond anyone's projections. Eventually, the Government of Belize supported enhancement of local natural resource management and production of systemic changes in conservation policy at the national level. However, it should also be noted that minimal government resources and political commitment (Catzim, 2002) have left many communities with insufficient support and no interconnected co-management system despite their interest in and will to proceed with conservation initiatives. Thus conservation contagion must be coupled with appropriate and sustainable support to achieve maximum impact.

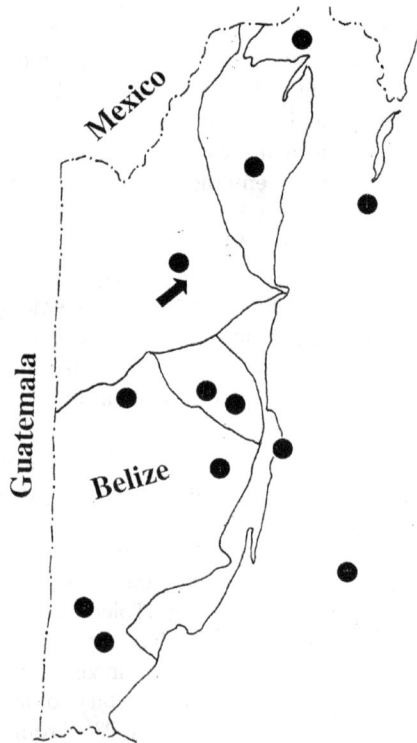

Fig. 1. Map of Belize showing distribution of the 12 community co-managed protected areas (dots) and the Community Baboon Sanctuary (arrow)

Figure 2 shows a synopsis of the catalyzing influences and social network of interactions that were occurring in Belize that led to the conservation contagion (for details, see Young & Horwich, 2007). The Community Baboon Sanctuary was initiated in 1985 in affiliation with the Belize Audubon Society, the latter having been commissioned by the Belizean government in 1984 to administer the existing park system. This partnership led to meetings between the Community Baboon Sanctuary Manager, the late Fallet Young, and the Director of the Cockscomb Basin Wildlife Sanctuary, Ernesto Saqui, an employee of the Belize Audubon Society. With the Community Baboon Sanctuary becoming widely publicized nationally and internationally, it had a conservation influence on rural communities country-wide (Government of Belize, 1998). In 1988, Young addressed the village of Monkey River, eventually catalyzing that community to move the government to create Payne's Creek Wildlife Sanctuary, that had a significant population of howlers (Horwich et al., 1993). By the early 1990s, three other projects were developed using our model: Slate Creek and Siete Milas village project in 1991 in the Mayan Mountains (Bevis & Bevis, 1991), a sea turtle nesting project initiated on Ambergris Cay by Greg Smith which was moved to Gales Point in 1992 when we initiated the Gales Point Manatee project (Lyon & Horwich, 1996); and, the Community Hicatee (turtle) Sanctuary along the Sibun River in 1994. Siete Milas later worked with Itzamna, a community-based organization co-managing the Elio Panti National Park in the Mayan Mountains.

Many rural communities in Belize, seeing the efforts of the Community Baboon Sanctuary villages, realized that they too could participate in the growing conservation/ecotourism movement and stimulated the government to create protected areas adjacent to their communities (Horwich & Lyon, 1999). St. Margaret's village initiated the establishment of Five Blues Lake National Park on Earth Day 1991 (Horwich & Lyon, 1999). We tried unsuccessfully to initiate a community-based project that included the Sarstoon-Temash mangrove forest area in 1989. Then, in 1994, it became a National Park without community input. The government held an ecotourism conference in 1994 and produced a movie and a booklet on community ecotourism (Horwich & Lyon, 1999). In 1995, the Association of Friends of 5 Blues Lake National Park signed a formal co-management agreement with the Government of Belize. In 1997, the Inuit Circumpolar Conference sponsored a co-management workshop in Toledo featuring the Community Baboon Sanctuary as an example of co-management. Later that year, *Community Conservation* initiated the process for a stakeholders workshop for the Sarstoon-Temash National Park (Produciones de la Hamaca & Community Conservation Consultants, 1998).

Stimulated by that conference, *Community Conservation* developed a proposal for a community co-managed park system that was to include Freshwater Creek Forest Reserve, Five Blues Lake National Park, Gales Point Manatee and Aguacaliente Wildlife Sanctuary (PACT, 1998). As a result of this project, when the Government of Belize created the Protected Areas System Plan in 2005 it had a section on community co-management. They presented the plan to the conservation community in early 2006 (Government of Belize, 2005). These actions were some that initiated the conservation contagion. Figure 2 gives a visual representation of a complex network that contributed to the conservation contagion that resulted throughout the small nation of Belize. Figure 2 demonstrates the complexity of social connections that occurred over the years contributing to the conservation contagion.

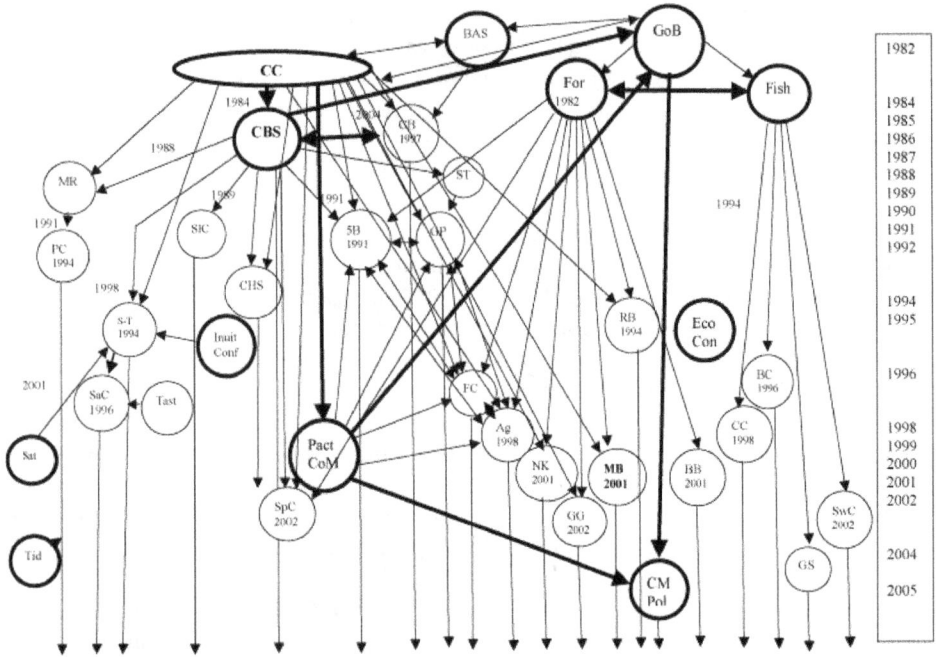

Government – GoB - Government of Belize, For - Forestry Department, Fish - Fishery Department, PACT Coman - Protected Areas Conservation Trust co-management grant, Cmpol – Co-management policy, Eco Con - Ecotourism Conference,

NGOs – CC – *Community Conservation*, BAS - Belize Audubon Society, Inuit Conf-Inuit Conference, Sat - SATIIM, Tast - Taste, Tid - TIDE,

CBOs/PAs - Ag – Aguacaliente Wildlife Sanctuary, BB – Billy Barquedier, BC - Bacalar Chico, CB – Cockscomb Basin Wildlife Sanctuary, CBS – Community Baboon Sanctuary, Cc – Cay Caulker, CHS – Community Hickatee Sanctuary, FC – Freshwater Creek Forest Reserve, 5B – 5 Blues Lake National Park, GG – Gra Gra Lagoon, GP – Gales Point, GS – Gladen Spit, MB - Mayflower Bocwina National Park, MR –Monkey River, NK -- Noj Kaax Meen Eligio Panti National Park, PC - Paynes Creek, RB - Rio Blanco National Park, SaC - Sapadilla Cayes, SwC – Swallow Cay, SlC- Slate Creek, SpC - Spanish Creek Wildlife Sanctuary, ST- Sea Turtle Project, S-T – Sarstoon-Temash National Park.

Fig. 2. Network connections of *Community Conservation* and their relationships with community-based organizations and Protected Areas, non-governmental organizations and government organizations.

4.2 Namibia

In the early 1980s in northern Namibia, wildlife was severely depleted by poaching (Hoole, 2010). At the same time that the Community Baboon Sanctuary was developed in Belize, a non-governmental organization in Namibia, the Namibian Wildlife Trust, appointed Garth Owen-Smith, a former government game ranger, to respond to the poaching crisis in the northern area of the country by collaborating with village headmen who shared concern for the loss of wildlife (Hoole, 2010; Jones, 2001). Using village contacts from his prior career, Owen-Smith began to establish relationships with local headmen. Working with the government conservator, Chris Eyre and in cooperation with village leaders, Owen-Smith instituted informal community game guard protection of wildlife, leading, over time, to increased population densities of large mammals threatened by poaching (Jones, 2001). In the mid-1980s, Owen-Smith and anthropologist Margaret Jacobsohn negotiated agreements with safari operators to pay the community a US$5 fee per tourist visiting the area to view game (Jones, 2001). In both Belize and Namibia, conservationists initiated their efforts by responding to environmental problems in discrete, manageable regions, engaging with stakeholders as allies having common interests and goals to preserve biodiversity. These associations also addressed resistance and other challenges arising within, between, and outside of village networks.

By 1989-1991 a community game guard program was firmly established, and Owen-Smith and Jacobsohn moved on to initiate a second program in northeastern Namibia (Jones, 2001). While projects in Belize had informal support from local politicians, there was no formal support from government because lands were privately owned by subsistence farmers. Their practices, especially "milpa" (slash-and-burn) or clear-cutting methods of land clearance, were often destructive to habitat and organisms inhabiting forests. In Namibia, before independence, the South African government viewed Owen-Smith's liaison with black communities suspiciously, ultimately terminating support for his programs (Jones, 2001). By 1990, however, the Namibian initiative proved to be on the right side of history (Westley et al., 2006) when the country gained independence, extending rights over wildlife to the majority black government, thereby terminating private ownership of land by white farmers (Hoole, 2010). The post-independence government engaged the non-governmental organization, Integrated Rural Development and Nature Conservation (IRDNC), formed by Owen-Smith and Jacobsohn, to develop a community-based natural resource management program (Hoole, 2010). This was similar to the process transpiring in Belize at approximately the same time.

The pioneering projects in Belize and Namibia had similar trajectories, with conservationists and small non-governmental organizations acting as catalysts, initiating locally-based programs and educating villagers. In both cases, through disseminated information initial projects led to broad-based regional conservation efforts. This seemingly paradoxical role is the foundation of reliable and replicable tactics and strategies of community conservation, including features characteristic of *Community Conservation's* 9-stage model.

Namibian independence in 1990 removed the obstructionist South African regime, and the new Namibian government invited Owen-Smith and Jacobsohn to resume and to expand their work. Namibia's community-based natural resource management program had been modeled after other African initiatives such as Zimbabwe's Communal Areas Program for Indigenous Resources (CAMPFIRE) and Zambia's Administrative Management Design for Game Management Areas program (ADMADE) (Hoole, 2010). Zimbabwe's conservation efforts, however, were obstructed by resistance at lower levels of government, a condition opposite to that in Namibia (Murphree, 2005). With major financial support from World Wildlife Fund (WWF) and the United States Agency for International Development (USAID, 2005) as well as backing from the central government, Namibia's Communal Conservancy Program caught on rapidly, exhibiting the conservation contagion seen in Belize and Assam, India (Horwich et al., 2010). By 2009, Namibia was managing 59 registered conservancies cumulatively, covering 12.2 million hectares (Figure 3; 2009 map from www.nacso.org.na). Wildlife populations in the northeast conservancy regions increased markedly and money was generated from eco-projects (NACSO, 2009) with long-term potential for financial sustainability. While the Namibian case demonstrates that our model for the implementation of reliable community conservation projects is not the only one capable of sustained success, the *Community Conservation* template is explicitly detailed by stages characteristic of all community-based conservation initiatives. We base this statement upon our documentation of the methods and outcomes of other programs compared and contrasted to our own model.

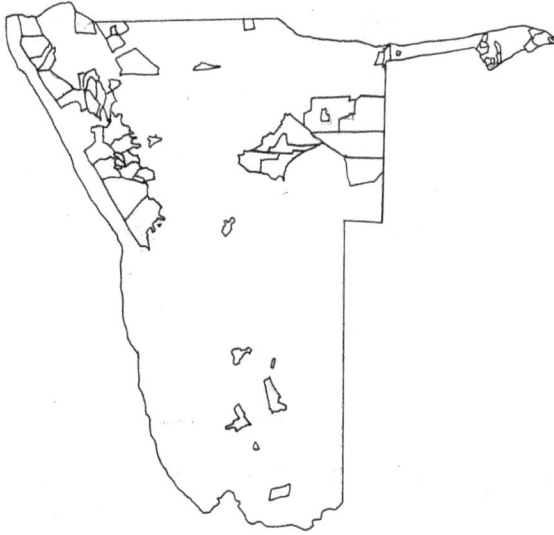

Fig. 3. Map of Namibia's 59 Community Conservancies in 2009 (redrawn from 2009 map on www.nacso.org.na).

4.3 Assam, India

The Golden Langur Conservation Project was initiated in 1998 to protect the Manas Biosphere Reserve and the golden langur (*Trachypithecus geei*), a folivorous monkey. Forests of the Manas Biosphere Reserve have been threatened by illegal logging since the early 1990s, and, in the last 15 years, approximately one third to one half of the three reserve forests (~200,000 ha) making up the reserve (Ripu, Chirrang, and Manas), were deforested by clear-cutting (Bose & Horwich, unpublished data). Based on transect surveys, including interviews with residents in and near the reserves, these three reserve forests, a group of southern isolated reserve forests, and the Royal Manas Sanctuary on Bhutan's northern border are presently the primary range of the golden langur. Understanding the potential of conservation contagion to create regional change, *Community Conservation* included local communities, non-governmental organizations, and agencies of the governing body of Assam in planning meetings and seminars to discuss the conservation project. While the Assam Forestry Department showed some interest in the effort, it was not until a new tribal government, the Bodoland Territorial Council (BTC), was formed in 2004, that community conservation was embraced, leading rapidly to the contagious spread of support for our proposals noted earlier, ultimately attracting crowds of 20,000 to 35,000 people (Figure 4) to our informational programs (Horwich et al., 2010).

Fig. 4. Crowd Attending the Manas Biosphere Celebration at Ultapani

An important aspect of the Assam project is the manner in which a second tier of catalysis arose from activities of community-based organizations and non-governmental organizations. In 2006, Forest Protection Forces were created within the Manas Biosphere Reserve with support from the Bodoland Territorial Council. Subsequently, conservation contagion gained momentum, culminating in cooperative management by 14 community groups forming the Unified Forest Conservation Network of Bodoland, supported by local and governmental networks (Horwich and Bose, personal observation). A schematic representation of connections between the three forests making up the reserve is presented in Figure 5. The lower circular configuration representing the Unified Forest Conservation Network of Bodoland that protects the Biosphere mimics the military squad network topology of Christakis and Fowler (2009). This ring network topology experimentally was shown to facilitate problem solving (Christakis & Fowler, 2009).

Similarly, contagion occurred around the Kakoijana Reserve Forest where community members inhabiting or proximal to the reserve forest, housing a small golden langur population, were attracted to the conservation project. Today, by contagious processes, 28 communities surrounding Kakoijana Reserve Forest (Figure 6) have created two federations (Nature Guard, Green Conservation Federation) to protect the 17km^2 reserve (Bose & Horwich, personal observation). These are represented by the two upper, left spheres in Figure 5. Most importantly, these community protection efforts resulted in an increased Indian golden langur population from 1500 to almost 5600 langurs (Figure 7). The Kakoijana Reserve Forest increased its forest from 5% to 70-80% canopy (Figure 8) accompanied by an increase of golden langurs from less than 100 to over 500 langurs (Horwich et al., 2010; Horwich et al., 2011, Bose and Horwich unpublished data).

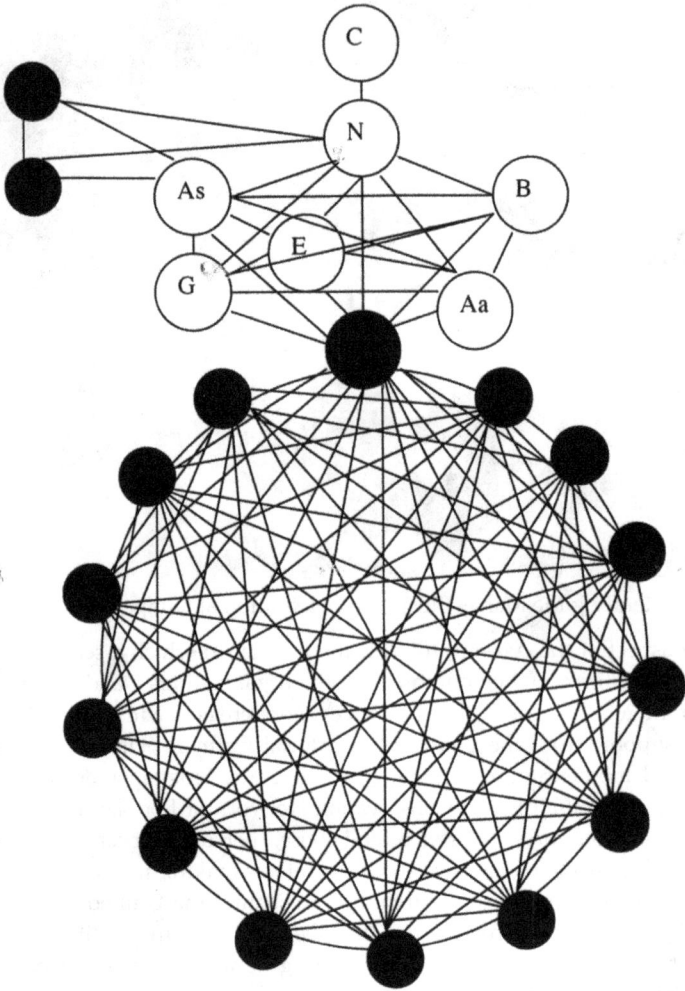

Government: As – Assam Forest Department, B – Bodoland Territorial Council
NGOs: Aa – Aaranyak, C – *Community Conservation*, G – Green Heart Nature Club, N – Natures Foster,
CBOs (black spheres): Biodiversity Conservation Society, Green Forest Conservation (larger black sphere), Manas Agrang Society, Manas Bhuyapara Conservation and Ecotourism Society, Manas Maozigendri Ecotourism Society, Manas Souci Khongar Ecotourism Society, New Horizon, Panbari Manas National Park Protection and Ecotourism Socierty, Raigajli Ecotourism and Social Welfare Society, Swarnkwr Mithinga Onsai Afut and four other unnamed community organizations.
CBOs around Kakoijana Reserve Forests (black spheres upper left) Green Conservation Federation, Nature Guard

Fig. 5. Network Connections Established by the Golden Langur Conservation Project in Assam, India Between Government, Non-Government and Community Organizations.

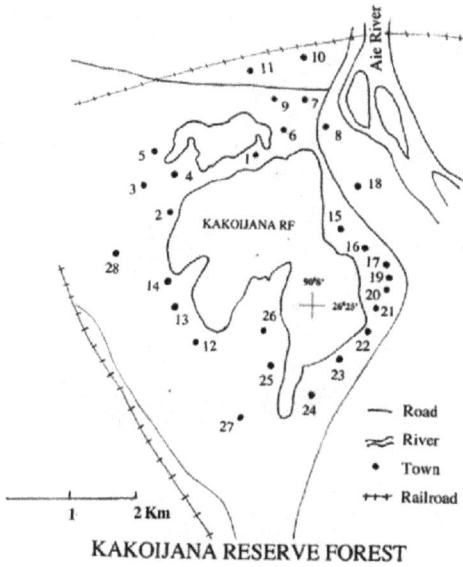

Fig. 6. Map of Kakoijana Reserve Forest surrounded by 28 villages composing the Green Conservation Federation and Nature Guard that protect the reserve forest

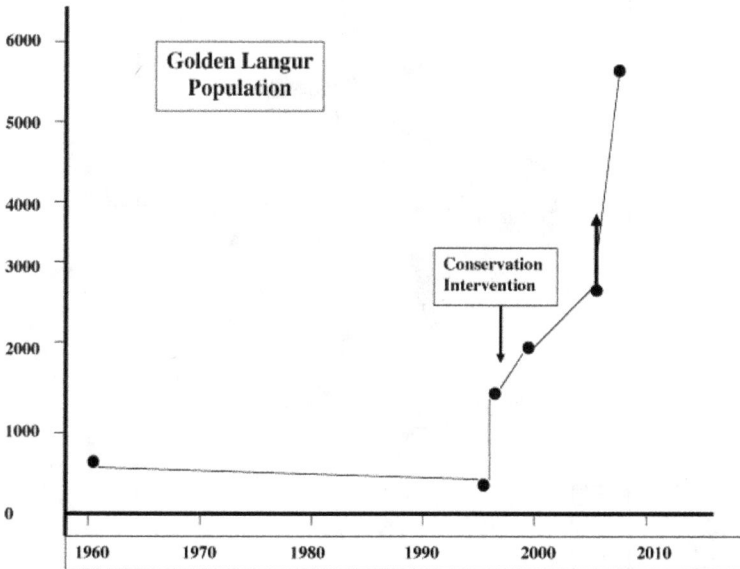

Fig. 7. Graph of Indian Golden Langur Population estimates indicating population increase following initiation of the Golden Langur Conservation Project (data from Gee, 1964 for 1960; Srivastava, et al. 2001 for 1997; Choudhury, 2002 for 2000; Ghosh, 2008a & b for 2008; Anonymous, 2009, Bose, 2007, 2008 (unpublished data), Ghosh, 2008a for 2009)

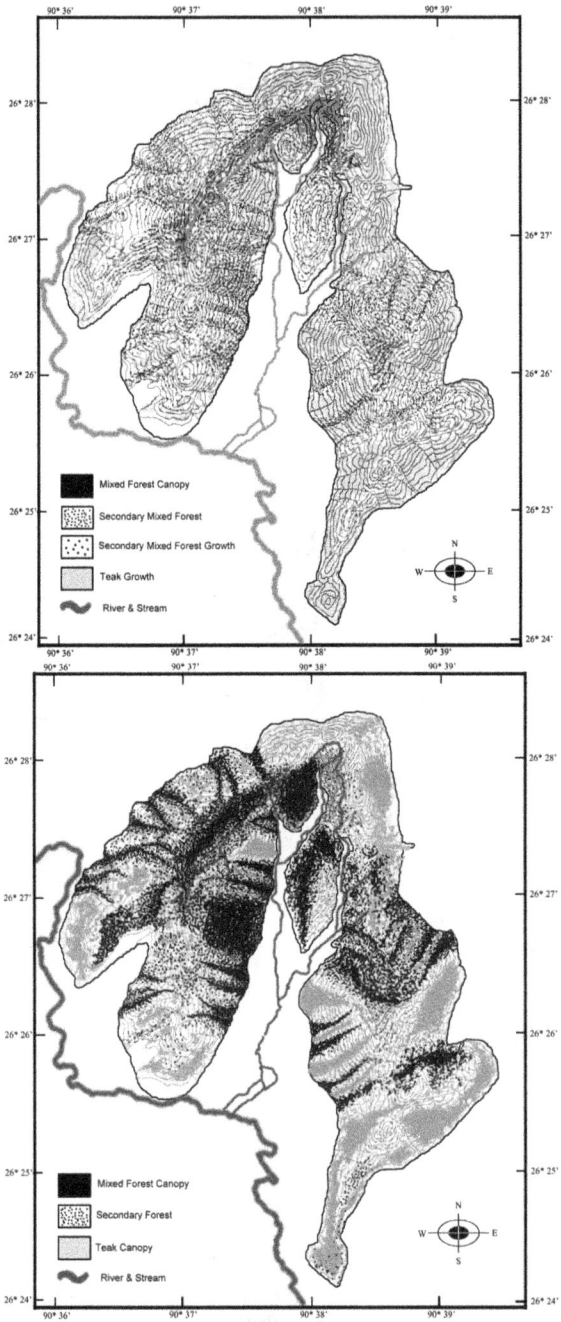

Fig. 8. Vegetation maps of Kakoijana Reserve Forest in 1996 and 2008 showing an increase in canopy cover from 5% to 70% as a result of community reforestation and protection.

In the Assam project, local villagers were trained as community organizers, community researchers, and community para-veterinarians (Figure 9). Through expansion of many programs and capacity-building, the project has influenced conservation efforts at broader regional scales, similar to the cases described for Belize and Namibia.

Fig. 9. Veterinarian Sarma training para-veterinarians near Kakoijana Reserve Forest

For example, one trained community organizer initiated self-help groups coordinating environmental awareness sessions around Manas National Park for the Pygmy Hog (*Porcula salvania*) Project (Bose and Horwich, personal observation). He is also organizing communities for the Assam Haathi Project whose goal is to mediate human-elephant conflict along the southern basin of the Brahmaputra River in Golpara District. Another local conservationist, trained in cutting-edge primate census methods by *Community Conservation* personnel, worked with the Tripura State Forest Department to census Phayre's langurs (*Trachypithecus phayrei*), leaf-eating monkeys related to golden langurs. Members of communities in these areas gained capacity and experience and were able to enhance their employability as semi-professional conservationists, disseminating project goals and procedures throughout the region. They currently perform a role similar to trained para-taxonomists in Papua New Guinea, in Guyana, and elsewhere (see Basset et al., 2000). In other locales, community members have been trained by a veterinarian as para-veterinarians (Figure 9), similar to bare-foot doctors supporting the Hen Can Change a Man Program organized to increase the income of villagers engaged in poultry husbandry around the Kakoijana Reserve Forest. These examples of effective educational initiatives reinforce our suggestion that all community-based conservation programs, *ceteris paribus*, are characterized by features similar to our 9-stage model because, due to psychological, economic, political, social and other comparable constraints, there are a restricted number of tactics and strategies likely to lead to sustained success. Training and capacity-building for both individuals and non-governmental organizations, then, have been integral and critical parts of expansion of community conservation programs on a range of fronts. Following the definition of catalysis presented above, energy inputs required to activate and effect initial stages of community-based conservation were decreased as a result of increased efficiency resulting from skill- and evidence-based organization. The broad regional success in Assam and elsewhere is dependent upon community conservation contagion traceable back to a modest, small-scale starting point in local areas. Our model, then, schematizes the trajectory of successful bottom-up conservation programs, creating stable "horizontal networks" capable of developing productive associations and activities with components of "vertical networks" (Berkes, 2004).

5. Lessons learned from practice

There is a current dearth of articles comparing and contrasting community-based conservation, community conservation, collaborative management, ICDPs, etc. These terms have, effectively, become buzz-words with biologists and sociologists referring to "new conservation" models incorporating community-based conservation practices. With so many different approaches, it has become difficult to know what works best and what doesn't. First hand practical experience with 23 small on-going community conservation projects over the past 26 years have given us insights about what factors are most critical for achieving program success (Table 1).

Rural communities are a major resource for a new breed of conservationists. Since they live where the highest concentration of natural resources exists, can easily exploit them with traditional practices, have additional knowledge about their habitats and people, and are experts in their own right, mainstream conservationists must create incentives for indigenous and other local people in order that their talents, knowledge, expectations, desires, and persistence can be utilized for their own benefit and that of the forests and wildlife. Conservationists can use existing rural institutions or can help rural people create new institutions to deal with modern problems (Agrawal and Gibson, 1999).

To be most effective, practitioners need to be catalysts of community conservation, helping community-based organizations to form in local communities, monitoring them, building capacity and re-catalyzing them when needed. When initiating a community conservation project, a well-crafted proposal may have major power to interest community members, government officials, donors, and other stakeholders to become involved. Enlisting a support coordinator and seed money are important to get a young project started and to help a fledgling community-based organization.

Seed money and money for simple project maintenance is essential. Looking for a simple mechanism for partial financial sustainability is very difficult, but, when found, adds a great deal toward continuity and longevity, preventing community discouragement. Using a guiding, entrance, or membership fee for protected areas can create a minimal budget to keep the project going. Community ecotourism, while a double-edged sword, can provide potential in this direction.

While financial sustainability is to be strived for, project continuity and longevity depend on creating social incentives. Properly equipped volunteers are very important to keep incentives going. Discontinuity of funds or other incentives discourages community members. Although projects may pass through ups and downs, if the original incentives were good, there is a recycling effect; a project is only a failure if it totally disappears. Establishing model projects is extremely important to encourage other projects and can have a regional effect. A replication of techniques from one project to another helps to propagate ideas and models.

Connection to land is important to rural residents and ownership contributes greatly to this. Thus, historical land tenure may be problematic or can be a source for positive conservation alternatives. While private landowners can make important contributions, encouraging communal and tribal or clan ownership, where in place, has a greater relevance for project longevity and long-term environmental protection, especially if there is a strong formal or informal institution in place. Since private landownership has its limitations, formal land protection mechanisms for private lands can create a long-lasting effect. There is a difference between land ownership and management and community groups can achieve some benefits from management alone. Land use planning provides an important vision for future conservation and protection.

While governments are often slow to move, may be corrupt, or can hamper a project, there is a great need for balanced community-government communication. Too much government reduces community initiative. However, effective, strong laws can provide strong legal land protection. Additionally, once government sees the advantage to involving communities, their tactical and financial support can provide project sustainability. Co-management of protected areas can thus be an important balanced conservation solution.

Non-governmental organizations have an important role (Agrawal & Gibson, 1999) as intermediaries and community trainers and educators. They can provide an initial open communicatory link between communities and government. Often, they can provide expertise and motivation that governments cannot. Non-governmental organizations can provide networking to connect communities to the resources they need to develop a stronger community-based organization. In any effective system that involves communities, non-governmental organizations, and government, there is a need to strengthen all partners, especially the weaker ones.

6. Policy implications for the future

Comprehensive, multi-scale, and resilient policies are required to respond effectively to social, political, and economic networks facilitating successful catalytic incidents and the emergence of conservation contagion. In this chapter, we have documented a resilient community-based model whose successes have resulted from the expansion of existing ground-level networks into a broader horizontal matrix by capitalizing on catalytic events and subsequent contagious diffusion. These tactics and strategies operate along with indigenous and other local values, beliefs, folkways, resource use, and political structures as well as social, cultural, and economic activities in order to capitalize on opportunities to build complex horizontal networks (see Berkes, 2004, Christakis and Fowler, 2009). In these cases, we have observed that a critical mass of complexity (e.g., by increasing inter-individual and multi-scale interactions) may lead to a threshold response at which a tipping point is reached, leading from one state of relative equilibrium to another (Gladwell, 2002). These non-linear events may have negative as well as positive outcomes from the perspective of conservation goals and objectives. Indeed, some studies suggest that increased complexity may lead some community members to resist or abandon active engagement with conservation initiatives (see Hoare and du Toit 1999 in Berkes, 2004), a dynamic state of affairs providing challenges to community-based organizations but capable of being addressed by components of our model applied to community-based organization programs by trained facilitators.

Community-based initiatives and their practitioners deliver specialized, reliable tactics and strategies to indigenous and other local communities for empowerment of community networks and, ultimately, transfer of power to these entities. Paramount to success is remembering that indigenous and other local communities are necessary components for solutions to the worldwide biodiversity crisis. We have demonstrated that application of our model and treating indigenous and other stakeholders with consistent and reliable respect and humility, valuing their folkways, habits, and cultures, their knowledge, leadership skills, and significant expertise without patronizing attitudes and behaviors, a majority of local stakeholders become willing and active participants in conservation programs (see Persha et al., 2011). If these factors are obtained, trusting, enduring relationships between horizontal and vertical networks can be built having the potential to establish multi-scale, multidimensional, complex, and dynamic associations, incorporating indigenous and other local people as co-conservationists (Haldane &May, 2011; Persha et al., 2011). In addition to consolidating horizontal and vertical networks, *Community Conservation's* mission and the successful implementation of our 9-stage model that we extend to other community-based organizations, has promoted the success of our programs by providing incentives, fostering pride, advancing program self-sufficiency, and effecting sustainable project ownership.

Our case studies demonstrate applications of the 9-stage paradigm that have influenced non-governmental organizations and/or central governments to adopt community-based conservation as policy, an outcome with the potential to impact conventional top-down environmental procedures linking horizontal with vertical networks. It is important to emphasize that successful implementation of our strategies depends upon careful, calculated planning resulting from the training and expertise of *Community Conservation* personnel. Equally important are the social and other skills, attendant traits, and motivation of indigenous leaders and other local individuals providing the initial commitment and impetus for successful implementation of our model and, later, cooperatively designed plans. Further, the flexibility and resilience of our 9-stage paradigm permits accommodation and adjustment to a range of local conditions, contingencies best evaluated stage by stage throughout the dynamic process of the multi-stage implementation of programs.

An example of the flexibility and resilience of our 9-stage plan involves the different contexts and challenges encountered when community members of low socioeconomic rank prove, initially, to be the most committed to implementation of conservation projects. In our experience, when the principal innovators and drivers of change in a target community are of relatively low socioeconomic or other status compared to, for example, group leaders, a longer period of time is required for the broader community and region to absorb, integrate, and respond to new ideas and ultimate project success. The cases we describe emphasize the need to adopt a flexible, resilient, and holistic systems perspective including social, economic, and political components of values and practices of indigenous and other local agents' relationships to their environment given the short-, mid- and long-range goals and objectives.

Currently, indigenous or other local people with lower socioeconomic status, including some men, most women, and youth, are assuming key roles in *Community Conservation's* community-based conservation projects. Large international non-governmental organizations may see the potential of using a catalyst method by adopting our 9-stage paradigm as a component of their conservation policy. Incorporating plans such as

Community Conservation's tactics and strategies would permit non-governmental organizations, central governments, and other entities to modify their methods, developing paradigms representative of their own circumstances and complying with the goals, objectives, and philosophies of their particular organizations. For example, World Wildlife Fund began innovative community conservation projects in southern Madagascar and Namibia. Conservation International is supporting the Tree Kangaroo Conservation in Papua New Guinea (Ancrenaz et al., 2007). The Nature Conservancy has shown innovative community projects in Papua New Guinea and the Solomon Islands (Mayer and Brown, 2007). Similarly, the Wildlife Conservation Society works with local communities to protect the critically endangered Cross River Gorilla (*Gorilla gorilla diehli*) limited to a restricted area along the Nigerian-Cameroon border (Nicholas et al., 2010). Successes can be documented; however, community conservation projects may not always be sufficiently comprehensive, ambitious, or efficient in time and energy to persuade non-governmental organizations with large budgets to invest in relatively small-scale conservation activities (Brockington et al., 2008). Furthermore, in our experience, large-scale non-governmental organizations rarely advertise their small community conservation accomplishments, and, if this policy were reversed, community-based conservation would likely be poised to gain a significant degree of legitimacy with non-governmental organizations and other entities in vertical networks.

Attempts to evaluate the outputs and successes of historical policies and practices of protective programs and the field of conservation biology suggest that these entities lag behind most other policy fields because of their resistance to and slow incorporation of participatory philosophy and models (Ferraro & Pattanayak, 2006; also see Milner-Gulland et al., 2010). Yet there is growing recent data that show the effects of community and indigenous projects in reducing deforestation. Recent studies by Porter-Bollard et al. (2011), comparing 40 tropical protected areas to 33 community managed forests, indicated that the community managed forests showed lower and less variable annual deforestation than the protected areas. Soares-Filho et al. (2010), focusing on Brazil's recent push to reduce Amazonian deforestation by expanding the Amazon protected area network of 1.9 million km2, showed a generalized inhibitory effect on deforestation in protected areas which would greatly reduce carbon emissions which in turn would effect climate change. However, they used a broad definition of protected areas. Looking closer at these 595 protected areas, only 90 or 15% were in the strictly protected category used by Porter-Bollard et al. (2011) while 494 or 83% were in the sustainable use (176) and indigenous lands (318) categories. Thus the positive results are mainly due to sustained use and indigenous lands, indicating the importance of community conservation.

New conservation policies incorporate and embrace community conservation as one approach to biodiversity and ecosystem conservation, necessitating involvement and active participation by horizontal networks, including community and other stakeholder entities, community-based organizations, regional and national agencies, central governments, non-governmental organizations, conservation biologists, and other conservation practitioners (e.g., forest guards and rangers; scientists conducting research in habitat countries and hotspots) and stakeholders. Connecting horizontal and vertical networks among conservation entities has the potential to increase network diversity, scale, and resilience, maximizing likelihoods of success of conservation programs. Such connectivity also has the potential to compensate for limitations of agents and entities in both horizontal and vertical networks since bottom-up entities have generally lacked the resources to invest in the creation, management, and sustainability of global organizations or in sustaining these networks if and when established. Higher-order, protectionist networks with "fortress conservation" policies have traditionally been averse to incorporating the interests, values, and bodies of indigenous and other local agents and units (informal and formal) into the goals and objectives of their mission statements and programs. This state of affairs has been maintained in large part because the morality, ethics, beliefs, attitudes, and philosophy of local people's relationships to their natural resources has conflicted with those of vertical networks' historical prioritization of preservation over sustainable use of forests, waterways, etc. (Berkes, 2004). When multi-level networks cooperate and establish meta-networks designed to resolve differences of vision and to consolidate communication networks and integrated representation of all stakeholders, differences and potential conflicts can be minimized through a multi-scale and multidimensional system sensitive to the interests and objectives of all network units over time and space.

7. Acknowledgements

We thank all of the community members and non-governmental organizations from the countries in which we worked, especially Belize and India. They are the heroes and doers. We are only the catalysts. We especially acknowledge the input and expertise of Dr. Colin, Jesse, and the late Fallett Young in Belize, whose support and encouragement were critical as we implemented our first international project. Our gratitude is also extended to Sadahib Senn, Raju Das and other members of Nature's Foster, the late Rajen Islari, and others from Green Forest Conservation, Bablu Dey and others from Green Heart Nature Club, Mahesh Moshahary and others from New Horizon, Firoz Ahmed and Aaranyak, and the members of the Biodiversity Conservation Society, the Raigajli Ecotourism and Social Welfare Society, the Panbari Manas National Park Protection and Ecotourism Society, the Swarnkwr Mithinga Onsai Afut, the Manas Maozigendri Ecotourism Society, Manas Bhuyapara Conservation and Ecotourism Society, the Manas Souchi Khongkar Ecotourism Society, the Manas Agrang Society, the Green Conservation Federation, and the Nature Guard of Chiponshila for making the Golden Langur Conservation Project a major success. We also are grateful to Minister of the Environment Kampa Borgoyari and the Bodoland Territorial Council for their support. The Margot Marsh Biodiversity Foundation, Conservation International, Primate Conservation, Inc., World Wildlife Fund, and the US Fish & Wildlife Service provided financial support, and we appreciate the generosity of these organizations. Thanks, also, to Ashley Morga for critiquing the manuscript.

8. References

Agrawal, A., & Gibson, C.C. (1999). Enchantment and disenchantment: the role of community in natural resource conservation. *World Development*, Vol.27, pp. 629-649, ISSN 0305-750X

Allendorf, T. (2007). Residents' attitudes toward three protected areas in southwestern Nepal. *Biodiversity and Conservation*, Vol.16, pp. 2087-2102, ISSN 0960-3115

Ancrenaz, M.; Dabek, L., & O'Neill, S. (2007). The costs of exclusion: recognizing a role for local communities in biodiversity conservation. *PloS Biology*, Vol.5, No.11, pp. 2443-2448, ISSN1544-9173

Anonymous. (2009). Comprehensive census reveals 4,231 golden langurs. In: *Assam Tribune* April 13, 2009, Available from: <http://ne.icindia.org/2009/04/13/comprehensive-census-reveals-4231-golden-langurs/> accessed 12/27/09

Anonymous. (2011). *India Tiger Estimate 2010*. Ministry of Environment and Forests, Government of India.

Ariely, D. (2010). *The Upside of Irrationality*. HarperCollins Publishers, New York, , *ISBN* 978-006-1995-03-3

Arnstein, S.A. (1969). A ladder of citizen participation. *Journal of the American Institute of Planners* ,Vol.35, pp. 216-224, ISSN 0002-8991

Balmford, A., & Whitten, T. (2003). Who should pay for tropical conservation, and how could the costs be met? *Oryx*, Vol.37, pp. 238-250, ISSN 0030-6053

Barrow, E. (1996). Frameworks for community participation. *1996 Pan Africa Symposium on Sustainable Use of Natural Resources and Community Participation*. Harare, IUCN, ISBN 079-7419-07-1

Basset, Y.; Novotny, V., Miller, S.E., & Pyle, R. (2000). Quantifying biodiversity: experience with parataxonomists and digital photography in Papua New Guinea and Guyana. *Bioscience*, Vol.50, No.10, pp. 899-908, ISSN 0006-3568

Belsky, J. M. (2000). The meaning of the manatee: an examination of community-based ecotourism discourse and practice in Gales Point, Belize. In: *People, Plants and Justice*, C. Zerner (Ed.), pp.(285-308), Columbia University Press, ISBN 023-1108-10-9, New York

Belsky, J.M. (1999). Misrepresenting communities: the politics of community-based rural ecotourism in Gales Point Manatee, Belize. *Rural Sociology*, Vol.64, pp. 641-666, ISSN 0036-0112

Berkes, F. (1994). Co-management: bridging the two solitudes. *Northern Perspectives*, Vol.22, pp. 18-20, ISSN 0380-5522

Berkes, F. (2004). Rethinking community-based conservation. *Conservation Biology*, Vol.18, pp. 621-630, ISSN 0888-8892

Bernstein, S.E. (2005). *Training to Develop Community-based Co-management Capacity in Belize*. M.S. Thesis, University of Wisconsin-Madison.

Bevis, J. & Bevis, M. (1991). *Slate Creek Preserve*. Unpublished manuscript.

Borrini-Feyerabend, G. (1996). *Collaborative Management of Protected Areas: Tailoring the Approach to the Context*. International Conservation Union, Gland, Switzerland

Borrini-Feyerabend, G..; Pimbert, M., Taghi Farvar, J.C., Kothari, A., & Renard, Y. (2004). *Sharing Power. Learning by Doing in Co-management of Natural Resources Throughout the World*. International Institute for Environment and Development (IIED) and IUCN, ISBN 184-3694-44-1, Cenesta, Tehran.

Brandon, K.; Redford, K.H., & Sanderson, S.E. (Eds.). (1998). *Parks in Peril: People, Politics and Protected Areas*. Island Press/The Nature Conservancy, ISBN 155-9636-07-6, Washington D.C.

Brechin, S.R.; Wilshusen, P.R., Fortwangler, C.L.,& West, P.C. (2002). Reinventing a square wheel: a critique of the new protectionist paradigm in international biodiversity conservation. *Society and Natural Resources*, Vol.15, pp. 41-64, ISSN 0894-1920

Broad, J. (1994). The poor and the environment: friends or foes? *World Development*, Vol.22, pp. 811-822, ISSN 0305-750X.

Brockington, D.; Duffy, R., & Igoe, J. (2008). *Nature Unbound*. Earthscan, ISBN 978-1-84407-440-2, London.

Brooks, T.M.; Mittermeier, R.A., da Fonseca, G.A.B., Gerlack, J., Hoffman, M., Lamoreux, J.F., Mittermeier, C.G., Pilgim, J.D., & Rodrigues, A.S.L. (2006). Global biodiversity conservation priorities. *Science* ,Vol.313, pp. 58-61, ISSN 0036-8075

Brosius, P.; Tsing, A.L., & Zerner, C. (1998). Representing communities: histories and politics of community-based natural resource management. *Society and Natural Resources*, Vol.11, pp. 157-168, ISSN 0899-1920

Caddy, E.; Ch'oc, G., Paul, S. (2000). The Sarstoon-Temash Institute of Indigenous Management: a grassroots initiative for social equity and sustainable development. In: *MS presented at the IUCN World Congress*, Amman, Jordan, Oct 10.

Catzim, A. (2002). *Project Systematization Main Report. PACT/GEF/UNDP The Community Co-managed Parks System Project 1999-2002*, PACT, Belmopan.

Choudhury, A. (2002). Golden langur *Trachypithecus geei* threatened by habitat fragmentation. *Zoo's Print Journal*, Vol.17, No.2, pp. 699-703

Christakis, N.A., Fowler, J.H. (2009). *Connected: The Surprising Power of Our Networks and How They Shape Our Lives*. Little, Brown and Company, ISBN 978-0-326-03614-6, New York

Davey, A. G. (1998). *National System Planning for Protected Areas*, IUCN, ISBN 283-1703-99-9, Gland, Switzerland, Cambridge, United Kingdom

Dowie, M. (2009). *Conservation Refugees*. MIT Press, ISBN 978- 0-26201-261-4, Cambridge, Massachusetts

Dupuis, E.M. & Vandergeest, P. (Eds.). (1996). *Creating the Countryside: the Politics of Rural and environmental Discourse*, Temple University Press, ISBN 156-6393-60-4, Philadelphia

Feeny, D.; Berkes, F., McCay, B.J. & Acheson, J.M. (1990). The tragedy of the commons: twenty-two years later. *Human Ecology*, Vol.18, pp. 1-19, ISSN 0300-7839

Ferraro, P.J. & Pattanayak, S.K. (2006). Money for Nothing? A Call for Empirical Evaluation of Biodiversity Conservation Investments, *PLoS Biology*, Vol.4, No.4, e105. doi:10.1371/journal.pbio.0040105, ISSN 1544-9173

Gadgil, M. & Guha, R. (1993). *This Fissured Land: an Ecological History of India*, University of California Press, ISBN 052-0076-21-4, Berkeley, California

Gee, E. P. (1964). *The Wildlife of India*, Collins, London.

Gezon, L. (1997). Institutional Structure and the Effectiveness of Integrated Conservation and Development Projects: Case Study from Madagascar, *Human Organization*, Vol.56, No.4, pp. 462-470, ISSN 0018-7259

Ghosh, S. (2008a). *Report on Population Estimation of Golden Langur (Southern Population) in Chakrashila Wildlife Sanctuary and Reserve Forests Under Kokrajhar, Dhubri and Bongaigaon Districts of Assam*, Report to the Assam Forest Department, Guwahati, Assam

Ghosh, S. (2008b). *Report of Wild Elephant (Elephas maximus) Population Estimation in Bodoland Territorial Council (20th-26th Feb 2008)*, Report to the Assam Forest Department, Guwahati, Assam

Gladwell, M. (2002). *The Tipping Point: How Little Things Can Make a Big Difference*, Little, Brown and Company, ISBN 031-6316-96-2, New York.

Government of Belize. (2005). *The Belize National Protected Areas System Plan*. Ministry of Natural Resources and the Environment, Belmopan, Belize

Government of Belize. (1998). *Belize's Interim National Report. Submitted to The Convention on Biological Diversity*. Ministry of Natural Resources, Belmopan, Belize

Guha, R. (1989). *The Unquiet Woods*, Oxford University Press, ISBN 052-0222-35-0, Delhi

Haldane, A.G., & May, R.M. (2011). Systematic risk in banking ecosystems. Nature, Vol.469, pp. 351-355, ISSN 0028-0836

Ham, S. M.; Sutherland, D.S., & Meganck, R.A. (1993). Applying environmental interpretation in protected areas of developing countries: problems in exporting a US model, *Environmental Conservation*, Vol.20, pp. 232-242, ISSN 0376-8929

Hardin, G. (1968). The tragedy of the commons, *Science*, Vol.162, pp. 1243-1248, ISSN 0036-8075

Hoole, A.F. (2010). Place – power –prognosis: community-based conservation, partnerships and ecotourism enterprise in Namibia. *Inernational Journal of the Commons*, Vol.4, No.1, pp. 78-99, ISSN 1875-0281

Horwich, R. H. & Lyon, J. (1988). Experimental Technique for the conservation of private lands. *Journal of Medical Primatology*, Vol.17, pp. 169-176, ISSN 0047-2565

Horwich, R. H. (1990a). How to develop a community sanctuary - an experimental approach to the conservation of private lands, *Oryx*, Vol.24, pp. 95-102, ISSN 0030-6053

Horwich, R.H. & Lyon, J. (1995). Multilevel conservation and education at the Community Baboon Sanctuary, Belize. In: *Conserving Wildlife: International Education and Communication Approaches*, Jacobson, S.K., (Ed.), pp. (235-253), Columbia University Press, ISBN 023-1079-67-2, New York

Horwich, R.H. & Lyon, J. (1998). Community development as a conservation strategy: the Community Baboon Sanctuary and Gales Point Manatee projects compared, In: *Timber, Tourists, and Temples: Conservation and Development in the Maya Forests of Belize, Guatemala and Mexico*, R.B. Primack, D.H. Bray, A. Galetti, A. & I. Ponciana, (Eds.), pp. (343-364), Island Press, ISBN 1-55963-541-X, Washington DC

Horwich, R.H. & Lyon, J. (1999). Rural ecotourism as a conservation tool. In: *Development of Tourism in Critical Environment*, T.V. Singh & S. Singh (Eds.), pp. (102-119), Cognizant Communication Corporation, ISBN 1-882345-19-3, New York

Horwich, R.H. & Lyon, J. (2007). Community conservation: practitioners' answer to critics. *Oryx*, Vol.41, No.3, pp. 376-385, ISSN 0030-6053

Horwich, R.H. (1990b). *A Biosphere Reserve for Toledo District*. Unpublished manuscript.

Horwich, R.H. (1998). Effective solutions for howler conservation, *International Journal of Primatology*, Vol.19, pp. 579-598, ISSN 0164-0291

Horwich, R.H. (2005). Communities saving Wisconsin birds: north and south. *Passenger Pigeon*, Vol.67, pp. 85-98, ISSN 0031-2703

Horwich, R.H., Lyon, J. & Bernstein, S.E. (2004). *An evaluation tool for internal and external assessments of community-based conservation projects*, Unpublished manuscript

Horwich, R.H.; Islari, R., Bose, A., Dey, B., Moshahary, M., Dey, N.K., Das, R. & Lyon, J. (2010). Community Protection of the the Manas Biosphere Reserve in Assam, India and the endangered golden Langur (*Trachypithecus geei*). *Oryx*, Vol.44, No. 2, pp. 252-260, ISSN 0030-6053

Horwich, R.H.; Lyon, J., & Bose, A. (2011). What Belize can teach us about grassroots conservation. *Solutions*, May-June, pp. 51-58, ISSN 2154-0896

Horwich, R.H.; Murray, D., Saqui, E., Lyon, J. & Godfrey, G. (1993). Ecotourism and community development : A view from Belize. In: *Ecotourism : A Guide For Planners and Managers*, K. Lindberg. & D. E. Hawkins,(Eds.), pp.(152-168), The Ecotourism Society, ISBN 096-3633-10-4, North Bennington, Vermont

Hutton, J.; Adams, W.M., & Murombedzi, J.C. (2005). Back to the barriers? Changing narratives in biodiversity conservation. *Forum for Developmental Studies NUPI*, No.2, pp. 341-370

Inamdar, A.; de Jode, H., Lindsay, K. & Cobb, S. (1999). Capitalizing on nature: protected area management. *Science*, Vol.283, pp. 1856-1857, ISSN 0036-8075

IUCN, (2003). *Policy Matters*, Vol.12

Johnson, M. (1992). *Lore: Capturing Traditional Environmental Knowledge*. International Development Research Centre, ISBN 088-9366-44-6, Ottawa, Ontario

Jones, B. (2001). The evolution of community-based approach to wildlife management at Kunene, Namibia. In: *African Wildlife and Livelihoods: The Promise and Performance of Community Conservation*, D. Hulme & M. Murphree (Eds.), pp. (38-58), James Currey, Ltd., ISBN 032-5070-60-1, Oxford

Kiss, A. (2004). Making biodiversity conservation a land-use priority. In: *Getting Biodiversity Projects to Work*, T.O. McShane & M.P.Wells (Eds.), pp. (98-123). Columbia University Press, ISBN 023-1127-64-2, New York

Kramer, R.; van Schaik, C., & Johnson, J. (Eds.) (1997). *Last Stand: Protected Areas and the Defense of Tropical Biodiversity*, Oxford University Press, ISBN 0-19-509554-5, New York

Lyon, J., & Horwich, R.H. (1996). Modification of tropical forest patches for wildlife protection and community conservation in Belize, In: *Forest Patches in Tropical landscapes*, J. Schelhas & R. Greenberg (Eds.), pp. (205-230), Island Press, ISBN 155-9634-25-1, Washington, D.C.

Mayer, E., Brown, S. (2007). *Community Resource Management at the Arnavons*, Accessed November 25, 2007, Available from:
<http://www.worldwildlife.org/bsp/bcn/learning/commsrcmgt/comms>

McGill, J.N.A. (1994). *Special Development Areas*. Consultancy Report No. 13, The Forest Planning and Management Project, Ministry of Natural Resources, Belmopan, Belize.

McShane, T.O. & Newby, S.A. (2004). Expecting the unattainable: the assumptions behind ICDPs, In: *Getting Biodiversity Projects to Work*, T.O. McShane & M.P. Wells (Eds.), pp. (49-74), Columbia University Press, ISBN 023-1127-64-2, New York

McShane, T.O., Wells, M.P. (Eds.). (2004). *Getting Biodiversity Projects to Work*, Columbia University Press, ISBN 023-1127-64-2, New York

Meerman, J.C. (2005). *Belize Protected Areas Policy and System Plan: Result 2: Protected Area System Assessment & Analysis PUBLIC DRAFT*. Unpublished Report to the Protected Areas Systems Plan Office.

Milner-Gulland, E.J.; Fisher, M., Browne, S., Redford, K.H., Spencer, M. & Sutherland, W.J. (2010). Do we need to develop a more relevant conservation literature? *Oryx*, Vol.44, pp.1-2, ISSN 0030-6053

Murphree, M.W. (2005). Congruent objectives, competing interests, and strategic compromise: concept and process in the evolution of Zimbabwe's CAMPFIRE, 1984-1996, In: *Communities and Conservation: History and Politics of Community-Based*

Management, J.P. Brosius, A.L. Tsing & C. Zerner, (Eds.), pp. (105-147), Altamira Press, ISBN 075-9105-05-7, New York

NACSO. (2009). *Namibia's communal conservancies: a review of progress 2008*, NACSO, Windhoek.

Nicholas, A.; Warren, Y., Bila, S., Ekinde, A., Ikfuingei, R. & Tampie, R. (2010). Successes in community-based monitoring of Cross River gorillas (*Gorilla gorilla diehli*) in Cameroon. *African Primates*, Vol.7, No.1, pp. 55-60, ISSN 1093-8966

Oates, J.F. (1999). *Myth and Reality in the Rain Forest: How Conservation Strategies Are Failing in West Africa*. University of California Press, ISBN 052-0217-82-9, Berkeley

PACT. (1998). *Creating a Co-managed Protected Areas System in Belize: A Plan for Joint Stewardship Between Government and Community*. Unpublished Manuscript

Pathak, N.; Bhatt, S., Tasneem, B., Kothari, A., & Borrini-Feyerbend, G. (2004). *Community Conservation Areas. A Bold Frontier for Conservation*. TILCEPA, IUCN, CENESTA, CMWG, and WAMIP, ISBN Tehran.

Persha L.; Agrawal A., & Chhatre A. (2011). Social and ecological synergy: local rulemaking, forest livelihoods, and biodiversity conservation, *Science*, Vol.331, pp. 1606-1608, ISSN 0036-8075

Poffenberger, M., & McGean, B. (Eds.). (1996). *Village Voices, Forest Choices, Joint Forest Management in India*, Oxford University Press, ISBN 019-5636-83-X, Delhi.

Porter-Bolland, L.,: Ellis, A.E., Guariguata, M.R., Ruiz-Mallén, I., Negrete-Yankelevich, S., & Reyes-García, V. 2011 (2011 In Press). Community managed forests and forest protected areas: an assessment of their conservation effectiveness across the tropics. *Forest Ecology and Management*. ISSN 0378-1127

Producciones de la Hamaca & Community Conservation Consultants. (1998). *Sarstoon-Temash National Park, Transcript of Stakeholders', Workshop*. Producciones de la Hamaca, Cay Caulker, Belize and Orang-utan Press, Gays Mills, Wisconsin

Rhoades, R.E. & Stalling, J. (2001). *Integrated Conservation and Development in Tropical America*, SANREP CRSP and CARE—SUBIR, ISBN 1-59111-009-2, Quito, Ecuador.

Robinson, J.G. & Redford, K.H. (2005). Jack of all trades, master of none: inherent contradictions among ICD approaches, In: *Getting Biodiversity Projects to Work*, T.O. McShane & M.P. Wells (Eds.), pp. (10-340, Columbia University Press, ISBN 023-1127-64-2, New York.

Robinson, J.G. (1993). The limits to caring: sustainable living and the loss of biodiversity. *Conservation Biology*, Vol.7, pp. 20-28, ISSN 0888-8892

Sayer, J. & Wells, M.P. (2004). The pathology of projects, In: *Getting Biodiversity Projects to Work*, T.O. McShane & M.P. Wells (Eds.), pp. (35-48), Columbia University Press, ISBN 023-1127-64-2, New York

Schipper, J. ; et al. (2008). The status of the world's land and marine mammals: diversity, threat, and knowledge. *Science*, Vol.322, pp. 225-230, ISSN 0036-8078

Schmitt, C.B.; et al. 2009 Global Analysis of the protection status of the world's forests. *Biological Conservation*, Vol.142, pp.2122-2130, ISSN 0006-3207

Shanee, N.; Shanee, S. & Maldonado, A.M. (2007). Conservation Assessment and Planning for the Yellow Tailed Woolly Monkey (*Oreonax flavicauda*) in Peru. *Wildlife Biology in Practice*, Vol.3, No.2, pp. 73-82, ISSN 1646-1509

Shepard, G. (2004). Poverty and forests: sustaining livelihoods in integrated conservation and development, In: *Getting Biodiversity Projects to Work*, T.O. McShane & M.P. Wells (Eds.), pp. (340-371), Columbia University Press, ISBN 0888-8892, New York.

Singleton, S. (1998). *Constructing Cooperation The Evolution of Institutions of Co-management.*,The University of Michigan Press, Ann Arbor

Soares-Filho, B. et al. (2010). Role of Brazilian Amazon protected areas in climate change mitigation. *PNAS*, Vol. 107, No. 24, pp.10821-10826

Srivastava, A.; Biswas, J., Das, J., & Bezbarua, P. (2001). Status and distribution of golden langurs (*Trachypithecus geei*) in Assam, India. *American Journal of Primatology*, Vol.55, pp.15-23., ISSN 0275-2565

SSC (Species Conservation Planning Task Force). (2008). *Strategic Planning for Species Conservation: A Handbook*, IUCN, Gland, Switzerland.

Stevens, S. (ed.), (1997). *Conservation Through Cultural Survival: Indigenous Peoples and Protected Areas*, Island Press, ISBN 155-9634-49-9, Washington, DC

Stolton, S. & Dudley, N. (1999). *Partnerships for Protection*, Earthscan Publications, Ltd., ISNBN 185-3836-14-1, London

Terborgh, J. (1999). *Requiem for Nature*. Island Press, ISBN 155-9635-87-8, Washington, DC.

Turner, W.R.; Brandon, K., Brooks, T.M., Costanza, R., da Fonseca, G.A.B., & Portella, R. (2007). Global conservation of biodiversity and ecosystem services. *Bioscience*, Vol.57, pp. 868-873, ISSN 0006-3568

Uphoff, N. & Langholz, J. (1998). Incentives for avoiding the Tragedy of the Commons? *Environmental Conservation*, Vol.25, pp. 251-261, ISSN 0376-8929

USAID. (2005). *USAID/Namibia Annual Report. FY2005. June 16, 2005*. Accessed on December 12, 2010. Available from: http://www.dec.org,

Vincent, K. (1994). *Report of the Community-Based Ecotourism Gathering*. Ministry of Tourism and the Environment and Belize Enterprise for Sustained Technology, Belmopan.

Wells, M.P.; McShane, T.O., Dublin, H.T., O'Connor, S., & Redford, K.H. (2004). The future of integrated conservation and development projects: building on what worked, In: *Getting Biodiversity Projects to Work*, T.O. McShane & M.P. Wells (Eds.), pp. (398-421), Columbia University Press, ISBN 023-1127-64-2, New York.

Westley, F.; Zimmerman, B., & Patton, M.Q. (2006). *Getting to Maybe How the World is Changed*, Random House, ISBN 978-0-679-31443-1, Canada.

Wilson, K. (2002). The new microfinance, an essay on the Self-Help group movement in India. *Journal of Microfinance*, Vol.4, pp. 217-245

Young, C. & Horwich, R.H. (2007). History of protected area designation, co-management and community participation in Belize, In: *Taking Stock: Belize at 25 Years of Independence*, B.S.Balboni. & J.O. Palacio (Eds.), pp. (123-150), Cubola Books, ISBN 978-976-8161-18-5, Benque Viejo del Carmen, Belize.

15

Agroforestry Systems and Local Institutional Development for Preventing Deforestation in Chiapas, Mexico

Lorena Soto-Pinto[1],
Miguel A. Castillo-Santiago[2] and Guillermo Jiménez-Ferrer[3]
[1]*Institut de Cíencia i Tecnologia Ambientals, Universitat Autonoma Barcelona,*
[2]*El Colegio de la Frontera Sur (ECOSUR). Unidad San Cristóbal, Chiapas,*
[3]*Veterinary Faculty, Universitat Autonoma Barcelona,*
[1,3]*Spain*
[2]*Mexico*

1. Introduction

The transformation of natural forest to secondary forest and pastures has been the most common process of land use change in tropical countries in recent decades (FAO, 2010). The main causes of deforestation include institutional factors, markets, public policies and global forces, which often act synergistically (Deininger and Minten, 1999; Bocco et al. 2001; Lambin et al., 2001).

Mexico is a country with $64,802 \times 10^3$ ha of forested land, and it is one of the ten countries with the largest area of primary forest (3% of total). The annual net loss of deforestation in Mexico has been estimated to be 0.52% for the period of 1990-2010, but the net loss, on average, has decreased over the past few years (FAO, 2010). The highest deforestation rates are concentrated in the south and central regions of the country, as documented elsewhere: 8.4% in el Nevado de Toluca, state of Mexico (1972-2000) (Maass et al., 2006); 8% in Patzcuaro, Michoacan (1960-1990) (Klooster, 2000); 6.9% in some areas of Campeche (Reyes-Hernández et al., 2003); 6.1% in the highlands of the state of Chiapas (Cayuela et al., 2006; Echeverría et al., 2007); and 2-6.7% in Selva Lacandona, also in Chiapas (Ortiz-Espejel & Toledo, 1998; de Jong et. al., 2000). Precisely, the states of Chiapas and Yucatan have registered the highest rate of forest conversion to grasslands and slash-and-burn cultivation over the past two decades, and Chiapas alone has contributed towards 12% of national deforestation during the period 1993-2007 (De Jong et al, 2010; Díaz-Gallegos et al., 2010).

In Mexico, deforestation occurs because forests become converted to agriculture, livestock and urban areas. But also because logging activities fail to meet the requirements of forest management plans. All these processes result in the loss of forest goods and services (Lambin et al. 2003), and they contribute to ecosystem fragmentation (Ochoa-Gaona & González Espinosa, 2000; Cayuela et al., 2006), biological invasions (Hobbs, 2000), greenhouse gas emissions (Watson et al., 2000), biodiversity loss (Lugo et al., 1993), soil degradation (Lal, 2004) and water siltation (Sweeney et al., 2004).

Deforestation is one of the key contributors to greenhouse gas (GHG) concentrations in the atmosphere (IPCC, 2000; Canadell & Raupach, 2008). Between 2003 and 2008, GHG deforestation-related emissions in Chiapas were estimated to be 16,477 (±7,299) Gg CO_2/year, while 414 Gg CO_2-eq were attributed to forest fires during the same period. These emissions represent 23.5% of national land-use change related emissions over the same period (Gobierno del Estado de Chiapas, 2011). In Chiapas, deforestation processes have affected highland, cloud and tropical forests. These forests have decreased in favour of agriculture, pastures and secondary vegetation. The original areas of some of these forest types have been reduced by 50% (de Jong et al, 2010). In particular, tropical mountain cloud forests and mangroves have been threatened by commercial agriculture, and considerable endemic biodiversity has subsequently been lost (Hirales-Cota, 2010; Toledo-Aceves et al., 2011). The Selva Lacandona (Lacandon rainforest) is located in the southern region of the country next to Guatemala, and it has been severely deforested during the last 40 years. The rainforest area of *Marques de Comillas* lost 81,080 ha of tropical forests between 1986 and 2005, which represents 48% of the original forest cover (Castillo-Santiago 2009).

Deforestation and degradation have occurred despite the fact that a variety of policy tools and approaches have been developed by national and regional governments, as well as by civil organisations, in order to guarantee forest and biodiversity conservation, including people-oriented conservation areas; community-based sustainable resource use management approaches; technological innovations for improved forest and agricultural management practices; and payments for ecosystem services (PES) (Deininger & Minten, 1999; Corbera, 2005; Cayuela et al, 2006). It is recognised that managed forests and agroforestry systems can contribute to conserve soil, regulate water flows, support biodiversity and sequester significant amounts of carbon by including timber-focused trees for durable products. Some of these land management systems can maintain biomass for longer, restore site capacity and increase economic benefits compared to a business-as-usual scenario (Kotto-Same et al., 1997; De Jong et al., 2000; Albrecht & Kandji, 2003; Montagnini & Nair, 2004; Soto-Pinto et al., 2010). The evidence presented before, however, has demonstrated that most of these policy and project-based approaches have been far from successful, and have been unable to halt land-use change in the region as a whole. There are of course apparently successful experiences, which should help us to learn lessons and identify avenues for improving the design and effectiveness of existing policies and instruments. One of these, institutional development, aims to encourage and facilitate local inputs and experimentation, interaction and consultation of local actors among themselves and to interact with relevant external agents to increase opportunities for the poor; it has a crucial importance in territorial development (Evans, 2004; Schejtman and Berdegué, 2004).

One of the latter experiences concerns the project being analysed in this chapter, which has been promoting agroforestry, reforestation and conservation activities for offsetting GHG emissions since 1994. The Scolel Té project has allowed carbon offsets to be sold through the voluntary market and will provide non-timber and timber products in the short, medium and long term. In the following sections, we examine the involvement in the project of three municipalities located in the northern-eastern tropical area of Chiapas, Mexico (i.e., Marqués de Comillas, Chilón and Salto de Agua) (Figure 1), and we discuss the potential of alternatives of avoided deforestation and agroforestry systems as well as institutional arrangements resulting from project development in these municipalities. The following section of the chapter describes the process of deforestation in Marqués de Comillas, a representative area of the state's ongoing deforestation and degradation processes. The central part of the chapter highlights Scolel Te's conservation and forest management approach in our selected municipalities, and it discusses the drivers and constraints for the successful realisation of environmental, economic and social benefits in the three selected cases. The last part of the chapter discusses the overall results in the context of evolving carbon forestry markets and the through the enhancement, conservation and sustainable management of forest stocks.

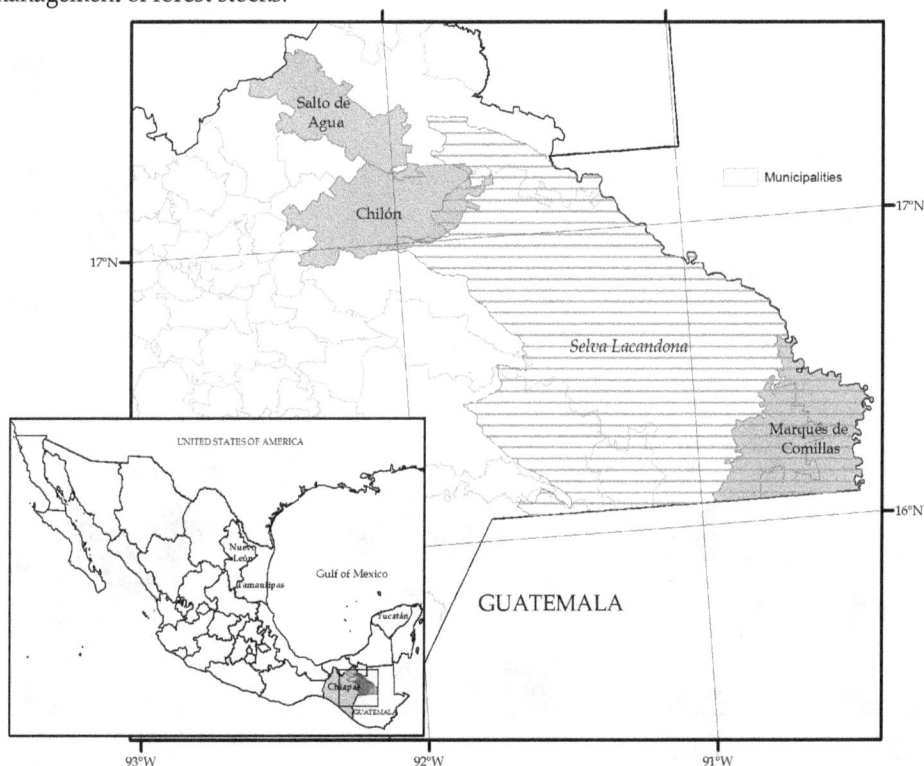

Fig. 1. Study area in Chiapas, Mexico.

2. The process of deforestation in Marqués de Comillas

Marqués de Comillas is one of the four municipalities of La Selva Lacandona and it comprises a total area of 203,200 ha. In the last few four decades, this municipality showed one of the highest deforestation rates in Mexico. Between 1986 and 2005 it lost approximately 81,080 ha of tropical rainforest, which represents 48% of its original forest cover (Figure 2). This loss contributed to 1.5% of the total CO_2 emissions from land use change in Mexico (Castillo-Santiago 2009).

Settlements in La Selva Lacandona were promoted during 1970-1980, and they have been recognised as the main driver of deforestation in Marqués de Comillas. Population was relocated to resolve conflicts of land scarcity and social rebellion in other Mexican states. Consequently, land in Marques de Comillas, which was completely forested at the beginning of the 1970s, was distributed in 37 ejidos[1] (Mariaca, 2002). The colonisation process was complemented by the construction of a main road, which allowed people and goods to flow between rural areas and the main cities of Tabasco and Chiapas (Harvey, 1998). Additionally, public policy favoured land use change from forest to maize agriculture, thereby financing deforestation in the 1970s. The process began with forest logging, followed by the use of fire for the cultivation of maize during three or four years. Once fertility was decreased and the land was weeded, due to intense cultivation, the natural steps were to intensify land use, fallow the land or let the grasses grow. Importantly, forested land was considered "idle land", and the process of deforestation was actually funded by government programs that encouraged cattle ranching. In 1978, the protected area of Montes Azules was established adjacent to Marques de Comillas, covering and area of 331,200 ha.

Subsequently, from 1992 to 1998, in Selva Lacandona six protected areas were established, five of them managed by the State, and other communal, with a total area of 123.660 ha: Chankin Protected Area (12.184 ha), Bonampak Natural Monument (4.357 ha), Biosphere Reserve Lacan-Tun (61.873 ha), areas of wildlife protection Nahá (3.847 ha), Metzabok (3.368 ha), Natural Monument Yaxchilan (2.621 ha), and by agreement of the Lacandon Community, the Communal Reserve of Sierra Cojolita with 35.410 ha (INE 2000).

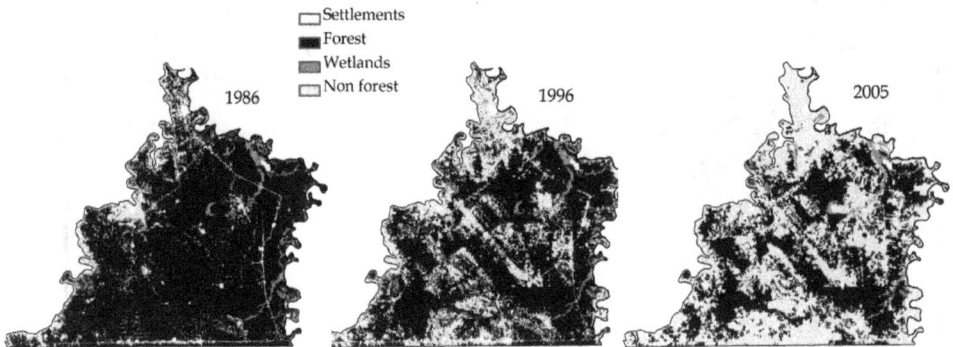

Fig. 2. Land-use changes from 1986 to 2005 in Marques de Comillas Chiapas, México.

In Marqués de Comillas, crops, such as cardamom, cocoa, rice, vanilla and rubber, were promoted by public and private investments to diversify the agricultural system and to employ local people who mainly participated as labourers in the 1980s and 1990s. Most of these projects were unsuccessful, due to their continuous requirements of external inputs, particularly technical support and capital. Forest initiatives were discouraging as well. In 1996, the Pilot Forest Plan was launched to promote timber extraction and establish the basis for a rational use of community forests. However, only few ejidos have maintained the deforestation rate under the regional average. More recently, other public initiatives have favoured the growth of oil palm for biofuels, which are rapidly growing in this

[1]Ejido is a particular form of land tenure in which the State has given the land to a group of people called "ejidatarios". The general use of forest and water is regulated by federal laws; land sale and use are decisions of the owner, this last is locally monitored and regulated by The Public Assembly of the community.

municipality, even at the expense of forests. Livestock has also gained ground. Livestock has very often represented the best economic option from the farmers' point of view because it has a relatively stable national market and a low level of investment is required. Livestock is thus still the dominant activity in the area and it occupies most of the landscape in Marques de Comillas at present (Table 1).

Type of vegetation	Area (ha)	Percentage
Tropical rainforest	69,360	34
Riparian forest	3,054	1
Wetlands	3,387	2
Secondary forest	16,544	8
Secondary vegetation (shrubs and herbaceous lands)	25,904	13
Pastures for livestock	56,339	28
Rainfed agriculture	24,324	12
Human settlements	2,404	1
Rivers	3,124	2
Total	204,440	100

Source: Castillo-Santiago, 2009

Table 1. Land use in Marqués de Comillas (Chiapas, México).

Livestock in Selva Lacandona has normally been characterised by its extensiveness, and it has been devoted mainly to grow calves that are fed with naturalised grasses of low nutritional value. Moreover, it has been characterised as having a low management level, high number of animals per land unity, low capital investments, poor infrastructure, scarce technical assistance and scarce financial support (Jiménez et al, 2008; Martínez & Ruiz de Oña, 2010). Table 2 shows the main features of the livestock system in communities of Selva Lacandona. Most of the farmers (95%) have designed their systems to produce calves to sell them to intermediaries in the local and regional markets. The trees located in the pastures are mainly tropical forest remnant trees tolerated for shading cattle, including *Blepharidium guatemalensis, Sabal mauritiformis, Vatairea lundellii, Guarea glabra, Albizia adinocephala, Bursera simaruba, Spondias mombin,* and *Swietenia macrophylla.* Specific timber species, such as "popiste" (*B. mexicanum*) and guanacastle (*Enterolobium cyclocarpum*), are sometimes favoured with the purpose of being used as a lumber source in rural construction (local market or self-supply).

Recently, several studies have highlighted the importance of silvopastoral systems and other agroforestry systems for conserving biodiversity and connecting countryside landscape with reserves (Harvey et al, 2006; Rice & Greenberg, 2004). Several institutions, including the Commission for Natural Protected Areas (CONANP), the Mesoamerican Biological Corridor (CBM) and the National Commission of Knowledge and Use of Diversity (CONABIO) have launched programs in Selva Lacandona, specifically in Marques de Comillas, to improve rural production and promote conservation. One of these initiatives is the Scolel Te' project, which began in 1994 as a pilot experience in Chiapas and it is managed by the local organisation AMBIO cooperative. It originally involved a few dozens of farmers in the central highlands of the state and it has now grown to encompass more than 700 participants and their families (3500 beneficiaries) in approximately 50 communities of Chiapas and the neighbouring states of Oaxaca and Tabasco. The project is built on a

Characteristics of livestock production units	La Siria, Ocosingo	Ach lum Monte Libano, Ocosingo	Amatitlan, Maravilla Tenejapa	La Corona, Marques de Comillas
Land use type	Ejido	Ejido	Ejido	Ejido
Ethnic group	Tseltal	Tseltal	Chol and Mestizo	Mestizo
Altitude (m a.s.l.)	150-200	300 - 500	275 - 590	75 – 125
Land use	Maize agriculture, Fruits, Livestock Forestry	Maize agriculture Livestock	Maize agriculture Livestock Forestry	Maize agriculture Livestock Forestry PES-Carbon
Average land area (ha/family)	15	20	10	45
Pasture area for cattle grazing (ha)	10	15	5	25
Stocking rate AU/ha	1.9	2.1	1.5	2.7
Management system	Livestock with improved pastures (*Brachiaria brizantha, B humidicola*). Rotations without technical assistance No supplement.	Livestock with native grasses and "estrella" grass (*Cynodon niemufensis*). Without technical assistance and financial support. Breeding and sales of young calves	Livestock with forest fallow grazing, and crop residues (maize stubbles)	Livestock on pastures with dispersed trees, live fences, and pastures with forest patches. Improved pastures (*B. decumbens, B. humidicola, Andropogon gayanus*). Growing and sale of calves recently wean
Product destination (mainly meat)	Local market consumption	Local market consumption	Local market consumption	Local and regional market consumption
Forage trees on pasture grazing areas	*G. sepium, Parmentiera aculeata, Brosimum allicastrum, Guazuma ulmifolia, Leucaena leuucocephala*	*Whiteringia meiantha, Thitonia diversifolia, G. sepium, G. ulmifolia, Eupatorium morifolium*	*G. ulmifolia, Diphysa americana, Spondias mombin, Bahuinia herrerae*	*G. sepium, Cecropia obtusifolia, Erythrina sp, L. leucocephala, P. aculeata*

Table 2. Socio-technical characteristics of livestock in four communities in Selva Lacandona (Chiapas, México).

participatory method that identifies rural development and forest management opportunities and constraints of each involved farmer and community. Subsequently, it helps farmers and communities to select and establish trees on individual or collective lands according to their preferences, providing technical support and paying participants for the provision of carbon offsets to national and mostly international buyers (Corbera 2010; Ruiz de Oña, 2011). Over time, the project has become a global landmark for the development of community-based payments for ecosystem services, and it has created its own design and implementation standard (www.planvivo.org). This approach has been extended to other similar projects in Uganda, Mozambique, Malawi and Cameroon

3. Avoiding deforestation in La Corona, Marqués de Comillas

In Ejido La Corona, a previous study estimated that 305 ha of forested land would have to be lost for agriculture purposes, as a baseline scenario during the period 2004-2009 (Quechulpa, et al., 2010). With the support of the AMBIO cooperative and financial support from Pro-Árbol (i.e., a governmental PES program developed by Mexico's National Forest Commission (CONAFOR) a participatory planning method was developed to design interventions for avoiding deforestation and increasing tree cover in deforested land. Concurrent financial resources were allocated from several programs, including Scolel Te, the Mexican Fund for Nature Conservation (FMCN) and Mesoamerican Biological Corridor (CBM). Financial resources were aimed to reduce pressure on forest land to avoid land-use change and to intensify cattle raising activities.

The following two approaches were suggested by community members and selected for project development: 1) management of secondary vegetation and forest conservation and 2) establishment of agrosilvopastoral and agroforestry systems in open and grazing areas. As a result, secondary vegetation was managed by pruning and thinning trees to eliminate competition and favour growth of the most commercially valuable species. Activities for forest conservation included the opening and maintenance of 22 km of fire protection rifts, acquiring equipment for fires, combating brigades trained for fire control, supervising and regulating agricultural burns. Communal forest conservation incorporated a wider vision of the territory and collective agreements related to resource access into the working plan, thereby establishing rules and monitoring to regulate land use change according to the plan. The establishment of 45.5 km of live fences was carried out in accordance to the work plan (Figure 3) using "cocoite" forage trees (*G. sepium*) and other timber species, such as *Tabebuia rosea*, *T. guayacan*, *C. odorata*, *S. macrophylla* and *Pachira aquatica*. All forage trees were native species produced in communal nurseries. Other activities included the use of forage grasses such as *Brachiaria*, the promotion of cattle-feed supplements with multi-nutritional blocks, improvement of cattle breeds, establishment of livestock infrastructure, establishment of technical training and creation of a farmer-to-farmer exchange system (Ambio Cooperative, 2010). In parallel, other projects particularly targeted towards women, such as the substitution of traditional open stoves by fuel wood-saving stoves, improvement of house conditions and improvement of other communal infrastructure, were also launched.

In the first five years of the project, all of these activities resulted in 179 ha of prevented deforestation out of the 305 ha previously estimated to be lost, which translated in reduced CO_2 emissions from land-use change. According to inventories from permanent plots, 407 ha of passively restored secondary forests accumulated biomass with a rate of 4.4 ton ha^{-1} year^{-1}

(approximately 3000 ton of CO_2 per year), and it was estimated that the establishment of live fences fixed approximately 3200 tonnes of CO_2. Other intangible benefits included the organisation of farmers for fire prevention; the organisation and training of brigades for carrying out forest inventories; and the improvement of productive systems. Above all, the recognition that forests may contribute to environmental and socioeconomic direct benefits was one of the main gains resulting from the project.

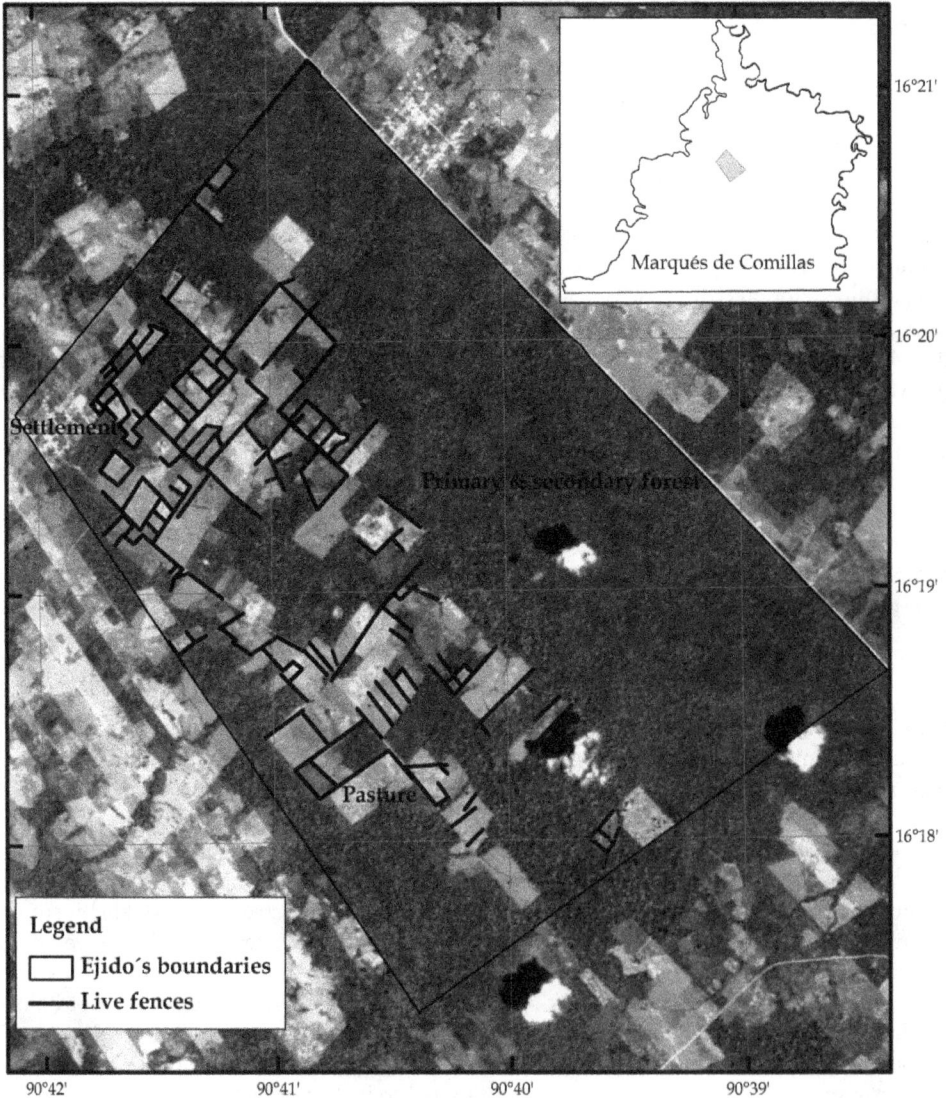

Fig. 3. Participatory planning for designing forestry and agroforestry interventions in La Corona and Marques de Comillas in Chiapas, Mexico.

As a result of this management, Ejido La Corona, with a total area of 2254 ha, conserves 68% of the total rain forest (1530 ha), which includes primary and old secondary tropical rain forests. Nonetheless, livestock (528 ha), agriculture (177 ha) and urbanisation (20 ha) coexist with forests and the biological reserve, and these combined areas support 292 people (Quechulpa 2010).

Subsequent studies have evaluated the performance of the systems established by producers in La Corona. An example is shown in Table 3, where the carbon components of pastures with and without trees are presented.

Carbon components	Pasture in monoculture	Pasture with live fences	Pasture with dispersed trees
Trees (Mg C ha⁻¹)	0.00	7.6	4.23
Herbs (including grasses Mg C ha⁻¹)	1.33	0.91	0.64
Total roots (fine and coarse Mg C ha⁻¹)	0.66	1.88	1.12
Live biomass (Mg C ha⁻¹)	1.99	10.4	5.99
Soil organic matter 0-40cm (Mg C ha⁻¹)	60.62	66.68	76.89
Total carbon (Mg C ha⁻¹)	64.62	87.5	88.89

Source: modified from Aguilar-Argüello, 2007.

Table 3. Carbon stocks in different components of monoculture pastures, live fences and dispersed trees in pastures in La Corona and Reforma Agraria in Chiapas (Mg C ha⁻¹).

4. Agroforestry in Chilón and Salto de Agua, Chiapas

Study communities of these municipalities are located in Chiapas tropical zone. Salto de Agua is located approximately 200 m a.s.l., and it has a warm and humid climate. In addition, Salto de Agua has a tropical rainforest. Meanwhile, Chilón is located in the intermediate tropical zone between 700 and 900 m a.s.l. It has a warm climate and abundant summer rains. Chilón has also a tropical rainforest, and the main soil types in this area are Regosols, Leptosols, and Cambisols (INEGI 1984; Soto-Pinto et al., 2010). Land is devoted mainly to agriculture, which is based on maize cultivation in both municipalities. While in Salto de Agua, maize cultivation is the key commercial and subsistence crop, Chilón farmers cultivate coffee as the main source of income. In both "ejidos", farmers organise around small activity groups for establishing agroforestry systems. In Chilón, farmers established maize associated to trees (Taungya rotational systems), improved fallows and shaded coffee systems; while farmers in Salto de Agua established taungya systems converted finally to silvopastoral systems. Improved fallow consists of enriching fallow lands with secondary vegetation with timber trees, so far as the latter are planted during the first five years of the fallow period. Taungya, in turn, consists of enriching maize cultivated plots with timber trees in a rotational pattern; and coffee systems were enriched with timber trees in association to other variety of previously existing native trees as shading cover. The project Scolel Té has contributed to organizing agroforestry practices, training and monitoring.

Previous evaluations by El Colegio de la Frontera Sur (ECOSUR) have shown that agroforestry systems provide multiple environmental services and can increase productivity, land and labour worth in relation to conventional land uses, such as extensive cattle farming and maize crops without trees (Soto-Pinto et al, 2010; Soto-Pinto, submitted). Coffee cultivation under the shade of trees conserves at least 40% of the total woody plant diversity in scarce neighbouring forests (Soto-Pinto et al, 2000; Romero-Alvarado et al, 2002; Peeters et al, 2003). Organic coffee cultivation translates into a higher carbon content in the upper soil layer (0-30 cm) than non-organic coffee systems, being able to store between 129.8 and 215.6 Mg C ha^{-1} in their components, including soil C (Soto-Pinto et al., 2010). However, the amount of aboveground biomass depends on the structure and composition of shade vegetation (Table 4). Economic analyses comparing conventional coffee with and without enrichment of timber trees and PES and organic coffee with timber trees and PES have shown positive benefit/cost ratios, as follows: 0.8, 1.2 and 1.8 for conventional management, conventional management enriched with timber trees, and organic coffee plus carbon sequestration and timber, respectively (Table 5).

Carbon components	Natural traditional polyculture coffee	Natural traditional polyculture coffee enriched with timber trees	Organic traditional polyculture coffee enriched with timber trees
Adult trees (≥10cm)	17.02±3.32	27.3±4.79	37.89±5.17
Tree saplings (<10cm)	0.14±0.05	0.36±0.14	0.85±0.26
Coffee shrubs (Mg C ha^{-1})	11.37±1.56	11.03±2.25	8.83±1.30
Herbs (Mg C ha^{-1})	0.44±0.09	0.122±0.03	0.24±0.07
Fallen tree branches (Mg C ha^{-1})	6.16±0.58	8.03±1.1	9.67±1.08
Live biomass (Mg C ha^{-1})	35.13±3.57	46.84±7.1	57.47±7.13
Roots (Mg C ha^{-1})	1.47±0.79	0.67±0.13	0.33±0.07
Litter (Mg C ha^{-1})	5.24±0.88	5.6±1.09	5.72±0.91
Dead organic matter (Mg C ha^{-1})	6.71±1.29	6.27±1.09	6.0±0.85
Soil organic matter (Mg C ha^{-1})	87.96±12.55	117.35±	152.12±10.26
Total Carbon (Mg C ha^{-1})	129.8±15.69	170.46±	215.64±12.16

Source: modified from Aguirre, 2006

Table 4. Carbon components in three types of coffee systems in Northern Chiapas (Mexico).

Indicator	Natural polyculture coffee	Natural polyculture coffee enriched with timber trees	Organic polyculture coffee enriched with timber trees
Current costs (USD)	105.87	129.83	204.42
Current benefits (USD)	87.33	263.1	384.34
Net present value (USD)	- 18.55	133.25	179.94
Annual net present value	- 927	6.66	8.99
Cost-benefit	0.8	2.0	1.9
Internal rate of return (%)		15.2	19.4

Source: modified from Aguirre, 2006

Table 5. Economic evaluation for three scenarios of coffee farms: natural polyculture coffee, natural polyculture enriched with timber trees and organic polyculture enriched with timber trees in Northern Chiapas, Mexico.

Improved fallows have demonstrated to be an adequate alternative to slash-and-burn agriculture, due to their capacity to increase biomass, productivity, economic value, complexity, carbon and diversity (Roncal 2007; Soto-Pinto et al., submitted). Tables 6 and 7 show structural and functional variables of these systems in the region. Results of these evaluations show that these systems may contribute to sedentarise the maize system, increase economic value and avoid deforestation while diversifying non-timber and timber related products.

Both systems have shown their multifunctionality for improving productivity and restoring site features, environmental conditions and livelihood conditions (Soto-Pinto et al., Submitted).

Land use systems	Adult tree density (trees ha-1)	Tree sapling density (100m2)	Tree Height (m)	Tree Diameter (cm)
Taungya system 9-13years	520±218	31.9±19.0	8.8±1.4	16.2±3.9
Improved Fallow at 9th year	623.3±106.3	4800±2361.0	8.26±1.3	16.88±3.16
Traditional Fallow >30 years	463.8±191.8	3616.7±2700	7.4±1.3	17.7±3.47
Inga-shaded organic coffee >10 years	79.3±79.3	25.1±7.5	5.6±1.0	19.2±13.0
Polyculture-shaded organic coffee >10 years	115.0±115.0	22.8±6.9	7.3±4.1	15.0±9.9
Polyculture-shaded non organic coffee >10 years	206.3±180.0	21.5±7.5	6.5±3.5	15.4±10.2
Pasture with dispersed trees >10 years	20.0±10.0	112.0±16.0	6.1±3.4	14.8±4.5
Pasture with live fences >10 years	56.0±37.3	116.0±68.7	4.5±1.5	10.3±4.5
Pasture in monoculture >10 years	0	0	0	0
Continuous maize 4-7 years	210.0±217.0	66.6±57.7	2.1±0.42	7.17±2.2

Source: Aguilar-Argüello 2007; Roncal –García et al., 2007; Aguirre 2006; Soto-Pinto et al. 2010; Soto-Pinto et al. submitted; and other original data

Table 6. Structural variables of agroforestry systems in Selva Lacandona (Mexico).

In Salto de Agua and Chilón the decisions on land use are taken individually, often under the consensus of the family and the work group. The impact seems to be centered at the plot level. However the impact on the territory has not been evaluated.

In each community, regional and local technicians act as training guides. Local decisions are taken individually or by group. Regional and state-level assemblies of technicians discuss, analyse problems, and propose solutions, new projects and financial supports to resolve specific problems. Civil and academic organizations and government dependencies accompany the process offering punctual technical assistance to resolve specific questions and financial support. Academy plays an important role in knowledge management, offering training and developing human resources. All of the sectors are involved in thematic networks contributing to the building of a public policy in the thematic of forestry programs, PES and territorial developing, among other issues.

Land use systems	Aboveground tree biomass (Mg ha-1)	Soil Carbon 0-30cm in depth (Mg C ha-1)	Tree Species Richness (number of species)
Taungya system 9-13years	44.4±25.7	104.7±30.1	3.4±2.3 (500m²)
Improved Fallow at 9th year	164.3±65.4	88.85±4.7	15.8±3.9 (1000m²)
Traditional Fallow >30 years	109.35±66.7	120.4±7.0	19.7±3.8 (1000m²)
Inga-shaded organic coffee >10 years	34.1±17.6	75.83±28.6	5.6±2.9 (1000m2)
Polyculture-shaded organic coffee >10 years	75.8±27.4	131.13±23.4	15.7±3.3 (1000m2)
Polyculture-shaded non organic coffee >10 years	54.6±25.9	101.13±27.9	8.0±3.3 (1000m2)
Pasture with dispersed trees >10 years	8.5±6.0	46.4±13.0	2.6±2.6
Pasture with live fences >10 years	15.2±10.7	40.5±9.8	1±0
Pasture in monoculture >10 years	0	50.2±14.6	0
Continuous maize 4-7 years	4.38±3.62	115.0±12.3	14.0±4.8 (1000m2)

Source: Aguilar-Argüello 2007; Roncal –García et al., 2007; Aguirre 2006; Soto-Pinto et al. 2010; Soto-Pinto et al. submitted; and other original data

Table 7. Functional variables of agroforestry systems in Selva Lacandona (Chiapas).

5. Discussion

In some places traditional communities have managed their resources sustainably for long time, even better than in many protected areas managed by the State, especially in Latin America (Bray et al., 2008; Porter-Bolland, et al., In Press). However, in the last years the effects of public policy, colonization process and urban development, among other land use change drivers led to high deforestation rates in sites which until four decades before were completely forested, this is the case of Selva Lacandona in Mexico where Marques de Comillas is a referent.

Enriched shaded coffee, alternative rotational and silvopastoral systems have demonstrated benefits in the topics of food production, biodiversity and economy of livelihoods. Results demonstrate the value of agroforestry systems as a potential strategy for tree cover recovery and carbon sequestration (Haile et al., 2008; Nair et al., 2010; Soto-Pinto et al., 2010) and suggest that adequate planning, incentives and capacity-building efforts can lead to better conservation practices (Berkes, 2007). In Ejidos La Corona and communities in Chilón and Salto de Agua people have improved their local organisational activities, either in groups or collectively through preventing deforestation; intensifying the agriculture process; reforesting the deforested and open areas; controlling fire; acquiring new abilities; creating norms, sanctions, work plans, and social rearrangements; and reinforcing old capacities for developing a forest culture as a part of a new institutional development and good governance (Evans 2004; Corbera, 2005, 2010). All this, coupled with the accompaniment of the civil society and academic institutions, and the involvement of government in a network of ecosystem services has been key in order to begin a governance development (Ruiz de Oña et al., 2011). However, the scaling up of this process is a challenge since it represents a greater organizational complexity and negotiation, a matter of governance (Swiderska et al., 2008).

In Marques de Comillas collective organization and decisions can impact broad and decisively on the territory in the short term, while in Chilón and Salto de Agua decisions taken by individuals or groups may be slower than collective decisions. Future studies on social resilience on both types of patterns should be of great importance. The lesson gained in this experience is that public programs must consider community priorities based on an integral regional vision, environmental education, strengthening of local capabilities and organisation. As a whole, public priorities may be a better appropriation and adaptation of programs in a new model of territorial development by considering institutional arrangements (Merino and Warnholtz, 2005). Farmers are the leading experts regarding their context and livelihood conditions and may or may not adopt and adapt programs that involve a territorial vision. The intense participation of communities may allow better monitoring, especially when these programs are integrated into their own community-management plans (Franzel and Scherr, 2002).

Territorial participatory planning of agroforestry and restoration systems, in addition to the institutional development have demonstrated their capacity to benefit environment, farmer's livelihoods and social capital, through their contribution to increase productivity, complexity, diversity, economic value, organization capacities and knowledge, which as a whole may contribute to avoid deforestation (Alburquerque, 2002; Bray, 2008; Swiderska et al., 2008). However, other products, service, process and management innovations are required to have a broader menu of options for farming systems and livelihoods to ensure the permanence of forest systems in a competitive context. Some of the needs to be resolved in order to conceive this process as a territorial development are: the best practices to achieve food self-sufficiency, development of market chains, education, training, infrastructure, financial support, policy arrangements and skills of competitiveness for production systems (Alburquerque, 2002; Ruiz de Oña et al., 2011). An integration of an agroecological matrix which enables agricultural production and natural resource conservation and a new political vision may help facilitate social and environmental synergies in rural areas (Perfecto and Vandermeer, 2010)

Although the experience run by the project Scolel te' in Marqués de Comillas is not precisely REDD+, may offer a set of lessons learned about the conditions required for a good management for reducing deforestation. Some elements which need to be considered in order to contribute to the design of PES programs are the following: 1) technology adaptation, 2) ordering of land use with a territorial development vision, 3) broad participation of all stakeholders, 4) institutional development at local, regional levels and, 5) involvement in networks to launch processes of governance for the management of an adequate public policy of PES. Although these systems may contribute to avoiding deforestation, their potential to become part of REDD programs need to be better discussed since the relationship among markets and actors' rights and duties, in addition to the uncertainty regarding the factors influencing effectiveness on deforestation is unclear (Coad et al., 2008).

In Chiapas, a process of institutional development around the issue of PES is emerging. In the last years groups working on ecosystem services (State Group of Ecosystem Services, GESE by its acronyms in Spanish) and deforestation and degradation (REDD) have matured in the state of Chiapas; representatives of the "Camara de Diputados y Senadores del Estado de Chiapas" (State House of Representatives and Senate) have approved the Law for Mitigation and Adaptation to Climate Change in the State; meanwhile, a group of research institutions and organizations of the civil society carry out an effort of monitoring, reporting and verification

(MRV) in rural communities as a basis for a REDD planning in Chiapas. Moreover, the Government of the States of Chiapas and Campeche has recently signed an agreement with Acre Brazil and the Government of California through The Governors' Climate and Forests Task Force (GCF) for REDD + (http://tropicalforestgroup.blogspot.com/2010/11/text-of-ca-chiapas-acre-mou-on-redd.html) with the aim to "developing a common subnational REDD+ framework or patform for adoption and implementtion in GCF states and provinces" and "to building databases, developing options for linkage arrangements for getting financial support, technical assistance, capacity building, and advancing stakeholder involvement" (http://www.gcftaskforce.org). For its part, the Chiapas Government has allocated funds from the vehicle ownership tax payment to farmers focused on a Biodiversity Hotspot, the Selva Lacandona (Lacandon Community). This region has been widely supported by conservation projects, but from the point of view of other authors it lacks of organizational conditions for effective management of forest carbon (Castillo-Santiago et al 2009).

At the national level, CONAFOR (Forestry National Comission) has convened a technical advisory board (Consejo Tecnico Consultivo CTC) aimed to "promote and deliver recommendations to the government institutions in order to influence the building of a functional mechanism for designing and implementing REDD+ in Mexico, guarantig the transference and maximization of environmental and social benefits" (http://www.reddmexico.org). However, groups of civil society organizations drew attention to the topic of indigenous peoples' rights. Hence, much effort is still needed to really get to have a shared vision.

Forest alternatives will only be adopted by farmers if they respond to expressed needs of population, reduce risks, alleviate constraints, and increase productivity. Deforestation process seems to take its course if drivers persist (Cortina et al., in press). Farmers will be interested since they participate and design innovations from the beginning of the project and also if the initiatives constitute a good deal from the set of alternatives offered in the territory (Franzel & Scherr, 2002; Merino & Warnholtz, 2005; Bray, 2008). Mere academic exercises of land use ordering may be at risk of abandonment if local priorities are left out, so that the participation of farmers (men and women) in planning, designing and applying technology, norms, rules and sanctions among other local institutions must be established with broad participation of all of the actors involved.

6. Acknowledgments

We would like to thank CONACYT for financial support (projects SEP-2004-C01-46244 and FORDECYT 116306), and we would also like to thank El Colegio de la Frontera Sur (ECOSUR), Esteve Corbera from Universitat Autònoma de Barcelona (UAB), Cooperative Ambio, and anonymous reviewers. Sandra Roncal, Cecilia Armijo, Victor Aguilar Argüello and Carlos Mario Aguirre helped for fieldwork.

7. References

Aguilar-Argüello V.H. (2007). Almacenamiento de carbono en sistemas de pasturas en monocultivo y silvopastoriles, en dos comunidades de la Selva Lacandona, Chiapas, México. MSc Thesis. Universidad Autónoma Chapingo. Centro de Agroforestería para el Desarrollo Sustentable, Chapingo Edo. Mexico, 89p.

Aguirre D.C.M. (2006). Servicios ambientales: Captura de carbono en sistemas de café bajo sombra en Chiapas, México. MSc Thesis. Universidad Autónoma Chapingo. Centro de Agroforestería para el Desarrollo Sustentable, Chapingo Edo. Mexico, 84p.

Albrecht A., S.T. Kandji. (2003). Carbon sequestration in tropical agroforestry systems. Agriculture, Ecosystems and Environment 99: 15–27.

Alburquerque L.F. (2002). Guia para agentes. Desarrollo economico territorial. Instituto de Desarrollo Regional. Fundación Universitaria. Sevilla Spain, 214 p.

Berkes F. 2007. Community-based conservation in a globalized world. PNAS 104 (39): 15188-15193.

Bocco G., M. Mendoza, O. Masera. (2001). La dinámica del cambio de uso del suelo en Michoacán. Una propuesta metodológica para el estudio de los procesos de deforestación. Investigaciones Geográficas. Boletín del Instituto de Geografía de la UNAM 44:18-38.

Bray D.B., E. Duran, V.H. Ramos, J.F. Mas, A. Velazquez, R.B. McNab, D. Barry, J. racachowsky. 2008. Tropical deforestation, community forests, and protected áreas in the maya forests. Ecology and Society 13(2): 56

Canadell JG, Raupach MR (2008) Managing forests for climate change mitigation. Science 320(5882):1456–1457.

Castillo-Santiago, M.A., (2009). Análisis con imágenes satelitales de los recursos forestales en el trópico húmedo de Chiapas. Un estudio de caso en Marqués de Comillas. Tesis de doctorado en Ciencias Biológicas, Universidad Nacional Autónoma de México. 123 p.

Cayuela L., J. M. Benayas, C. Echeverria. (2006). Clearance and fragmentation of tropical montane forests in the Highlands of Chiapas, Mexico (1975–2000). Forest Ecology and Management 226, 208–218.

Coad, L., Campbell, A., Clark, S., Bolt, K., Roe, D., Miles, L. (2008). *Protecting the future: Carbon, forests, protected areas and locallivelihoods.* Revised May 2008. UNEP World Conservation Monitoring Centre, Cambridge, U.K.

Corbera E. (2005). Bringing development into carbon forestry markets: Challenges and outcomes of small-scale carbon forestry activities in Mexico. In: D. Murdiyarso y H. Herawati.pp. 42-56. Carbon forestry: ¿who will benefit? CIFOR (Center for Intnal. Forestry Res.), Bogor, Indonesia.

Corbera E. (2010). Mexico's PES-carbon programme: a preliminary assessment and impacts on rural livelihoods. In: L. Tacconi, S. Maharty and H. Sulch, 54-81pp. Payments for environmental services, forest conservation and climate change. Edmund Elgar, Cheltenham UK.

Cortina V.S., H. Plascencia, R. Vaca, G. Schroth, Y. Zepeda, L. Soto-Pinto. Resolving the conflict between ecosystem protection and land use in protected areas of the Sierra Madre de Chiapas, Mexico. Environmental Management. In Press.

de Jong B.H., Ochoa S, Cairns MA. (2000). Carbon flux and patterns of land-use/land-cover change in the Selva Lacandona, Mexico. Ambio 29(8):504–511.

de Jong B.H., V. Maldonado, F. Rojas G., M. Olguín A., V. de la Cruz A., F. Paz P., G. Jiménez Ferrer, M. A. Castillo-Santiago. (2010). Sector uso del suelo, cambio de uso de suelo y silvicultura del inventario estatal de gases de efecto invernadero del Estado de Chiapas. Gobierno del Estado de Chiapas. Tuxtla Gutiérrez, Chiapas México. 71p.

Deininger K.W., B. Minten. (1999). Poverty, Policies, and Deforestation: The Case of Mexico. Economic Development and Cultural Change, 47 (2):313-344.

Díaz-Gallegos J.R., J.F. Mas, A. Velázquez. (2010). Trends of tropical deforestation in Southeast Mexico. Singapore Journal of Tropical Geography, 31 (2): 180-196.

Echeverría C., L. Cayuela, R.H. Manson, D.A. Coomes, A. Lara, J.M. Rey-Benayas, A.C. Newton. (2007). Spatial and temporal patterns of forest loss and fragmentation in Mexico and Chile. In: A.C. Newton (Ed.) pp. 14-42. Biodiversity and conservation in fragmented forest landscapes. CABI, Oxford UK.

Evans P. (2004). Development as institutional change: the pitfalls of monocropping and the potentials of deliberation. Studies in Comparative International Development. 38 (4): 30-52.

FAO (2007). Paying Farmers for Environmental Services, Map 2, page 18. Rome.

FAO (2010). Global forest resources assessment. (2010). Main report. Food and Agriculture Organization of the United Nations. FAO Forestry Paper 163. Rome Italy, 340p.

Franzel S., S.J. Scherr. (2002). Trees on the farm. Assessing the adoption potential of agroforestry practices in Africa. CABI. New York, 185p.

Gobierno del Estado de Chiapas. (2011). Inventario de gases efecto invernadero del estado de Chiapas.
        http://www.cambioclimaticochiapas.org/portal/descargas/igei/resumen_igei.pdf

Haile S., P.K.R. nair, V.D. Nair. (2008). Carbon storage of different soil-size fractions in Florida silvopastoral systems. J. Environ. Qual. 37: 1789-1797

Harvey C.A., A. Medina, D. Merlos S., S. Vilchez, B. Hernández, J.C. Saez, J.M. Maes, F. Casnoves, F.L. Sinclair. (2006). Patterns of animal diversity in different forms of tree cover in agricultural landscapes. Ecological Applications 16 (5): 1986-1999.

Harvey, N. (1998). El fin del "desarrollo" en Marqués de Comillas: discurso y poder en el último rincón de la Selva lacandona. In: Reyes R., M.E., Moguel, R., y van der Haar, G. (cords.), Espacios disputados: transformaciones rurales en Chiapas, Universidad Autónoma Metropolitana y El Colegio de la Frontera Sur, México.

Hirales-Cota M., J. Espinoza-Avalos, B. Schmook, A. Ruiz-Luna, R. Ramos-Reyes. (2010). Drivers of mangrove deforestation in Mahahual-Xcalak, Quintana Roo, southeast Mexico Ciencias Marinas 36 (2):147-159.

Hobbs R.J. (2000). Land use changes and invassions. In Mooney H. A. and R.J. Hobbs (Eds.). Pp. 55-64. Invasive species in a changing world. Island Press, Washington DC.

INEGI (Instituto Nacional de Estadística, Geografía e informática). (1984.) Carta de uso del suelo y vegetación 1:25000. Maps E 15–8, E 15–11. Instituto Nacional de Estadística e Informática, Aguascalientes, México.

INE (Instituto Nacional de Ecología). 2000. Programa de manejo de la Reserva de la Biosfera Montes Azules. México. 256 p.

IPCC (Intergovernmental Panel for Climate Change). (2000). Intergovernmental Panel for Climate Change special report on land use, land-use change and forestry. Special Report. IPCC. Cambridge University Press, Cambridge, USA (http://www.grida.no/publications/other/ipcc_sr/).

Jiménez-Ferrer G., Aguilar AV., Soto-Pinto L. (2008). Livestock and carbon sequestration in the Lacandon rainforest, Chiapas México. In : Livestock and Global Climate Change. BSAS,Cambridge Press, UK, 195-197.

Klooster D. (2000). Beyond deforestation: the social context of forest change in two indigenous communities in Highland Mexico. Yearbook, Conference of Latin Americanists Geographers, 26: 47-59.

Kotto-Same J, Woomer PL, Appolinaire M, Louis Z. (1997). Carbon dynamics in slash-and-burn agriculture and land use alternatives of the humid forest zone in Cameroon. Agric Ecosyst Environ 65:245–256

Lal R. (2004). Soil carbon sequestration to mitigate climate change. Geoderma 123 (1-2): 1-22.

Lambin E.F., B. L. Turner, H. J. Geist, S. B. Agbola, A. Angelsen, J. W. Bruce, O. T. Coomes, R. Dirzo, G. Fischer, C. Folke, P. S. George, K. Homewood, J. Imbernon, R. Leemans, X. Li, E. F. Moran, M. Mortimore, P. S. Ramakrishnan, J. F. Richards, H. Skånes, W. Steffen, G. D. Stone, U. Svedin, T.A. Veldkamp, C. Vogel and J. Xu. (2001). The causes of land-use and land-cover change: moving beyond the myths. Global Environmental Change, Vol. 11 (4): 261-269.

Lugo E, J. A. Parrotta, S. Brown. (1993). Loss in species caused by tropical deforestation and their recovery through management. Ambio 22 (2/3): 106-109.

Maass S.F., H.H. Regil G., J. A. Ordoñez. (2006). Dinámica de población-recuperación de las zonas forestadas en el Parque Nacional Nevado de Toluca. Madera y Bosques 12(1):17-28

Mariaca, R. (2002). Marqués de Comillas, Chiapas: procesos de inmigración y adaptabilidad en el trópico cálido húmedo de México. Tesis de Doctorado en Antropología Social, Universidad Iberoamericana. 396 p.

Martínez M.P. y C. Ruiz de Oña P. (2010). Planeación comunitaria empleando la herramienta plan vivo. Consultancy Report to Corredor Biológico Mesoamericano. San Cristóbal de las Casas, Chiapas, México 43p.

Merino-Pérez, L., G. Segura-Warnholtz. (2005). Las políticas forestales y de conservación y sus impactos en las comunidades forestales en México. In: Bray, D. B., L. Merino-Pérez, D. Barry (Eds), pp. 49-70. The Community forests of Mexico: managing for sustainable landscapes. University of Texas Press. Austin.

Montagnini F., P. K. R. Nair. (2004). Carbon sequestration: An underexploited environmental benefit of agroforestry systems. Agroforestry Systems 61: 281–295, 2004.

Nair P.K.R., V.D. Nair, B. M. Kumar, J.M. Showatter. 2010. Carbon sequestration in agroforestry systems. Advances in agrofomy 108:237-307.

Ochoa-Gaona, S., González-Espinosa, M. (2000). Land use and deforestation in the highlands of Chiapas, Mexico. Appl. Geogr. 20, 17–42.

Ortiz-Espejel B. y V.M. Toledo. (1998). Tendencias en la deforestación de la Selva Lacandona (Chiapas Mexico): el caso de Las Cañadas. Interciencia 23(6): 318-327.

Peeters l.Y.K, L. Soto-Pinto, H. Perales, G. Montoya, M. Ishiki. (2003). Coffee production, timber and firewood in Sothern México. Agric. Ecosyst. and Environmt 95 (2-3):481-493

Perfecto, I., J. Vandermeer. 2010. The agroecological matrix as alternative to the land-sparing/agriculture intensification model. PNAS 107 (13): 5786-5791.

Porter-Bolland L., E. A. Ellis, M. R. Guariguata, I-Ruiz-Mallén, S. Negrete-yankelevich, V. Reyes-García. 2011. Community managed forest and forest protected areas: an assesment of their conservation effectiveness across the tropics. Forest Ecology and Management. In Press.

Quechulpa, S., Castillo-Santiago, M.A., Esquivel, E., Hernández, M.A., Trujillo, J.M. (2010). Captura y reducción de emisiones de bióxido de carbono en el ejido La Corona, municipio de Marqués de Comillas, Chiapas. Informe para La Comisión Nacional Forestal, México. 33 p.

Reyes-Hernández H., S. Cortina-Villar, H. perales-Rivera, E. Kauffer-Michel, J,M. Pat Fernandez. (2003). Efecto de los subsidios agropecuarios y apoyos gubernamentales sobre la deforestación durante el período 1990-2000 en la región de Calakmul, Campeche, México. Investigaciones Geográficas, Boletín del Instituto de Geografia, UNAM 51: 88-106.

Rice R., R. Greenberg. (2004). Silvopastoral systems and socioeconomic benefits and migratory bird conservation. In: Schroth G., G.A.B. da Fonseca, C. Harvey, C. Gascon, H.L. Vasconcelos, A.N. Izac (Eds). Agroforestry and biodiversity. Conservation in tropical landscapes pp. 453-472. Island Press, Washington D.C.

Romero-Alvarado, Y., L. Soto-Pinto, L.E. García-Barrios y J. F. Barrera Gaytán. (2002). Coffee yields and soil nutrients under the shades of Inga sp. Vs. multiple species in Chiapas, Mexico. Agroforestry Systems, 54:215-224.

Roncal G. S. (2007). Almacenamiento de carbono en sistemas agroforestales en Chiapas, México. Tesis de MC. El Colegio de la Frontera Sur (Mex), 75 p.

Ruiz De Oña P.C., L. Soto-Pinto, S. Paladino, F. Morales, E. Esquivel. (2011). Constructing public policy in a participatory manner: from local carbon sequestration projects to network governance in Chiapas, Mexico. In: B. M. Kumar and P. K. R. Nair (Eds). Pp. 247-262. Carbon Sequestration in Agroforestry: Processes, Policy, and Prospects. Springer. Dordrecht.

Schejtman, A. and J. A. Berdegué. (2004). Desarrollo territorial rural. Serie Debates y Temas rurales 1. Rimisp-Centro Latinoamericano para el Desarrollo Rural. Santiago, Chile.

Soto-Pinto L., Perfecto I., Castillo-Santiago H.J. and Caballero N.J. (2000). Shade effect on coffee production at the northern Tzeltal zone of state of Chiapas, Mexico. Agricult. Ecosyst. Environ. 80: 61–69.

Soto-Pinto L., Anzueto-Martinez M., Mendoza V. J., Jimenez-Ferrer G., B. de Jong B. 2010. Carbon sequestration through agroforestry in indigenous communities of Chiapas, Mexico. Agroforestry Systems 78 (1):39-51.

Soto Pinto M.L., G. Montoya Gómez, J. F. Hernádez, T. Morales C, A. Flores, G. Jiménez Ferrer, E. Pineda D.B. (2011). Elaboración de la primera fase del sistema de evaluación de prácticas agroforestales en la Selva Lacandona. Technical Report of Consultancy to Corredor Biológico Mesoamericano 116 p.

Soto-Pinto L., S. Roncal, M. Anzueto. Improved fallows as alternative to shifting cultivation in Chiapas México. Submitted.

Swiderska K., D. Row, L. Siegele, M. Grieg-Gran. (2008). The governance of nature and the nature of governance: policy that works for biodiversity and livelihoods. IIED. London, UK, 184p.

Sweeney B.W., T. Bott, J. K. Jackson, L.A. Kaplan, J. D. Newbold, L.J. Standley, W. C. Hession and R.J. Horwitz. (2004). Riparian deforestation, stream narrowing, and loss of stream ecosystem services. PNAS 101 (39) : 14132–14137.

Toledo-Aceves, T., J. A. Meave, M. González-Espinosa, N. Ramírez-Marcial. (2011). Tropical montane cloud forests: Current threats and opportunities for their conservation and sustainable management in Mexico. Journal of Environmental Management, 92 (3): 974-981.

Watson R.T, I. R. Noble, B. Bolin, N. Ravindranath H., D. J. Verardo, D.J. Dokken. (2000). Land use, lande-use change and forestry: a special report of the Intergovernmental Panel on Climate Change. World Bank, Washington D.C. 388p.

16

Efficiency of the Strategies to Prevent and Mitigate the Deforestation in Costa Rica

Óscar M. Chaves

Instituto para la Conservación y el Desarrollo Sostenible (INCODESO), San José
Costa Rica

1. Introduction

During the 2000-2010 period, worldwide deforestation, mainly due to conversion of forests to agricultural lands, was responsible for the loss of 5.2 million ha of forest per year or 140 km^2 of forest per day (Food and Agriculture Organization of the United Nations [FAO], 2010a). During this same period, Central America loss 248,000 ha of forest per year and, with the exception of Costa Rica, the forest cover continues decreasing in the area (Food and Agriculture Organization of the United Nations [FAO], 2010a). Tropical forest deforestation and the consequent habitat loss and forest fragmentation are among the main causes of the global decline in biodiversity (Brook *et al.*, 2003; Sehgal, 2010). For this reason, studies evaluating the efficiency of the strategies contributing to prevent, minimize, and revert the forest covert loss are critical for the long-term survival of many species of plants and animals (including human beings).

Costa Rica is often hailed as a model for how developing nations can balance the protection of the nature and the economic development. The country is recognized to devoted *ca.* 25% of its territory to forest conservation and for the income that the country generates from different sustainable activities (Buchsbaum, 2004; Food and Agriculture Organization of the United Nations [FAO], 2010a).

However, the "conservationist" reputation of the country is relatively recent and largely contrasts from the unsustainable practices driving the economic model of Central America during the second half the XX century. Overall, the land speculation by cattle ranchers in combination with interest rate subsidies (i.e., financial incentives widely provided by government entities and public banks) is pointed as the main cause of deforestation in Costa Rica and in the rest of Latin America (Kull *et al.*, 2007; Roebeling & Hendrix, 2010). Thus, since 1960 most government policies, national and international investment, and international cooperation programs have all promoted the deforestation, colonization, and "hamburgerization" (i.e., the pasture expansion for the beef industry proliferation for export to USA) of the Central American land (Harrison, 1989; Myers, 1981). As result of these "depredatory" politics, Costa Rican's forest cover decreased from 2.71 million ha in 1950 to 1.01 million ha in 1983, while the deforestation rates increased from 36,018 ha/year during the 1950-1961 period to 97,317 ha/year during 1977-1983 (reviewed by Harrison, 1989). However, the economic balance for the beef industry was extraordinary positive because the number of head cattle exported to USA explosively increased from 6903 in 1955 to 148,882 in 1973. Even after 1983, the Costa Rican's government across the Instituto de Desarrollo

Agrario (IDA) continued promoting the deforestation and clearing of 500,000 ha of "underutilized lands" (i.e., forested or partially forested lands). Most of these lands were distributed among poor "campesinos"and/or small farmers to incentive the agriculture (Harrison, 1989).

Although there is a complex network of environmental, social, economic, and politic factors influencing the changes in forest cover in the country (Calvo-Alvarado *et al.*, 2009; Kull *et al.*, 2007; Sánchez-Azofeifa *et al.*, 2001; Sánchez-Azofeifa & Van Laake, 2004), most studies highlight the importance of the ecotourism, the payments for environmental services, the private nature reserves, and the environmental education as strategies to prevent, minimize, and/or revert the deforestation in the country. These four factors are described below.

1.1 Ecotourism

Ecotourism is often perceived as an excellent tool for sustainable development in developing countries (Almeyda *et al.*, 2010; Gossling, 1999). Among the main benefits of this "green" economic strategy frequently are mentioned: (1) the protection of forest resources and hence the prevention of deforestation in general, and (2) the economic benefits for local communities (Gossling, 1999; Horton, 2009).

Since the boom of the ecotourism in Costa Rica, at the end of 1980s and into the 1990s, the activities related with ecotourism (e.g., proliferation of eco-lodges and/or private reserves, nature-tours, and agro-ecotourism), represent one of the main economic activities of the country (Koens *et al.*, 2009; Weaver, 1999). Corresponding with this "boom", the hotel sector in the country has grown over 400% from 1987 to 2000 (Instituto Centroamericano de Administración de Negocios [INCAE], 2000). As result, the agro-export economic model based in the extensive plantations of coffee and banana (i.e., the main economic activity since the nineteenth century), was displaced to a secondary place as source of economic resources (Iveniuk, 2006). Thus, annually *ca.* 1.2 million tourists visit the country, which result in an annual turnover range from US$ 1200 million in 2003 to US$ 1983 million in 2010 (Instituto Costarricense de Turismo [ICT], 2011a).

Although the above, some researchers question the contribution of the ecotourism to conservation and community development due to some potential negative impacts such as habitat destruction, waste generation, visitor impacts, and socio-cultural ills (Stem *et al.*, 2003; Wearing & Neil, 2009). Below I discuss this topic in the light of the evidence provided by different case studies in Costa Rica.

1.2 Payments for Environmental Services (PES)

The payment for environmental services is an economic incentive received to the landowners for the services and goods provided for their forested lands or forest patches (Sánchez-Azofeifa *et al.*, 2007). In Costa Rica the current PES system was created with the 1996 Forestry Law 7575, which recognizes four main environmental services: mitigation of green house gas emissions, hydrological services, biodiversity conservation, and provision of scenic beauty for recreation and ecotourism (Kull *et al.*, 2007; Sánchez-Azofeifa *et al.*, 2007).

The Forestry Law 7575 provides the legal and regulatory basis to contract with landowners for the environmental services provided by their lands, empowers the National Forestry Financing Fund (FONAFIFO) to issue such contracts, and establishes a financing mechanism for this purpose (Sánchez-Azofeifa *et al.*, 2007; Sierra & Russman, 2006). Previously to receive the PES economic benefits, the landowners need sign a 5-10 year contract with the government, agreeing to either protect forest cover or engage in reforestation. Funding for PES program comes from a 3.5% fossil fuel tax, private-sector and international donor contributions, a World Bank loan, and the sale of carbon offsets to industrialized countries established in the Kyoto Protocol (Kull *et al.*, 2007).

The percent of the country territory receiving PES increase from 5.5% (representing 4400 contracts) in 2001 to 7.3% (representing 4600 contracts) in 2010 (National Forestry Financing Fund [FONAFIFO], 2011). According with the same source, currently PES covers 373,074 ha of forests in different succession stages and protects *ca.* 2.7 million of trees throughout the 7 provinces of the country. Although this program might facilitate field abandonment in some marginal zones already affected by agricultural liberalization (Sierra & Russman, 2006), its relative efficiency as deforestation-avoiding strategy is even polemic.

1.3 Private nature reserves

Traditionally, conservation efforts to preserve the Costa Rican' natural legacy (i.e., ecosystems and species) have been focused on the establishment of government-controlled areas such as national parks, wildlife refuges, biological reserves, and other management categories (Food and Agriculture Organization of the United Nations [FAO], 2010b). However, with the boom of the ecotourism at the end of 1980s and into the 1990s (see above), this fact has changed noticeably. Thus, the number of private reserves registered increases from 65 reserves in 2005 to *ca.* 200 reserves in 2011 (Red Costarricense de Reservas Naturales, 2011).

Overall, private reserves range from 20 to 1500 ha in size (Herzog & Vaughan, 1998; Red Costarricense de Reservas Naturales, *pers. comm.*) and together cover over 81,429 ha of forests along the country (Red Costarricense de Reservas Naturales, 2011). This fact highlights the importance of private landholdings in the fate of the Costa Rican forests, particularly in the trends in the forest cover along the time.

1.4 Environmental Education (EE)

Environmental education is the process of recognizing values and clarifying concepts in order to develop skills and attitudes necessary to understand and appreciate the interrelatedness among men, his culture and his biophysical surroundings (International Union for the Conservation of Nature [IUCN], 1970). Overall, acording to Jacobson (1991) a well-designed and well-implemented EE program have the potential to increase ecological awareness, foster favorable attitudes toward the environment, and promote the conservation of the natural resources (including discourage of deforestation and/or selective logging). Even, some authors consider that an appropriate EE might produce significant behavioral changes in the people and hence, may be more crucial to successful long-term conservation than a strictly scientific work (Blum, 2008; Jacobson, 1991; Palmer, 1998).

In this sense, since 1980s Costa Rica has been one of the world-leaders in efforts to promote environmental learning and national policies integrating education, conservation, and different sustainable activities related to ecotourism (Blum, 2008). For instance, as early as 1975, the Universidad Nacional (one of the most important public universities in the country) established a School of Environmental Sciences which included an environmental education program. Similarly in 1994, the National Council of Vice-Chancellors created an Inter-University Commission for Environmental Education which works to 'environmentalise' all of the state universities (Oficina de Educación Ambiental, 2002). Currently, most of the environmental education in Costa Rican schools and/or communities is supported by the Ministry of Environment and diverse non-governmental organizations (Blum, 2008).

Nevertheless, to date the controversy on the relative efficiency of the aforementioned factors as deforestation-avoiding strategies still persist. Some authors concludes that the current expansive network of efforts in conservation, environmental management, education and ecotourism contribute directly to prevent and minimize the deforestation in the country (e.g., Blum, 2008). Conversely, other authors support that the forest cover recovery in some areas of Costa Rica have been result of a process referred as "forest transition" (i.e., deforestation trends are replaced with reforestation following the trends in the economic development and urbanization), rather a direct effect of the environmental policies (Calvo-Alvarado et al., 2009; Schelhas & Sánchez-Azofeifa, 2006). Therefore, in this chapter I reviewed the available literature to assess the relative efficiency of the ecotourism, PES, private reserves and environmental education to prevent, mitigate, and/or reverts the deforestation in different regions of Costa Rica. Finally, I synthesize the major strengths and weaknesses of these strategies and suggest future directions to minimize the deforestation and to improve the environmental performance in the country.

2. Methods

2.1 Literature review

To achieve the main objectives of this study, I conducted a systematic review of published articles, book chapters and dissertations up to August 2011 using ISI Web of Science, Biological Abstracts, Google Scholar, and Costa Rican environmental agencies online data bases. Overall, I amassed a total of 90 studies on the study topic distributed as follow: 36 studies on ecotourism, 17 studies on payment for environmental services, 8 studies on private nature reserves, and 29 studies on environmental education.

2.2 Efficiency indicators

To determine the efficiency of the above mentioned strategies I used changes in deforestation rates and/or forest cover before the implementation of each strategy and 4-20 years after its implementation in the same region. As sources of the information on the deforestation rates and/or forest cover in the sites in which a particular strategy was implemented, I used the information available in the literature and in online data bases from the Ministerio del Ambiente, Energía y Tecnología (MINAET) de Costa Rica (http://www.minae.go.cr/), Sistema de Infomación de Recursos Forestales (http://www.sirefor.go.cr), Fondo Nacional de Financiamiento Forestal (http://www.fonafifo.com/), Oficina Naciona Forestal (ONF), Tribunal Ambiental Administrativo (http://www.tribunalambiental.org), and Centro de Derecho Ambiental y de Recursos Naturales (http://www.cedarena.org/).

2.3 Statistical analysis

To evaluate if forest area covers by private and public forests differed between years, I used generalized linear models (GLM: Lehman et al., 2005). I constructed the following model: AREA = FOREST TYPE (PRIVATE AND PUBLIC) nested within YEAR + FOREST TYPE*YEAR. Data were first arcsine transformed, and tested for a normal distribution with a Shapiro-Wilk test (passed, $P>0.05$). I then selected Normal distribution with an identity link-function to the response variable. I performed the statistical analyses using JMP (version 7.0, SAS Institute, Cary, NC).

Case study	Location (coordinates)	Positive aspects	Negative aspects	Ref.
Monte Verde Could Forest Reserve	Puntarenas province (10°18'25.14"N, 84°48'35.03"W)	1) Protection over 10,500 ha of could forest, including different threatened species of animals and plants. 2) Protection of an important water reservoir for the region. 3) Promotion of environmental education and sustainable management of resources across the non-profit organization Monte Verde Institute 4) Generation of feeds for local inhabitants inside and outside the reserve. 5) Creation of a women s handicraft cooperative with nearly 100 members.	1) Trail and soil erosion with the increase in the number of visitors. 2) Potential pollution and/or over-exploitation of the water reservoir	Buckley (2003)
Rara Avis Reserve	Heredia province (10°16'52.23"N, 84° 2'43.09"W)	1) The reserve directly conserves 485 ha of primary rainforest and has indirectly conserved an additional 1000 ha. 2) Protection of more than 500 species of trees and palms, and over 362 species of birds. 3) By protecting the forest, the reserve has avoided the emission of over 337,000 of CO_2, which would have been emitted had the forest been cut, as was planned by the previous owners. 4) Promotion of environmental education and alternative sustainable uses of the rainforest. 5) Low-moderate number of visitors along the year, which minimizes the human disturbance in the area. 6) Promotion of the scientific research on the birds, amphibians and trees.	1) No clear economic benefits for the neighbor communities.	Buckley (2003)
Punta Islita Eco-lodge	Nicoya, Guanacaste province (9°51'23.80"N, 85°23'55.89"W)	1) The eco-lodge property increased in forest cover from 4% to 76% from 1975 to 2008. 2) Protection of 27 ha of tropical dry forest. 3) Provide employment for 175 local inhabitants and indirect employment for an undetermined number of permanent residents. 4) By generation employment for local inhabitants, has contributed to reduce the problems such as alcoholism, drug addiction, and prostitution.	1) Future increase in deforestation rates due to uncontrolled development of hotels and housing projects.	Almeyda et al. (2010)

Table 1. Three study case on ecotourism efficiency in different top tourism areas of Costa Rica.

3. Results and discussion

3.1 Relative efficiency of ecotourism

Overall, the findings for the three analyzed case studies (Table 1) concur with previous studies suggesting that ecotourism represent a valuable tool for conservation and sustainable development (e.g., Almeyda et al., 2010; Ceballos-Lascurain, 1998; Wearing & Neil, 2009). Although in the three case studies there are some negative aspects, the general balance undoubtedly can favors the forest protection and the ecosystem regeneration (Table 1). Similarly, there is evidence indicating that the expansion of the ecotourism plays a relevant role in the reduction of the deforestation in Costa Rica. Thus, considering forests on farmlands and non-farmland areas, the deforestation rates decreased from 47,219 ha/year during the 1950-1961 decade to 15,677 ha during 1973-1984 (reviewed by Harrison, 1989), which correspond to the boom of ecotourism in the country (see above).

However, other studies do not support that ecotourism per se have a relevant role as deforestation-avoiding strategy. For instance, Kruger (2005) in his review of 251 ecotourism case studies found that ecotourism did not create enough revenues to prevent 'consumptive' land use (e.g., forest conversion to agriculture or pasture) among households. Furthermore, the increasing development in the ecotourism "hotspots" of Costa Rica (e.g., Nicoya, Monte Verde, Siquirres, Motezuma, Quepos) might result in a higher deforestation in theses areas. Thus, Almeyda et al. (2010) mention that uncontrolled development of standard hotel operations and large condo developments in Nicoya, seeks to capitalise on the region's natural beauty and may reverse land cover trends if they are not accompanied by adequate forest conservation strategies and government monitoring.

3.2 Relative efficiency of the PES

Since the implementation of PES the amount of industrial wood extracted from forests gradually decreased from 248,362 m^3 in 1998 to ca. 50,000 m^3 in the years 1999-2006 (Figure 1). Conversely, wood derived from agroforestry increased from 458,538 m^3 in 1998 to a maximum value of 673,426 m^3 in 2001 and two years latter this value drop to 205,401 m^3 and remained relatively constant until 2007 (Figure 1). Furthermore, during this period occurred a noticeable increase in the amount of wood derived from forest plantations (mainly Smelina arborea and Tectona grandis, National Forestry Financing Fund, pers. comm.) (Figure 1). Similarly, during the period 2000-2010 (i.e., 15 year after the implementation of PES), the forest area of Costa Rica increased from 2.37 to 2.60 million hectares (Food and Agriculture Organization of the United Nations [FAO], 2010a). However, the history of the forest cover was quite different during the 1980s when the nation was losing 4% of its forest cover annually – the highest deforestation rate in the western hemisphere at the time (Carriere, 1991). Indeed, between 1970 and 1980 more than 7000 km^2 were cleared, and by 1987 total forest cover had been reduced to only 31% of the land mass or approximately 16,000 km^2 (Carriere, 1991).

Currently there are 4599 PES contracts covering 373,074 ha of lands along the country. Most of the area covered by PES (70%) was devoted to forest protection. Similarly, from the four main PES categories, most contracts were devoted to forest protection (54.2%), following by reforestation (29.8%), protection of wildlife areas (6.4%), and forest management (1.5%) (Table 2). This suggests that PES can promotes the protection of the forest cover in different private nature lands, at least during the contract period (i.e., 5-years). However, there was a large variation in the number of contracts and area covered by PES among provinces (Table 2). Alajuela province presented the larger number of contracts (1045 contracts) and cumulative number of hectares covered by PES (73,516 ha), while the lower values were found in Cartago province (130 contracts and 28,822 ha, respectively) (Table 2).

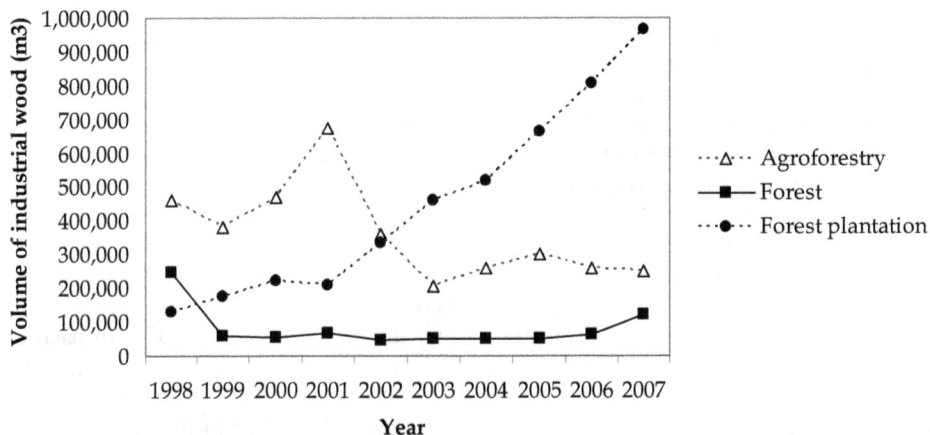

Fig. 1. Amount of industrial wood extracted from agroforestry, forests, and forest plantations during a 10-year period. Source: Oficina Nacional Forestal (ONF) (2009).

Province	No. contracts	Total area (ha)	Hectares covered by PES category (No. contracts)[a]				
			Forest protection	Reforestation	PWA	FM	Total
Alajuela	1045	73,516.67	38,292.08 (447)	21,255.07 (443)	6786.8 (63)	2351.82 (31)	68,685.77 (984)
Limón	652	70,894.68	52,346.37 (389)	3099.17 (149)	9645.7 (44)	587.32 (10)	65,678.56 (592)
Puntarenas	670	69,662.64	56,742.44 (422)	3705.7 (146)	5928 (62)	0	66,376 (630)
Guanacaste	1027	63,290.57	46,582.38 (541)	11,130.89 (375)	547.5 (10)	0	58,260.77 (926)
Heredia	484	35,722.00	18,989.42 (213)	3885.51 (113)	8745.9 (83)	1792.66 (30)	33,413.49 (439)
San José	590	31,165.95	24,462.32 (392)	2430.23 (122)	1785.6 (18)	0	28,678.15 (532)
Cartago	130	28,822.38	24,074.58 (90)	848.4 (21)	3429.1 (14)	0	28,352.08 (124)
Total	**4599**	**373,074.90**	**261,489.59(2494)**	**46,354.98 (1369)**	**36,868.60 (294)**	**4731.8 (71)**	**349,444.96**

[a]PWA= Protection of wildlife area, FM= forest management.

Table 2. Current area covered by PES category in the 7 provinces of Costa Rica. Data come from the National Forestry Financing Fund of Costa Rica, FONAFIFO, (2011). Data are ordered according to the total area protected.

The aforementioned trends in forest cover recovery might be related to the different environmental restrictions and government incentives related to the PES established in the Forestry Law 7575. However, other authors suggest that the increase in the forest cover observed in different areas of Costa Rica are in fact a result of a process of "forest transition" (Calvo-Alvarado *et al.*, 2009; Schelhas & Sánchez-Azofeifa, 2006). For instance, Calvo-Alvarado *et al.* (2009) analyzed the process of deforestation and restoration of the tropical dry forest in Guanacaste, Costa Rica, using socioeconomic data and satellite images of the forest cover from 1960 to 2005. They concluded that the restoration of the Guanacaste's forest cover after the 1980s was the result of multiple socioeconomic factors rather than the efficiency of the PES. Similarly, Almeyda *et al.* (2010) found that the proportion of forest cover in the Nicoya peninsula decrease from 0.51 in 1987 to 0.36 in 2008, indicating a poor efficiency of the PES or any other potential deforestation-avoiding strategies implemented by the government during that period.

For this reason, Calvo-Alvarado *et al.* (2009) mentions that neither the PES nor the other conservation policies implemented in the country are enough to protect this tropical dry forest at the long-term. This statement might be particularly true if we take in consideration the multiple financial limitations of the MINAET and the limited number of environmental inspectors of the Secretaría Técnica Nacional Ambiental, the public institution responsible for authorizing the construction and monitoring the environmental performance of the diverse public and private infrastructures along the country (MINAET, *pers. comm.*).

3.3 Relative efficiency of private nature reserves

During the decades 1990s and 2000s the forest area on control of private landholdings was noticeable greater than the forest area on control of the government (Figure 2). Nevertheless, the latter forest area was duplicated from 1990 to 2005 (Figure 2). However, until today, most of the Costa Rican forests (55%) are on control of private landholdings (Food and Agriculture Organization of the United Nations [FAO], 2010b). Overall, the forest cover protected by the private reserves increased *ca.* 2.5 times from 1995 (32,895 ha) to 2010 (81,429 ha) (Programa Estado de la Nación, 2011). Interestingly, from the 200 private reserves recorded until 2010 only a minimal part is managed by individual landowners. Thus, 52% of private reserves are managed by non-government organizations (mainly public universities and non-profit conservationist institutions), 46% are managed by profit organizations, and only 2% are managed by individual landowners (Programa Estado de la Nación, 2011). This fact undoubtedly can benefit the long-term conservation of these areas because contrasting with individual landowners, the activities of the non-profit organizations and universities are frequently monitored by different government institutions and hence, it is more difficult to they change the land use of the forested areas. The deforestation and/or selective logging in reserves management by individual landowners is hard to detect and penalize (particularly when there is not a formal denounce) (Tribunal Ambiental Administrativo, 2011).

On the other hand, considering both private reserves and mixed public-private areas, they cover over 13% of the total continental area of the country while the government-protected forests cover 21% of this area (Table 3). Additionally, as in most America Latina, the main uses of these reserves are ecotourism and investigation (Mesquita, 1999). Furthermore, since private reserves often are located in ecoregions poorly represented in the government system and in regions of the country without existing reserves, they might also contributes to protects a number of threatened animals such as agouti, peccary, jaguar, and puma

(Herzog & Vaughan, 1998). These facts strongly suggest that the private reserves play an important role in the protection of the forest cover in Costa Rica. Even, some of these reserves are also important research centers and frequently receive researches and students of diverse countries of the world. For instance, the La Selva Biological Station protects 1600 ha of primary tropical rainforest, (including 1000 tree species and 420 bird species), and also is one of the most well-studied and recognized tropical rainforest around the world with hundreds of scientific publications in high-impact international journals.

Management category	N	Area (ha)	%STA[a]	% of the country[b]
Goverment-protected areas				
National Parks	28	629,121	37.1	12.3
Biological Reserve	8	22,036	1.3	0.4
Wildlife Refugie	12	61,708	3.6	1.2
Absoult Natural Reserve	2	1,369	0.8	0
Indigenous Reserves	24	335,851	19.8	6.6
National Monument and other reserves	3	23,768	1.4	0.5
subtotal	77	1,073.853	64	21
Mixed (Public-Private)				
Wildlife Refugie	25	106,572	6.3	2.1
Protected Zone	31	157,715	9.3	3.1
Forest Reserve	9	221,239	13.0	4.3
Wetland	13	68,543	4.0	1.3
subtotal	78	554,069	32.66	10.8
Private Reserves				
Wildlife Refuges	34	68,447	4.03	1.3
Natural Reserves	72	52,156	NA	1
subtotal	106	120,603	4.03	2.3

aPercent of total protected area.
bPercent of the continental area of the country (i.e., 51,100 km2).

Table 3. Area covered by public and private protected areas in Costa Rica during 2007. Modified from Food and Agriculture Organization of the United Nations FAO (2010b).

3.4 Relative efficiency of environmental education
The findings indicated that most environment education programs (EEP) in Costa Rica are linked to different ecotourism activities (79% of 29 analyzed studies), following by activities promoted by government and/or non-profit organizations. Currently, most Costa Rican eco-lodges, private reserves, and ecotourism companies have EEPs for employees, visitants and/or local inhabitants (Instituto Costarricense de Turismo [ICT], *pers. comm.*). Even, the EEP is an essential requirement for all those companies and institutions procuring the Certification for Sustainable Tourism (CST, see below). Currently, 178 eco-lodges, 59 tourism agencies, and 4 car rental agencies are certificated (Instituto Costarricense de Turismo [ICT], 2011b), and hence with some type of EEP in course.

Unfortunately, in most cases it is not possible to evaluate the relative efficiency of environmental education as deforestation-avoiding strategy due that its effect is difficult to separate from other complementary conservation strategies. Nevertheless, evidence suggests that it contributes significantly to reinforce conservation attitudes of the people respect to the deforestation and other environmental problems (Blum, 2008; Koens *et al.*, 2009; Palmer, 1998). Further studies in particular communities of Costa Rica are necessary to evaluate if the environmental education per se is able to minimize, or even revert, the deforestation at the long-term.

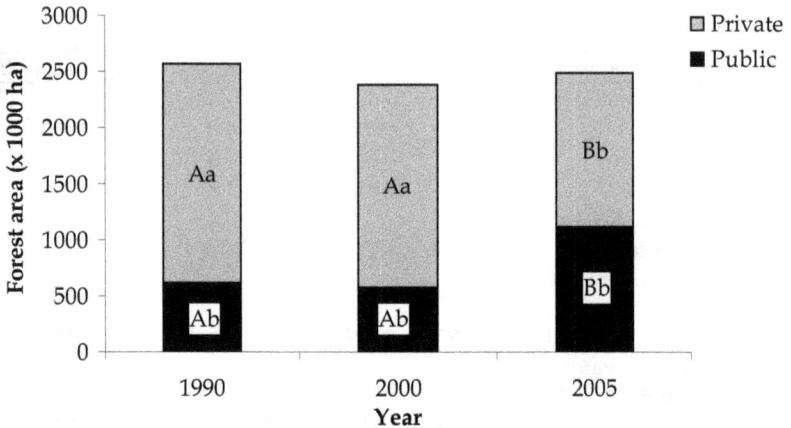

Fig. 2. Changes in the number of forest hectares under government and private management in Costa Rica. Different capital letters indicate significant differences among years, and different lowercase letters indicate differences among reserve type (contrast tests, $P<0.05$).

4. Conclusion and recommendations

Although some limitations (e.g., scarce information on changes in forest cover, lack of quantitative analyses), the information provided in this chapter indicates that, in general, the four strategies analyzed have contributed to discourage the deforestation along the country and contributes to reinforce conservation attitudes in the people.

Thanks to these strategies, the conceptualization of the nature as an obstacle to development has experienced a noticeable change, at least when comparing with the beliefs of some decades ago (see Introduction). However, further studies are critical to improve our understanding on how ecotourism, PES, private reserves, and environmental education contribute to avoid the deforestation and the long-term conservation of the ecosystems. In this sense, field studies analyzing changes in forest cover and/or deforestation rates begin and after the implementation of a determined strategy (using satellite imagery, remote sensing imagery, or other similar methodology) can be particularly useful. Nevertheless, there are indications that one or few isolated conservation strategies are insufficient to deter further deforestation in Costa Rica. Probably the most efficient deforestation-avoiding strategy is not an isolated one, but an integrated conservation program that considers simultaneously the main environmental, social, politic, and economic aspects involving in the phenomenon of deforestation as suggested by Calvo-Alvarado *et al.* (2009). This strategy must take in consideration the appropated planning of multple land uses within the heterogenous landscapes predominant along the country. Government, non-profit organizations, and civil society in general, should priorizing the protection of the more vulnerable forested areas (i.e., those that contain threated species of animals and plants and/or water reservoir). When necessary, the government should consider the expropriation of some of these areas to guarantee their ecological integrity.

Furthermore, to a more efficient control of the deforestation and/or logging along the country, it is important increase the resources invested in the environmental monitoring. In this sense, are necessary more private and public initiatives to monitoring and penalize the deforestation and other environmental crimes and/or reinforce those initiatives that have showed be successful. For instance, one successful initiative is the 2008' public program namely "barridas ambientales" from the Tribunal Ambiental Admistrativo. In this initiative, a multidisciplinary team of biologists, forest engineers, chemistries, and lawyers carried out sudden visits (5-10 times/year) to different areas of the country and, with the cooperation of the local inhabitants, they detect, record, and sanction any environmental crime (including illegal logging). As result, the number of denounces has been increased by a 100% (Tribunal Ambiental Administrativo, 2011). Other interesting initiative is the program Certification for Sustainable Tourism (CST), from the Instituto Costarricense de Turismo. This program was designed to differentiated businesses of the tourism sector, based on way that they interact with the nature, local communities, and social resources in general (Instituto Costarricense de Turismo [ICT], 2011b).

5. Acknowledgements

I thank Júlio César Bicca-Marques for logistical support for the redaction of this chapter in Brazil. The Ministerio del Ambiente, Energía y Telecomunicaciones de Costa Rica (MINAET) provided important information on changes in the forest cover in different regions of Costa Rica.

6. References

Almeyda, A. M., Broadbent, E. N., Wyman, M. S., & Durham, W. H. (2010). Ecotourism Impacts in the Nicoya Peninsula, Costa Rica. *International Journal of Tourism Research,* 12:803-819

Blum, N. (2008). Environmental education in Costa Rica: building a framework for sustainable development? *International Journal of Educational Development* 28:348-358.

Brook, B. W., Sodhi, N. S., & Ng, P. K. L. (2003). Catastrophic extinctions follow deforestation in Singapore. *Nature,* Vol. 424, pp. 420-423.

Buckley R. (2003). *Case studies in ecotourism,* CABI Publishing, Cambridge, 264 p

Buchsbaum, B. D. (2004). *Ecotourism and sustainable development in Costa Rica.* M.Sc. dissertation, Virginia Polytechnic Institute and State University, Virginia, 59 p.

Calvo-Alvarado, J., McLennan, B., Sánchez-Azofeifa, A., & Garvin, T. (2009). Deforestation and forest restoration in Guanacaste, Costa Rica: putting conservation policies in context. *Forest Ecology and Management,* Vol. 258, pp. 931-940

Carriere, J. (1991). The crisis in Costa Rica: an ecological perspective. In: *Environment and development in Latin America: the politics of sustainability,* Redclift, M. & Goodman, D., (Eds.), pp. 184-204, Manchester University Press, Manchester

Ceballos-Lascurain, H. (1998). Ecotourism as a worldwide phenomenon. In: *Ecotourism: a guide for planners and managers*, Lindberg, K. & Hawkins, D. (Eds.), The Ecotourism Society, North Bennington

Food and Agriculture Organization of the United Nations, FAO. (2010a). *Global forest resource assessment 2010*, FAO, Rome, Retrieved from <http://www.fao.org/forestry/fra/en/>

Food and Agriculture Organization of the United Nations, FAO. (2010b). *Evaluación de los recursos forestales mundiales 2010: informe Costa Rica*, FAO, Rome, Retrieved from <http://www.fao.org/forestry/20308-0982e92a29545eed28d27cd4149dce658.pdf>

Gossling, S. (1999). Ecotourism: a means to safeguard biodiversity and ecosystem functions? *Ecological Economics*, Vol. 29, pp. 303-320

Harrison, S. (1989). *Population, land use, and deforestation in Costa Rica, 1950-1983*, Standford University, California, 75 p

Herzog, P., & Vaughan, C. (1998). Conserving biological diversity in the tropics: the role of private nature reserves in Costa Rica. *Revista de Biologia Tropical*, Vol. 46, pp. 183-190

Honey, M. (1999). *Ecotourism and Sustainable Development. Who owns Paradise?*, Island Press, Washington D.C.

Horton, L. R. (2009). Buying up nature economic and social impacts of Costa Rica's ecotourism boom. *Latin American Perspectives*, Vol. 36, pp. 93-107

Instituto Centroamericano de Administración de Negocios, INCAE. (2000). Tourism in Costa Rica: a competitive challenge, INCAE, San José, Costa Rica

Instituto Costarricense de Turismo, ICT. (2011a). *Cifras turísticas*. Accessed: 10 October 2011. Available from: <http://www.visitcostarica.com/ict/paginas/cifras_turisticas/Agosto_2011/ictblt1.html>

Instituto Costarricense de Turismo, ICT. (2011b). *Sustainability CST*. Accessed: 15 December 2011. Available from:
<http://www.visitcostarica.com/ict/paginas/sostenibilidad.asp?tab=0>

IUCN. (1970). *International Working Meeting on Environmental Education in the School Curriculum*, UNESCO, Paris

Iveniuk, J. (2006). The consumption of conservation ecotourism in Costa Rica. *Nexus*, Vol. 19, pp. 102-125

Jacobson, S. K. (1991). Evaluation model for developing, implementing, and assessing conservation education programs: examples from Belize and Costa Rica. *Environmental Management*, Vol. 15, pp. 143–150

Koens, J. F., Dieperink, C., & Miranda, M. (2009). Ecotourism as a development strategy: experiences from Costa Rica. *Environment, Development and Sustainability*, Vol. 11, pp. 1225-1237

Kruger, O. (2005). The role of ecotourism in conservation: Panacea or Pandora's box? *Biodiversity and Conservation*, Vol. 14, pp. 579-600

Kull, C. A., Ibrahim, C. K., & Meredith, T. C. (2007). Tropical forest transitions and globalization: neo-liberalism, migration, tourism, and international conservation agendas. *Society & Natural Resources*, Vol. 20, pp. 723-737

Lehman, A., O'Rourke, N., Hatcher, L., & Stepanski, E. J. (2005). *JMP® for basic univariate and multivariate statistics: a step-by-step guide*, SAS Institute, North Carolina, USA

Liverman, D. M., & Vilas, S. (2006). Neoliberalism and the environment in Latin America. *Annual Review of Environment and Resources*, Vol. 31, lines 327-363

Mesquita, C. A. B. (1999). *Caracterización de las reservas naturales privadas en América Latina.* CATIE, Turrialba, Costa Rica

Myers, N. (1981). The hamburger connection: how Central America's forests become North America's hamburgers. *Ambio*, Vol. 10, pp. 3-8

National Forestry Financing Fund. (2011). In: *Fondo Nacional de Financiamiento Forestal, FONAFIFO, San José, Costa Rica.* Accessed: 10 October 2011. Available from: < http://www.fonafifo.go.cr/paginas_espanol/consultas_psa/e_cp_mapas.htm>

Oficina de Educación Ambiental. (2002). *Marco conceptual, legal y áreas de acción de la Oficina de Educación Ambiental,* Oficina de Educación Ambiental, Ministerio de Educación Pública, San José, Costa Rica

Oficina Nacional Forestal, ONF. (2009). Registros de madera en rollo procesada por la industria, periodo 1998-2007. Ministerio del Ambiente, Energía y Tecnología (MINAET), San José, Costa Rica

Palmer, J. A. (1998). *Environmental education in the 21st century: theory, practice, progress and promise.* Routledge, New York

Programa Estado de la Nación. (2011). Decimoséptimo Informe Estado de la Nación en Desarrollo Humano Sostenible. Aspectos ambientales. Consejo Nacional de Rectores, San José, Costa Rica

Red Costarricense de Reservas Naturales (2011). *Red Costarricense de Reservas Naturales, San José, Costa Rica.* Accessed: 5 October 2011. Available from: <http://www.reservasprivadascr.org/ver3/ndex.php?x=5>

Roebeling, P. C., & Hendrix, E. M. T. (2010). Land speculation and interest rate subsidies as a cause of deforestation: the role of cattle ranching in Costa Rica. *Land Use Policy*, Vol. 27, pp. 489-496

Sánchez-Azofeifa, G. A., & Van Laake, P. E. (2004). Focus on deforestation: zooming in on hot spots in highly fragmented ecosystems in Costa Rica. *Agriculture Ecosystems & Environment*, Vol. 102, pp. 3-15

Sánchez-Azofeifa, G. A., Pfaff, A., Robalino, J. A., & Boomhower, J. P. (2007). Costa Rica's payment for environmental services program: Intention, implementation, and impact. *Conservation Biology*, Vol. 21, pp. 1165-1173

Sánchez-Azofeifa, G., Harriss, R. C., & Skole, D. L. (2001). Deforestation in Costa Rica: a quantitative analysis using remote sensing imagery. *Biotropica*, 33:378-384

Schelhas, J., & Sánchez-Azofeifa, A. (2006). Post-Frontier forest change adjacent to Braulio Carrillo National Park, Costa Rica. *Human Ecology*, Vol. 34, pp. 407-431

Sehgal, R. N. M. (2010). Deforestation and avian infectious diseases. *Journal of Experimental Biology*, Vol. 213, pp. 955-960

Sierra, R., & Russman, E. (2006). On the efficiency of environmental service payments: a forest conservation assessment in the Osa Peninsula, Costa Rica. *Ecological Economics*, Vol. 59, pp. 131-141

Stem, C.J., Lassoie, J. P., Lee, D. R., Deshler, D. J. (2003). How 'eco' is ecotourism? A comparative case study of ecotourism in Costa Rica. *Journal of Sustainable Tourism*, Vol. 11, pp. 322-347

Tribunal Ambiental Administrativo. (2011). In: *Tribunal Ambiental Administrativo,* MINAET, San José, Costa Rica. Accessed: 5 December 2011. Available from: <http http://www.tribunalambiental.org>

Wearing S, Neil J. (2009). *Ecotourism: impacts, potentials, and possibilities,* Elsevier, New York, 282 p

Weaver, D. B. (1999). Magnitude of ecotourism in Costa Rica and Kenya. *Annals of Tourism Research,* Vol. 26, pp. 792-816

17

Economic Models
of Shifting Cultivation: A Review

Yoshito Takasaki
University of Tsukuba
Japan

1. Introduction

Shifting cultivation is a dominant agricultural system in tropical forests. Shifting cultivators transform nutrients stored in standing forests to soils by slashing, felling, and burning forests (i.e., slash-and-burn); they regularly shift crop lands by replacing depleted plots with cleared forest lands (Denevan & Padoch, 1987; Kleinman et al., 1995; Ruthenberg, 1980). Approximately 300–500 million people practice slash-and-burn agriculture on almost one third of the planet's 1,500 million ha of arable land (Giaradina et al., 2000; Goldammer, 1993). Shifting cultivation is central to the poverty-environment nexus in the tropics. On one hand, shifting cultivation is a dominant livelihood activity among small-scale tropical farmers with various cultural, ethnic, and social backgrounds, and thus it is tightly linked with poverty and development (Angelsen & Wunder, 2003; Byron & Arnold, 1999; Reardon & Vosti, 1995; Sunderlin et al., 2005; Wunder, 2001). On the other hand, not only is shifting cultivation one of the major causes of tropical deforestation, but also, the associated forest-cover change leads to multiple environmental problems, such as soil degradation, biodiversity loss, and reduced carbon sequestration (e.g., Chazdon et al., 2009; Dent & Wright, 2009; Kleinman et al., 1995; Lawrence et al., 2005; Myers, 1992). As such, shifting cultivation can conflict with various conservation efforts, such as maintaining protected areas, engaging in community-based conservation, sustaining integrated conservation-development programs (ICDPs), making payments for environmental services (PES), and reducing emissions from deforestation and forest degradation (REDD) (e.g., Angelsen, 2008; Wilshusen et al., 2002; Wunder, 2006). A win-win goal of poverty alleviation and rainforest conservation in shifting cultivation systems is a global challenge of the first order. To design an effective policy mix, it is crucial to develop a better understanding of shifting cultivators' decision making; to that end, economic modeling is a powerful tool.

This chapter reviews economic models of shifting cultivation and those of deforestation and soil conservation related to shifting cultivation developed by economists over the last two decades. My goal is not to offer a comprehensive review, but to highlight key modeling approaches (what is modeled and what is not, and with what assumptions), clarify how they are useful and incomplete in efforts to examine shifting cultivators' behaviors, and point to promising directions for future modeling. I encourage readers to see other reviews on economic models, such as Kaimowitz and Angelsen (1998) and Barbier and Burgess (2001) for deforestation and Barbier (1997) for land degradation in developing countries. As far as I know, no other reviews on economic models of shifting cultivation are available.

I focus on farm-level models that characterize individual farmers' behaviors (endogenous variables) under certain environmental and institutional conditions, such as resource stock, markets, and property rights (Binswanger & McIntire, 1987).[1] Farm models allow modelers to examine how farmers' behaviors are affected by policy parameters (exogenous variables). Modelers usually focus on individual farmers' key decisions that directly or indirectly determine environmental outcomes of interest (e.g., forest clearing in deforestation models). Although no models fully capture the complexity of the real world, economic models highlight key aspects of the reality to better understand causal mechanisms.

1.1 Modeling approach

Three important choices in modeling approaches require attention: static vs. dynamic modeling, market conditions, and policies. Economic models are generally classified into static or dynamic models; whereas static models capture economic agents' decisions at a point in time, dynamic models consider the potentially changing path of their behaviors. The choice depends on whether agents' decisions at a point in time affect their future decisions. This dynamic linkage is described by state equations, i.e., the law of motion of state variables, which can be the outcome of interest. Although static models characterize agents' optimal decisions at a given point in time, dynamic models characterize the over-time path of their optimal decisions (control variables) and corresponding state variables. For example, in a soil-conservation model, the state variable can be soil stock (or fertility) and the control variables can be farmers' choices that affect soil fertility, such as cultivation intensity and soil conservation input. The simplest dynamic model is a two-period model, although most dynamic models discussed below consider an infinite time horizon, while in this chapter, models are considered to be static when agents make current decisions based only on the present value of the net benefit/cost stream.

Although perfect markets enable an efficient allocation of resources, market imperfection is the norm in developing countries, where most tropical forests are situated. Better understanding market imperfection and non-market institutions has been a central theme of development economics over the last three decades (Bardhan & Udry, 1999; Ray, 1998). Although many shifting cultivation, deforestation, and soil conservation models in the literature assume perfect markets to examine price policies, such as those related to taxes and subsidies, some models consider imperfect factor markets. In particular, although with a perfect labor market a market price (wage) supports a separation of farm households' consumption (labor supply) and production (labor demand) decisions, market imperfection can break this separation (Singh et al., 1986); here wage represents the opportunity cost of

---

[1] Kaimowitz and Angelsen (1998) review deforestation models other than farm-level models, such as regional-level models and national and macro-level models, including general equilibrium models (see also Angelsen & Kaimowitz, 1999). Although tropical forests are often common property, soils are individual farmers' private property; most soil conservation models are farm-level models.

labor in the form of returns to any non-farm activities (Benjamin, 1992). Not surprisingly, market imperfection commonly gives rise to ambiguous policy impacts. In contrast, some models employ a framework that does not involve any factor markets (e.g., models focusing on fallow-cultivation cycle).

Most models examine farm output price (mostly food price) and wage (opportunity cost of labor), which can be altered by various macroeconomic policies; some models also examine input price other than wage, technological progress, and property rights.[2] Many dynamic models highlight the role of the discount rate, which can be altered by credit policies. Some models that consider farmers' decisions with uncertainty – especially in production and price – focus on the roles of risk and risk aversion. Most deforestation models show that promoting farming through price and technology leads to greater forest clearing as the farmers augment farm production; in contrast, promoting non-farm activities discourages forest clearing. Most dynamic models reveal that a lower discount rate encourages investment not only in soils (soil conservation), but also in land holdings (forest clearing). Other policy impacts are generally mixed, depending on modeling specification (assumption). Specific theoretical predictions of each model are not reviewed in this chapter.

1.2 Organization of the chapter

The remainder of the chapter is organized as follows. Sections 2, 3, and 4 review deforestation, soil conservation, and shifting cultivation models, respectively. The main papers cited in these sections are listed in chronological order in Tables 1, 2, and 3, respectively, which summarize decision variables, outcome variables, policy parameters, modeling frameworks (static vs. dynamic), and factor markets (perfect vs. imperfect vs. not modeled).

The tables also report whether the modeling work is accompanied with a substantial empirical analysis; an empirical analysis can be a case study, a descriptive analysis of micro data, simulation work based on micro data, or a regression analysis (to test theoretical hypotheses). Whereas some models – especially those accompanied with an empirical analysis – consider specific empirical contexts (e.g., colonists in Amazonia), others are developed in general contexts. Although this distinction is not always clear, it is clarified when needed. In some models I show mathematical equations to highlight their key features in a concrete way; when I do so, I change original notations (and functions in some cases) to uniform notations for clarity and clear comparisons across models.

Based on these reviews, Section 5 discusses major lacunae in extant shifting cultivation models and promising avenues for future modeling. Section 6 concludes.

2. Deforestation models

Most farm-level deforestation models examine forest-clearing labor as a key decision variable. Assuming a simple function of forest clearing with labor as a unique input (which is valid among small-scale farmers who do not use chainsaws), cleared forest is directly captured by forest-clearing labor.

---

[2]Welfare-augmenting policies are usually considered. It is a straightforward process to examine welfare impacts of specific policies in dynamic models by applying the procedure developed by Caputo (1990) (see Takasaki, 2006 for an example).

	Main decision variables	Main outcome variables	Main policy parameters	Static vs. dynamic	Factor markets	Empirics
Southgate (1990)	Forest-clearing labor, soil conservation labor	Forest-clearing labor, soil conservation labor	Output price, wage, interest rate	Static	Perfect	None
Larson (1991)	Forest-clearing labor, soil conservation labor	Forest-clearing labor, soil conservation labor	Output price, wage, interest rate, technological progress	Static	Perfect	None
DeShazo and DeShazo (1995)	On-farm labor	Forest clearing (land value)	Output price, input price, wage, cost of land clearing	Static	Perfect	None
Bluffstone (1995)	Labor for firewood collection	Firewood collection, forest stock	Wage	Dynamic	Perfect, imperfect	Nepal (simulation)
Angelsen (1999)	Cleared forest (distance)	Cleared forest (distance)	Output price, wage, transport cost, discount rate, population	Static	None, perfect	None
Barrett (1999)	Forest-clearing labor	Forest-clearing labor	Output price - mean and standard deviation	Static	Perfect	Madagascar (case study)
Barbier (2000)	Forest-clearing labor	Cultivated land	Output price, wage	Dynamic	Perfect	Mexico, Ghana (case study)
Pendleton and Howe (2002)	Forest-clearing labor	Cleared forest	Market integration (generated from price and wage), technological progress	Dynamic (2 periods)	Perfect	Bolivia (regression)
van Soest et al. (2002)	Forest-clearing labor	Cleared forest	Technological progress, output price	Static	Perfect, imperfect	None
Takasaki (2007)	Forest-clearing labor	Cleared forest	Output price, wage, land price, discount rate	Dynamic (2 periods)	Perfect, imperfect	None
Delacote (2007)	Proportion of land cultivated	Proportion of land cultivated	Risk, risk aversion, population, forest profitability	Static	Not modeled	None

Table 1. Deforestation models

	Main decision variables	Main outcome variables	Main policy parameters	Static vs. dynamic	Factor markets	Empirics
McConnell (1983)	Soil loss, non-soil input	Soil depth	Tenure	Dynamic	Perfect	None
Barbier (1990)	Soil-degrading input, soil-conserving input	Soil depth	Output price, input price	Dynamic	Perfect	Indonesia (descriptive)
Barrett (1991)	Soil loss, non-soil input	Soil depth	Output price	Dynamic	Perfect	None
Clarke (1992)	Farm input, soil investment	Soil quality	Output price, input price, discount rate	Dynamic	Perfect	None
LaFrance (1992)	Cultivation input, soil-conservation input	Soil stock	Output price	Dynamic	Perfect	None
Krautkraemer (1994)	Soil loss	Soil fertility	Population	Dynamic	Perfect	None
Barrett (1996)	Soil loss, soil-conservation input	Soil depth	Output price, discount rate	Dynamic	Perfect	None
Grepperud (1997a)	Farming labor, soil-conservation labor	Farming labor, soil-conservation labor	Farming support, soil-conservation support, off-farm support	Static	Perfect	None
Grepperud (1997b)	Farm input, investment in soil-conservation structure	Soil stock	Output price, discount rate	Dynamic	Perfect	None
Bulte and van Soest (1999)	Soil loss, farming labor	Soil depth	Output price	Dynamic	Perfect, imperfect	None
Grepperud (2000)	Farming intensity (soil depleting/ conserving)	Soil fertility	Risk aversion	Dynamic	Perfect	None
Lichtenberg (2006)	Soil loss, farming labor	Soil depth	Output price	Dynamic	Perfect	None
Graff-Zivin and Lipper (2008)	Soil carbon-sequestration investment	Soil carbon-sequestration investment	Sequestration cost, output price, discount rate, risk aversion	Dynamic	Perfect	None

Table 2. Soil-conservation models

	Main decision variables	Main outcome variables	Main policy parameters	Static vs. dynamic	Factor markets	Empirics
Barrett (1991)	Cultivation length, fallow length	Fallow-cultivation cycle	Output price	Dynamic	Not modeled	None
Jones and O'Neill (1993)	Proportion of land cultivated	Fallow length	Output price, wage, discount rate, population	Static	Perfect	None
López (1997)	Cleared forest	Cleared forest	Output price	Dynamic	Perfect	Ghana (regression)
Tachibana et al. (2001)	Proportion of upland land cultivated, upland forest cleared	Proportion of upland cultivated, shifting cultivation area, upland forest cleared	Lowland technological progress, lowland farm area, output price, forest-clearing cost, tenure security	Dynamic	Perfect	Vietnam (regression)
Batabyal and Lee (2003)	Fallow length	Fallow length	Return to fallow, discount rate	Dynamic	Not modeled	None
Sylwester (2004)	Proportion of land cultivated	Land quality	Income transfer, output price, population	Dynamic	Not modeled	None
Willassen (2004)	Fallow-cultivation cycle	Fallow-cultivation cycle, soil fertility (present value of gross output)	Output price	Dynamic	Not modeled	None
Takasaki (2006)	Proportion of land cleared	Proportion of land cleared	Output price, wage, discount rate, soil-regeneration rate, soil erosivity	Dynamic	Perfect	None
Pascual and Barbier (2006)	Farming labor (clearing and on-farm labor with a fixed proportion)	Fallow soil fertility, forest clearing	Population	Dynamic	Perfect	Mexico (simulation)
Pascual and Barbier (2007)	Farming labor (clearing and on-farm labor with a fixed proportion)	Fallow soil fertility, forest clearing	Output price	Dynamic	Perfect	Mexico (simulation)
Balsdon (2007)	Cultivation length	Cultivation length	Output price, non-farm income	Dynamic	Not modeled	None
Brown (2008)	Proportion of land cultivated	Proportion of land cultivated	Preference, spatial dependency	Dynamic	Perfect	Cameroon (regression, simulation)

Table 3. Shifting cultivation models

2.1 Static deforestation models

Early deforestation models are static. Southgate (1990), which is elaborated by Larson (1991), considers not only forest-clearing labor, but also soil-conservation labor among colonists in the forest frontier;[3] these two labors separately determine the present value of agricultural production (cropping and livestock) and soil conservation. DeShazo and DeShazo (1995) apply an agricultural household model (Singh et al., 1986) to forest clearing with a perfect labor market, though they capture forest clearing through the value of land (rent), not forest clearing itself. van Soest et al. (2002) directly extend the agricultural household model to forest clearing, comparing effects of farm technological progress on forest clearing under perfect and no labor-market conditions.

Barrett (1999) and Delacote (2007), respectively, examine influences of price and production risk in farming on forest clearing in their static models; Delacote (2007) also addresses effects of risk aversion and returns to standing forest in the form of non-timber forest products (NTFPs).[4]

2.2 Discrete dynamic deforestation models

Static deforestation models effectively treat cleared land as a variable input (produced by labor) for farming. This setup is valid if tropical farmers replace their old infertile plots with newly cleared forest lands every agricultural season or do not consider future production on their cleared lands because of insecure tenure. This is not a common practice among shifting cultivators, because (1) forest clearing is very costly to them (especially with no use of chainsaws), (2) they can employ a variety of traditional soil management techniques (in particular fallowing), and (3) forest clearing and cultivation often give them some claims to the land (Takasaki, 2007). Instead, shifting cultivators crop their cleared lands for more than one agricultural season over time.

Takasaki (2007) treats forest clearing as both an input for current production and an investment in future production in his two-period model. Quality-adjusted land for cultivation at period t is given by:

$$A_1 = a(L_1) \tag{1.1}$$

$$A_2 = (1-\rho)A_1 + a(L_2) \tag{1.2}$$

where L_t is labor allocated to clear forest at period t, a is forest-clearing function, and ρ captures fertility decline through cultivation (depreciation rate). van Soest et al. (2002) use the same forest-clearing function as in equation (1.1); equation (1.2) is a state equation of

---

[3]Although conflicts over property rights are central issues among colonists in the forest frontier (e.g., Alston et al., 2000; Anderson & Hill, 1990; Hotte, 2001; Mueller, 1997), related theoretical modeling is not reviewed in this chapter.

[4]The potential role of NTFPs for sustainable development and poverty alleviation in the tropics is often emphasized (e.g., Arnold & Perez, 2001; Coomes et al., 2004; Wunder, 2001); at the same time, overexploitation of forest resources as local commons among poor populations has been a major concern (i.e., poverty-environment trap) (Barbier, 2010; Dasgupta, 1993, 2001; Jodha, 1986). In particular, firewood collection and associated forest degradation have received much attention. Bluffstone (1995), for example, examines firewood/fodder collection and forest biomass evolution.

crop land. Takasaki (2007) considers not only labor-market conditions, but also land-market conditions, comparing four distinct market institutions (Latin America vs. Sub-Saharan Africa), including the effects of land price.

Some static models, such as Southgate (1990), Larson (1991), and Angelsen (1999), jointly address input and investment aspects of forest clearing by considering the benefit/cost stream over time generated by current forest clearing; such models capture neither farmers' behaviors over time nor the evolution of land assets.[5]

Pendleton and Howe (2002) develop a two-period model for Amerindians in Bolivia, capturing forest clearing in the dry season (period 1) for production in the wet season (period 2). Distinct from other modeling works, Pendleton and Howe (2002) distinguish between primary and secondary forests; they also construct a measure of market integration from market prices.

2.3 Continuous dynamic deforestation models

Following a standard capital model, dynamic farm-level deforestation models consider forest clearing as a pure investment in land capital for future production. This modeling is commonly used to examine a society's optimal deforestation – i.e., exploitation of tropical forests as the commons – in the literature (e.g., Barbier & Burgess, 1997; Ehui et al., 1990; López, 1994; López & Niklitschek, 1991); most models employ control theory in a continuous time framework (e.g., Kamien & Schwartz, 1991; Seietstad & Sydsaeter, 1987).

Assuming that a fixed proportion of arable land (δ) is fallowed in each time period, Barbier (2000) considers the following state equation:

$$\dot{A} = a(L) - \delta A \qquad (2)$$

where time index is suppressed and $\dot{A} = dA/dt$. The depreciation rate δ is effectively the same as ρ in equation (1) in the discrete-time framework.

3. Soil-conservation models

Soil-management measures are classified into two groups based on their costs: one with reduced current output levels, such as less intensified cultivation, forest fallowing, and perennial systems, and the other with input use, which can take various forms, such as mulching, composting, terracing, and creating hedgerows, depending on agroecological conditions in specific locales. Although fertilizer is an essential input in other agricultural systems, fertilizer use is very limited in shifting cultivation that relies heavily on forest-based measures (forest clearing and fallowing) (Nicholaides et al., 1983; Sanchez et al., 1982). Grepperud (1997a) examines how programs supporting farming, soil conservation, and non-farm activities affect labor allocations for these three activities in his static model, in the same spirit as Southgate (1990) and Larson (1991).

---

[5]The key decision variable in Angelsen's model (1999) is the distance to forest cleared. Such spatial modeling, which is common among geographers, is not reviewed in this chapter (other examples of spatial farm-level deforestation models developed by economists include Angelsen, 1994; Chomitz & Gray, 1996; Mendelsohn, 1994). Angelsen (1999) compares four models under distinct modeling assumptions and property rights, not market conditions, in a unified framework.

All soil conservation models developed in the literature examine continuous cultivation with fixed land size.

3.1 Canonical soil dynamics

McConnell (1983) models the dynamics of soil depth x as follows:

$$\dot{x} = \alpha - s \tag{3}$$

where α is natural soil regeneration and s is soil loss associated with cultivation; farm output is a function of soil loss, soil depth (fertility), and non-soil inputs (evaluated at factor price).[6] This model captures only the adjustment of cultivation intensity among soil-management measures.

3.2 Input-based soil-conservation models

Economists have extended McConnell's (1983) dynamic model by incorporating input-based soil-conservation measures in various ways. Clarke (1992) adds soil investment as a choice variable to equation (3); Barbier (1990) and LaFrance (1992) consider inputs for (soil degrading) cultivation and soil conservation separately; Barrett (1996) adds a soil-conservation measure as a function of conservation input to equation (3); and Grepperud (1997b) considers an investment in soil-conservation structure, such as terraces, modeling the joint evolution of soil stock and conservation structure.

Bulte and van Soest (1999) examine the soil dynamics with no labor market, using the following state equation:

$$\dot{x} = \alpha(l) - s \tag{4}$$

where l is labor for soil conservation. Equation (4) captures labor-intensive soil conservation.[7]

Grepperud (2000) examines how risk aversion influences soil conservation with production and price uncertainty. Graff-Zivin and Lipper (2008) examine the farmer's decision on investment in soil carbon sequestration by explicitly modeling soil carbon as well as soil fertility with production risk; they examine effects of sequestration cost and risk aversion, as well as output price and discount rate.

3.3 Continuous vs. cyclical farming

Assuming stock-dependent soil regeneration (cf. equations 3 and 4),

$$\dot{x} = \alpha(x) - s \tag{5}$$

Krautkraemer (1994) shows that in the presence of nonconvexity in the net benefit function, a non-continuous farming strategy – periodic cycles of cultivation and fallow – can be an

---

[6]Barrett (1991) compares McConnell's (1983) models with and without non-soil inputs.
[7]Using equation (4), Bulte and van Soest (2001) examine an environmental Kuznets curve for land degradation with no labor market. Lichtenberg (2006) demonstrates that ambiguous impacts of output price found by Bulte and van Soest (1999) is not attributable to labor-market failure, but can occur depending on the labor supply's wage elasticity.

equilibrium (Lewis & Schmalensee, 1977, 1979) and that population growth leads to a shift from cyclical cultivation to continuous cultivation (*sensu* Boserup, 1965).

4. Shifting cultivation models

Shifting cultivation models in the economics literature can be classified into four: the fallow-cultivation cycle model, the forest-fallow model, the cultivation-intensity model, and the land-replacement model.[8] Almost all models are dynamic; all models except for Tachibana et al. (2001) assume a fixed land size.

4.1 Fallow-cultivation cycle models

Fallow-cultivation cycle models focus on fallow and/or cultivation length as decision variables, ignoring all other decisions, such as labor allocation. Barrett (1991) extends the optimal forest-rotation problem (Faustmann, 1995) to fallow-cultivation cycles by treating both fallow and cultivation lengths as choice variables. This rotation problem does not explicitly capture soil dynamics. In contrast, Willassen (2004) models the cyclical evolution of soil fertility in the cultivation and fallow phases; the farmer chooses only the phase – binary choice $q = 0$ (fallow) or 1 (cultivation) – over time, and distinct from soil conservation models (e.g., equation 3), soil dynamics under cultivation as well as fallow are assumed to be determined by soil fertility level x only.

In these cyclical models, the farmer does no cultivation in the fallow phase. This simplification is for analytical tractability. Of course, in practice, shifting cultivators mix different stages of cultivation and fallow across plots.

Assuming fixed fallow length and on-farm soil dynamics characterized by equation (5), Balsdon (2007) focuses on cultivation length as a choice variable; distinct from other cyclical models, the termination of the cultivation phase in one plot is instantly followed by cultivation on the next plot. Batabyal and Lee (2003), in contrast, focus on the choice of fallow length.

4.2 Cultivation-intensity models

Cultivation-intensity models capture soil degradation resulting from shortened fallow through the cultivation-intensity measure without explicitly modeling fallow dynamics. Although cultivation-intensity models differ depending on their focus, their common feature is to capture cultivation intensity through the proportion of land cultivated (b). For a given land size, $1 - b$ is the proportion of fallow land and $1/b$ represents fallow length. For example, for $b = .1$, fallow length is 10 (years).

4.2.1 Early cultivation-intensity models

Larson and Bromley (1990) develop a dynamic model with a fixed cultivation intensity. Jones and O'Neill (1993) develop a static model using cultivation intensity b as a key decision variable.[9]

---

[8]Batabyal and Beladi (2004) and Batabyal and Nijkamp (2009) apply stochastic modeling to shifting cultivation, which is not reviewed in this chapter.
[9]Jones and O'Neill (1993) extend their model to a spatial model.

4.2.2 Cultivation-intensity models with soil dynamics

In Sylwester's (2004) model, the soil dynamics under cultivation follows equation (5), with soil loss s replaced with a function of cultivation intensity b; distinct from other cultivation-intensity models, Sylwester does not model factor markets as in fallow-cultivation cycle models.

Whereas Brown (2008) considers a binary choice between cultivation and fallow – on each plot over time – as in fallow-cultivation cycle models, he solves the dynamic problem by treating this binary variable q as continuous; that is, he effectively uses cultivation intensity b as a choice variable. His focus is to examine the roles of preference (measured by the revealed preference approach) and spatial dependency in farmers' forest clearing using simulation (see also Brown, 2006).

4.2.3 Cultivation-intensity models with land dynamics

Tachibana et al. (2001) develop a cultivation-intensity model that endogenizes the evolution of upland holdings (T) among Vietnamese farmers who combine upland shifting cultivation and lowland paddy cultivation:

$$\dot{T} = a - \delta(b)bT \tag{6}$$

where a is (upland) forest cleared and endogenized depreciation rate $\delta(b)$ (cf. equation 2) captures soil degradation through shortened fallow (higher b captures depriving intensification). Note that distinct from equation (2), T is total land holdings, consisting of cultivated land bT (=A) and fallow land $(1-b)T$ (= $T - A$). Furthermore, fallow land is under the risk of being grabbed by neighbors. Tachibana et al. (2001) examine how the proportion of cultivated upland land (inverse of fallow length), shifting cultivation area, and upland forest clearing are affected by a rich set of policies, such as lowland technological progress, lowland farm area, forest clearing cost, and upland tenure security, as well as output price.

4.3 Forest-fallow models
4.3.1 Forest-fallow models with communal fallow forest

Forest-fallow models endogenize the dynamics of biomass accumulation in fallow forest as a soil builder. Fallow forest is explicitly or implicitly assumed to be communally owned by villagers. López (1997) introduces the following dynamics of fallow biomass density η:

$$\dot{\eta} = \gamma - \frac{\sum_i a_i}{Q}\eta \tag{7}$$

where γ is the intrinsic growth of secondary vegetation, a_i is cleared forest by household i, and Q is total land area under both cultivation and fallow – of the village. Equation (7) assumes that fallow biomass density is determined by the proportion of cleared forest land for cultivation, i.e., village-level cultivation intensity.[10]

Assuming equation (7) and a simple conversion of biomass to soil fertility on cleared fallow forest, Pascual and Barbier (2006; 2007) derive the dynamics of soil fertility on cleared forest (Pascual & Barbier, 2006, equation 5). They assume that in each period of time the farmer

---

[10]In the forest-fallow model, adding NTFPs collected from secondary fallow forest as an additional benefit of fallowing is a straightforward extension.

cultivates only the cleared land; then, on-farm soil conservation is irrelevant. In Pascual and Barbier (2006; 2007), the only decision variable is farm labor, which is assumed to be allocated between forest clearing and cultivation with a fixed proportion. Pascual and Barbier (2006; 2007) examine impacts of population density (n/Q, where n is the number of households in the village) and output price on forest clearing and fallow soil fertility.

4.3.2 Forest-fallow models with private fallow forest
Shifting cultivators commonly have usufruct of not only the cultivated land they have cleared, but also their fallow land; customary tenure of fallow land tends to be insecure, however, and this tenure insecurity influences their forest clearing and fallowing decisions (Otsuka & Place, 2001; Place & Otsuka, 2001; Tachibana et al., 2001). It is straightforward to revise equation (7) to characterize such an alternative customary tenure setting; then, soil fertility of cleared fallow forest is effectively determined by fallow length or the inverse of cultivation intensity, $1/b$. In this way, the fallow-forest model with private fallow forest directly corresponds to the cultivation-intensity model; a key difference is that the former focuses on fallow dynamics and the latter highlights other dynamics, such as on-farm soil or land holdings.

4.4 Land-replacement models
Fallow-cultivation cycle models assume a cyclical switch of the whole land between cultivation and fallow; fallow-forest models assume that the farmer cultivates cleared forest land only in each period of time. In practice, shifting cultivators replace some depleted plots with cleared forest land each time, while continuing to cultivate the remaining plots; replacing all plots simultaneously is a polar case.

This aspect is explicitly captured in the land-replacement model (with fixed land size) introduced by Takasaki (2006). The key choice variable is the proportion of cultivated land, not total land, replaced with cleared forest land (c). This modeling approach highlights the tension between replaced (cleared) and non-replaced (remaining) plots – the former is more fertile but clearing is costly. It also directly captures new soils on cleared forest land added to soils on remaining plots. Specifically, the dynamics of on-farm soil stock is obtained by extending equation (3):

$$\dot{x} = \varphi c + \alpha (1 - c) - s \tag{8}$$

where φ is soil stock (per unit of land) of cleared forest (see Takasaki, 2006, Figure 1 for derivation). Note that for $c = 0$ (continuous cultivation), equation (8) is the same as (3); for $c = 1$ (complete replacement), equation (8) corresponds to forest-fallow models, though fallow dynamics is not modeled (φ is not endogenized). Takasaki (2006) examines effects on forest clearing (measured by c) of soil-regeneration rate α and soil erosivity altered by soil conservation programs, as well as output price, wage, and discount rate.

5. Discussion

5.1 Primary vs. secondary forests
The review in the last section indicates two significant lacunae in the extant shifting cultivation models. The first lacuna is that the extant models do not distinguish between

primary and secondary forests.[11] This distinction is critically important for both environmental and economic reasons. First, in general, protecting primary forest with greater biodiversity needs to be given a higher priority than secondary forest protection. At the same time, as primary forest becomes scarce in the tropics, researchers and practitioners pay greater attention to secondary fallow forest (Coomes et al., 2000). In particular, short fallow results in less matured secondary forest with limited biomass accumulation and poor protection of erodible soils, as well as low biodiversity, weak carbon sequestration, and limited timber and NTFPs (Brown & Lugo, 1990; Chazdon et al., 2009; Dalle & de Bois, 2006; Dent & Wright, 2009; Lawrence et al., 2005). Shifting cultivation models need to jointly address cleared primary forest and fallow length of secondary forest as key environmental outcomes.

Second, the choice between primary and secondary forest is determined by farmers' decisions under specific environmental and economic conditions: In particular, secondary forest is less fertile but easier to clear than primary forest (Scatena et al., 1996), and this comparison depends on fallow length (farmer's decision) (Dvořàk, 1992) and the availability of primary forest (determined by population growth, etc.). This choice also has a direct implication for asset accumulation: Although clearing secondary forest does not alter total land holdings (only the plot phase changes from fallow to cultivation), clearing new primary forest augments land holdings. That is, although secondary forest brings fertile soil, primary forest brings both more fertile soil and new land itself. Shifting cultivation models need to capture these key differences.

Pendleton and Howe (2002) address the choice between primary and secondary forests as a pure forest-clearing problem; they neither model the role of secondary fallow forest as a soil builder nor consider soil addition through primary forest clearing. No other deforestation models distinguish or specify the type of cleared forest; this is also true in dynamic deforestation models, which necessarily involve land accumulation (Barbier, 2000; Takasaki, 2007). Not only all soil conservation models but also most shifting cultivation models assume fixed land holdings, and thus implicitly focus on secondary forest; Tachibana et al. (2001) do not distinguish or specify the type of cleared forest, either.

This lacuna in the theoretical literature is in contrast to the considerable number of empirical studies on primary and secondary forests. Smith et al. (1999), for example, show that the relative importance of secondary forest to primary forest increases over time among Amazonian colonists; Coomes et al. (2000; 2011) also find this pattern over a longer time span among Amazonian peasants (in their study village in Peru, primary forest has virtually disappeared).

5.2 On-farm soil conservation in shifting cultivation

Supporting non-farm activities discourages farming, thereby releasing pressure on forests. This policy option becomes available and significant only after non-agricultural sectors sufficiently develop, often following massive deforestation and forest degradation. What policies can slow down this trend along the development path?

---

[11] Primary forest "has had little or no anthropogenic intervention" and secondary forest is "woody successional vegetation that regenerates after the original forest cover has been removed for agriculture or cattle ranching" (Smith et al., 1999, p.86).

The second lacuna not only in the extant theoretical works on shifting cultivation, but also in related empirical works is the investigation into potential roles of on-farm soil conservation. Among poor shifting cultivators, forest-based soil-management options (forest clearing and fallowing) outweighs on-farm soil conservation (Barbier, 1997); when degraded land can be easily replaced, farmers have little incentive to adopt expensive input-based soil-conservation measures. Then, the question is whether policy makers can alter shifting cultivators' benefit-cost calculations by introducing effective soil-conservation programs, as discussed by Takasaki (2006) (see also Grepperud, 1997a).

Although developing locally adoptable, effective soil-conservation measures in tropical forests has been a daunting task (Lal, 1995), soil scientists' recent growing interest in biochar in Amazonia may lead to significant improvement in soil fertility and soil carbon sequestration in shifting cultivation systems (Glaser, 2007; Marris, 2006; Steiner et al., 2004). Biochar, also known as black carbon, is the residue of organic matter that has been pyrolyzed (partially combusted in a low-oxygen environment). Research indicates that Amazonian black carbon (*terra preta*) has, on average, three times more soil organic matter (SOM) content, higher nutrient levels, and a better nutrient retention capacity than surrounding infertile soils (Glaser, 2007). How the labor-intensive alternative "slash-and-char" system, combined with sustainable charcoal production, can be promoted among poor shifting cultivators is still an open question, however (Swami et al., 2009) (see Coomes & Burt, 2001 for charcoal production among Amazonian peasants).

Soil-conservation models extensively developed in the literature can well capture various input-based soil-conservation measures; in particular, equation (4) or its variant can be applied to labor-intensive conservation like biochar.

5.3 Shifting cultivation regimes

It is very useful to differentiate two regimes of shifting cultivation. In regime 1, where primary forest is available, farmers choose to clear primary or secondary forest. Although the extant deforestation and shifting cultivation models effectively capture primary forest clearing and secondary fallow forest clearing (cyclical cultivation), respectively, neither of them addresses the choice of these two. As primary forest becomes scarce (deforestation), cultivation shifts to regime 2, in which only secondary forest is cleared; in another words, primary forest has been so degraded that clearing primary forest is too costly or simply not an available option. Policies effectively protecting primary forest (in particular, protected areas with compliance) can also make this regime shift.[12] Although the extant shifting

---

[12] Migration can also significantly affect the regime shift. Coomes et al. (2011) find that urban migration plays an important role in lowering pressure on diminishing forest land among shifting cultivators in their study village. The extensive migration option in the forest frontier, however, may allow farmers to clear forest – both primary and secondary – without employing fallowing practices; this is possible among colonists in land-abundant areas in Latin America, especially in locales where selling cleared lands is an additional motive for forest clearing (Barbier, 2004; Binswanger, 1991; Takasaki, 2007). Conceptually, further regime shifts following regime 2 can be considered. Once shifting cultivators start to employ continuous cultivation on some plots, regime 3 emerges; in this new regime, in addition to forest fallow management, farmers make a key choice between shifting and continuous. Lastly, regime 3 is followed by the complete shift to continuous cultivation, i.e., abandonment of shifting cultivation (Krautkraemer, 1994).

cultivation models essentially focus on regime 2, protecting remaining primary forest and promoting sustainable secondary forest management (long fallow) in regime 1 should be given a higher priority for conservation and development in shifting cultivation systems.

5.4 Future modeling

It is now clear that a promising avenue for future modeling of shifting cultivation is to extend extant models for secondary fallow forest in regime 2 by adding primary forest clearing to capture regime 1 and by endogenizing on-farm soil conservation to examine its effects on forest outcomes. That is, a unified farm model of primary forest clearing, forest fallowing, and on-farm soil conservation is needed to examine effective policies for protecting primary forest and maintaining sustainable long fallow.

Two extensions toward such a unified model are suggested. The first is to augment a cultivation-intensity model so that it captures the dynamics of both on-farm soil and land holdings (through primary forest clearing). Such an augmented model could explicitly capture the mechanism of depriving intensification embedded in $\delta(b)$ in equation (6).

The second extension is to augment Takasaki's (2006) land-replacement model by endogenizing cultivation intensity and capturing acquisition of new land and soil through primary forest clearing. The proportion of total land, not cultivated land, replaced with fallow forest is bc, and fallow length $1/bc$ determines the soil stock of cleared fallow forest φ in equation (8).

5.5 Hypothetical effects of on-farm soil conservation

How does better on-farm soil conservation affect forest outcomes? On one hand, in regime 2 with no primary forest clearing, it is expected that shifting cultivators intensify on-farm soil conservation and rely less on fallow soils (less frequent clearing), resulting in longer fallow. On the other hand, in regime 1, better on-farm soil conservation encourages shifting cultivators to clear more primary forest with increased returns to farming; at the same time, primary forest clearing (land accumulation) is balanced with secondary forest clearing (fallow management). A well-designed soil conservation program might result in longer fallow at the cost of primary forest; then, it becomes crucial to combine the soil program with other measures to protect primary forest, such as protected areas.

The unified farm model proposed above can dissect shifting cultivators' benefit-cost calculations, shedding light on an effective policy mix for conservation and development and pointing to promising avenues for empirical research.

6. Conclusion

This chapter reviewed farm-level economic models of shifting cultivation, as well as those of deforestation and soil conservation related to shifting cultivation. Although economists have made significant progress in modeling shifting cultivation over the last two decades, extant economic models neither clearly distinguish between primary and secondary forests nor address potential roles of on-farm soil conservation in shifting cultivation. Developing a unified farm model of primary forest clearing, forest fallowing, and on-farm soil conservation is needed to examine effective policies for protecting primary forest and maintaining sustainable secondary fallow forest. The chapter pointed to promising avenues for future modeling.

7. Acknowledgment

This chapter has benefited significantly from the comments and suggestions of Oliver Coomes. This research has been made possible through financial support provided by the Japan Society for the Promotion of Science and the Ministry of Education, Culture, Sports, Science and Technology in Japan. Any errors of interpretation are solely the author's responsibility.

8. References

Alston, L. J., Libecap, G. D. & Mueller, B. (2000). Land reform policies, the sources of violent conflict, and implications for deforestation in the Brazilian Amazon. *Journal of Environmental Economics and Management,* Vol. 39, No., 162-188.

Anderson, T. L. & Hill, P. J. (1990). The race for property rights. *Journal of Law and Economics,* Vol. 33, No. 1, 177-197.

Angelsen, A. (1994). Shifting cultivation expansion and intensity of production: the open economy case. Working Paper 3, Chr. Michelsen Institute, Bergen.

Angelsen, A. (1999). Agricultural expansion and deforestation: modeling the impact of population, market forces and property rights. *Journal of Development Economics,* Vol. 58, No. 1, 185-218.

Angelsen, A. (Ed.), (2008). *Moving ahead with REDD: Issues, options and implications,* Center for International Forestry Research, Bogor.

Angelsen, A. & Kaimowitz, D. (1999). Rethinking the causes of deforestation: lessons from economic models. *The World Bank Research Observer,* Vol. 14, No. 1, 73-98.

Angelsen, A. & Wunder, S. (2003). Exploring the forest-poverty link: key concepts, issues and research implications. CIFOR Occasional Paper No. 40, Center for International Forestry Research, Bogor.

Arnold, J. E. M. & Perez, M. R. (2001). Can non-timber forest products match tropical forest conservation and development objectives? *Ecological Economics,* Vol. 39, No. 3, 437-447.

Balsdon, E. M. (2007). Poverty and the Management of Natural Resources: A Model of Shifting Cultivation. *Structural Change and Economic Dynamics,* Vol. 18, No. 3, 333-347.

Barbier, E. B. (1990). The farm-level economics of soil conservation: the uplands of Java. *Land Economics,* Vol. 66, No. 2, 198-211.

Barbier, E. B. (1997). The economic determinants of land degradation in developing countries. *Philosophical Transactions of the Royal Society of London, Series B,* Vol. 352, No. 1356, 891-899.

Barbier, E. B. (2000). Links between economic liberalization and rural resource degradation in the developing regions. *Agricultural Economics,* Vol. 23, No. 3, 299-310.

Barbier, E. B. (2004). Agricultural expansion, resource boom and growth in Latin America: implications for long-run economic development. *World Development,* Vol. 32, No. 1, 137-157.

Barbier, E. B. (2010). Poverty, Development, and Environment. *Environment and Development Economics,* Vol. 15, No. 6, 635-660.

Barbier, E. B. & Burgess, J. C. (1997). The economics of tropical forest land use options. *Land Economics*, Vol. 73, No. 2, 174-195.

Barbier, E. B. & Burgess, J. C. (2001). The economics of tropical deforestation. *Journal of Economic Surveys*, Vol. 15, No. 3, 413-433.

Bardhan, P. & Udry, C. (1999). *Development Microeconomics*, Oxford University Press, Oxford.

Barrett, C. B. (1999). Stochastic food prices and slash-and-burn agriculture. *Environment and Development Economics*, Vol. 4, No. 2, 161-176.

Barrett, S. (1991). Optimal soil conservation and the reform of agricultural pricing policies. *Journal of Development Economics*, Vol. 36, No. 2, 167-187.

Barrett, S. (1996). Microeconomic responses to macroeconomic reforms in the optimal control of soil erosion. In: *The Environment and Emerging Development Issues*, P. Dasgupta & K. G. Mäler, (Eds.), Clarendon Press, Oxford.

Batabyal, A. A. & Beladi, H. (2004). Swidden agriculture in developing countries. *Review of Development Economics*, Vol. 8, No. 2, 255-265.

Batabyal, A. A. & Lee, D. M. (2003). Aspects of land use in slash and burn agriculture. *Applied Economics Letters*, Vol. 10, No. 13, 821-824.

Batabyal, A. A. & Nijkamp, P. (2009). The Fallow and the Non-fallow States in Swidden Agriculture: A Stochastic Analysis. *Letters in Spatial and Resource Sciences*, Vol. 2, No. 1, 45-51.

Benjamin, D. (1992). Household composition, labor markets and labor demand: testing for separation in agricultural household models. *Econometrica*, Vol. 60, No. 2, 287-322.

Binswanger, H. P. (1991). Brazilian policies that encourage deforestation in the Amazon. *World Development*, Vol. 19, No. 7, 821-829.

Binswanger, H. P. & McIntire, J. (1987). Behavioral and material determinants of production relations in land-abundant tropical agriculture. *Economic Development and Cultural Change*, Vol. 36, No. 1, 73-99.

Bluffstone, R. (1995). The effect of labor market performance on deforestation in developing countries under open access: an example from rural Nepal. *Journal of Environmental Economics and Management*, Vol. 29, No. 1, 42-63.

Boserup, E. (1965). *The Conditions of Agricultural Growth: The Economics of Agrarian Change under Population Pressure*, Aldine, New York.

Brown, D. R. (2006). Personal preferences and intensification of land use: their impact on southern Cameroonian slash-and-burn agroforestry systems. *Agroforestry Systems*, Vol. 68, No., 53-67.

Brown, D. R. (2008). A spatiotemporal model of shifting cultivation and forest cover dynamics. *Environment and Development Economics*, Vol. 13, No. 5, 643-671.

Brown, S. & Lugo, A. E. (1990). Tropical secondary forests. *Journal of Tropical Ecology*, Vol. 6, No. 1, 1-32.

Bulte, E. H. & van Soest, D. P. (1999). A note on soil depth, failing markets and agricultural pricing. *Journal of Development Economics*, Vol. 58, No. 1, 245-254.

Bulte, E. H. & van Soest, D. P. (2001). Environmental degradation in developing countries: households and the (reverse) Environmental Kuznets Curve. *Journal of Development Economics*, Vol. 65, No. 1, 225-235.

Byron, N. & Arnold, M. (1999). What future for the people of the tropical forests? *World Development*, Vol. 27, No. 5, 789-805.

Caputo, M. R. (1990). How to do comparative dynamics on the back of an envelope in optimal control theory. *Journal of Economic Dynamics and Control*, Vol. 14, No. 3-4, 655-683.

Chazdon, R. L., Peres, C. A., Dent, D., Sheil, D., Lugo, A. E., Lamb, D., Stork, N. E. & Miller, S. E. (2009). The potential for species conservation in tropical secondary forests. *Conservation Biology*, Vol. 23, No. 6, 1406-1417.

Chomitz, K. M. & Gray, D. A. (1996). Roads, Land Use, and Deforestation: A Spatial Model Applied to Belize. *World Bank Economic Review*, Vol. 10, No. 3, 487-512.

Clarke, H. R. (1992). The supply of non-degraded agricultural land. *Australian Journal of Agricultural Economics*, Vol. 36, No. 1, 31-56.

Coomes, O. T., Barham, B. L. & Takasaki, Y. (2004). Targeting conservation-development initiatives in tropical forests: insights from analyses of rain forest use and economic reliance among Amazonian peasants. *Ecological Economics*, Vol. 51, No. 1-2, 47-64.

Coomes, O. T. & Burt, G. J. (2001). Peasant charcoal production in the Peruvian Amazon: rainforest use and economic reliance. *Forest Ecology and Management*, Vol. 140, No. 1, 39-50.

Coomes, O. T., Grimard, F. & Burt, G. J. (2000). Tropical forests and shifting cultivation: secondary forest fallow dynamics among traditional farmers of the Peruvian Amazon. *Ecological Economics*, Vol. 32, No. 1, 109-124.

Coomes, O. T., Takasaki, Y. & Rhemtulla, J. (2011). Land-use poverty traps identified in shifting cultivation systems shape long-term tropical forest cover. *Proceedings of the National Academy of Sciences of the United States of America*, Vol. 108, No. 34, 13925-13930.

Dalle, S. P. & de Bois, S. (2006). Shorter fallow cycles affect the availability of noncrop plant resources in a shifting cultivation system. *Ecology and Society*, Vol. 11, No. 2, 2.

Dasgupta, P. (1993). *An Inquiry into Well-Being and Destitution*, Clarendon Press, Oxford.

Dasgupta, P. (2001). *Human Well-being and the Natural Environment*, Oxford University Press, Oxford.

Delacote, P. (2007). Agricultural expansion, forest products as safety nets, and deforestation. *Environment and Development Economics*, Vol. 12, No. 2, 235-249.

Denevan, W. M. & Padoch, C. (Eds.), (1987). *Swidden-Fallow Agroforestry in the Peruvian Amazon*, The New York Botanical Garden, Bronx.

Dent, D. H. & Wright, S. J. (2009). The future of tropical species in secondary forests: a quantitative review. *Biological Conservation*, Vol. 142, No. 12, 2833-2843.

DeShazo, R. P. & DeShazo, J. R. (1995). An economic model of smallholder deforestation: a consideration of the shadow value of land on the frontier. In: *Management of tropical forests: towards an integrated perspective*, O. Sandbukt, (Ed.), Center for Development and the Environment, University of Oslo, Oslo.

Dvořàk, K. A. (1992). Resource management by West African farmers and the economics of shifting cultivation. *American Journal of Agricultural Economics*, Vol. 74, No. 3, 809-815.

Ehui, S., Hertel, T. W. & Preckel, P. V. (1990). Forest resource depletion, soil dynamics, and agricultural productivity in the tropics. *Journal of Environmental Economics and Management*, Vol. 18, No. 2, 136-154.

Faustmann, M. (1995). Calculation of the value which forest land and immature stands possess for forestry. *Journal of Forest Economics*, Vol. 1, No. 1, 7-44.

Giaradina, P. M., Sanford, R. L., Dockersmith, I. C. & Jaramillo, V. J. (2000). The effects of slash burning on ecosystem nutrients during the land preparation phase of shifting cultivation. *Plant Soil*, Vol. 220, No. 1/2, 247-260.

Glaser, B. (2007). Prehistorically modified soils of central Amazonia: a model for sustainable agriculture in the twenty-first century. *Philosophical Transactions of The Royal Society B*, Vol. 362, No., 187-196.

Goldammer, J. G. (1993). Historical biogeography of fire: tropical and subtropical. In: *Fire in the environment: the ecological, atmospheric, and climatic importance of vegetation fires*, P. J. Crutzen & J. G. Goldammer, (Eds.), Wiley, New York.

Graff-Zivin, J. & Lipper, L. (2008). Poverty, Risk, and the Supply of Soil Carbon Sequestration. *Environment and Development Economics*, Vol. 13, No. 3, 353-373.

Grepperud, S. (1997a). Soil conservation and government policies in tropical areas: does aid worsen the incentives for arresting erosion? *Agricultural Economics*, Vol. 12, No. 2, 129-140.

Grepperud, S. (1997b). Soil conservation as an investment in land. *Journal of Development Economics*, Vol. 54, No. 2, 455-467.

Grepperud, S. (2000). Optimal Soil Depletion with Output and Price Uncertainty. *Environment and Development Economics*, Vol. 5, No. 3, 221-240.

Hotte, L. (2001). Conflicts over property rights and natural-resource exploitation at the frontier. *Journal of Development Economics*, Vol. 66, No. 1, 1-21.

Jodha, N. S. (1986). Common property resources and rural poor in dry region of India. *Economic and Political Weekly*, Vol. 21, No. 27, 1169-1181.

Jones, D. W. & O'Neill, R. V. (1993). Human-environmental influences and interactions in shifting agriculture when farmers form expectations rationally. *Environment and Planning A*, Vol. 25, No. 1, 121-136.

Kaimowitz, D. & Angelsen, A. (1998). *Economic Models of Tropical Deforestation: A Review*, Center for International Forestry Research, Bogor.

Kamien, M. I. & Schwartz, N. L. (1991). *Dynamic Optimization: The Calculus of Variations and Optimal Control in Economics and Management*, Elsevier, Amsterdam.

Kleinman, P. J. A., Pimentel, D. & Bryant, R. B. (1995). The ecological sustainability of slash-and-burn agriculture. *Agriculture, Ecosystems and Environment*, Vol. 52, No. 2-3, 235-249.

Krautkraemer, J. A. (1994). Population growth, soil fertility, and agricultural intensification. *Journal of Development Economics*, Vol. 44, No. 2, 403-428.

LaFrance, J. T. (1992). Do increased commodity prices lead to more or less soil degradation? *Australian Journal of Agricultural Economics*, Vol. 36, No. 1, 57-82.

Lal, R. (1995). *Sustainable management of soil resources in the humid tropics*, United Nations University Press, Tokyo.

Larson, B. A. (1991). The causes of land degradation along "spontaneously" expanding agricultural frontier in the third world: comment. *Land Economics*, Vol. 67, No. 2, 260-266.

Larson, B. A. & Bromley, D. W. (1990). Property rights, externalities, and resource degradation. *Journal of Development Economics*, Vol. 33, No. 2, 235-262.

Lawrence, D., Suma, V. & Mogea, J. P. (2005). Changes in species composition with repeated shifting cultivation: limited role of soil nutrients. *Ecological Applications*, Vol. 15, No. 6, 1953-1967.

Lewis, T. R. & Schmalensee, R. (1977). Nonconvexity and optimal exhaustion of renewable resources. *International Economic Review*, Vol. 18, No. 3, 535-552.

Lewis, T. R. & Schmalensee, R. (1979). Nonconvexity and optimal harvesting strategies for renewable resources. *Canadian Journal of Economics*, Vol. 12, No. 4, 677-691.

Lichtenberg, E. (2006). "A note on soil depth, failing markets and agricultural pricing": Comments. *Journal of Development Economics*, Vol. 81, No. 1, 236-243.

López, R. (1994). The environment as a factor of production: the effects of economic growth and trade liberalization. *Journal of Environmental Economics and Management*, Vol. 27, No. 2, 163-184.

López, R. (1997). Environmental externalities in traditional agriculture and the impact of trade liberalization: the case of Ghana. *Journal of Development Economics*, Vol. 53, No. 1, 17-39.

López, R. & Niklitschek, M. (1991). Dual economic growth in poor tropical areas. *Journal of Development Economics*, Vol. 36, No. 2, 189-211.

Marris, E. (2006). Putting the carbon back: black is the new green. *Nature*, Vol. 442, No., 624-626.

McConnell, K. E. (1983). An economic model of soil conservation. *American Journal of Agricultural Economics*, Vol. 65, No. 1, 83-89.

Mendelsohn, R. (1994). Property rights and tropical deforestation. *Oxford Economic Papers*, Vol. 46, No. Oct., 750-756.

Mueller, B. (1997). Property rights and the evolution of a frontier. *Land Economics*, Vol. 73, No. 1, 42-57.

Myers, N. (1992). *The Primary Source: Tropical Forests and Our Future*, W. W. Norton, New York.

Nicholaides, J. J. I., Sanchez, P. A., Bandy, D. E., Villachica, J. H., Coutu, A. J. & Valverde, C. S. (1983). Crop production systems in the Amazon basin. In: *The Dilemma of Amazonian Development*, E. F. Moran, (Ed.), Westview Press, Boulder.

Otsuka, K. & Place, F. (Eds.), (2001). *Land Tenure and Natural Resource Management*, The Johns Hopkins University Press, Baltimore and London.

Pascual, U. & Barbier, E. B. (2006). Deprived land use intensification in forest-fallow agricultural systems under population pressure. *Agricultural Economics*, Vol. 34, No. 2, 155-165.

Pascual, U. & Barbier, E. B. (2007). On Price Liberalization, Poverty, and Shifting Cultivation: An Example from Mexico. *Land Economics*, Vol. 83, No. 2, 192-216.

Pendleton, L. H. & Howe, E. L. (2002). Market integration, development, and smallholder forest clearance. *Land Economics*, Vol. 78, No. 1, 1-19.

Place, F. & Otsuka, K. (2001). Population, tenure, and natural resource management: the case of customary land area in Malawi. *Journal of Environmental Economics and Management*, Vol. 41, No. 1, 13-32.

Ray, D. (1998). *Development Economics*, Princeton University Press, Princeton.

Reardon, T. & Vosti, S. A. (1995). Links between rural poverty and the environment in developing countries: asset categories and investment poverty. *World Development*, Vol. 23, No. 9, 1495-1506.

Ruthenberg, H. (1980). *Farming Systems in the Tropics*, Clarendon Press, Oxford.

Sanchez, P. A., Bandy, D. E., Villachica, J. H. & Nicholaides, J. J. (1982). Amazon basin soils: management for continuous crop production. *Science*, Vol. 216, No. 4548, 821-827.

Scatena, F. N., Walker, R. T., Homma, A. K. O., de Conto, A., Ferreira, C. A. P., Carvalho, R. d. A., de Rocha, A. C. P. N., dos Santos, A. I. M. & de Oliverira, P. M. (1996). Cropping and fallowing sequences of small farmers in the "terra firme" landscape of the Brazilian Amazon: a case study from Santarem, Para. *Ecological Economics*, Vol. 18, No. 1, 29-40.

Seietstad, A. & Sydsaeter, K. (1987). *Optimal Control Theory with Economic Applications*, Elsevier, Amsterdam.

Singh, I., Squire, L. & Strauss, J. (Eds.), (1986). *Agricultural Household Models: Extensions, Applications, and Policy*, The John Hopkins University Press, Baltimore.

Smith, J., van de Kop, P., Reategui, K., Lombardi, I., Sabogal, C. & Diaz, A. (1999). Dynamics of secondary forests in slash-and-burn farming: interactions among land use types in the Peruvian Amazon. *Agriculture Ecosystems & Environment*, Vol. 76, No. 2-3, 85-98.

Southgate, D. (1990). The causes of land degradation along "spontaneously" expanding agricultural frontiers in the third world. *Land Economics*, Vol. 66, No. 1, 93-101.

Steiner, C., Teixeira, W. G. & Zech, W. (2004). Slash and char: an alternative to slash and burn practiced in the Amazon basin. In: *Amazonian Dark Earths: Origins, Properties, and Management*, J. Lehman, D. C. Kern, B. Glaser & J. Woods, (Eds.), Kluwer Academic Publishers, Dordrecht.

Sunderlin, W. D., Angelsen, A., Belcher, B., Burgess, P., Nasi, R., Santos, L. & Wunder, S. (2005). Livelihoods, forests, and conservation in developing countries: an overview. *World Development*, Vol. 33, No. 9, 1383-1402.

Swami, S. N., Steiner, C., Teixeira, W. G. & Lehmann, J. (2009). Charcoal making in the Brazilian Amazon: economic aspects of production and carbon conversion efficiencies of kilns. In: *Amazonian Dark Earths: Wim Sombroek's Vision*, W. I. Woods, (Ed.), Springer, Heidelberg.

Sylwester, K. (2004). Simple Model of Resource Degradation and Agricultural Productivity in a Subsistence Economy. *Review of Development Economics*, Vol. 8, No. 1, 128-40.

Tachibana, T., Nguyen, T. M. & Otsuka, K. (2001). Agricultural intensification versus extensification: a case study of deforestation in the northern-hill region of Vietnam. *Journal of Environmental Economics and Management*, Vol. 41, No. 1, 44-69.

Takasaki, Y. (2006). A model of shifting cultivation: can soil conservation reduce deforestation? *Agricultural Economics*, Vol. 35, No. 2, 193-201.

Takasaki, Y. (2007). Dynamic household models of forest clearing under distinct land and labor market institutions: can agricultural policies reduce tropical deforestation? *Environment and Development Economics*, Vol. 12, No. 3, 423-443.

van Soest, D. P., Bulte, E. H., Angelsen, A. & van Kooten, G. C. (2002). Technological change and tropical deforestation: a perspective at the household level. *Environment and Development Economics*, Vol. 7, No. 2, 269-280.

Willassen, Y. (2004). On the economics of the optimal fallow-cultivation cycle. *Journal of Economic Dynamics and Control*, Vol. 28, No. 8, 1541-1556.

Wilshusen, P. R., Brechin, S. R., Fortwangler, C. L. & West, P. C. (2002). Reinventing the square wheel: critique of a resurgent "protection paradigm" in international biodiversity conservation. *Society and Natural Resources*, Vol. 15, No. 1, 17-40.

Wunder, S. (2001). Poverty alleviation and tropical forests - what scope for synergies? *World Development*, Vol. 29, No. 11, 1817-1833.

Wunder, S. (2006). The efficiency of payments for environmental services in tropical conservation. *Conservation Biology*, Vol. 21, No. 1, 48-58.

Permissions

The contributors of this book come from diverse backgrounds, making this book a truly international effort. This book will bring forth new frontiers with its revolutionizing research information and detailed analysis of the nascent developments around the world.

We would like to thank Paulo Moutinho, for lending his expertise to make the book truly unique. He has played a crucial role in the development of this book. Without his invaluable contribution this book wouldn't have been possible. He has made vital efforts to compile up to date information on the varied aspects of this subject to make this book a valuable addition to the collection of many professionals and students.

This book was conceptualized with the vision of imparting up-to-date information and advanced data in this field. To ensure the same, a matchless editorial board was set up. Every individual on the board went through rigorous rounds of assessment to prove their worth. After which they invested a large part of their time researching and compiling the most relevant data for our readers. Conferences and sessions were held from time to time between the editorial board and the contributing authors to present the data in the most comprehensible form. The editorial team has worked tirelessly to provide valuable and valid information to help people across the globe.

Every chapter published in this book has been scrutinized by our experts. Their significance has been extensively debated. The topics covered herein carry significant findings which will fuel the growth of the discipline. They may even be implemented as practical applications or may be referred to as a beginning point for another development. Chapters in this book were first published by InTech; hereby published with permission under the Creative Commons Attribution License or equivalent.

The editorial board has been involved in producing this book since its inception. They have spent rigorous hours researching and exploring the diverse topics which have resulted in the successful publishing of this book. They have passed on their knowledge of decades through this book. To expedite this challenging task, the publisher supported the team at every step. A small team of assistant editors was also appointed to further simplify the editing procedure and attain best results for the readers.

Our editorial team has been hand-picked from every corner of the world. Their multi-ethnicity adds dynamic inputs to the discussions which result in innovative outcomes. These outcomes are then further discussed with the researchers and contributors who give their valuable feedback and opinion regarding the same. The feedback is then collaborated with the researches and they are edited in a comprehensive manner to aid the understanding of the subject.

Apart from the editorial board, the designing team has also invested a significant amount of their time in understanding the subject and creating the most relevant covers. They scrutinized every image to scout for the most suitable representation of the subject and create an appropriate cover for the book.

The publishing team has been involved in this book since its early stages. They were actively engaged in every process, be it collecting the data, connecting with the contributors or procuring relevant information. The team has been an ardent support to the editorial, designing and production team. Their endless efforts to recruit the best for this project, has resulted in the accomplishment of this book. They are a veteran in the field of academics and their pool of knowledge is as vast as their experience in printing. Their expertise and guidance has proved useful at every step. Their uncompromising quality standards have made this book an exceptional effort. Their encouragement from time to time has been an inspiration for everyone.

The publisher and the editorial board hope that this book will prove to be a valuable piece of knowledge for researchers, students, practitioners and scholars across the globe.

List of Contributors

Rachindra Mawalagedara and Robert J. Oglesby
University of Nebraska, Lincoln, USA

Vani Starry Manoharan
Environmental Sciences Division, Argonne National Laboratory, Argonne, USA

John Mecikalski, Ronald Welch and Aaron Song
Department of Atmospheric Sciences, University of Alabama in Huntsville, USA

Ranyére Silva Nóbrega
University Federal of Pernambuco, Brazil

Serafeim Polyzos and Dionysios Minetos
University of Thessaly, Department of Planning and Regional Development, Pedion Areos, Volos, Greece

Maria Anete Lallo
Universidade Paulista (UNIP), São Paulo, Brazil

Diandong Ren
Australian Sustainability Institute, Curtin University, Perth, Australia

Lance M. Leslie
School of Meteorology, University of Oklahoma, Norman, USA

Qingyun Duan
Beijing Normal University, People's Republic of China

Andrej Kranjc
Slovenian Academy of Sciences and Arts, Slovenia

K.P. Chethan, Jayaraman Srinivasan, Kumar Kriti and Kaki Sivaji
TCS Innovation Labs Bangalore, Tata Consultancy Services, India

Luzia Alice Ferreira de Moraes
Federal University of the State of Rio de Janeiro UNIRIO, Brasil

Nathaniel O. Adeoye
Department of Geography, Obafemi Awolowo University, Ile-Ife, Nigeria

Albert A. Abegunde and Samson Adeyinka
Department of Urban & Regional Planning, Obafemi Awolowo University, Ile-Ife, Nigeria

Antonia Macedo-Cruz
Hidrociencias, Campus Montecillo, Colegio de Postgraduados, Montecillo, Estado de México, México

M. Santos-Peñas and G. Pajares-Martinsanz
Facultad de Informática, Universidad Complutense de Madrid, España

I. Villegas-Romero
Universidad Autónoma Chapingo, México

Chris Goodman
Object Consulting Pty Ltd, Australia

Anthony Stocks
Idaho State University, USA

Andrew Noss, Malgorzata Bryja and Santiago Arce
Wildlife Conservation Society – WCS, USA

Robert H. Horwich, Jonathan Lyon, Arnab Bose and Clara B. Jones
Community Conservation, USA

Jonathan Lyon
Merrimack College, USA

Arnab Bose
Natures Foster, India

Lorena Soto-Pinto
Institut de Cíencia i Tecnologia Ambientals, Universitat Autonoma Barcelona, Spain

Miguel A. Castillo-Santiago
El Colegio de la Frontera Sur (ECOSUR)., Unidad San Cristóbal, Chiapas, Mexico

Guillermo Jiménez-Ferrer
Veterinary Faculty, Universitat Autonoma Barcelona, Spain

Óscar M. Chaves
Instituto para la Conservación y el Desarrollo Sostenible (INCODESO), San José, Costa Rica

Yoshito Takasaki
University of Tsukuba, Japan